# STUDENT SOLUTIONS MANUAL

TO ACCOMPANY

# ELEMENTARY AND INTERMEDIATE ALGEBRA: DISCOVERY AND VISUALIZATION
THIRD EDITION

# STUDENT SOLUTIONS MANUAL

TO ACCOMPANY

# ELEMENTARY AND INTERMEDIATE ALGEBRA: DISCOVERY AND VISUALIZATION
## THIRD EDITION

Hubbard/Robinson

# Emily J. Keaton

Houghton Mifflin Company   Boston   New York

Sponsoring Editor: Paul Murphy
Editorial Associate: Marika Hoe
Senior Project Editor: Nancy Blodget
Senior Manufacturing Coordinator: Marie Barnes
Marketing Manager: Ben Rivera

Copyright © 2002 by Houghton Mifflin Company. All rights reserved.

No part of this work may be reproduced or transmitted in any form or by any means, electronic or mechanical, including photocopying and recording, or by any information storage or retrieval system without the prior written permission of Houghton Mifflin Company unless such copying is expressly permitted by federal copyright law. Address inquiries to College Permissions, Houghton Mifflin Company, 222 Berkeley Street, Boston, MA 02116-3764.

Printed in the U.S.A.

ISBN: 0-618-12994-4

3 4 5 6 7 8 9 – BB – 05 04 03 02

# TABLE OF CONTENTS

**Chapter 1:   The Real Number System**
    1.1 The Real Numbers      1
    1.2 Addition      3
    1.3 Subtraction      5
    1.4 Multiplication      7
    1.5 Division      9
    1.6 Exponents, Square Roots, Order of Operations      11
    1.7 Properties of the Real Numbers      14
    Chapter Project: Air Pressure and Altitude      17
    Chapter 1 Review Exercises      17
    Looking Back      19
    Chapter 1 Test      19

**Chapter 2:   Expressions and Functions**
    2.1 Algebraic Expressions and Formulas      21
    2.2 Simplifying Expressions      25
    2.3 The Coordinate Plane      28
    2.4 Evaluating Expressions Graphically      32
    2.5 Relations and Functions      36
    2.6 Functions: Notation and Evaluation      40
    2.7 Analysis of Functions      46
    Chapter Project: Bicycle Helmet Usage      51
    Chapter 2 Review Exercises      52
    Looking Back      55
    Chapter 2 Test      56
**Cumulative Test: Chapters 1 – 2**      57

**Chapter 3:   Linear Equations and Inequalities**
    3.1 Introduction to Linear Equations      59
    3.2 Solving Linear Equations      63
    3.3 Formulas      72
    3.4 Modeling and Problem Solving      77
    3.5 Applications      86
    3.6 Linear Inequalities      91
    3.7 Compound Inequalities      97
    3.8 Absolute Value: Equations and Inequalities      104
    Chapter Project: Health Insurance      110
    Chapter 3 Review Exercises      110
    Looking Back      117
    Chapter 3 Test      117

**Chapter 4:   Properties of Lines**
    4.1 Linear Equations in Two Variables      119
    4.2 Slope of a Line      124
    4.3 Applications of Slope      129
    4.4 Equations of Lines      134
    4.5 Graphs of Linear Inequalities      140
    Chapter Project: Bald Eagle Population      147
    Chapter 4 Review Exercises      147

| | | |
|---|---|---|
| | Looking Back | 152 |
| | Chapter 4 Test | 152 |

**Chapter 5: Systems of Linear Equations**
    5.1 Graphing and Substitution Methods     155
    5.2 The Addition Method     163
    5.3 Systems of Equations in Three Variables     172
    5.4 Matrices     185
    5.5 Applications     189
    5.6 Systems of Linear Inequalities     197
    Chapter Project: Office Communications     200
    Chapter 5 Review Exercises     200
    Looking Back     207
    Chapter 5 Test     208

**Cumulative Test: Chapters 3 – 5**     210

**Chapter 6: Exponents and Polynomials**
    6.1 Properties of Exponents     213
    6.2 Zero and Negative Exponents     216
    6.3 Scientific Notation     219
    6.4 Polynomials     221
    6.5 Multiplication and Special Products     225
    Chapter Project: Microgravity and Space Flight     228
    Chapter 6 Review Exercises     229
    Looking Back     231
    Chapter 6 Test     231

**Chapter 7: Factoring**
    7.1 Common Factors and Grouping     233
    7.2 Special Factoring     235
    7.3 Factoring Trinomials of the Form $ax^2 + bx + c$     238
    7.3 Supplementary Exercises     241
    7.4 Solving Equations by Factoring     243
    7.5 Division     249
    Chapter Project: Foreign Currency Exchange     254
    Chapter 7 Review Exercises     255
    Looking Back     257
    Chapter 7 Test     258

**Chapter 8: Rational Expressions**
    8.1 Introduction to Rational Expressions     259
    8.2 Multiplication and Division     263
    8.3 Addition and Subtraction     267
    8.4 Complex Fractions     272
    8.5 Equations with Rational Expressions     277
    8.6 Applications     283
    Chapter Project: Major League Salaries     288
    Chapter 8 Review Exercises     288
    Looking Back     292
    Chapter 8 Test     292

**Cumulative Test: Chapters 6 – 8**     294

**Chapter 9:    Radical Expressions**
     9.1 Radicals     296
     9.2 Rational Exponents     298
     9.3 Properties of Rational Exponents     301
     9.4 The Product Rule for Radicals     304
     9.5 The Quotient Rule for Radicals     307
     9.6 Operations with Radicals     311
     9.7 Equations with Radicals and Exponents     318
     9.8 Complex Numbers     324
     Chapter Project: Visitors to National Parks and Monuments     328
     Chapter 9 Review Exercises     328
     Looking Back     331
     Chapter 9 Test     332

**Chapter 10:    Quadratic Equations**
     10.1 Special Methods     334
     10.2 Completing the Square     338
     10.3 The Quadratic Formula     345
     10.4 Equations in Quadratic Form     352
     10.5 Applications     360
     10.6 Quadratic Inequalities     367
     10.7 Quadratic Functions     374
     Chapter Project: Heart Attack Risks     380
     Chapter 10 Review Exercises     380
     Looking Back     384
     Chapter 10 Test     385

**Chapter 11:    Exponential and Logarithmic Functions**
     11.1 Algebra of Functions     387
     11.2 Inverse Functions     392
     11.3 Exponential Functions     397
     11.4 Logarithmic Functions     402
     11.5 Properties of Logarithms     407
     11.6 Exponential and Logarithmic Equations     410
     Chapter Project: Telecommuting     416
     Chapter 11 Review Exercises     416
     Chapter 11 Test     420
     **Cumulative Test: Chapters 9 – 11**     421

# Chapter 1

# The Real Number System

## Section 1.1  The Real Numbers

1.  The decimal name of a rational number is a terminating or repeating decimal.

3.  $\frac{3}{8} = 0.375$ is a terminating decimal.

5.  $\frac{35}{99} = 0.\overline{35}$ is a repeating decimal.

7.  Integers that are not whole numbers are negative integers.

9.  Rational numbers that are integers are numbers that can be written as $p/q$ such that $q = 1$.

11. The statement "Every integer is also a whole number" is false. For example, $-1$ is an integer but not a whole number.

13. The statement "Every whole number is also an integer" is true. The integers are made up of the whole numbers and their opposites.

15. The statement "Every nonterminating decimal repeats" is false. A nonterminating decimal neither terminates nor repeats.

17. The number 1.75 is a terminating decimal, and the digit 5 is followed by zeros. The number $1.\overline{75}$ is a repeating decimal, and the block of digits 75 repeats without end.

19. The calculator gives 0.9459459459 for $\frac{35}{37}$. Since this suggests that $\frac{35}{37} = 0.\overline{945}$, we have that the number is rational.

21. The calculator gives $-1.047197551$ for $-\frac{\pi}{3}$. Since this is not a repeating decimal, this decimal representation suggests that $-\frac{\pi}{3}$ is irrational.

23. The calculator gives 0.5 for $\frac{\sqrt{2}}{\sqrt{8}}$. Since this is a terminating decimal, this decimal representation suggests that $\frac{\sqrt{2}}{\sqrt{8}}$ is rational.

25. The calculator gives $-1.118033989$ for $-\frac{\sqrt{5}}{2}$. Since this is not a repeating decimal, this decimal representation suggests that $\frac{\sqrt{7}}{5}$ is irrational.

27. (a) Natural Numbers: $\sqrt{16}$
    (b) Whole Numbers: $0, \sqrt{16}$
    (c) Integers: $-4, 0, -17, \sqrt{16}$
    (d) Rational Numbers: $-4, 0, \frac{3}{5}$, $0.25, -17, 0.\overline{63}, \sqrt{16}$
    (e) Irrational Numbers: $\sqrt{7}, \pi$

**29.** $\{1, 2, 3, -2, -4\}$

**31.** $\left\{\dfrac{1}{2}, 0.75, -\dfrac{4}{3}, 0, -1.2\right\}$

**33.** $\left\{-3, -\dfrac{1}{2}, 3, -\pi, 4.3, \sqrt{4}, -\sqrt{4}, -2\dfrac{1}{3}\right\}$

**35.** $-12 < -5$

**37.** $-\dfrac{15}{7} < -\dfrac{7}{15}$

**39.** $\dfrac{22}{7} > \pi$

**41.** $\sqrt{7} > 2.\overline{645}$

**43.** $|-4| > -4$

**45.** The opposite of $-6$ is $6$.

**47.** The opposite of $0$ is $0$.

**49.** The symbols $|7|$ and $|-7|$ represent the distances from 0 to 7 and from 0 to $-7$, respectively. Because these distances are the same, $|7| = |-7|$.

**51.** If the number $n$ is 5 units from 0, then $n = 5$ or $n = -5$

**53.** If, when increased by 2, $n$ is 4 units from 0, then $n = -6$ or $n = 2$.

**55.** $|-6| = 6$

**57.** $-|7| = -7$

**59.** $|0| = 0$

**61.** $|-2| - |2| = 2 - 2 = 0$

**63.** $|-6| - |-5| = 6 - 5 = 1$

**65.** $1440 is equivalent to
$$\$1440 \cdot \dfrac{1 \text{ pound}}{\$1.60} = 900 \text{ pounds}.$$
If 600 pounds are spent in England, you return with $900 - 600 = 300$ pounds, which is equivalent to
$$300 \text{ pounds} \cdot \dfrac{\$1.50}{1 \text{ pound}} = \$450.$$

**67.** The wholesale price per pound is
$$\dfrac{\$75}{50 \text{ pounds}} = \$1.50 \text{ per pound}.$$
To make a profit of 40 cents per pound, you should charge $\$1.50 + \$0.40 = \$1.90$ per pound.

**69.** If two-thirds of the teachers were over 40 years of age, then one-third is 40 or younger. That is,
$$\dfrac{1}{3}(5601) = 1867 \text{ teachers under the}$$
age of 40.

**71.** About 63% of frozen desserts consumed consists of ice cream. Since the total amount of frozen desserts consumed is 5 gallons per person per year, we have that 63% of 5 gallons = $(0.63)(5) = 3.15$ gallons of ice cream per person per year.

**73.** About 3% of frozen desserts consumed consists of sherbet. Since the total amount of frozen desserts consumed is 5 gallons per person per year, we have that 3% of 5 gallons = $(0.03)(5) = 0.15$ gallons of sherbet per person per year. To find the number of ounces, multiply by 128 ounces per gallon: $128(0.15) = 19.2$ ounces.

**75.** If $a < 0$, then $-a\ \underline{>}\ 0$.

**77.** If $a < 0$, then $|a|\ \underline{>}\ a$.

**79.** $2.7 = \dfrac{27}{10}$

Section 1.2 Addition

**81.** $0.\overline{3} = \dfrac{1}{3}$

## Section 1.2 Addition

1. Add the absolute values and retain the common sign.

3. $8+(-5)=3$ (Unlike signs)

5. $-6+2=-4$ (Unlike signs)

7. $-3+(-7)=-10$ (Like signs)

9. $-6+(-2)=-8$ (Like signs)

11. $-16+16=0$

13. $-3+5=2$

15. $12+(-4)=8$

17. $-6+14=8$

19. $-7+0=-7$

21. $0+(-55)=-55$

23. $-9+17=8$

25. $-12+12=0$

27. $-11+(-11)=-22$

29. $100+(-101)=-1$

31. $-14+3=-11$

33. $-14+(-2)+(-1)=-16+(-1)=-17$

35. $-3+[6+(-9)]=-3+(-3)=-6$

37. $6+(-7)+2=-1+2=1$

39. $-12+(-2)+(-5)+10=-14+(-5)+10$
    $=-19+10$
    $=-9$

41. $-6+8+[(-5)+3]=-6+8+(-2)$
    $=2+(-2)$
    $=0$

43. Because the addends have the same sign (both negative), the sum should be negative: $-2.3+(-4.5)=-6.8$.

45. Because $|9.5|>|-6.7|$, the sum should be positive: $-6.7+9.5=2.8$.

47. Because the addends have the same sign (both negative), the sum should be negative: $-45.84+(-5.73)=-51.57$.

49. Because $|24.47|<|-34.78|$, the sum should be negative: $24.47+(-34.78)=-10.31$.

51. The value of 11 more than $-6$ is $-6+11=5$.

53. If 26, $-73$, and 40 are the addends, the sum is $26+(-73)+40=-7$.

55. If 12 is added to the sum of $-5$ and 2, the result is
    $12+(-5+2)=12+(-3)=9$.

57. $-\dfrac{1}{3}+\dfrac{5}{3}=\dfrac{4}{3}$

59. $-\dfrac{2}{3}+\dfrac{3}{5}=-\dfrac{10}{15}+\dfrac{9}{15}=-\dfrac{1}{15}$

61. $\dfrac{5}{16}+\left(-\dfrac{3}{8}\right)=\dfrac{5}{16}+\left(-\dfrac{6}{16}\right)=-\dfrac{1}{16}$

63. $-5+\left(-\dfrac{2}{3}\right)=-\dfrac{15}{3}+\left(-\dfrac{2}{3}\right)=-\dfrac{17}{3}$

65. Because $-7+2=-5$, the unknown number is $-7$.

67. Because $-12+(-1)+4=-9$, the third addend is $-12$.

**69.** $8 + (-5) = 3$; $8 + (-5.1) = 2.9$
Because $3 > 2.9$, we have:
$8 + (-5) > 8 + (-5.1)$

**71.** $6 + |-6| = 12$; $-6 + |-6| = 0$
Because $12 > 0$, we have:
$6 + |-6| > -6 + |-6|$

**73.** $17 + (-8) = 9$

**75.** $6 + (-8) + (-2) = -2 + (-2) = -4$

**77.** $-6 + |-8| = -6 + 8 = 2$

**79.** $187.24 + (-187.24) = 0$

**81.** $-6 + \dfrac{3}{5} = -\dfrac{30}{5} + \dfrac{3}{5} = -\dfrac{27}{5}$

**83.** $-50 + 5 = -45$

**85.** $0 + (-18.3) = -18.3$

**87.** $-12 + (-3) + 10 + (-2) = -15 + 10 + (-2)$
$\qquad\qquad\qquad\qquad\quad = -5 + (-2)$
$\qquad\qquad\qquad\qquad\quad = -7$

**89.** $|-7| + (-7) = 7 + (-7) = 0$

**91.** $-\dfrac{3}{4} + \left(-\dfrac{21}{8}\right) = -\dfrac{6}{8} + \left(-\dfrac{21}{8}\right) = -\dfrac{27}{8}$

**93.** $-\$26.84 + \$35.25 = \$8.41$
The checkbook balance after the deposit and the payment of the electric bill is $\$8.41$.

**95.**
| Week | Pounds | Weight |
|---|---|---|
| 1 | 5 (loss) | $216 - 5 = 211$ |
| 2 | 1 (gain) | $211 + 1 = 212$ |
| 3 | 2 (loss) | $212 - 2 = 210$ |
| 4 | 1 (loss) | $210 - 1 = 209$ |
| 5 | 4 (gain) | $209 + 4 = 213$ |
| 6 | 3 (loss) | $213 - 3 = 210$ |
| 7 | 2 (gain) | $210 + 2 = 212$ |

His weight was:
$216 + (-5) + 1 + (-2) + (-1) + 4 + (-3)$
$+ 2 = 212$ pounds at the end of seven weeks.

**97.** Fenway: $-38 + 1950 = 1912$
$\qquad\quad 2000 - 1912 = 88$
Wrigley: $-36 + 1950 = 1914$
$\qquad\quad 2000 - 1914 = 86$
Yankee: $-27 + 1950 = 1923$
$\qquad\quad 2000 - 1923 = 77$
Dodger: $+12 + 1950 = 1962$
$\qquad\quad 2000 - 1962 = 38$
Turner: $+47 + 1950 = 1997$
$\qquad\quad 2000 - 1997 = 3$

**99.** The percent of cardholders who use their cards fewer than four times per month is the sum of the percents who use their cards 0, 1, or 2-3 times per month: $6\% + 13\% + 22\% = 41\%$.

**101.** The number who use their cards 2 times per month is not necessarily equal to the number who use their cards 3 times per month.

**103.** The number $|a|$ is positive and the number $b$ is positive, so $|a| + b$ is positive.

**105.** The number $a$ is negative and the number $(-b)$ is negative, so $a + (-b)$ is negative.

**107.** To advance one step, the person takes 3 steps (one step backward and two steps forward). The person takes a total of $3(27) = 81$ steps.

## Section 1.3 Subtraction

1. Change the minus sign to a plus sign and change 8 to $-8$: $-5+(-8)$.

3. $4-15 = 4+(-15) = -11$

5. $-7-3 = -7+(-3) = -10$

7. $9-(-4) = 9+4 = 13$

9. $-2-(-8) = -2+8 = 6$

11. $-5-8 = -5+(-8) = -13$

13. $0-(-52) = 0+52 = 52$

15. $13-(-5) = 13+5 = 18$

17. $-12-12 = -12+(-12) = -24$

19. $-6-(-3) = -6+3 = -3$

21. $24-(-24) = 24+24 = 48$

23. $-4-16 = -4+(-16) = -20$

25. $-8-(-5) = -8+5 = -3$

27. $15-20-5 = 15+(-20)+(-5)$
    $= -5+(-5)$
    $= -10$

29. $-20-(-14)-(-1) = -20+14+1$
    $= -6+1$
    $= -5$

31. $10-(6-9) = 10-(-3) = 10+3 = 13$

33. $-8-10-(-3)+7 = -8+(-10)+3+7$
    $= -18+3+7$
    $= -15+7$
    $= -8$

35. $-3+(-2)-(-1+10) = -3+(-2)-(9)$
    $= -3+(-2)+(-9)$
    $= -5+(-9)$
    $= -14$

37. $16-[(-3)+10-(-9)]$
    $= 16-[7-(-9)]$
    $= 16-(7+9)$
    $= 16-16$
    $= 16+(-16)$
    $= 0$

39. The difference between $-17$ and $-18$ is $-17-(-18) = -17+18 = 1$.

41. Nine less than $-1$ is
    $-1-9 = -1+(-9) = -10$.

43. $-\dfrac{1}{3}-\dfrac{4}{3} = \dfrac{-1+(-4)}{3} = -\dfrac{5}{3}$

45. $-\dfrac{3}{5}-\left(-\dfrac{1}{2}\right) = -\dfrac{3}{5}+\dfrac{1}{2}$
    $= -\dfrac{6}{10}+\dfrac{5}{10}$
    $= \dfrac{-6+5}{10}$
    $= -\dfrac{1}{10}$

47. $-5-\left(-\dfrac{1}{3}\right) = -5+\dfrac{1}{3}$
    $= -\dfrac{15}{3}+\dfrac{1}{3}$
    $= \dfrac{-15+1}{3}$
    $= -\dfrac{14}{3}$

49. $\dfrac{7}{16}-\left(-\dfrac{5}{8}\right) = \dfrac{7}{16}+\dfrac{5}{8}$
    $= \dfrac{7}{16}+\dfrac{10}{16}$
    $= \dfrac{7+10}{16}$
    $= \dfrac{17}{16}$

51. The opposite of the difference of $a$ and $b$ is the difference of $b$ and $a$.

**53.** Using a calculator,
$-22-14-(-13)-4+(-7) = -34$

**55.** Using a calculator,
$13.84-(-26.19)-(-27.5)-25-(-31.7) = 74.23$

**57.** Because $3-5=-2$, the unknown number is 5.

**59.** Because $6-7=-1$, the unknown number is 6.

**61.** Use the information given in the figure:
$AB = |-7-(-1)| = |-7+1| = |-6| = 6$
$BC = |9-(-1)| = |9+1| = |10| = 10$
$AC = |9-(-7)| = |9+7| = |16| = 16$
$AB + AC = 6 + 10 = 16 = AC$

**63.** $-7-(-2) = -7+2 = -5$

**65.** $-47.87-(-6.74) = -47.87+6.74 = -41.13$

**67.** $7-(-36) = 7+36 = 43$

**69.** $|-8|-|-3| = 8-3 = 5$

**71.**
$-\dfrac{2}{3} - \dfrac{4}{5} = -\dfrac{2}{3} + \left(-\dfrac{4}{5}\right)$
$= -\dfrac{10}{15} + \left(-\dfrac{12}{15}\right)$
$= \dfrac{-10+(-12)}{15}$
$= -\dfrac{22}{15}$

**73.** $-6-10+(-5) = -6+(-10)+(-5)$
$= -16+(-5)$
$= -21$

**75.** $-16-16 = -16+(-16) = -32$

**77.**
$-\dfrac{3}{4} - 4 = -\dfrac{3}{4} + (-4)$
$= -\dfrac{3}{4} + \left(-\dfrac{16}{4}\right)$
$= \dfrac{-3+(-16)}{4}$
$= -\dfrac{19}{4}$

**79.** $0-(-32) = 0+32 = 32$

**81.** $23.89-(-35.87) = 23.89+35.87 = 59.76$

**83.** $-45-0 = -45$

**85.** $4-9-(-5) = 4+(-9)+5$
$= -5+5$
$= 0$

**87.** $|4|-|-12| = 4-12 = 4+(-12) = -8$

**89.** $-24-56 = -24+(-56) = -80$

**91.** $9-27+2-(-1) = 9+(-27)+2+1$
$= -18+2+1$
$= -16+1$
$= -15$

**93.** Let negative values be associated with below sea level. The difference between the two altitudes is: $14{,}495 - (-280) = 14{,}495 + 280 = 14{,}775$ feet.

**95.** (a) 
| Category | Difference |
|---|---|
| Always | $32\% - 20\% = +12\%$ |
| Sometimes | $33\% - 35\% = -2\%$ |
| Seldom | $19\% - 25\% = -6\%$ |
| Never | $16\% - 20\% = -4\%$ |

(b) The sum of the differences is 0. The total is 100% for both years. Thus, the increase in the Always category is offset by decreases in the other categories.

(c) "At least sometimes" means Always or Sometimes: 65%.

**97.**
| State | Difference in Altitude |
|---|---|
| CO | $14{,}431 - 2500 = 11{,}931$ feet |
| FL | $325 - 2500 = -2175$ feet |
| IL | $1241 - 2500 = -1259$ feet |

Section 1.4   Multiplication

RI   $805 - 2500 = -1695$ feet
TN   $6642 - 2500 = 4142$ feet
WA   $14{,}408 - 2500 = 11{,}908$ feet

**99.** There are 5280 feet in 1 mile. Mt. Rainer is $14{,}408 \div 5280 \approx 2.73$ miles high.

**101.** Because $a$ is negative and $b$ is positive, $a - b = a + (-b)$ represents the sum of two negative numbers. Thus $a - b$ is negative.

**103.** Because $a$ is a negative number, $|a| = -a$. So, $a - |a| = a - (-a) = a + a = 2a$. Because $2a$ is a negative number, $a - |a|$ is a negative number.

**105.** The statement $|a - b| = |b - a|$ is true for all real numbers $a$ and $b$ because the distance between $a$ and $b$ on the number line is the same regardless of the starting point.

**107.** Of the nine numbers, five are odd and four are even. Regardless of the signs, the result will be an odd number.

## Section 1.4  Multiplication

**1.** Determine the product of the absolute values. The answer is positive.

**3.** $7 \cdot (-4) = -28$   (Unlike signs)

**5.** $-5 \cdot 2 = -10$   (Unlike signs)

**7.** $-3 \cdot (-4) = 12$   (Like signs)

**9.** $-6 \cdot (-2) = 12$

**11.** $-10 \cdot 10 = -100$

**13.** $11 \cdot (-4) = -44$

**15.** $-7 \cdot 1 = -7$

**17.** $4 \cdot (-6) = -24$

**19.** $0 \cdot (-45) = 0$

**21.** Because the factors have like signs, the product is positive. $-2.3 \cdot (-2.7) = 6.21$

**23.** Because the factors have unlike signs, the product is negative. $2.1 \cdot (-2.3) = -4.83$

**25.** Because the factors have like signs, the product is positive. $-26.4 \cdot (-57.8) = 1525.92$

**27.** A numerical expression for the product of $-1.8$ and $1.8$ is $(-1.8) \cdot (1.8)$. Evaluation: $(-1.8) \cdot (1.8) = -3.24$

**29.** A numerical expression for $-\frac{3}{4}$ of 32 is $-\frac{3}{4} \cdot (32)$. Evaluation: $-\frac{3}{4} \cdot (32) = -24$

**31.** A numerical expression for twice the product of 5 and $-3$ is $2 \cdot [5 \cdot (-3)]$. Evaluation: $2 \cdot [5 \cdot (-3)] = -30$

**33.** $-\dfrac{1}{3} \cdot \dfrac{3}{5} = -\dfrac{3}{15} = -\dfrac{1}{5}$

**35.** $-\dfrac{2}{5} \cdot \left(-\dfrac{5}{8}\right) = \dfrac{10}{40} = \dfrac{1}{4}$

**37.** $-5 \cdot \left(-\dfrac{2}{3}\right) = -\dfrac{5}{1} \cdot \left(-\dfrac{2}{3}\right) = \dfrac{10}{3}$

**39.** $-\dfrac{34}{5} \cdot \left(-\dfrac{25}{4}\right) = \dfrac{850}{20} = \dfrac{85}{2}$

**41.** The result is positive because the number of negative factors is even.

**43.** $(-2)(-3)(-4) = -24$

**45.** $-5(7)(-2) = 70$

**47.** $(-2)(-1)(3)(2) = 12$

**49.** $6(-1)(4)(-2) = 48$

**51.** $(-1)(-2)(3)(1)(-6) = -36$

**53.** $(-2)(-2)(-2)(-2)(-2) = -32$

**55.** $(-3)(-3)(-3)(-3) = 81$

**57.** The left side is negative and the right side is positive. So the order is $2(-3) \underline{<} -2(-3)$. The actual values, with order, are $-6 < 6$.

**59.** The left side is negative (odd number of negative factors) and the right side is positive (even number of negative factors). So, the order is $(-1)^5 \underline{<} (-1)^4$. The actual values, with order, are $-1 < 1$.

**61.** Because $(-3) \cdot (-8) = 24$, the unknown number is $-3$.

**63.** Because $6 \cdot (-3) = -18$, the unknown number is 6.

**65.** $\dfrac{11}{4} \cdot \left(-\dfrac{4}{3}\right) = -\dfrac{44}{12} = -\dfrac{11}{3}$

**67.** $-5.76 \cdot (-4.95) = 28.512$

**69.** $-5|-4| = -5(4) = -20$

**71.** $5(-2)(4) = -40$

**73.** $36.47 \cdot (-4.47) = -163.0209$

**75.** $(-2)(-1)(-2)(-3) = 12$

**77.** $|9| \cdot |-8| = 72$

**79.** $\dfrac{3}{5} \cdot \dfrac{5}{7} \cdot \left(-\dfrac{7}{9}\right) = -\dfrac{1}{3}$

**81.** $-4 \cdot (-24) = 96$

**83.** $-14 \cdot 10 = -140$

**85.** The numbers 2 and 3 are such that their sum is 5 and their product is 6.

**87.** The numbers 4 and $-3$ are such that their sum is 1 and their product is $-12$.

**89.** The numbers $-2$ and $-6$ are such that their sum is $-8$ and their product is 12.

**91.** The numbers $-8$ and 1 are such that their sum is $-7$ and their product is $-8$.

**93.** The product $7(-2)$ can be interpreted
$7(-2) = -2 + (-2) + (-2) + (-2) + (-2) + (-2) + (-2) = -14$

**95.** The golfer has one $-2$ score, three $-1$ scores, nine 0 scores, two $+1$ scores, and three $+2$ scores. The total is
$1(-2) + 3(-1) + 9(0) + 2(1) + 3(2) = -2 + (-3) + 0 + 2 + 6 = 3$. Because par for the course is 72, the golfer's score is $72 + 3 = 75$ for the 18 holes.

**97.** (a) To find total annual consumption in billions of gallons, convert the population in millions to population in billions, then multiply by the annual per-person consumption times 8 ounces (per serving), and divide by 128 ounces (per gallon). Total Annual Consumption (billions of gallons):

Brazil: $\dfrac{0.162(122)(8)}{128} \approx 1.235$

China: $\dfrac{1.221(4)(8)}{128} \approx 0.305$

Germany: $\dfrac{0.082(201)(8)}{128} \approx 1.030$

Mexico: $\dfrac{0.094(322)(8)}{128} \approx 1.892$

Russia: $\dfrac{0.147(6)(8)}{128} \approx 0.055$

United States: $\dfrac{0.263(343)(8)}{128} \approx 5.638$

Zimbabwe: $\dfrac{0.011(60)(8)}{128} \approx 0.041$

(b) Because its annual per-person consumption of 8-ounce servings is nearest to 365, the United States is the country in which nearly every

Section 1.5   Division

the country in which nearly every person consumes nearly one serving of Coca-Cola per day.

**99.** To find the total national debt for 1999, multiply the average cost per citizen in 1999 by the population in 1999.
$-20{,}700(274 \text{ million}) = -\$5.6718$ trillion, or approximately $-\$5.7$ trillion.

**101.** Using the result of $-\$399$ billion in interest on the debt in 1999, the average cost per citizen to pay the interest on the national debt in 1999 is
$$\frac{-\$399 \text{ billion}}{274 \text{ million}} = -\$1456 \text{ per person}.$$

**103.** Because $a$ and $b$ are both negative, $ab$ is positive.

**105.** Because $a$ is negative and $|b|$ and $c$ are positive, $a \cdot |b| \cdot c$ is negative.

**107.** Because $a$ is negative, $a \cdot a$ is positive.

## Section 1.5  Division

**1.** Divide the absolute values. The result is positive.

**3.** $8 \div (-4) = -2$   (Unlike signs)

**5.** $\dfrac{15}{-5} = -3$   (Unlike signs)

**7.** $-40 \div (-5) = 8$   (Like signs)

**9.** $-16 \div (-2) = 8$   (Like signs)

**11.** $\dfrac{-80}{-5} = 16$

**13.** $-21 \div 7 = -3$

**15.** $\dfrac{-20}{5} = -4$

**17.** $-7 \div 1 = -7$

**19.** $\dfrac{-2}{0}$ is undefined

**21.** $0 \div (-35) = 0$

**23.** $-6 \div 3 \div (-4) = -2 \div (-4) = \dfrac{1}{2}$

**25.** $24 \div [-6 \div (-2)] = 24 \div 3 = 8$

**27.** $-16 \cdot 2 \div 8 = -32 \div 8 = -4$

**29.** $18 \div (-6 \cdot 3) = 18 \div (-18) = -1$

**31.** A nonzero number divided by itself is 1.

**33.** Because the dividend and divisor have like signs, the quotient is positive.
$-5.9 \div (-2.7) \approx 2.19$

**35.** Because the dividend and divisor have unlike signs, the quotient is negative.
$-4.7 \div 9.3 \approx -0.51$

**37.** A numerical expression for the quotient of $-16.4$ and $-4.1$ is $-16.4 \div (-4.1)$.
Evaluation: $-16.4 \div (-4.1) = 4$

**39.** A numerical expression for 0 divided by $-10$ is $0 \div (-10)$. Evaluation:
$0 \div (-10) = 0$

**41.** A numerical expression for $-20$ divided into $-10$ is $\dfrac{-10}{-20}$. Evaluation:
$\dfrac{-10}{-20} = \dfrac{1}{2}$

**43.** $-\dfrac{1}{3} \div \dfrac{5}{9} = -\dfrac{1}{3} \cdot \dfrac{9}{5} = -\dfrac{9}{15} = -\dfrac{3}{5}$

**45.** $-\dfrac{2}{5} \div \left(-\dfrac{3}{10}\right) = -\dfrac{2}{5} \cdot \left(-\dfrac{10}{3}\right) = \dfrac{20}{15} = \dfrac{4}{3}$

**47.** $-5 \div \left(-\dfrac{2}{3}\right) = -\dfrac{5}{1} \cdot \left(-\dfrac{3}{2}\right) = \dfrac{15}{2}$

**49.** $\left(-\dfrac{3}{4}\right) \div 8 = \left(-\dfrac{3}{4}\right) \cdot \dfrac{1}{8} = -\dfrac{3}{32}$

**51.** (a) The opposite of 5 is $-5$.
(b) The reciprocal of 5 is $\dfrac{1}{5}$.

**53.** (a) The opposite of $-\dfrac{3}{4}$ is $\dfrac{3}{4}$.
(b) The reciprocal of $-\dfrac{3}{4}$ is $-\dfrac{4}{3}$.

**55.** (a) The opposite of 0 is 0.
(b) 0 has no reciprocal because division by 0 is undefined.

**57.** Because the numerators are the same and the denominator on the left is less than the denominator on the right, the order should be $\dfrac{3}{7} > \dfrac{3}{8}$. Calculator approximations, with the order, give $0.43 > 0.38$.

**59.** Because the divisors are the same and the dividend on the left is less than the dividend on the right, the order should be $-8 \div 9 \ < \ -\dfrac{7}{9}$. Calculator approximations, with the order, give $-0.89 < -0.78$.

**61.** Because $14 \div (-7) = -2$, the unknown number is 14.

**63.** Because $-40 \div 10 = -4$, the unknown number is $-40$.

**65.** $-26.4 \div (-57.8) \approx 0.46$

**67.** $-\dfrac{2}{3} \div \dfrac{8}{27} = -\dfrac{2}{3} \cdot \dfrac{27}{8} = -\dfrac{54}{24} = -\dfrac{9}{4}$

**69.** $-|-5| \div |-5| = -5 \div 5 = -1$

**71.** $-56 \div 4 \cdot 2 = -14 \cdot 2 = -28$

**73.** $\dfrac{9}{-3} = -3$

**75.** $18 \div (-2) \div (-3) = -9 \div (-3) = 3$

**77.** $12 \div (-2 \cdot 3) = 12 \div (-6) = -2$

**79.** $0 \div (-1.274) = 0$

**81.** $-20 \div (-10) \div (-6) = 2 \div (-6) = -\dfrac{1}{3}$

**83.** $27 \div (-3) = -9$

**85.** $72 \div (-6 \div 3) = 72 \div (-2) = -36$

**87.** $\dfrac{0}{0}$ is undefined because division by 0 is undefined.

**89.** Division by 0 is not defined.

**91.** $\dfrac{-5}{-8} = \dfrac{5}{8}$

**93.** $-\dfrac{-1}{-6} = \dfrac{-1}{6} = \dfrac{1}{-6} = -\dfrac{1}{6}$

**95.** $-\dfrac{-4}{7} = \dfrac{4}{7}$

**97.** (a) Average score: $\dfrac{5+1+7+4+5+2}{6} = \dfrac{24}{6} = 4$
(b) Average percentage of correct answers: $\dfrac{20+(-4)}{20} = \dfrac{16}{20} = 0.8 = 80\%$

**99.** (a) In 2020, the 65-74 age group will have approximately 30 million, the 75-84 age group will have approximately 15 million, and the 85 and over age group will have approximately 7 million.
(b) Because the 65-74 age group was approximately 18 million in 1990 and will have nearly 35 million in 2050, it is the group expected to almost double. Because the 75-84 age group was approximately 10 million in 1990 and will have around 26 million in 2050, it is the group expected to almost triple.
(c) The 85 and over age group has the greatest projected increase from

Section 1.6  Exponents, Square Roots, Order of Operations

2020 to 2050 from around 7 million to around 19 million, an increase of 12 million, as compared to an increase of nearly 5 million for the 65-74 age group and an increase of 11 million for the 75-84 age group.

**101.** The difference in temperature from sea level to an altitude of 36,000 feet is: $-67-60=-127°F$.

**103.** Using the result (from Exercise 102) of a $-3.5°F$ change in temperature for each 1000-feet increase in altitude, the temperature at 22 thousand feet would be: $60°F + (22)(-3.5°F) = 60°F - 77°F = -17°F$.

**105.** Because $a$ is negative and $b$ is positive, $\dfrac{a}{b}$ is negative.

**107.** Because $-a$ is positive and $b$ is positive, $\dfrac{-a}{b}$ is positive.

**109.**
$$\dfrac{\frac{14}{3}}{\frac{-7}{6}} = \dfrac{14}{3} \div \left(-\dfrac{7}{6}\right)$$
$$= \dfrac{14}{3} \cdot \left(-\dfrac{6}{7}\right)$$
$$= -\dfrac{84}{21} = -4$$

## Section 1.6  Exponents, Square Roots, Order of Operations

**1.** In $3^4$, 3 is the base. In $\sqrt{3}$, 3 is the radicand.

**3.** $7 \cdot 7 \cdot 7 \cdot 7 \cdot 7 \cdot 7 = 7^6$

**5.** $-[(-5) \cdot (-5) \cdot (-5)] = -(-5)^3$

**7.** $a \cdot a \cdot a = a^3$

**9.** $(2b)(2b)(2b)(2b) = (2b)^4$

**11.** $\left(\dfrac{7}{9}\right)^4 = \left(\dfrac{7}{9}\right) \cdot \left(\dfrac{7}{9}\right) \cdot \left(\dfrac{7}{9}\right) \cdot \left(\dfrac{7}{9}\right)$

**13.** $2x^2 = 2 \cdot x \cdot x$

**15.** $5^2 y^3 = 5 \cdot 5 \cdot y \cdot y \cdot y$

**17.** $5^2 = 5 \cdot 5 = 25$

**19.** $(-2)^3 = (-2)(-2)(-2) = -8$

**21.** $-7^2 = -1(7)(7) = -49$

**23.** $2^5 = 2 \cdot 2 \cdot 2 \cdot 2 \cdot 2 = 32$

**25.** $\left(\dfrac{2}{3}\right)^2 = \left(\dfrac{2}{3}\right) \cdot \left(\dfrac{2}{3}\right) = \dfrac{4}{9}$

**27.** $(-1)^6 = (-1) \cdot (-1) \cdot (-1) \cdot (-1) \cdot (-1) \cdot (-1) = 1$

```
12^5
            248832
(-3)^11
           -177147
(3.47)^7
           6057.68
```

**Figure for Exercises 29, 31, and 33**

**29.** $12^5 = 248,832$  (see first entry)

**31.** $(-3)^{11} = -177,147$  (see second entry)

**33.** $(3.47)^7 \approx 6057.68$  (see third entry)

35. Because $6^2 = 36$ and $(-6)^2 = 36$, the two square roots of 36 are 6 and $-6$. However, the radical symbol refers to the principal (positive) square root. Thus, $\sqrt{36} = 6$.

37. Because $2^2 = 4$ and $(-2)^2 = 4$, the square roots of 4 are 2 and $-2$.

39. Because $6^2 = 36$ and $(-6)^2 = 36$, the square roots of 36 are 6 and $-6$.

41. There is no real number which when squared gives $-25$. There are no real number square roots of $-25$.

43. Because $23^2 = 529$ and $(-23)^2 = 529$, the square roots of 529 are 23 and $-23$.

45. The radical sign indicates the principal square root. So, $\sqrt{4} = 2$.

47. Because $\sqrt{36} = 6$, $-\sqrt{36} = -6$.

49. The radical sign indicates the principal square root. So, $\sqrt{\dfrac{4}{9}} = \dfrac{2}{3}$.

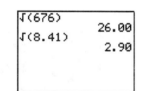

**Figure for Exercises 51 and 53**

51. $\sqrt{676} = 26$ (see first entry)

53. $\sqrt{8.41} = 2.9$ (see second entry)

**Figure for Exercises 55 and 57**

55. $\sqrt{534} \approx 23.11$ (see first entry)

57. $\sqrt{63{,}876} \approx 252.74$ (see second entry)

59. For $(-6)^2$, multiply $(-6)(-6)$ to obtain 36. For $-6^2$, multiply $-1(6)(6)$ to obtain $-36$.

61. $6 + 4(-3) = 6 + (-12) = -6$

63. $\dfrac{240}{6 \cdot 4} = \dfrac{240}{24} = 10$

65. $3 \cdot 6 - 4 \cdot 2 = 18 - 8 = 10$

67. $-2|-4| + |5| = -2(4) + 5$
$= -8 + 5$
$= -3$

69. $6 - 4 \cdot 2 + 20 \div 5 = 6 - 8 + 4$
$= -2 + 4$
$= 2$

71. $\dfrac{5^2 + 7}{11 - 3^2} = \dfrac{25 + 7}{11 - 9} = \dfrac{32}{2} = 16$

73. $\dfrac{9 - 3}{3 - 9} = \dfrac{9 - 3}{3 + (-9)} = \dfrac{6}{-6} = -1$

75. $12 - (5 + 3) \div 2 + 6 = 12 - 8 \div 2 + 6$
$= 12 - 4 + 6$
$= 12 + (-4) + 6$
$= 8 + 6$
$= 14$

77. $5 - [3(2 - 1) - 4(3 - 5)]$
$= 5 - [3(1) - 4(-2)]$
$= 5 - (3 + 8)$
$= 5 - 11$
$= 5 + (-11)$
$= -6$

79. *Verbal description:* four less than the product of $-2$ and 7.
*Numerical expression:* $-2 \cdot 7 - 4$
*Evaluation:* $-2 \cdot 7 - 4 = -14 - 4 = -18$

**81.** *Verbal description:* add −3 to 7 and divide the sum by 2. Then subtract −4 from the quotient.

*Numerical expression:* $\dfrac{-3+7}{2}-(-4)$

*Evaluation:*
$\dfrac{-3+7}{2}-(-4)=\dfrac{4}{2}+4=2+4=6$

**83.** (a) $-3\cdot(4-5)=-3\cdot(-1)=3$
(b) $(-3\cdot 4)-5=-12-5=-17$
(c) $-(3\cdot 4-5)=-(12-5)=-7$

**Figure for Exercises 85, 87, and 89**

**85.** $-4\cdot 3^2=-36$   (see first entry)

**87.** $-(-2)^4+3\sqrt{16}=-4$ (see second entry)

**89.** $5-4\sqrt{3\cdot 7-5}=-11$   (see third entry)

**Figure for Exercises 91 and 93**

**91.** $(5-2)^2+(-3-1)^2=25$   (see first entry)

**93.** $(-3)^2-4(2)(-1)=17$   (see second entry)

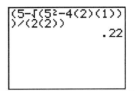

**Figure for Exercise 95**

**95.** $\dfrac{5-\sqrt{5^2-4(2)(1)}}{2(2)}\approx 0.22$   (see screen)

**97.**

| $x$ | −4 | −1 | 0 | 1 | 4 | 9 | 16 | 25 |
|---|---|---|---|---|---|---|---|---|
| $\sqrt{x}$ | X | X | 0 | 1 | 2 | 3 | 4 | 5 |
| $x^2$ | 16 | 1 | 0 | 1 | 16 | 81 | 256 | 625 |

**99.** $2^6\ 2^5\ 2^4\ 2^3\ 2^2\ 2^1\ 1$
↓ ↓ ↓ ↓ ↓ ↓ ↓
1  1  0  1  1  0  1

$=1\cdot 2^6+1\cdot 2^5+0\cdot 2^4+1\cdot 2^3+1\cdot 2^2$
$+0\cdot 2^1+1\cdot 1$
$=64+32+0+8+4+0+1$
$=109$

**101.** A numerical expression for the total cost of a one-day, 160-mile trip is $26+160\cdot 0.12$. The multiplication would be performed first.

**103.** From the bar graph, the percentage of children under age 6 who received assistance was 5.5% + 41.2% = 46.7%. Multiply the total number of children 18 years of age or younger receiving aid by this percentage to find the total number of children under age 6 who received aid: (8.8 million) · 0.467 ≈ 4.1 million.

**105.** (a) Because $x^2$ is always nonnegative, the value of $-x^2$ is positive for no values of $x$.
(b) The value of $-x^2$ is 0 when $x=0$.
(c) The value of $-x^2$ is negative when $x\neq 0$.

**107.** $a > 0$, $b > 0$, and $c < 0$
  (a) $a - c = a + (-c)$ is the sum of two positive numbers. The expression is positive
  (b) $a + c$ is the sum of a positive and a negative number. The expression could be either positive or negative.
  (c) $ab$ is the product of two positive numbers. The expression is positive.
  (d) $ac$ is the product of a positive and a negative number. The expression is negative.
  (e) $bc - a = bc + (-a)$ is the sum of two negative numbers. The expression is negative.
  (f) $a^2 c$ is the product of a positive and a negative number. The expression is negative.
  (g) $ac^2$ is the product of two positive numbers. The expression is positive.
  (h) $\dfrac{a+b}{c}$ is the quotient of a positive and a negative number. The expression is negative.

**109.** $3^3 + 3 = 27 + 3 = 30$

## Section 1.7 Properties of the Real Numbers

**1.** When $-9 - 4x + x^2$ is written as $-9 + (-4x) + x^2$, the Commutative Property of Addition can be applied to write the expression as $x^2 + (-4x) + (-9) = x^2 - 4x - 9$.

**3.** $3 + (2 + b) = (3 + 2) + b$ illustrates the Associative Property of Addition.

**5.** $4x + 0 = 4x$ illustrates the Additive Identity Property.

**7.** $4(x - 5) = 4x - 20$ illustrates the Distributive Property.

**9.** $\dfrac{2}{3}x - \dfrac{2}{3} = \dfrac{2}{3}(x - 1)$ illustrates the Distributive Property.

**11.** $5 \cdot 1 = 5$ illustrates the Multiplicative Identity Property.

**13.** $-1 \cdot (2y) = -2y$ illustrates the Multiplication Property of $-1$.

**15.** $5x - 3x = (5 - 3)x$ illustrates the Distributive Property.

**17.** $x \cdot (-7) = -7x$. So, $x \cdot (-7)$ is matched with (i).

**19.** $y + (2 + 9) = (y + 2) + 9$. So, $y + (2 + 9)$ is matched with (b).

**21.** $1(2x - y) = 2x - y$. So, $1(2x - y)$ is matched with (c).

**23.** $2(3x) = (2 \cdot 3)x$. So, $2(3x)$ is matched with (f).

**25.** Using the Associative Property of Addition: $(x + 4) + 3 = \underline{x + (4 + 3)}$.

**27.** Using the Associative Property of Multiplication: $6(2x) = \underline{(6 \cdot 2)x}$.

**29.** Using the Additive Identity Property: $(x + 0) + 6 = \underline{x + 6}$.

Section 1.7  Properties of the Real Numbers

**31.** Using the Property of Multiplicative Inverses: $\left(3 \cdot \dfrac{1}{3}\right)x = \underline{1x}$.

**33.** Using the Distributive Property: $12z + 15 = \underline{3(4z+5)}$.

**35.** Suppose that 0 had a reciprocal $r$. Then, according to the Property of Multiplicative Inverses, $0 \cdot r = 1$. But the Multiplication Property of 0 states that $0 \cdot r = 0$. Therefore, 0 does not have a reciprocal.

**37.** Using the Property of Multiplicative Inverses, $5 \cdot \dfrac{1}{\underline{5}} = 1$.

**39.** Using the Multiplicative Identity Property, $-\dfrac{2}{3} \cdot \underline{1} = -\dfrac{2}{3}$.

**41.** Using the Distributive Property, $\underline{-1}(2x-5) = -2x+5$.

**43.** $(y+0)+(-y)$
$= (0+y)+(-y)$    (a) Commutative Property of Addition
$= 0+[y+(-y)]$    (b) Associative Property of Addition
$= 0+0$    (c) Property of Additive Inverses
$= 0$    (d) Additive Identity Property

**45.** $(a-b)+b$
$= [a+(-b)]+b$    (a) Definition of Subtraction
$= a+[(-b)+b]$    (b) Associative Property of Addition
$= a+0$    (c) Property of Additive Inverses
$= a$    (d) Additive Identity Property

**47.** $x+(3+2) = x+5$

**49.** $2z+(7-2) = 2z+5$

**51.** $(-5 \cdot 3)x = -15x$

**53.** $\left[-\dfrac{5}{6}\left(-\dfrac{9}{10}\right)\right]x = \dfrac{3}{4}x$

**55.** $5(4+3b) = 5 \cdot 4 + 5 \cdot (3b) = 20+15b$

**57.** $-\dfrac{3}{5}(5x+20) = -\dfrac{3}{5} \cdot 5x + \left(-\dfrac{3}{5}\right) \cdot (20)$
$= -3x-12$

**59.** $-3(5-2x) = -3 \cdot 5 + (-3) \cdot (-2x)$
$= -15+6x$

**61.** $x(y+z) = xy+xz$

**63.** $5x+5y = 5 \cdot x + 5 \cdot y = 5(x+y)$

**65.** $3x+12 = 3 \cdot x + 3 \cdot 4 = 3(x+4)$

**67.** $5y+5 = 5 \cdot y + 5 \cdot 1 = 5(y+1)$

**69.** $6-3x = 3 \cdot 2 + 3 \cdot (-x) = 3(2-x)$

**71.** $-(x-4) = -1 \cdot (x-4) = -x+4$

**73.** $-(2x+5y-3) = -1(2x+5y-3)$
$= -2x-5y+3$

**75.** $(3x) \cdot 4 = 4 \cdot (3x) = (4 \cdot 3) \cdot x = 12x$

**77.** $(-6x)(2x) = (-6x)(x \cdot 2)$
$= -6 \cdot (x \cdot x) \cdot 2$
$= -6 \cdot x^2 \cdot 2$
$= -6 \cdot 2 \cdot x^2$
$= -12x^2$

**79.** $(5+x)+3 = (x+5)+3$
$= x+(5+3)$
$= x+8$

**81.** $-5x + (3 + 5x) = -5x + (5x + 3)$
$= (-5x + 5x) + 3$
$= 0 + 3$
$= 3$

**83.** $\dfrac{2}{9} + \dfrac{3}{7} + \dfrac{5}{9} + \dfrac{4}{7} + \dfrac{2}{9}$
$= \left(\dfrac{3}{7} + \dfrac{4}{7}\right) + \left(\dfrac{2}{9} + \dfrac{5}{9} + \dfrac{2}{9}\right)$
$= \left(\dfrac{7}{7}\right) + \left(\dfrac{9}{9}\right)$
$= 1 + 1$
$= 2$

**85.** $2 \cdot (-87) \cdot 5 \cdot (-1)$
$= (2 \cdot 5)[-87(-1)]$
$= 10 \cdot 87$
$= 870$

**87.** $7 \cdot 108 = 7 \cdot (100 + 8)$
$= 7 \cdot 100 + 7 \cdot 8$
$= 700 + 56$
$= 756$

**89.** $15 \cdot 98 = 15 \cdot (100 - 2)$
$= 15 \cdot 100 - 15 \cdot 2$
$= 1500 - 30$
$= 1470$

**91.** If $n$ is the number,
$9n = (10 - 1)n = 10n - n$.

**93.** The additive inverse of $\dfrac{5}{7}$ is $-\dfrac{5}{7}$. The multiplicative inverse of $\dfrac{5}{7}$ is $\dfrac{7}{5}$.

**95.** The additive inverse of $\dfrac{2}{3}x$ is $-\dfrac{2}{3}x$. The multiplicative inverse of $\dfrac{2}{3}x$ is $\dfrac{3}{2x}$.

**97.** The additive inverse of $x + 3$ is $-x - 3$. The multiplicative inverse of $x + 3$ is $\dfrac{1}{x + 3}$.

**99.** The additive inverse of $4 - 3x$ is $3x - 4$. The multiplicative inverse of $4 - 3x$ is $\dfrac{1}{4 - 3x}$.

**101.** The Property of Additive Inverses guarantees that the sum of your monthly income and expenses is 0.

**103.** The Commutative Property of Addition guarantees that the sum is the same added from top to bottom or from bottom to top.

**105.** The statement $a - b = b - a$ is sometimes true. It is true if $a = b$. For example, if $a = b = 3$, this is $3 - 3 = 3 - 3 = 0$. Otherwise, it is false. For example, if $a = 4$ and $b = 3$, we have $4 - 3 = 1$ for $a - b$, but $3 - 4 = -1$ for $b - a$.

**107.** The statement $|a + b| = |a| + |b|$ is sometimes true. It is true if $a$ and $b$ have like signs. For example, if $a = 2$ and $b = 3$, we have $|2 + 3| = 2 + 3 = |2| + |3| = 5$. Otherwise, it is false. For example, if $a = -2$ and $b = 3$, we have $|-2 + 3| = |1| = 1$ for $|a + b|$ but $|-2| + |3| = 2 + 3 = 5$ for $|a| + |b|$.

**109.** The statement $(a + b)^2 = a^2 + b^2$ is always false for nonzero numbers.

**111.** $\dfrac{a + b}{3} = \dfrac{1}{3}(a + b)$
$= \dfrac{1}{3}a + \dfrac{1}{3}b$
$= \dfrac{a}{3} + \dfrac{b}{3}$

## Chapter 1 Project

**1.** A bar graph of the information given in the table is shown below.

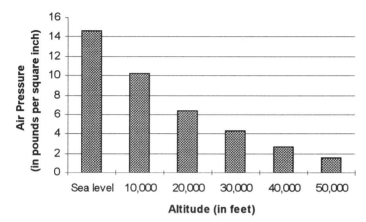

**3.** For the six highest points listed in Exercises 97-100 in Section 1.3, the following table gives the air pressure as estimated from the bar graph and curve from Exercises 1 and 2.

| Highest Point | Altitude (in feet) | Estimated Air Pressure |
|---|---|---|
| Mount Elbert, CO | 14,431 | 8.5 |
| Iron Mountain, FL | 325 | 14.5 |
| Charles Mound, IL | 1241 | 14.1 |
| Durfee Hill, RI | 805 | 14.3 |
| Clingmans Dome, TN | 6642 | 11.7 |
| Mount Rainier, WA | 14,408 | 8.5 |

**5.** The air pressure drops approximately 0.2 pounds per square inch for every 1000 feet increase in altitude between 20,000 and 30,000 feet, $\frac{6.4 - 4.3}{10} \approx 0.2$. An air pressure of 4.5 pounds per square inch is closest to a value of 4.3 (for 30,000 feet) in the table. Because $4.5 - 0.2 = 4.3$, subtract 1000 feet from the altitude of 30,000. An estimate of the altitude of Mount Everest, if the air pressure is 4.5 pounds per square inch, is 29,000 feet.

## Chapter 1 Review Exercises

**1.** The statement that $-8$ is a natural number is false. The number $-8$ does not belong to the set $\{1, 2, 3, 4, \ldots\}$.

**3.** The statement that every whole number is also an integer is true.

**5.** The statement that the value of $\sqrt{4.8}/\sqrt{2.7}$ is an irrational number is false. Because $\sqrt{4.8}/\sqrt{2.7} = 4/3$, it is a rational number.

**7.** The statement that for any real number $n$, $-n$ represents a negative real number is false. For example, if $n = -5$, then $-n = 5$.

**9.** Because $0.5 = \frac{1}{2}$, the statement $0.5 < \frac{1}{2}$ is false. However, $0.5 \leq \frac{1}{2}$ is true because $\leq$ means less than or equal to.

**11.** (a) Because $1 - 3 = -2$ and $1 + 3 = 4$, the number $n$ could be either $-2$ or $4$.
(b) Because $|7| = 7$ and $|-7| = 7$, the number n could be either $-7$ or $7$.

**13.** Addends are the numbers that are added. In $4 + 6$, 4 and 6 are addends.

**15.** $8 + (-11) = -3$

**17.** $-\dfrac{2}{3} + \dfrac{3}{4} = -\dfrac{8}{12} + \dfrac{9}{12} = \dfrac{1}{12}$

**19.** $-|-2| + |-5| = -2 + 5 = 3$

**21.** Because $11 + (-5) = 6$, the unknown number is 11.

**23.** The average temperature for the day is the sum of the high and low temperatures, divided by 2:
$\dfrac{26 + (-4)}{2} = \dfrac{22}{2} = 11°F$

**25.** In $6 - 2$, 6 is the minuend and 2 is the subtrahend.

**27.** $7 - (-13) = 7 + 13 = 20$

**29.** $-\dfrac{2}{5} - \dfrac{3}{4} = -\dfrac{8}{20} - \dfrac{15}{20} = -\dfrac{23}{20}$

**31.** $-|-3| - |-7| = -3 - 7$
$= -3 + (-7)$
$= -10$

**33.** Because $-5 - 4 = -9$, 4 must be subtracted from $-5$ to obtain $-9$.

**35.** The difference in elevation between the highest point in Montana and the mean elevation of the United States is $12{,}850 - 2500 = 10{,}350$ feet.

**37.** The factors are the numbers that are multiplied.

**39.** $5 \cdot (-4) = -20$

**41.** $-6 \cdot \left(-\dfrac{5}{12}\right) = -\dfrac{6}{1} \cdot \left(-\dfrac{5}{12}\right) = \dfrac{30}{12} = \dfrac{5}{2}$

**43.** $-\dfrac{20}{33} \cdot \left(-\dfrac{22}{15}\right) \cdot \left(-\dfrac{3}{16}\right) = -\dfrac{1320}{7920} = -\dfrac{1}{6}$

**45.** $(-1)(-3)(0)(4) = 0$

**47.** Because $-4.5 \cdot (-2) = 9$, $-4.5$ must be multiplied by $-2$ to obtain 9.

**49.** The tank is decreasing in volume at a rate of $100 - 60 = 40$ gallons per minute. The number of minutes from noon to 1:15 P.M. is $60 + 15 = 75$ minutes. So, the amount of water in the tank is $40(75) = 3000$ gallons less at 1:15 P.M. than at noon.

**51.** For $10 \div 2$, 10 is the dividend and 2 is the divisor.

**53.** $\dfrac{10}{-2} = -5$

**55.** $0 \div (-9) = 0$

**57.** $-\dfrac{4}{9} \div (-9) = -\dfrac{4}{9} \cdot \left(-\dfrac{1}{9}\right) = \dfrac{4}{81}$

**59.** Because $6 \div \left(-\dfrac{1}{2}\right) = \dfrac{6}{1} \cdot \left(-\dfrac{2}{1}\right) = -12$, if the quotient is $-12$ and the dividend is 6, the divisor is $-\dfrac{1}{2}$.

**61.** For $-5^2$, multiply $-1(5)(5)$ to obtain $-25$. For $(-5)^2$, multiply $(-5)(-5)$ to obtain 25.

**63.** $(-3)(-3)(-3)(-3)(-3) = (-3)^5 = -243$

**65.** $2^4 = 2 \cdot 2 \cdot 2 \cdot 2 = 16$

**67.** $0^{33} = 0$

**69.** $(-1)^{36} = 1$

**71.** $\sqrt{49} = 7$

73. $5+7\cdot 2-6\div 3+8 = 5+14-2+8$
    $= 19-2+8$
    $= 17+8$
    $= 25$

75. $-6^2-4(5-2^3) = -36-4(5-8)$
    $= -36-4(-3)$
    $= -36+12$
    $= -24$

77. $18-(7+2)-8\div 2 = 18-9-4$
    $= 9-4$
    $= 5$

79. The Multiplication Property of $-1$ states that $-1x = -x$.

81. $3+(5+c) = 3+(c+5)$ illustrates the Commutative Property of Addition.

83. $4\cdot(3\cdot x) = (4\cdot 3)\cdot x$ illustrates the Associative Property of Multiplication.

85. $6+0 = 6$ illustrates the Additive Identity Property.

87. Using the Commutative Property of Multiplication, $(c+5)\cdot 7 = \underline{7\cdot(c+5)}$.

89. Using the Distributive Property, $5\cdot x+5\cdot 8 = \underline{5(x+8)}$.

91. Using the Multiplicative Identity Property, $-5\left(-\dfrac{1}{5}\right)\cdot 1 = \underline{-5\left(-\dfrac{1}{5}\right)}$.

## Chapter 1 Looking Back

1. $-3-2(-5+1) = -3-2(-4)$
   $= -3+8$
   $= 5$

3. $\sqrt{3(0.9)+4} = \sqrt{2.7+4}$
   $= \sqrt{6.7}$
   $\approx 2.59$

5. $I = Pr$, $P = \$4500$, $r = 0.07$
   $I = 4500(0.07) = 315$
   The simple interest on $4500 at 7% interest is $315.

7. $3(2x-7) = 3\cdot(2x)+3\cdot(-7) = 6x-21$

9. $ax+3x = a\cdot x+3\cdot x = (a+3)x$

11. $d = \left|-7-3\right| = \left|-10\right| = 10$

## Chapter 1 Test

1. Of the elements in the set $\{-3.7, -2, -0.\overline{14}, 0, \frac{2}{3}, \pi, \sqrt{7}, 6, \sqrt{9}\}$, the elements $-2, 0, 6$, and $\sqrt{9}$ are integers.

3. Of the elements in the set $\{-3.7, -2, -0.\overline{14}, 0, \frac{2}{3}, \pi, \sqrt{7}, 6, \sqrt{9}\}$, the elements $-3.7, -2, -0.\overline{14}, 0, \frac{2}{3}, 6$, and $\sqrt{9}$ are rational numbers.

5. $15-(-8) = 15+8 = 23$
   The distance between the points whose coordinates are $-8$ and $15$ on the number line is $23$.

7. $-7-(-3) = -7+3 = -4$

9. $\dfrac{2-(-3)}{3(7)-2(-8)} = \dfrac{2+3}{21+16} = \dfrac{5}{37}$

11. $\begin{aligned}\dfrac{-3+\sqrt{3^2-4(2)}}{2} &= \dfrac{-3+\sqrt{9-8}}{2}\\ &= \dfrac{-3+\sqrt{1}}{2}\\ &= \dfrac{-3+1}{2}\\ &= \dfrac{-2}{2}\\ &= -1\end{aligned}$

13. $-\dfrac{3}{4}+\left(-\dfrac{1}{2}\right) = -\dfrac{3}{4}+\left(-\dfrac{2}{4}\right) = -\dfrac{5}{4}$

15. $(-2)(5)(-4)(0)(6) = 0$

17. (a) $\dfrac{0}{-3} = 0$

(b) It is not possible to evaluate $\dfrac{-3}{0}$. Division by zero is undefined.

19. Using the Commutative Property of Addition, $4-x = \underline{-x+4}$.

21. The opposite of $-|-4| = -4$ is 4.

23. 
```
(-3.7+(-4.13))/(
4-(3.1)²)
        1.395721925
```

$\dfrac{-3.7+(-4.13)}{4-(3.1)^2} \approx 1.40$

25. If a temperature of $-2.45°C$ is decreased by $1.04°C$, the new temperature is $-2.45°C - 1.04°C = -3.49°C$.

# Chapter 2

# Expressions and Functions

## Section 2.1  Algebraic Expressions and Formulas

1. For $x = 2$, the denominator is 0. Division by 0 is undefined.

3. Evaluate $5 - x$ for $x = -3$.
   $5 - x = 5 - (-3) = 5 + 3 = 8$

5. Evaluate $-4x$ for $x = -1$.
   $-4x = -4(-1) = 4$

7. Evaluate $5 - 3(x + 4)$ for $x = 2$.
   $5 - 3(x + 4) = 5 - 3(2 + 4)$
   $= 5 - 3(6)$
   $= 5 - 18$
   $= -13$

9. Evaluate $2x^2 - 4x - 9$ for $x = 3$.
   $2x^2 - 4x - 9 = 2(3)^2 - 4(3) - 9$
   $= 2(9) - 4(3) - 9$
   $= 18 - 12 - 9$
   $= 6 - 9$
   $= -3$

11. Evaluate $5 - |t + 3|$ for $t = -5$.
    $5 - |t + 3| = 5 - |-5 + 3|$
    $= 5 - |-2|$
    $= 5 - 2 = 3$

13. Evaluate $\dfrac{x + 3}{5 - x}$ for $x = 5$.
    $\dfrac{x + 3}{5 - x} = \dfrac{5 + 3}{5 - 5} = \dfrac{8}{0}$
    The expression is undefined for $x = 5$.

15. Evaluate $\dfrac{2x - 3}{3 - 2x}$ for $x = 1$.
    $\dfrac{2x - 3}{3 - 2x} = \dfrac{2(1) - 3}{3 - 2(1)} = \dfrac{-1}{1} = -1$

17. Evaluate $\sqrt{11 - 2x}$ for $x = 1$.
    $\sqrt{11 - 2x} = \sqrt{11 - 2(1)}$
    $= \sqrt{11 - 2}$
    $= \sqrt{9}$
    $= 3$

19. Evaluate $2x + 4y$ for $x = 3, y = -2$.
    $2x + 4y = 2(3) + 4(-2) = 6 - 8 = -2$

21. Evaluate $2x - (x - y)$ for $x = 0, y = -5$.
    $2x - (x - y) = 2(0) - [0 - (-5)]$
    $= 0 - (5)$
    $= -5$

23. Evaluate $|2s - t|$ for $s = -4, t = 2$.
    $|2s - t| = |2(-4) - 2|$
    $= |-8 - 2|$
    $= |-10|$
    $= 10$

25. Evaluate $\dfrac{2x + y}{2x - y}$ for $x = 3, y = 6$.
    $\dfrac{2(3) + 6}{2(3) - 6} = \dfrac{6 + 6}{6 - 6} = \dfrac{12}{0}$
    $\dfrac{2x + y}{2x - y}$ is undefined $x = 3$ and $y = 6$ because these values make the denominator 0.

**27.** Evaluate $\dfrac{2x+3y}{x^2+y^2}$ for $x=-6$, $y=4$.

$$\dfrac{2x+3y}{x^2+y^2} = \dfrac{2(-6)+3(4)}{(-6)^2+(4)^2}$$
$$= \dfrac{-12+12}{36+16}$$
$$= \dfrac{0}{52} = 0$$

**29.** Evaluate $\dfrac{(x-4)^2}{16}+\dfrac{(y+2)^2}{25}$ for $x=2$, $y=3$.

$$\dfrac{(x-4)^2}{16}+\dfrac{(y+2)^2}{25}$$
$$= \dfrac{(2-4)^2}{16}+\dfrac{(3+2)^2}{25}$$
$$= \dfrac{(-2)^2}{16}+\dfrac{(5)^2}{25}$$
$$= \dfrac{4}{16}+\dfrac{25}{25}$$
$$= \dfrac{1}{4}+1 = \dfrac{5}{4}$$

**31.** $a=2$, $b=-3$, $c=-2$

$$\dfrac{-b+\sqrt{b^2-4ac}}{2a}$$
$$= \dfrac{-(-3)+\sqrt{(-3)^2-4(2)(-2)}}{2(2)}$$
$$= \dfrac{3+\sqrt{9+16}}{4}$$
$$= \dfrac{3+5}{4}$$
$$= 2$$

**33.** $a=9$, $b=12$, $c=4$

$$\dfrac{-b+\sqrt{b^2-4ac}}{2a}$$
$$= \dfrac{-(12)+\sqrt{(12)^2-4(9)(4)}}{2(9)}$$
$$= \dfrac{-12+\sqrt{144-144}}{18}$$
$$= \dfrac{-12}{18} = -\dfrac{2}{3}$$

**35.** $x_1=-2$, $x_2=-4$, $y_1=3$, $y_2=-1$

$$\dfrac{y_2-y_1}{x_2-x_1} = \dfrac{-1-3}{-4-(-2)} = \dfrac{-4}{-2} = 2$$

**37.** $x_1=0$, $x_2=5$, $y_1=2$, $y_2=0$

$$\dfrac{y_2-y_1}{x_2-x_1} = \dfrac{0-2}{5-0} = \dfrac{-2}{5} = -\dfrac{2}{5}$$

**39.** Evaluate $3x-2$ for $x=-2.47$.

```
-2.47→X
              -2.47
3X-2
              -9.41
```

**41.** Evaluate $-y+3z$ for $y=2.1$, $z=-3.7$.

```
2.1→Y
              2.10
-3.7→Z
              -3.70
-Y+3Z
              -13.20
```

**43.** Evaluate $\sqrt{13y}+\sqrt{15x}$ for $x=8$ and $y=5$.

```
8→X
              8.00
5→Y
              5.00
√(13Y)+√(15X)
              19.02
```

**45.** Evaluate $\sqrt{x^2-3x+2}$ for $x=2$.

```
2→X
              2.00
√(X²-3X+2)
              0.00
```

**47.** Evaluate $|3-5x|-|4-7y|$ for $x=4$ and $y=2$.

```
4→X
              4.00
2→Y
              2.00
abs(3-5X)-abs(4-7Y)
              7.00
```

**49.** For $t=5$, $4-t=-1$. Because $\sqrt{-1}$ is not a real number, your calculator should respond with an error message.

Section 2.1  Algebraic Expressions and Formulas

51. If 5 is stored for $x$, then 4 must be stored for $y$ because $5 + 4 = 9$.

53. If 5 is stored for $x$, then $\frac{9}{5}$ must be stored for $y$ because $5 \cdot \left(\frac{9}{5}\right) = 9$.

55. If 5 is stored for $x$, then $-4$ or 14 must be stored for $y$ because $|-4-5| = 9$ and $|14-5| = 9$.

57. When we evaluate an expression, we perform all indicated operations to determine the value of the expression. To verify that a number is a solution of an equation, we substitute the number for the variable in the equation and then evaluate the two sides to determine whether they are equal.

59. For $7 - 3x = 1$ and the set $\left\{-5, -3, -\frac{1}{2}, 2, \frac{5}{2}, 4\right\}$, the true statement $1 = 1$ is the result when $x$ is replaced by 2. Replacing $x$ with any other member of the set results in a false statement. Thus, the solution is $x = 2$.

61. For $3x - 7 = 5$ and the set $\left\{-5, -3, -\frac{1}{2}, 2, \frac{5}{2}, 4\right\}$, the true statement $5 = 5$ is the result when $x$ is replaced by 4. Replacing $x$ with any other member of the set results in a false statement. Thus, the solution is $x = 4$.

63. For $13 - 4x = 3$ and the set $\left\{-5, -3, -\frac{1}{2}, 2, \frac{5}{2}, 4\right\}$, the true statement $3 = 3$ is the result when $x$ is replaced by $\frac{5}{2}$. Replacing $x$ with any other member of the set results in a false statement. Thus, the solution is $x = \frac{5}{2}$.

65. $3x - 12 = -3(4 - x)$
$3(0) - 12 = -3(4 - 0)$
$-12 = -12$
The last statement is true. Yes, 0 is a solution.

67. $x(x - 3) = 2$
$2(2 - 3) = 2$
$2(-1) = 2$
$-2 = 2$
The last statement is false. No, 2 is not a solution.

69. $x^2 - x - 6 = 0$
$(-2)^2 - (-2) - 6 = 0$
$4 + 2 - 6 = 0$
$6 - 6 = 0$
The last statement is true. Yes, $-2$ is a solution.

71. $\frac{x}{2} - 3 = 5$
$\frac{4}{2} - 3 = 5$
$2 - 3 = 5$
$-1 = 5$
The last statement is false. No, 4 is not a solution.

73. $a = 4$, $b = 5$, $c = 7$
$a^2 + b^2 = c^2$
$4^2 + 5^2 = 7^2$
$16 + 25 = 49$
$41 = 49$
No, the triangle is not a right triangle.

75. $a = 8$, $b = 15$, $c = 17$
$a^2 + b^2 = c^2$
$8^2 + 15^2 = 17^2$
$64 + 225 = 289$
$289 = 289$
Yes, the triangle is a right triangle.

**77.** $a = 9$, $b = 12$, $c = 15$
$$a^2 + b^2 = c^2$$
$$9^2 + 12^2 = 15^2$$
$$81 + 144 = 225$$
$$225 = 225$$
Yes, the triangle is a right triangle.

**79.** $a = 1$, $b = 1$, $c = \sqrt{2}$
$$a^2 + b^2 = c^2$$
$$1^2 + 1^2 = (\sqrt{2})^2$$
$$1 + 1 = 2$$
$$2 = 2$$
Yes, the triangle is a right triangle.

**81.** $a = 60$, $b = 60$, $c = 90$
$$a^2 + b^2 = c^2$$
$$60^2 + 60^2 = 90^2$$
$$3600 + 3600 = 8100$$
$$7200 = 8100$$
The infield is not perfectly square (if it were, the triangle formed by the first- and second-base lines and the line from home plate to second base would be a right triangle.) The distance from home plate to second base should be:
$$a^2 + b^2 = c^2 \Rightarrow c = \sqrt{a^2 + b^2}$$
$$c = \sqrt{60^2 + 60^2}$$
$$= \sqrt{3600 + 3600}$$
$$= \sqrt{7200}$$
$$\approx 84.9 \text{ feet}$$

**83.** No, the wall is not perfectly vertical. The ladder, building, and ground form a triangle whose sides are 6, 22, and 24. Because $6^2 + 22^2 \ne 24^2$, the wall is not perfectly vertical.

**85.**
$$C = \frac{5}{9}(F - 32)$$
$$C = \frac{5}{9}(50 - 32)$$
$$C = \frac{5}{9}(18)$$
$$C = 10$$
The Celsius temperature is 10°C.

**87.** Evaluate $A = P + Prt$, $P = \$1000$, $r = 0.06$, $t = 5$.
$$A = 1000 + (1000)(0.06)(5)$$
$$= 1000 + 300$$
$$= 1300$$
The value of the $1000 investment after 5 years at 6% interest is $1300.

**89.**

| State | High (in °C) | Low (in °F) |
|---|---|---|
| Hawaii | 37.8° | 14.0° |
| Oklahoma | 49.4° | −40.0° |
| Florida | 42.8° | −2.2° |
| Wisconsin | 45.6° | −54.4° |

**Hawaii:**

High: $C = \dfrac{5}{9}(100 - 32) \approx 37.8$

Low: $F = \dfrac{9}{5}(-10) + 32 = 14.0$

**Oklahoma:**

High: $C = \dfrac{5}{9}(121 - 32) \approx 49.4$

Low: $F = \dfrac{9}{5}(-40) + 32 = -40.0$

**Florida**

High: $C = \dfrac{5}{9}(109 - 32) \approx 42.8$

Low: $F = \dfrac{9}{5}(-19) + 32 = -2.2$

**Wisconsin:**

High: $C = \dfrac{5}{9}(114 - 32) \approx 45.6$

Low: $F = \dfrac{9}{5}(-48) + 32 = -54.4$

**91.** To obtain approximations of the percentages for the years 1992, 1994, 1996, and 1998 (as shown in the figure) substitute the values $t = 0, 2, 4, 6$.

**93.** Let $t = 5$:
$$5.35t + 18.7 = 5.35(5) + 18.7$$
$$= 26.75 + 18.7$$
$$= 45.45$$
In 1997, the percent of riders who wore helmets was about 45%.

**95.** The pattern for $a^2 + b^2 = c^2$ is that $a$ is increasing by 2, $b$ is increasing by multiples of 4, and $c$ is one more than $b$. Suppose that $n$ is the line number in

b. Suppose that $n$ is the line number in the table. Then for $a^2 + b^2 = c^2$, $a = 2n+1$, $b = n(a+1)$, and $c = b+1$. The next two lines are $9^2 + 40^2 = 41^2$ and $11^2 + 60^2 = 61^2$.

## Section 2.2  Simplifying Expressions

1. Suppose that $A$ is an algebraic expression with no explicitly written constant term. Because $A = A + 0$ by the Additive Identity Property, the constant term of the expression is 0.

3. $2x - y + 5$
   Terms: $2x$, $-y$, $5$
   Coefficients: $2, -1, 5$

5. $3a^3 - 4b^2 + 5c - 6d$
   Terms: $3a^3$, $-4b^2$, $5c$, $-6d$
   Coefficients: $3, -4, 5, -6$

7. $7y^2 - 2(3x-4) + \dfrac{2}{3}$
   Terms: $7y^2$, $-2(3x-4)$, $\dfrac{2}{3}$
   Coefficients: $7, -2, \dfrac{2}{3}$

9. $\dfrac{x+4}{5} - 5x + 2y^3$
   Terms: $\dfrac{x+4}{5}$, $-5x$, $2y^3$
   Coefficients: $\dfrac{1}{5}, -5, 2$

11. The expression $2x + 3$ contains one plus sign that separates the terms $2x$ and $3$. The plus sign in the expression $2(x+3)$ is inside parentheses and is not considered when identifying terms. Thus, this expression has only one term.

13. $2x + 1 - 3x + 2 = (2-3)x + (1+2)$
    $= -x + 3$

15. $2x - 3y + x - 5 = (2+1)x - 3y - 5$
    $= 3x - 3y - 5$

17. $2x + 4x - 7y = (2+4)x - 7y = 6x - 7y$

19. $-x + 3 - 4x + 1 + 5x$
    $= (-1 - 4 + 5)x + (3+1)$
    $= 4$

21. $\dfrac{3}{2}x + \dfrac{2}{3}y - \left(-\dfrac{1}{3}y\right) - \dfrac{5}{2}x$
    $= \dfrac{3}{2}x + \dfrac{2}{3}y + \dfrac{1}{3}y - \dfrac{5}{2}x$
    $= \left(\dfrac{3}{2} - \dfrac{5}{2}\right)x + \left(\dfrac{2}{3} + \dfrac{1}{3}\right)y$
    $= -x + y$

23. $8x^2 + 9x^2 - 7x + 10x$
    $= (8+9)x^2 + (-7+10)x$
    $= 17x^2 + 3x$

25. $3ac^2 - 7ac + 7ac^2 - 6ac$
    $= (3+7)ac^2 + (-7-6)ac$
    $= 10ac^2 - 13ac$

27. $5.23x^2 - 5.23x^3 + 7.98x^3 - 9.63x^2$
    $= (-5.23 + 7.98)x^3 + (5.23 - 9.63)x^2$
    $= 2.75x^3 - 4.40x^2$

29. We can write $-(2x-5)$ as $-1(2x-5)$, and we can write $3 - (4a+1)$ as $3 + (-1)(4a+1)$. In both cases, distributing $-1$ changes the signs of the terms inside the grouping symbols.

31. $-(a-b) = -1(a-b) = -a + b$

**33.** $5(2x-3) = 10x-15$

**35.** $-(2a-3b+7d) = -1(2a-3b+7d)$
$= -2a+3b-7d$

**37.** $3(x-2y)-(z-2) = 3x-6y-z+2$

**39.** $2(x-3)-5x = 2x-6-5x$
$= (2-5)x-6$
$= -3x-6$
The answer is given in (d).

**41.** $-x^2 + (-x)^2 = -x^2 + x^2 = 0$
The answer is given in (f).

**43.** $-[-(1-x)] = -[-1(1-x)]$
$= -(-1+x)$
$= -(x-1)$
The answer is given in (a).

**45.** $2x-(3x+1) = 2x-3x-1 = -x-1$

**47.** $3(x-4)-2x = 3x-12-2x = x-12$

**49.** $-(2a+4b)-(5a-2b)$
$= -2a-4b-5a+2b$
$= -7a-2b$

**51.** $3x+2(x-4)-5x = 3x+2x-8-5x$
$= -8$

**53.** $3(x+5)+2(1-3x) = 3x+15+2-6x$
$= -3x+17$

**55.** $-(x-1)+2x+1 = -x+1+2x+1$
$= x+2$

**57.** $(x+1)+(2x-1)+5 = x+1+2x-1+5$
$= 3x+5$

**59.** $5(2t+1)-2(t-4) = 10t+5-2t+8$
$= 8t+13$

**61.** $3x-5y+2x-(3x-5y+4)$
$= 3x-5y+2x-3x+5y-4$
$= 2x-4$

**63.** $6c-5[6c-5(c-5)] = 6c-5[6c-5c+25]$
$= 6c-5(c+25)$
$= 6c-5c-125$
$= c-125$

**65.** $2(3x-5)-3(4x-7)-8(x+6)$
$= 6x-10-12x+21-8x-48$
$= -14x-37$

**67.** $7+3[-(x-2)-(3-x)]$
$= 7+3(-x+2-3+x)$
$= 7+3(-1)$
$= 7-3$
$= 4$

**69.** $2x^2y^3 - 5x^3y^2 + 8x^3y^2 - x^2y^3$
$= x^2y^3 + 3x^3y^2$

**71.** $-(x^5y^2 - x^2y^5) - (x^2y^5 - x^5y^2)$
$= -x^5y^2 + x^2y^5 - x^2y^5 + x^5y^2$
$= 0$

**73.** The quotient of twice the number and 12 is represented by $\dfrac{2n}{12} = \dfrac{n}{6}$.

**75.** Three times the number plus the number itself is represented by $3n+n = 4n$.

**77.** Five less than a number is represented by $n-5$. The answer is given in (b).

**79.** The difference between a number and 5 is represented by $n-5$. The answer is given in (b).

**81.** Five less than a number is 3 is represented by $x-5 = 3$.

**83.** The product of 8 and a number is 11 is represented by $8x = 11$.

**85.** Twice a number is 3 less than the number is represented by $2x = x-3$.

**87.** Decreasing a number by 40% of the number results in 50 is represented by $x-0.4x = 50$.

**89.** If you take the difference between a number and twice another number, you obtain $-5$ is represented by $x-2y = -5$.

Section 2.2  Simplifying Expressions

**91.** The perimeter of a rectangle is twice the width plus twice the length. If the width is half the length,
$2W + 2L = 2(\frac{1}{2}L) + 2L = L + 2L = 3L$.

**93.** Because a quarter is worth 25 cents and a dime is worth 10 cents, the value of $q$ quarters and $d$ dimes is $25q + 10d$.

**95.** The cost of online service if $x$ hours are used is $2.5x + 15$.

**97.** The sum of three consecutive integers, the smallest being $n$, is
$n + (n+1) + (n+2) = n + n + 1 + n + 2$
$= 3n + 3$

**99.** The annual interest earned on $d$ dollars with simple interest rate 8% is $0.08d$.

**101.** The hypotenuse of a right triangle if one leg is 2 units less than the other leg, whose length is $a$, is $\sqrt{a^2 + (a-2)^2}$.

**103.** A person's salary after a 5% raise if the old salary was $s$ is $s + 0.05s = 1.05s$.

**105.** The phrase is ambiguous because we can read it as "(the product of 3 and a number) increased by 7" or as "the product of 3 and (a number increased by 7)." The corresponding translations are $3x + 7$ and $3(x + 7)$.

**107.** (a) To rewrite the model, apply the distributive property:
$1000(x + 6) = 1000x + 6000$
(b) Let $x = 3$:
$1000(x + 6) = 1000(3 + 6)$
$= 1000(9)$
$= 9000$
$1000x + 6000 = 1000(3) + 6000$
$= 3000 + 6000$
$= 9000$
(c) In 1998 ($x = 3$), the estimated number of airline complaints is 9000.
(d) Let $x = 7$:
$1000x + 6000 = 1000(7) + 6000$
$= 13,000$

**109.** If $d$ dollars out of a total of 25 billion dollars were spent on bait, the algebraic expression representing the percent of the total spent on bait is $\dfrac{d}{250,000,000}$.

**111.** If $x = 26$, then the amount spent on fishing equipment is
$0.26(\$25,000,000,000) =$
$\$6,500,000,000$. If $\$10,000,000$ is spent on bait, then the total amount spent on both fishing gear and bait is
$\$10,000,000 + \$6,500,000,000 =$
$\$6,510,000,000$, or $\$6.51$ billion.

**113.** $(-x)^2 = x^2$, so the expression is equivalent to (a).

**115.** $|x^2| = x^2$, so the expression is equivalent to (a).

**117.** $-|x|^2 = -x^2$, so the expression is equivalent to (b).

**119.** $x \cdot |x| + x^2$
If $x > 0$, then the first addend is positive, the second addend is positive, and $x \cdot |x| + x^2 = x^2 + x^2 = 2x^2$.

## Section 2.3 The Coordinate Plane

1. Starting at the origin, move 2 units to the left along the *x*-axis and then 4 units vertically upward. The destination is the point $A(-2, 4)$.

3.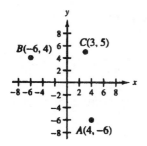

5.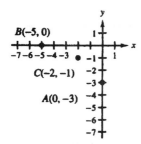

7. $\{(-2, -2), (0, -1), (2, 0), (4, 1), (6, 2), \ldots\}$. If the pattern continues, the next point should be 2 units to the right and 1 unit above the previous point. The next point should be $(8, 3)$.

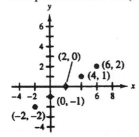

9. $\{(-3, 5), (-2, 5), (-1, 5), (0, 5), (1, 5), \ldots\}$. If the pattern continues, the next point should be 1 unit to the right of the previous point and have a *y*-coordinate of 5. The next point should be $(2, 5)$.

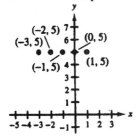

11. $\{(-4, 0), (-2, 3), (0, 0), (2, 3), (4, 0), \ldots\}$. If the pattern continues, the next point should be 2 units to the right of the previous point and have a *y*-coordinate of 3. The next point should be $(6, 3)$.

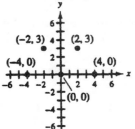

13. $P\left(-2, -\dfrac{4}{5}\right)$ is in Quadrant III because both coordinates are negative.

15. $P(2.4, 6.5)$ is in Quadrant I because both coordinates are positive.

17. $A(\sqrt{5}, -7)$ is in Quadrant IV because the *x*-coordinate is positive and the *y*-coordinate is negative.

19. $A(-6, 4)$

21. $C(0, 0)$

23. $E(5, 0)$

25. Points whose *x*-coordinates are 0 belong to the *y*-axis.

27. If $x > 0$, the point represented by $(x, y)$ would lie in Quadrant I, Quadrant IV, or on the *x*-axis.

29. If $y < 0$, the point represented by $(x, y)$ would lie in Quadrant III, Quadrant IV, or on the *y*-axis.

31. If the coordinates are $(x, 0)$, the point would lie on the *x*-axis.

33. If the coordinates are $(-3, y)$, the point could lie in Quadrant II, Quadrant III, or on the *x*-axis.

Section 2.3   The Coordinate Plane

35. If $xy < 0$, the point represented by $(x, y)$ would have either a negative $x$-coordinate or a negative $y$-coordinate, but not both. Therefore, the point could lie in Quadrant II or Quadrant IV.

37. If $P(x, y)$ lies on the $x$-axis, then $y = 0$. If $P(x, y)$ lies to the left of the origin, then $x < 0$.

39. If $P(x, y)$ lies on the $y$-axis, then $x = 0$. If $P(x, y)$ lies above the origin, then $y > 0$.

41. (a) $(+, -)$ if $P(a, b)$ lies in Quadrant IV.
    (b) $(+, +)$ if $P(a, b)$ lies in Quadrant I.
    (c) $(-, +)$ if $P(a, b)$ lies in Quadrant II.
    (d) $(-, -)$ if $P(a, b)$ lies in Quadrant III.

43. (a) For $(a, 0)$, $a > 0$, the point must lie on the positive $x$-axis.
    (b) For $(0, b)$, $b < 0$, the point must lie on the negative $y$-axis.
    (c) For $(a, -a)$, $a > 0$, the point must lie in Quadrant IV because the $x$-coordinate is positive and the $y$-coordinate is negative.
    (d) For $(a, a)$, $a \neq 0$, the point must lie in Quadrant I or Quadrant III because the coordinates are either both positive or both negative.

45. $P(0, 13)$

47. $P(9, 9)$

49. $P(-12, 6)$

51. Starting at $A(12, -8)$, 16 units to the left and up 14 units gives $(12 - 16, -8 + 14) = (-4, 6)$.

53. Starting at $A(12, -8)$, 5 units to the right and up 8 units gives $(12 + 5, -8 + 8) = (17, 0)$.

55. Find the square of the difference of the $x$-coordinates and the square of the difference of the $y$-coordinates. Add these quantities and take the square root of the result.

57. $P(5, 2)$, $Q(-12, 2)$
$$d = \sqrt{(x_2 - x_1)^2 + (y_2 - y_1)^2}$$
$$d = \sqrt{(-12 - 5)^2 + (2 - 2)^2}$$
$$= \sqrt{(-17)^2 + (0)^2}$$
$$= \sqrt{289}$$
$$= 17$$

59. $P(3, 8)$, $Q(3, -8)$
$$d = \sqrt{(x_2 - x_1)^2 + (y_2 - y_1)^2}$$
$$d = \sqrt{(3 - 3)^2 + [8 - (-8)]^2}$$
$$= \sqrt{(0)^2 + (16)^2}$$
$$= \sqrt{256}$$
$$= 16$$

61. $P(-3, 2)$, $Q(5, 8)$
$$d = \sqrt{(x_2 - x_1)^2 + (y_2 - y_1)^2}$$
$$d = \sqrt{[5 - (-3)]^2 + (8 - 2)^2}$$
$$= \sqrt{(8)^2 + (6)^2}$$
$$= \sqrt{64 + 36}$$
$$= \sqrt{100}$$
$$= 10$$

63. $P(-7, -2)$, $Q(-3, -5)$
$$d = \sqrt{(x_2 - x_1)^2 + (y_2 - y_1)^2}$$
$$d = \sqrt{[-3 - (-7)]^2 + [-5 - (-2)]^2}$$
$$= \sqrt{(4)^2 + (-3)^2}$$
$$= \sqrt{16 + 9}$$
$$= \sqrt{25} = 5$$

65. $P(-4, 3)$, $Q(-6, -7)$
$$d = \sqrt{(x_2 - x_1)^2 + (y_2 - y_1)^2}$$
$$d = \sqrt{[-4 - (-6)]^2 + [3 - (-7)]^2}$$
$$= \sqrt{(2)^2 + (10)^2}$$
$$= \sqrt{4 + 100}$$
$$= \sqrt{104}$$
$$\approx 10.20$$

**67.** $A(2, 4)$, $B(0, -2)$, $C(-1, -5)$

$$AB = \sqrt{(2-0)^2 + (-2-4)^2} = \sqrt{4+36}$$
$$= \sqrt{40} = 2\sqrt{10}$$
$$AC = \sqrt{(-1-2)^2 + (-5-4)^2} = \sqrt{9+81}$$
$$= \sqrt{90} = 3\sqrt{10}$$
$$BC = \sqrt{(-1-0)^2 + [-2-(-5)]^2} = \sqrt{1+9}$$
$$= \sqrt{10}$$

Because the sum of the two shorter distances equals the longer distance, the points are collinear.

**69.** $A(-4, 1)$, $B(-1, 2)$, $C(4, -1)$

$$AB = \sqrt{[-1-(-4)]^2 + (2-1)^2} = \sqrt{9+1}$$
$$= \sqrt{10}$$
$$AC = \sqrt{(-4-4)^2 + (-1-1)^2} = \sqrt{64+4}$$
$$= \sqrt{68} = 2\sqrt{17}$$
$$BC = \sqrt{(-1-4)^2 + (-1-2)^2} = \sqrt{25+9}$$
$$= \sqrt{34}$$

Because the sum of the two shorter distances does not equal the longer distance ($\sqrt{10} + \sqrt{34} \neq \sqrt{68}$), the points are not collinear.

**71.**

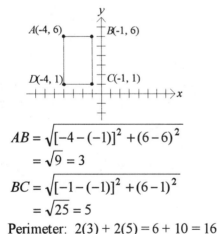

$$AB = \sqrt{[-4-(-1)]^2 + (6-6)^2}$$
$$= \sqrt{9} = 3$$
$$BC = \sqrt{[-1-(-1)]^2 + (6-1)^2}$$
$$= \sqrt{25} = 5$$

Perimeter: $2(3) + 2(5) = 6 + 10 = 16$
Area: $3(5) = 15$

**73.**

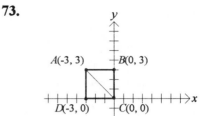

Length of side:
$$s = \sqrt{(0-0)^2 + (3-0)^2} = \sqrt{9} = 3$$
Perimeter: $4s = 4(3) = 12$
Area: $s^2 = 3^2 = 9$

**75.** Endpoints: $(4, -3)$, $(4, -1)$

$$x_m = \frac{4+4}{2} = 4,$$
$$y_m = \frac{-3+(-1)}{2} = -2$$

The midpoint is $(4, -2)$.

**77.** Endpoints: $(3, 7)$, $(-1, 3)$

$$x_m = \frac{-1+3}{2} = 1,$$
$$y_m = \frac{7+3}{2} = 5$$

The midpoint is $(1, 5)$.

**79.** Endpoints: $\left(2, \frac{1}{2}\right)$, $(-2, 3)$

$$x_m = \frac{-2+2}{2} = 0,$$
$$y_m = \frac{\frac{1}{2}+3}{2} = 1.75$$

The midpoint is $(0, 1.75)$.

**81.**

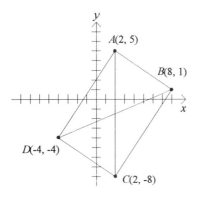

Midpoint of $AC$:
$$\left(\frac{2+2}{2}, \frac{5+(-8)}{2}\right) = (2, -1.5)$$

Midpoint of $BD$:
$$\left(\frac{-4+8}{2}, \frac{-4+1}{2}\right) = (2, -1.5)$$

Because the midpoints are the same, the diagonals bisect each other.

**83.**

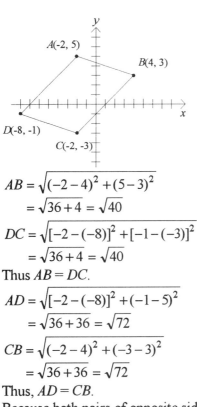

$$AB = \sqrt{(-2-4)^2 + (5-3)^2}$$
$$= \sqrt{36+4} = \sqrt{40}$$
$$DC = \sqrt{[-2-(-8)]^2 + [-1-(-3)]^2}$$
$$= \sqrt{36+4} = \sqrt{40}$$
Thus $AB = DC$.
$$AD = \sqrt{[-2-(-8)]^2 + (-1-5)^2}$$
$$= \sqrt{36+36} = \sqrt{72}$$
$$CB = \sqrt{(-2-4)^2 + (-3-3)^2}$$
$$= \sqrt{36+36} = \sqrt{72}$$
Thus, $AD = CB$.
Because both pairs of opposite sides are equal in length, the figure is a parallelogram.

**85.** $A(-2, 4)$, $B(0, 9)$, $C(3, 2)$
$$(AB)^2 = (-2-0)^2 + (4-9)^2$$
$$= 4 + 25 = 29$$
$$(BC)^2 = (3-0)^2 + (9-2)^2$$
$$= 9 + 49 = 58$$
$$(AC)^2 = (-2-3)^2 + (4-2)^2$$
$$= 25 + 4 = 29$$
Because $(AB)^2 + (AC)^2 = (BC)^2$, that is, $29 + 29 = 58$, the three points form a right triangle with the right angle at $A$.

**87.** $A(2, 4)$, $B(-2, 1)$, $C(3, -5)$
$$(AB)^2 = (-2-2)^2 + (4-1)^2$$
$$= 16 + 9 = 25$$
$$(BC)^2 = (-2-3)^2 + (-5-1)^2$$
$$= 25 + 36 = 61$$
$$(AC)^2 = (3-2)^2 + (-5-4)^2$$
$$= 1 + 81 = 82$$
Because $(AB)^2 + (BC)^2 \neq (AC)^2$, the three points do not form a right triangle.

**89.** Distance between accident $(3, -2)$ and house $(-1, 4)$:
$$d = \sqrt{[3-(-1)]^2 + [4-(-2)]^2}$$
$$= \sqrt{16+36}$$
$$= \sqrt{52}$$
$$\approx 7.2$$
Because the accident is located 7.2 miles away from your home (which is greater than the 6 mile radius given in the evacuation orders), you do not need to evacuate.

**91.** Distance between church $(0, 0)$ and school $(3, 2)$ in miles:
$$d = \sqrt{(3-0)^2 + (2-0)^2}$$
$$= \sqrt{9+4}$$
$$= \sqrt{13}$$
$$\approx 3.606 \text{ miles}$$
Because 6 inches = 1 mile, the church and school are $6 \cdot (3.606) \approx 21.6$ inches apart on the map.

93. (a) The highest bar on the graph belongs to the category of mothers employed full-time with one child.
    (b) The lowest bar on the graph belongs to the category of mothers employed full-time with four or more children.

95. (a) A graph of the data given in the table is shown below.

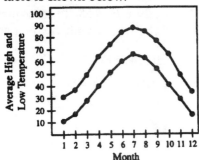

   (b) Choose the two points representing the average high and the average low for any month. If the $y$-coordinates of those points are $y_H$ and $y_L$, then the distance $y_H - y_L$ is the difference between the average high and low temperatures for that month.

97. The ordered pairs represented in the graph for the South are (1970, 63), (1980, 75), (1990, 85), (2000, 96).

99. A reasonable interpretation of the midpoint (1985, 79.5) of the line segment whose endpoints are (1970, 63) and (2000, 96) is that the estimated population in 1985 was 79.5 million.

101. (a) $AB = \sqrt{(x_1 - 0)^2 + (0 - 0)^2} = x_1$
    $CD = \sqrt{[(x_1 + x_2) - x_2]^2 + (y_1 - y_1)^2}$
    $= x_1$
    Thus, $AB = CD = x_1$

   (b) $AC = \sqrt{(x_2 - 0)^2 + (y_1 - 0)^2}$
    $= \sqrt{x_2^2 + y_1^2}$
    $BD = \sqrt{[(x_1 + x_2) - x_1]^2 + (y_1 - 0)^2}$
    $= \sqrt{x_2^2 + y_1^2}$
    Thus, $AC = BD = \sqrt{x_2^2 + y_1^2}$.

   (c) $ABCD$ is a parallelogram.

   (d) The midpoint of $\overline{AD}$ is
    $\left(\dfrac{x_1 + x_2 + 0}{2}, \dfrac{y_1 + 0}{2}\right) = \left(\dfrac{x_1 + x_2}{2}, \dfrac{y_1}{2}\right)$

   (e) The midpoint of $\overline{BC}$ is
    $\left(\dfrac{x_1 + x_2}{2}, \dfrac{y_1 + 0}{2}\right) = \left(\dfrac{x_1 + x_2}{2}, \dfrac{y_1}{2}\right)$

   (f) Because the diagonals have the same midpoint, they bisect each other.

   (g) The diagonals of a parallelogram bisect each other.

## Section 2.4 Evaluating Expressions Graphically

1. The advantage of using the **Y** screen rather than the home screen to evaluate an expression repeatedly is that entering the expression each time it is to be evaluated is not necessary.

3. ```
   Plot1 Plot2 Plot3
   \Y1■X-6
   \Y2=
   ```
   ```

Section 2.4  Evaluating Expressions Graphically

```
-1→X:Y₁
         -7
0→X:Y₁
         -6
4→X:Y₁
         -2
```

```
11→X:Y₁
         5
```

5.
```
Plot1 Plot2 Plot3
\Y₁■X²+4X-21
\Y₂=
```

```
-9→X:Y₁
         24
-4→X:Y₁
         -21
3→X:Y₁
         0
```

```
5→X:Y₁
         24
```

7.
```
Plot1 Plot2 Plot3
\Y₁■abs(2X+1)
\Y₂=
```

```
-5→X:Y₁
         9
-1→X:Y₁
         1
-0.5→X:Y₁
         0
```

```
8→X:Y₁
         17
```

9.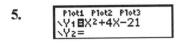

11. 
| X | Y1 |
|---|---|
| -8 | 0 |
| -1 | 14 |
| 2 | 20 |
| 10 | 36 |
Y₁■2(X+8)

13.
| X | Y1 |
|---|---|
| -5 | .4 |
| 1 | 1 |
| 25 | .52 |
| 50 | .51 |
Y₁■(X+1)/(2X)

15.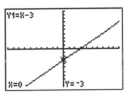

17.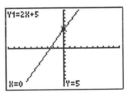

19.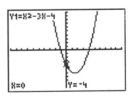

21. You obtain an error message for $x = 1$ because division by 0 is undefined.

23. From the graph, when $x = 3$, $y = 4$.

25. From the graph, when $x = -1$, $y = 1$.

27.

By tracing the graph, we find that the value of the expression is $-6$ when $x = 2$. Similarly, the expression has a value of 0 when $x = 5$ and a value of 4 when $x = 7$.

**29.**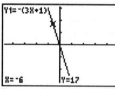

By tracing the graph, we find that the value of the expression is $-5$ when $x = -30$. Similarly, the expression has a value of 2 when $x = -9$ and a value of 9 when $x = 12$.

**31.**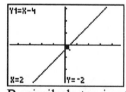

By tracing the graph, we find that the value of the expression is 17 when $x = -6$. Similarly, the expression has a value of $-1$ when $x = 0$ and a value of $-16$ when $x = 5$.

**33.** When $x = -13$, $-x + 5$ has a value of 18. Thus, the graph is of the expression $-x + 5$.

**35.** The first coordinate, 3, is the value of the variable, and the second coordinate, 7, is the value of the expression for $x = 3$.

**37.**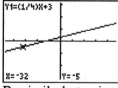

By similarly tracing the graph of $y_1 = x - 4$ to obtain missing coordinates, we obtain the following table.

| x | 2 | 4 | 0 | 5 | −1 | −2 |
|---|---|---|---|---|---|---|
| x − 4 | −2 | 0 | −4 | 1 | −5 | −6 |

**39.** 

By similarly tracing the graph of $y_1 = \frac{1}{4}x + 3$ to obtain missing coordinates, we obtain the following table.

| x | −32 | −8 | −12 | 16 | 6 | 28 |
|---|---|---|---|---|---|---|
| $\frac{1}{4}x + 3$ | −5 | 1 | 0 | 7 | 4.5 | 10 |

**41.**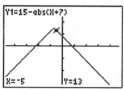

By similarly tracing the graph of $y_1 = 15 - |x + 7|$ to obtain missing coordinates, we obtain the following table.

| x | y |
|---|---|
| −5 | 13 |
| −15 | 7 |
| −30 | −8 |
| 15 | −7 |
| 20 | −12 |

**43.**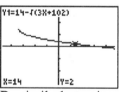

By similarly tracing the graph of $y_1 = 14 - \sqrt{3x + 102}$ to obtain missing coordinates, we obtain the following:

| x | y |
|---|---|
| 14 | 2 |
| −22 | 8 |
| −31 | 11 |
| −34 | 14 |
| 41 | −1 |

**45.**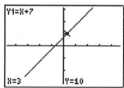

(a) By tracing the graph, we find that the expression has a value of 10 when $x = 3$.

(b) By tracing the graph, we find that the expression has a value of 16 when $x = 9$.

(c) By tracing the graph, we find that the expression has a value of 2 when $x = -5$.

Section 2.4  Evaluating Expressions Graphically

(d) By tracing the graph, we find that the expression has a value of −3 when $x = -10$.

47.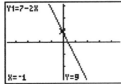

(a) By tracing the graph, we find that the expression has a value of 9 when $x = -1$.
(b) By tracing the graph, we find that the expression has a value of 3 when $x = 2$.
(c) By tracing the graph, we find that the expression has a value of −3 when $x = 5$.
(d) By tracing the graph, we find that the expression has a value of −9 when $x = 8$.

49.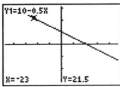

(a) By tracing the graph, we find that the expression has a value of 21.5 when $x = -23$.
(b) By tracing the graph, we find that the expression has a value of 7 when $x = 6$.
(c) By tracing the graph, we find that the expression has a value of 0 when $x = 20$.
(d) By tracing the graph, we find that the expression has a value of −6.5 when $x = 33$.

51.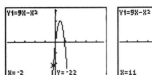

(a) By tracing the graph, we find that the expression has a value of −22 when $x = -2$ and $x = 11$.
(b) By tracing the graph, we find that the expression has a value of −10 when $x = -1$ and $x = 10$.
(c) By tracing the graph, we find that the expression has a value of 8 when $x = 1$ and $x = 8$.
(d) By tracing the graph, we find that the expression has a value of 14 when $x = 2$ and $x = 7$.

53.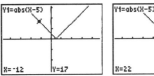

(a) By tracing the graph, we find that the expression has a value of 17 when $x = -12$ and $x = 22$.
(b) By tracing the graph, we find that the expression has a value of 8 when $x = -3$ and $x = 13$.
(c) By tracing the graph, we find that the expression has a value of 0 when $x = 5$.
(d) By tracing the graph, we find that the expression has a value of 4 when $x = 1$ and $x = 9$.

55. Because $2x + 1 = 2(1) + 1 = 3$ and $4 - x = 4 - (1) = 3$, the expressions have the same value when $x = 1$.

57. Because $3x = 3(2) = 6$ and $8 - x = 8 - (2) = 6$, the expressions have the same value when $x = 2$.

59.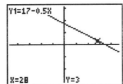

Because $y = 3$ when $x = 28$, the number of pizzas left when 28 people attend the reunion is 3.

61. (a) The variable represents the number of home runs hit.

(b)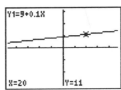

Because $y = 11$ (hundred thousand) when $x = 20$, the player's income is $1.1 million when he hits 20 home runs.

(c) If the player's income was $1,030,000, trace the graph to where $y = 10.3$ and find the corresponding $x$-coordinate.

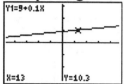

When the player's income is $1,030,000, he hit 13 home runs.

63.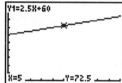

According to the graph, in 2001 (when $t = x = 5$), the number of game players is 72.5 million ($y = 72.5$).

65.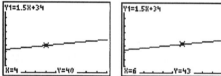

(a) According to the graph, there were approximately 43 million wireless device users in 2001.

(b) Because there were 40 million wireless device users in 1999, the percent increase from 1999 to 2001 is $\frac{43 - 40}{40} \times 100\% \approx 7.5\%$.

67.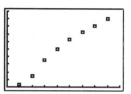

69.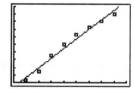

By inspecting the graph, the model appears to underestimate the number of elected officials from 1975 through 1990.

71. For the expression $2x + 1$, the $y$-coordinate is 1 more than twice the $x$-coordinate.

73. For the expression $2(x + 1)$, the $y$-coordinate is twice the sum of the $x$-coordinate and 1.

75.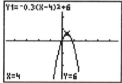

(a) When $y = 6$, there is only one $x$-value.
(b) When $y < 6$, there are two $x$-values.
(c) When $y > 6$, there are no $x$-values.

## Section 2.5 Relations and Functions

1. A relation is a set of ordered pairs. A function is a relation in which each first coordinate is paired with exactly one second coordinate. Every function is a relation, but not every relation is a function.

3. The set of ordered pairs with first coordinate an integer between $-3$ and 3, inclusive, and with second coordinate the square of the first coordinate is $\{(-3, 9), (-2, 4), (-1, 1), (0, 0), (1, 1), (2, 4), (3, 9)\}$.

Domain: $\{-3, -2, -1, 0, 1, 2, 3\}$
Range: $\{0, 1, 4, 9\}$

5. The set of ordered pairs with first coordinate the length of the side of a square, where the length is a positive integer not exceeding 5, and with second coordinate the perimeter of the square is $\{(1, 4), (2, 8), (3, 12), (4, 16), (5, 20)\}$.
Domain: $\{1, 2, 3, 4, 5\}$
Range: $\{4, 8, 12, 16, 20\}$

Section 2.5   Relations and Functions

7. The set of ordered pairs with first coordinate the number of hours worked (10, 20, 30, 40) and with second coordinate the salary at $7 per hour is {(10, 70), (20, 140), (30, 210), (40, 280)}.
   Domain: {10, 20, 30, 40}
   Range: {70, 140, 210, 280}

9. {(1, 2), (3, 4), (5, 6), (7, 8), (9, 10)}
   Domain: {1, 3, 5, 7, 9}
   Range: {2, 4, 6, 8, 10}
   The relation is a function.

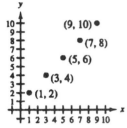

11. {(3, 1), (4, 1), (5, 1)}
    Domain: {3, 4, 5}
    Range: {1}
    The relation is a function.

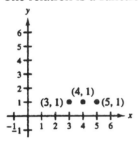

13. {(−2, 4), (−2, 5), (−2, 7)}
    Domain: {−2}
    Range: {4, 5, 7}
    The relation is not a function.

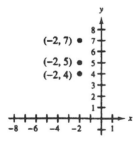

15. {(−2, 2), (−2, −3), (3, 4), (2, −3)}
    Domain: {−2, 2, 3}
    Range: {−3, 2, 4}
    The relation is not a function.

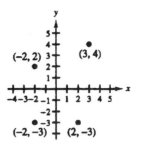

17. {(−3, −4), (−4, −5), (−5, −6), . . . }
    Domain: {. . . , −5, −4, −3}
    Range: {. . . , −6, −5, −4}
    The relation is a function.

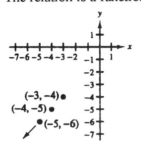

19. The domain of a relation can be estimated from a graph by determining the *x*-coordinates of the points of the graph.

21. For a graph to represent a function, no vertical line can intersect the graph in more than one point. If a vertical line intersects a graph in more than one point, there are two ordered pairs with the same first coordinate. This violates the definition of a function.

23. $\{x \mid x < -4\}$ in interval notation is $(-\infty, -4)$.

25. $\{x \mid -3 < x \leq 7\}$ in interval notation is $(-3, 7]$.

27. $[-3, 7)$ in set-builder notation is $\{x \mid -3 \leq x < 7\}$.

29. $(-\infty, 0]$ in set-builder notation is $\{x \mid x \leq 0\}$.

31. From the given figure, we see that the graph includes the points whose first coordinates extend from $-7$ to $1$, inclusive, and whose second coordinates extend from $-6$ to $2$, inclusive. The domain is $[-7, 1]$ and the range is $[-6, 2]$. The graph does not pass the Vertical Line Test: not a function.

33. From the given figure, we see that the graph includes points whose first coordinates are the integers from $-5$ to $3$, inclusive, and whose second coordinate is $3$. The domain is $\{-5, -4, -3, -2, -1, 0, 1, 2, 3\}$ and the range is $\{3\}$. The graph passes the Vertical Line Test: function.

35. From the given figure, we see that the graph includes the points whose first coordinates extend from $-\infty$ to $0$ (inclusive) and whose second coordinates extend from $-\infty$ to $\infty$. The domain is $(-\infty, 0]$ and the range is $\mathbf{R}$. The graph does not pass the Vertical Line Test: not a function.

37. From the given figure, we see that the graph includes points whose first coordinates extend from $-\infty$ to $\infty$ and whose second coordinates extend from $3$ (inclusive) to $\infty$. The domain is $\mathbf{R}$ and the range is $[3, \infty)$. The graph passes the Vertical Line Test: function

39. From the given figure, we see that the graph includes points whose first coordinates extend from $-\infty$ to $\infty$ and whose second coordinates extend from $-\infty$ to $\infty$. The domain is $\mathbf{R}$ and the range is $\mathbf{R}$. The graph passes the Vertical Line Test: function.

41. From the given figure, we see that the graph includes points whose first coordinates extend from $-\infty$ to $\infty$ and whose second coordinates extend from $-3$ (inclusive) to $\infty$. The domain is $\mathbf{R}$ and the range is $[-3, \infty)$: function.

43. $x = y + 3$
For each real number $x$, the expression $y + 3$ has exactly one value. Thus $x = y + 3$ represents a function.

45. $y = x^2 + 5$
For each real number $x$, the expression $x^2 + 5$ has exactly one value. Thus $y = x^2 + 5$ represents a function.

47. $x = (y + 1)^2$
For $x = 4$, for example, the equation $4 = (y + 1)^2$ is true for two values of $y$: $1$ and $-3$. Thus $x = (y + 1)^2$ does not represent a function.

49. $x = y^3$
For each real number $x$, the expression $y^3$ has exactly one value. Thus $x = y^3$ represents a function.

51. $x = |y|$
For $x = 3$, for example, the equation $3 = |y|$ is true for two values of $y$: $3$ and $-3$. Thus $x = |y|$ does not represent a function.

53. $y = \sqrt{x}$
For each nonnegative real number $x$, the expression $\sqrt{x}$ has exactly one value. Thus $y = \sqrt{x}$ represents a function.

55. $x^2 + y^2 = 4$
For $x = 0$, for example, the equation $0^2 + y^2 = 4$ is true for two values of $y$: $2$ and $-2$. Thus $x^2 + y^2 = 4$ does not represent a function.

57. $x^2 = y^2$
For $x = 2$, for example, the equation $2^2 = 4 = y^2$ is true for two values of $y$: $2$ and $-2$. Thus $x^2 = y^2$ does not represent a function.

**59.** $x = \sqrt[3]{y}$

For each real number $x$, the expression $\sqrt[3]{y}$ has exactly one value. Thus $x = \sqrt[3]{y}$ represents a function.

**61.** $y = x - 8$

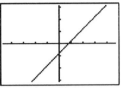

Domain: **R**; Range: **R**

**63.** $y = 9 - x^2$

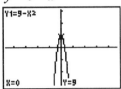

Domain: **R**; Range: $(-\infty, 9]$

**65.** $y = \sqrt{9 - x}$

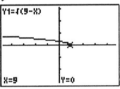

Domain: $(-\infty, 9]$; Range: $[0, \infty)$

**67.** $y = x^2 + 4x + 3$

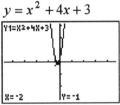

Domain: **R**; Range: $[-1, \infty)$

**69.** $y = |x - 2|$

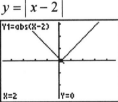

Domain: **R**; Range: $[0, \infty)$

**71.** $c = 420/n$

(a) Domain: $\{1, 2, 3, 4, 5, 6\}$;
Range: $\{70, 84, 105, 140, 210, 420\}$

(b) $\{(1, 420), (2, 210), (3, 140), (4, 105), (5, 84), (6, 70)\}$

**73.** (a) A function for the area of the base is $A = (20 - x)x = 20x - x^2$.

(b) Using a graph of the area function, we see that the largest possible area of the base is 100 square inches.

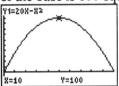

(c) There is no $y$-value larger than 100.

(d) The largest possible area of the base occurs when the dimensions of the base are 10 inches by 10 inches.

(e) Using a graph, we see that the area of 75 square inches is obtained when the dimensions of the box are 5 inches by 15 inches.

(f) The domain of the function is $\{x \,|\, 0 < x < 20\}$.

**75.** (a) {(Birds, 12.6), (Dogs, 52.9), (Rabbits, 5.7), (Cats, 59)}

(b) Yes, this relation represents a function because no two ordered pairs have the same first coordinate.

**77.** No, this mapping diagram does not represent a function because red and silver would appear as first coordinates in more than one pair.

**79.** $A =$ {(New Orleans, 256), (Tampa, 246), (Jackson, 231), (Mobile, 217), (Casper, 217)}. The set $A$ is a function.

**81.** No, the relation {(city, cost)} would not be a function because each ordered pair would have the same first coordinate.

**83.** $A = \{(x, y) \,|\, y \text{ is an ancestor of } x\}$

No, this is not a function because each person $x$ has many ancestors, starting with a mother and a father.

**85.** $C = \{(x, y) | y \text{ is the mother of } x\}$
Yes, this is a function because each person $x$ has only one biological mother.

**87.** $E = \{(x, y) | y \text{ is an aunt of } x\}$
No, this is not a function because a person $x$ could have more than one aunt.

**89.** $y = \dfrac{|x|}{x}$

From the graph, we see that the domain is $\{x | x \neq 0\}$ and the range is $\{-1, 1\}$.

**91.** $y = \dfrac{1}{4}x^4 - \dfrac{1}{3}x^3 - 6x^2$

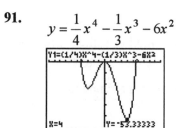

From the graph, we see that the domain is **R** and the range is $[-53.33, \infty)$.

## Section 2.6 Functions: Notation and Evaluation

**1.** To evaluate $f(3)$, replace $x$ in the expression $2x+1$ with 3 and perform the indicated operations.
$f(3) = 2(3) + 1 = 7$

**3.** $f(x) = 5x + 4$; $f(3) = 5(\underline{\ 3\ }) + 4$

**5.** $R(x) = 4 - x^2$; $R(2) = 4 - (\underline{\ 2\ })^2$

**7.** $h(x) = -x^2 + 2x + 5$;
$h(-2) = -(\underline{\ -2\ })^2 + 2(\underline{\ -2\ }) + 5$

**9.** $f(x) = 2x^3 - 4x + 5$;
$f(\underline{\ 2\ }) = 2(2)^3 - 4(2) + 5$

**11.** $h(x) = \sqrt{3x+4}$; $h(-1) = \sqrt{3(\underline{\ -1\ }) + 4}$

**13.** $a(x) = |x - 5|$; $a(\underline{\ 1\ }) = |1 - 5|$

**15.** $g(x) = 2x - 7$;
$g(3 + t) = 2(\underline{\ 3+t\ }) - 7$

**17.** $h(x) = -4x^2 + 3x - 2$;
$h(\underline{\ a\ }) = -4a^2 + 3a - 2$

**19.** $f(x) = 6 - x$
$f(3) = 6 - 3 = 3$
$f(-2) = 6 - (-2) = 6 + 2 = 8$

**21.** $g(x) = 2x + 9$
$g(-3) = 2(-3) + 9 = -6 + 9 = 3$
$g(4) = 2(4) + 9 = 8 + 9 = 17$

**23.** $R(x) = -x^2 + x + 3$
$R(-2) = -(-2)^2 + (-2) + 3$
$= -4 - 2 + 3 = -3$
$R(3) = -(3)^2 + (3) + 3 = -3$

**25.** Because $f(4) = -7$, a point of the graph is $(4, -7)$. The $x$-coordinate of each point of the graph is the value of the variable, and the $y$-coordinate is the value of the function.

Section 2.6  Functions: Notation and Evaluation

27. $f(x) = 7x - 2$, $g(x) = 2x - 5$
   (a) $g(-4) = 2(-4) - 5 = -8 - 5 = -13$
   (b) $f(-3) = 7(-3) - 2 = -21 - 2 = -23$
   (c) $g(3) - f(2) = 2(3) - 5 - [7(2) - 2]$
       $= 6 - 5 - [14 - 2]$
       $= 6 - 5 - (12)$
       $= -11$
   (d) $f(0) + g(1) = 7(0) - 2 + 2(1) - 5$
       $= 0 - 2 + 2 - 5$
       $= -5$

29. $g(x) = -x^2 - 3x$, $h(x) = x^4 - x$
   (a) $g(-1) = -(-1)^2 - 3(-1) = -1 + 3 = 2$
   (b) $h(-1) = (-1)^4 - (-1) = 1 + 1 = 2$
   (c) $h(-2) = (-2)^4 - (-2) = 16 + 2 = 18$
   (d) $g(-3) = -(-3)^2 - 3(-3) = -9 + 9 = 0$

31. $h(x) = \sqrt{3x + 4}$, $g(x) = |3 - 2x|$
   (a) $h(-1) = \sqrt{3(-1) + 4} = \sqrt{-3 + 4}$
       $= \sqrt{1} = 1$
   (b) $h(4) = \sqrt{3(4) + 4} = \sqrt{12 + 4} = 4$
   (c) $g(-2) = |3 - 2(-2)| = |3 + 4| = 7$
   (d) $g(2) = |3 - 2(2)| = |3 - 4| = 1$

33. Store the value for which the function is to be evaluated. Then enter the expression for the function on the home screen. The calculator returns the value of the function for the stored value of the variable.

35. Enter $9 - x^2$ as $Y_1$. The following screens demonstrate the home screen, Y-screen, and function notation methods for evaluating $Q(-3)$ and $Q(0)$, respectively.

37. Enter $x^3 + 2x^2 - x$ as $Y_1$. The following screens demonstrate the home screen, Y-screen, and function notation methods for evaluating $P(-1)$ and $P(1)$, respectively.

39. Enter $\dfrac{1}{3}x - \dfrac{x^2}{2}$ as $Y_1$. The following screens demonstrate the home screen, Y-screen, and function notation methods for evaluating $P(6)$ and $P(1)$, respectively.

41.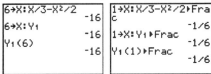

43.

**45.** $h(x) = \sqrt{3x-6} \div 1 - x$,
$k(x) = \sqrt{3x-6} \div (1-x)$

(a) $h(5) = \sqrt{3(5)-6} \div 1 - 5$
$= \sqrt{15-6} \div 1 - 5$
$= \sqrt{9} \div 1 - 5$
$= 3 - 5$
$= -2$

$k(5) = \sqrt{3(5)-6} \div (1-5)$
$= \sqrt{15-6} \div (-4)$
$= \sqrt{9} \div (-4)$
$= 3 \div (-4)$
$= -\dfrac{3}{4}$

(b) The parentheses in function $k$ require the subtraction to be performed before the division.

**47.** $f(x) = x^2 - (4-x)$
$f(-2) = (-2)^2 - [4-(-2)]$
$= 4 - (6)$
$= -2$

Calculator verification:
```
-2→X
           -2
X²-(4-X)
           -2
```

**49.** $h(c) = 5 - 5(5-c)$
$h(5) = 5 - 5(5-5)$
$= 5 - 5(0)$
$= 5 - 0$
$= 5$

Calculator verification:
```
5→C
           5
5-5(5-C)
           5
```

**51.** $k(m) = 7(3m-1) - (8-m)$
$k(-2) = 7[3(-2)-1] - [8-(-2)]$
$= 7[-6-1] - 10$
$= 7(-7) - 10$
$= -49 - 10$
$= -59$

Calculator verification:
```
-2→M
           -2
7(3M-1)-(8-M)
          -59
```

**53.** $m(s) = [4-(s-1)] + 8s$
$m(3) = [4-(3-1)] + 8(3)$
$= [4-(2)] + 24$
$= 2 + 24$
$= 26$

Calculator verification:
```
3→S
           3
(4-(S-1))+8S
          26
```

**55.** $G(x) = 3\{x - [x-(x-1)]\} - 3x$

(a) $G(1) = 3\{1 - [1-(1-1)]\} - 3(1)$
$= 3\{1 - [1-(0)]\} - 3$
$= 3\{1 - (1)\} - 3$
$= 3(0) - 3$
$= 0 - 3$
$= -3$

$G(-4) = 3\{-4 - [-4-(-4-1)]\} - 3(-4)$
$= 3\{-4 - [-4-(-5)]\} + 12$
$= 3\{-4 - 1\} + 12$
$= 3(-5) + 12$
$= -15 + 12$
$= -3$

$G(2.3) = 3\{2.3 - [2.3-(2.3-1)]\} - 3(2.3)$
$= 3\{2.3 - [2.3-1.3]\} - 6.9$
$= 3\{2.3 - 1\} - 6.9$
$= 3(1.3) - 6.9$
$= 3.9 - 6.9$
$= -3$

Section 2.6   Functions: Notation and Evaluation

(b) The results in part (a) suggest that $G(x) = -3$ for all $x$.

(c) $G(x) = 3\{x - [x - (x - 1)]\} - 3x$
$G(x) = 3\{x - [x - x + 1]\} - 3x$
$G(x) = 3\{x - (0 + 1)\} - 3x$
$G(x) = 3\{x - 1\} - 3x$
$G(x) = 3x - 3 - 3x$
$G(x) = -3$

**57.** (a) $f(x) = 2x - 3\left[x - \sqrt{x - 1}\right]$
$f(1) = 2(1) - 3\left[1 - \sqrt{1 - 1}\right]$
$= 2 - 3\left[1 - \sqrt{0}\right]$
$= 2 - 3$
$= -1$

(b) 2X – 3(X – √ (X – 1))

(c)
```
1→X
           1
2X-3(X-√(X-1))
          -1
```

**59.** (a) $K(x) = 3 - 5 \cdot |x - 3|$

3 – 5 * ABS(X – 3)

(b)
```
9→X
           9
3-5*abs(X-3)
         -27
```

**61.** $h(x) = \sqrt{3x + 4}$

(a) $h(20.8) = \sqrt{3(20.8) + 4}$
$= \sqrt{62.4 + 4}$
$= \sqrt{66.4}$
$\approx 8.15$

(b) $h(0.9) = \sqrt{3(0.9) + 4}$
$= \sqrt{2.7 + 4}$
$= \sqrt{6.7}$
$\approx 2.59$

**63.** $k(x) = \sqrt{x^2 - 2x - 15}$

(a) $k(11.3) = \sqrt{(11.3)^2 - 2(11.3) - 15}$
$= \sqrt{127.69 - 22.6 - 15}$
$= \sqrt{90.09}$
$\approx 9.49$

(b) $k(-4.5) = \sqrt{(-4.5)^2 - 2(-4.5) - 15}$
$= \sqrt{20.25 + 9 - 15}$
$= \sqrt{14.25}$
$\approx 3.77$

**65.** $c(x) = |x^2 + 2x - 7|$

(a) $c(106) = |(106)^2 + 2(106) - 7|$
$= |11{,}236 + 212 - 7|$
$= |11{,}441|$
$= 11{,}441$

(b) $c(-89) = |(-89)^2 + 2(-89) - 7|$
$= |7921 - 178 - 7|$
$= |7736|$
$= 7736$

**67.** $f(x) = \dfrac{x + 1}{2x - x^2}$

(a) $f(x) = \dfrac{x + 1}{2x - x^2}$
$f(-1) = \dfrac{-1 + 1}{2(-1) - (-1)^2}$
$= \dfrac{0}{2(-1) - (-1)^2}$
$= 0$

(b) $f(0.7) = \dfrac{0.7 + 1}{2(0.7) - (0.7)^2}$
$= \dfrac{1.7}{1.4 - 0.49}$
$= \dfrac{1.7}{0.91}$
$\approx 1.87$

**69.** $f(x) = -x^2$, $g(x) = x^2$, $h(x) = (-x)^2$

(a) $f(7) = -7^2 = -49$

$g(7) = 7^2 = 49$

$h(7) = (-7)^2 = 49$

(b) $f(-7) = -(-7)^2 = -(49) = -49$

$g(-7) = (-7)^2 = 49$

$h(-7) = [-(-7)]^2 = 7^2 = 49$

(c) $f$: B, A; $g$: B; $h$: A, B

(d) Functions $g$ and $h$ have the same graph because $x^2 = (-x)^2$.

(e) $f(x) = -x^2$

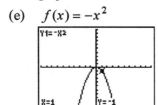

$g(x) = x^2$

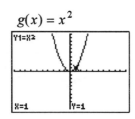

$h(x) = (-x)^2$

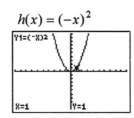

**71.** $B(x) = \sqrt{x^2 + 25}$, $C(x) = x + 5$

(a) $B(0) = \sqrt{0^2 + 25} = 5$

$C(0) = 0 + 5 = 5$

(b) $B(3) = \sqrt{3^2 + 25}$

$= \sqrt{9 + 25}$

$= \sqrt{34}$

$C(3) = 3 + 5 = 8$

(c) $B(-6) = \sqrt{(-6)^2 + 25}$

$= \sqrt{36 + 25}$

$= \sqrt{61}$

$C(-6) = -6 + 5 = -1$

(d) The two functions do not have the same graph because $B(x) \neq C(x)$ for all $x$.

**73.** $f(x) = |x + 7|$, $g(x) = |x| + 7$

(a) If we replace $x$ with only positive values, then $f(x) = |x + 7| = x + 7$ and $g(x) = |x| + 7 = x + 7$. Yes, the graphs of the two functions would be the same.

(b) If, for example, we replace $x$ with $-2$, $f(-2) = |-2 + 7| = 5$ but $g(-2) = |-2| + 7 = 2 + 7 = 9$. So, if we replace $x$ with *any* real number, the graphs of the two functions would not necessarily be the same.

(c) $f(x) = |x + 7|$

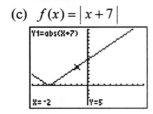

$g(x) = |x| + 7$

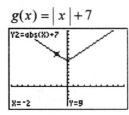

**75.** $f(x) = 5x - 7$

(a) $f(2a) = 5(2a) - 7$

$= 10a - 7$

(b) $f(a + 2) = 5(a + 2) - 7$

$= 5a + 10 - 7$

$= 5a + 3$

Section 2.6  Functions: Notation and Evaluation

**77.** $h(x) = x^2 - 5x + 6$
(a) $h(t) = t^2 - 5t + 6$
(b) $h(2t) = (2t)^2 - 5(2t) + 6$
$= 4t^2 - 10t + 6$

**79.** $f(x) = \sqrt{x^2 + 5}$
(a) $f(a) = \sqrt{a^2 + 5}$
(b) $f(-2a) = \sqrt{(-2a)^2 + 5}$
$= \sqrt{4a^2 + 5}$

**81.** $f(x) = |x^3 - 2|$
(a) $f(-s) = |(-s)^3 - 2|$
$= |-s^3 - 2|$
(b) $f(3s) = |(3s)^3 - 2|$
$= |27s^3 - 2|$

**83.** $f(x) = 2x - 5$
(a) $f(3) = 2(3) - 5 = 6 - 5 = 1$
(b) $f(3 + h) = 2(3 + h) - 5$
$= 6 + 2h - 5$
$= 2h + 1$
(c) $f(3+h) - f(3) = 2h + 1 - 1$
$= 2h$
(d) $\dfrac{f(3+h) - f(3)}{h} = \dfrac{2h}{h} = 2$

**85.** $f(x) = 5 - 3x$
(a) $f(-2) = 5 - 3(-2) = 5 + 6 = 11$
(b) $f(-2 + h) = 5 - 3(-2 + h)$
$= 5 + 6 - 3h$
$= 11 - 3h$
(c) $f(-2 + h) - f(-2) = 11 - 3h - 11$
$= -3h$
(d) $\dfrac{f(-2+h) - f(-2)}{h} = \dfrac{-3h}{h}$
$= -3$

**87.** $p(m) = 60 - \sqrt{2m}$
$p(15) = 60 - \sqrt{2(15)}$
$= 60 - \sqrt{30}$
$\approx 55$

The person's pace rate halfway through a 30-minute workout is 55 paces per minute.

**89.** $B(x) = -2x^2 - 10x + 1500$
$B(8) = -2(8)^2 - 10(8) + 1500$
$= -128 - 80 + 1500$
$= 1292$

There are 1292 bacteria left 24 hours after giving 8 mg of the drug.

**91.**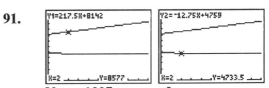
Year: 1997    x: 2
Semi-private: 8577
Private: 4734

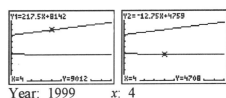
Year: 1999    x: 4
Semi-private: 9012
Private: 4708

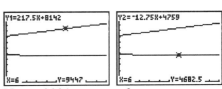
Year: 2001    x: 6
Semi-private: 9447
Private: 4683

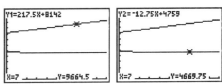
Year: 2002    x: 7
Semi-private: 9665
Private: 4670

**93.**

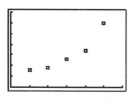

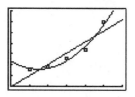

**95.**

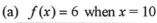

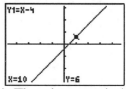

| Year | $f$ Difference | $g$ Difference |
|---|---|---|
| 1960 | $800 - 500 = 300$ | $800 - 841 = -41$ |
| 1970 | $900 - 1020 = -120$ | $900 - 842 = 58$ |
| 1980 | $1300 - 1540 = -240$ | $1300 - 1183 = 117$ |
| 1990 | $1700 - 2060 = -360$ | $1700 - 1864 = -164$ |
| 2000 | $3000 - 2580 = 420$ | $3000 - 2885 = 115$ |

Yes, the function $g$ still appears to be the better model.

**97.** $f(x) = \dfrac{3}{x}$

(a) (i) $f(1) = \dfrac{3}{1} = 3$

(ii) $f(0.1) = \dfrac{3}{0.1} = 30$

(iii) $f(0.01) = \dfrac{3}{0.01} = 300$

(iv) $f(0.001) = \dfrac{3}{0.001} = 3000$

(v) $f(0.0001) = \dfrac{3}{0.0001} = 30{,}000$

b) As $x$ becomes smaller, $3/x$ becomes larger.

c) The result of evaluating $f(0)$ is an error message because division by 0 is undefined.

## Section 2.7 Analysis of Functions

**1.** (a) To estimate the $y$-intercept of a graph with your calculator, trace to the point whose $x$-coordinate is 0 and read the corresponding $y$-coordinate.

(b) To estimate the $y$-intercept of a graph algebraically, evaluate the function for $x = 0$.

**3.** If the graph of a relation has two $y$-intercepts, the relation contains two ordered pairs with the same first coordinate. The graph would not pass the Vertical Line Test. Thus the relation is not a function.

**5.** (a) $f(x) = 6$ when $x = 10$

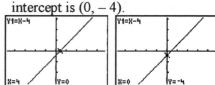

(b) The $x$-intercept is $(4, 0)$ and the $y$-intercept is $(0, -4)$.

## Section 2.7  Analysis of Functions

7. (a) $f(x) = -5$ when $x = 9$

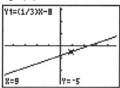

(b) The $x$-intercept is $(24, 0)$ and the $y$-intercept is $(0, -8)$.

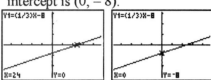

9. (a) $f(x) = 2$, $x < 0$ when $x \approx -3.45$

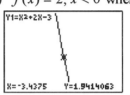

(b) The $x$-intercepts are $(1, 0)$ and $(-3, 0)$. The $y$-intercept is $(0, -3)$.

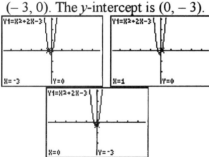

11. (a) $f(x) = 3$, $x > 0$ when $x \approx 1.45$

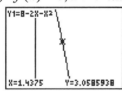

(b) The $x$-intercepts are $(2, 0)$ and $(-4, 0)$. The $y$-intercept is $(0, 8)$.

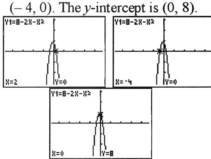

13. (a) $h(-15) = -15$

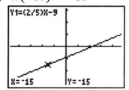

(b) $h(x) = -5$ when $x = 10$

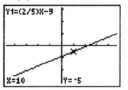

(c) $h(x) = 11$ when $x = 50$

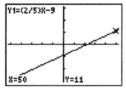

15. (a) $g(7) = 27$

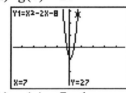

(b) $g(x) = 7$ when $x = 5$ and $x = -3$

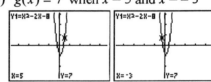

(c) $g(x) = -10$ for no values of $x$. The minimum point on the graph is $(1, -9)$.

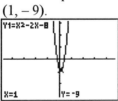

17. (a) $f(8) = 2$

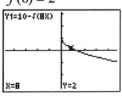

(b) $f(x) = -2$ when $x = 18$
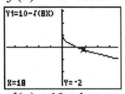

(c) $f(x) = 10$ when $x = 0$
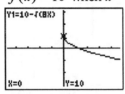

19. (a) $f(6) = 11$
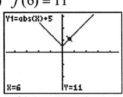

(b) $f(x) = 19$ when $x = 14$ or $x = -14$
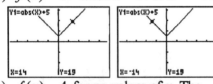

(c) $f(x) = 4$ for no values of $x$. The minimum point on the graph is $(0, 5)$.
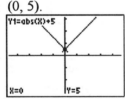

21. No absolute maximum; the absolute minimum is 0.
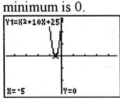

23. The absolute maximum is 9; no absolute minimum.
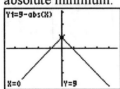

25. No absolute maximum; the absolute minimum is 6.
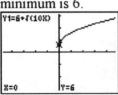

27. (a) The local minimum in Quadrant IV is approximately $(0.82, -1.09)$.
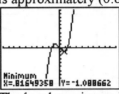

(b) The local maximum in Quadrant II is approximately $(-0.82, 1.09)$.
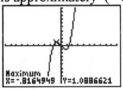

29. (a) The local minimum in Quadrant IV is approximately $(2.63, -9.71)$.
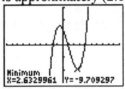

(b) The local maximum in Quadrant II is approximately $(-0.63, 7.71)$.
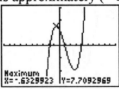

31. Local maximum: $(-1.43, 16.90)$

Local minimum: $(2.10, -5.05)$
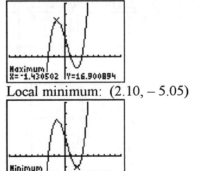

33. Because $f(x) = -(2 + |x|)$ is always negative, the graph of $f$ has no $x$-intercepts.

35. The following graph has an absolute maximum and an absolute minimum.

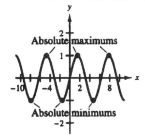

37. The following graph has an absolute minimum but no absolute maximum.

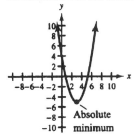

39. The following graph has three $x$-intercepts: $(-3, 0)$, $(1, 0)$, and $(5, 0)$. The graph also has local maximums at $(-1, 4)$ and $(3, 7)$, and a local minimum at $(5, 3)$.

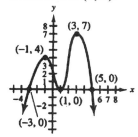

41.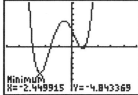

    (a) The graph has 4 $x$-intercepts and 1 $y$-intercept.
    (b) The graph has 2 local minimums and 1 local maximum.

(c) The domain is **R**; the range is approximately $[-4.84, \infty)$.

(d) The $x$-intercept farthest to the right is $(1, 0)$.

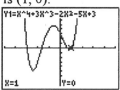

43. The value $-5$ is not in the domain of $g(x) = \sqrt{2x + 3}$ because if $x = -5$, then $g(-5) = \sqrt{-7}$, which is not a real number.

45. (a) There are no $x$-intercepts; the $y$-intercept is $(0, 5)$.

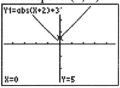

(b) The absolute minimum is 3; there is no absolute maximum.

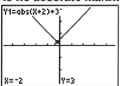

(c) The domain is **R**; the range is $[3, \infty)$.

47. (a) The $x$-intercept is $(1, 0)$; there is no $y$-intercept.

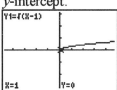

(b) The absolute minimum is 0; there is no absolute maximum.

(c) The domain is $[1, \infty)$; the range is $[0, \infty)$.

**49.** $f(x) = x + 1$, $g(x) = 2x - 3$

(a) $(3, 4)$ belongs to the graph of $f$.

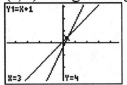

(b) $(0, -3)$ belongs to the graph of $g$.

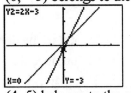

(c) $(4, 5)$ belongs to the graphs of $f$ and $g$.

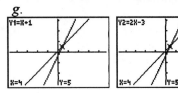

**51.** $f(x) = \frac{1}{2}x$, $g(x) = |x|$

(a) $(10, 5)$ belongs to the graph of $f$.

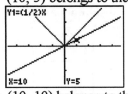

(b) $(10, 10)$ belongs to the graph of $g$.

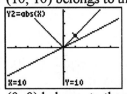

(c) $(0, 0)$ belongs to the graphs of $f$ and $g$.

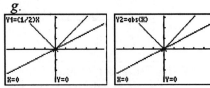

**53.** (a) The pair $(2, 8)$ belongs to the area graph. When the width is 2 feet, the area is 8 square feet.

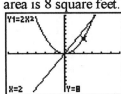

(b) The pair $(2, 12)$ belongs to the perimeter graph. When the width is 2 feet, the perimeter is 12 feet.

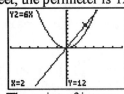

(c) The point of intersection is $(3, 18)$.

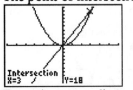

(d) From the $x$-coordinate of the point of intersection, we know that the board should be 3 feet wide. The length of the board is twice the width, or $2(3) = 6$ feet long.

(e) From the $y$-coordinate of the point of intersection, we know that the area of the board (in square feet) and the perimeter of the board (in feet) are both 18.

**55.**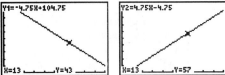

(a) From the graph, we see that the percentage of male shoppers in 2003 was 43%.

(b) From the graph, we see that the year in which female shoppers account for 57% of all shoppers is 2003.

**57.**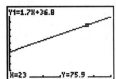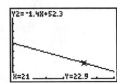

a) From the graph, we see that the year in which approximately 76% of homes heated with gas is 2003.

b) From the graph, we see that the percentage of homes heated with electricity in 2001 was about 23%.

Section 2.7 Analysis of Functions    51

**59.**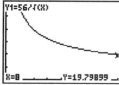

For the period 1998–2003, the absolute minimum of the function is in 2003 when the cost is about 20 cents per minute.

**61.** In 2005, $x = 10$:

$$C(x) = \frac{56}{\sqrt{x}}$$

$$C(10) = \frac{56}{\sqrt{10}} \approx 18$$

If the cost continues to change according to the model, the cost will be about 18 cents per minute in 2005.

**63.** $y = x - 3$

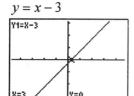

(a) When $y \leq 0$, $x \leq 3$.
(b) When $y \geq 0$, $x \geq 3$.

**65.** $y = x^2 - 9$

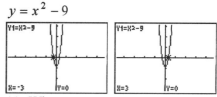

(a) When $y \leq 0$, $-3 \leq x \leq 3$.
(b) When $y \geq 0$, $x \leq -3$ or $x \geq 3$.

**67.** $y = \sqrt{x}$

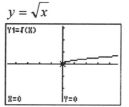

(a) When $y \leq 0$, $x = 0$.
(b) When $y \geq 0$, $x \geq 0$.

**69.** (a) The domain is **R**; Range is $\{5\}$.
(b) When $x_2 > x_1$, $g(x_2) = 5$ and $g(x_1) = 5$. Thus $g(x_2) \geq g(x_1)$ for all $x$ in the domain of $g$.

## Chapter 2 Project

**1.**

| $t$ | Actual Percent $A$ | Modeled Percent $M$ | $M - A$ | $(M - A)^2$ |
|---|---|---|---|---|
| 0 | 18% | 18.7% | + 0.7 | 0.49 |
| 2 | 30% | 29.4% | − 0.6 | 0.36 |
| 4 | 41% | 40.1% | − 0.9 | 0.81 |
| 6 | 50% | 50.8% | + 0.8 | 0.64 |

**3.** (i) $6t + 18$

| $t$ | Actual Percent $A$ | Modeled Percent $M$ | $M - A$ | $(M - A)^2$ |
|---|---|---|---|---|
| 0 | 18% | 18% | 0 | 0 |
| 2 | 30% | 30% | 0 | 0 |
| 4 | 41% | 42% | 1 | 1 |
| 6 | 50% | 54% | 4 | 16 |

(ii) $5.5t + 19$

| $t$ | Actual Percent $A$ | Modeled Percent $M$ | $M - A$ | $(M - A)^2$ |
|---|---|---|---|---|
| 0 | 18% | 19% | 1 | 1 |
| 2 | 30% | 30% | 0 | 0 |
| 4 | 41% | 41% | 0 | 0 |
| 6 | 50% | 52% | 2 | 4 |

(iii) $5t + 20$

| $t$ | Actual Percent $A$ | Modeled Percent $M$ | $M - A$ | $(M - A)^2$ |
|---|---|---|---|---|
| 0 | 18% | 20% | 2 | 4 |
| 2 | 30% | 30% | 0 | 0 |
| 4 | 41% | 40% | $-1$ | 1 |
| 6 | 50% | 50% | 0 | 0 |

(iv) $5.75t + 18$

| $t$ | Actual Percent $A$ | Modeled Percent $M$ | $M - A$ | $(M - A)^2$ |
|---|---|---|---|---|
| 0 | 18% | 18% | 0 | 0 |
| 2 | 30% | 29.5% | $-0.5$ | 0.25 |
| 4 | 41% | 41% | 0 | 0 |
| 6 | 50% | 52.5% | 2.5 | 6.25 |

5.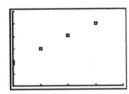

7. Answers will vary.

# Chapter 2 Review Exercises

1. Evaluate $3^2 + 3 + 1$ or store 3 for $x$ and evaluate $x^2 + x + 1$.

3. Evaluate $\dfrac{2(x+1)}{x}$ for $x = -2$.

   $\dfrac{2(x+1)}{x} = \dfrac{2(-2+1)}{-2}$
   
   $= \dfrac{2(-1)}{-2}$
   
   $= \dfrac{-2}{-2}$
   
   $= 1$

5. Evaluate $2x^2 - 3y + 4xy$ for $x = -3$ and $y = 2$.

   $2x^2 - 3y + 4xy$
   
   $= 2(-3)^2 - 3(2) + 4(-3)(2)$
   
   $= 2(9) - 3(2) + 4(-3)(2)$
   
   $= 18 - 6 + (-24)$
   
   $= -12$

7. For $6x + 5 = 1$ and the set $\left\{-7, -4, -\dfrac{2}{3}, 3, \dfrac{5}{4}, 6\right\}$, the true statement $1 = 1$ is the result when $x$ is

replaced by $-\frac{2}{3}$. Replacing $x$ with any other member of the set results in a false statement. Thus, the solution is $x = -\frac{2}{3}$.

9. If $-3$ is stored for $x$, you would need to store $-1$ for $y$ in order for $x/y$ to have a value of 3.

11. (a) $a = 3$, $b = 4$, $c = 5$
$$a^2 + b^2 = c^2$$
$$3^2 + 4^2 = 5^2$$
$$9 + 16 = 25$$
$$25 = 25$$
The triangle is a right triangle.
(b) $a = 12$, $b = 15$, $c = 18$
$$a^2 + b^2 = c^2$$
$$12^2 + 15^2 = 18^2$$
$$144 + 225 = 324$$
$$369 = 224$$
The triangle is not a right triangle.

13. The exponents on $x$ and $y$ are not the same.

15. $3x - 2y - z$
Terms: $3x$, $-2y$, $-z$
Coefficients: $3, -2, -1$

17. $7cb$ and $7cd$ are not like terms.

19. $-(2a + 3b - 4c)$
$= -1(2a + 3b - 4c)$
$= -1(2a) + (-1)(3b) + (-1)(-4c)$
$= -2a - 3b + 4c$

21. $14x^2 - 3x^3 + 2x^2 - 7x^3 + x^3$
$= -9x^3 + 16x^2$

23. (a) The product of $-5$ and the sum of 4 and 3 times the number is represented by
$-5(4 + 3n) = -20 - 15n$.

(b) Ten more than the number less 4 times the number is represented by
$10 + n - 4n = 10 - 3n$.

25. $3 - 3(2x - 1) + 4x = 3 - 6x + 3 + 4x$
$\phantom{3 - 3(2x - 1) + 4x} = -2x + 6$

27. $A(-4, 3)$

29. $C(-2, -3)$

31. $E(3, -4)$

33. Because $-\pi < 0$ and $\pi > 0$, the point $(-\pi, \pi)$ lies in Quadrant II.

35. Because $a < 0$ and $b > 0$, the point $(a, b)$ lies in Quadrant II.

37. If $c = 0$, the point $P(c, d)$ lies on the $y$-axis.

39. $\{(-3, 5), (0, -6), (2, -4), (-5, -3), (7, 0)\}$

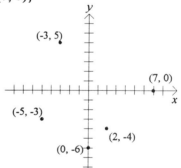

41. $A(-5, 4)$, $B(-8, -5)$
(a) Distance:
$$d = \sqrt{(x_2 - x_1)^2 + (y_2 - y_1)^2}$$
$$d = \sqrt{(-5 + 8)^2 + (4 + 5)^2}$$
$$= \sqrt{(3)^2 + (9)^2}$$
$$= \sqrt{9 + 81}$$
$$= \sqrt{90}$$
$$= 3\sqrt{10} \approx 9.49$$
(b) The midpoint is $(-6.5, -0.5)$:
$$x_m = \frac{-5 - 8}{2} = -6.5,$$
$$y_m = \frac{4 - 5}{2} = -0.5$$

**43.**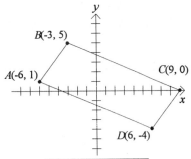

$AB = \sqrt{(-6+3)^2 + (1-5)^2} = \sqrt{25} = 5$

$BC = \sqrt{(-3-9)^2 + (5-0)^2} = \sqrt{169} = 13$

Perimeter: $2(5) + 2(13) = 10 + 26 = 36$

**45.**

**47.**

By tracing the graph, we find the value of the expression is $-6$ when $x = 18$.

**49.** The $y$-coordinate will be $-3$ because the $y$-coordinate is the value of $7 - 2x$ when $x = 5$.

**51.** $\{(0, 1), (3, 5), (-4, 7), (5, -6)\}$
Domain: $\{-4, 0, 3, 5\}$
Range: $\{-6, 1, 5, 7\}$
The relation is a function.

**53.** $y = |x+2|$

For each real number $x$, the expression $|x+2|$ has exactly one value. Thus $y = |x+2|$ represents a function.

**55.** $y^2 = x + 1$

For $x = 3$, for example, the equation $y^2 = 4$ is true for two values of $y$: 2 and $-2$. Thus $y^2 = x + 1$ does not represent a function.

**57.** $y = \sqrt{x+2}$

Domain: $[-2, \infty)$
Range: $[0, \infty)$

**59.** A line containing $(-3, 2)$ and $(4, 0)$ represents a function because the line passes the Vertical Line Test. A circle with its center at the origin does not represent a function because it does not pass the Vertical Line Test. Thus, only (i) represents a function.

**61.** $f(x) = -x^2 - 5x + 2$

$f(-2) = -(-2)^2 - 5(-2) + 2$
$= -4 + 10 + 2$
$= 8$

**63.** $h(c) = \sqrt{c^2 + 4c}$

$h(3) = \sqrt{3^2 + 4(3)}$
$= \sqrt{9 + 12}$
$= \sqrt{21} \approx 4.58$

**65.** $Q(a) = 5a - 3(5a - 3) - a$

$Q(-1) = 5(-1) - 3[5(-1) - 3] - (-1)$
$= -5 - 3(-5 - 3) + 1$
$= -5 - 3(-8) + 1$
$= -5 + 24 + 1$
$= 20$

**67.** $f(x) = |2x - 5|$, $g(x) = \dfrac{3x}{x^2 + 1}$

(a) $f(1) + g(0) = |2(1) - 5| + \dfrac{3(0)}{0^2 + 1}$

$= |2 - 5| + \dfrac{0}{0 + 1}$

$= |-3| + \dfrac{0}{1}$

$= 3 + 0$

$= 3$

(b) $g(1) - f(-2) = \dfrac{3(1)}{1^2+1} - |2(-2) - 5|$

$= \dfrac{3}{1+1} - |-4-5|$

$= \dfrac{3}{2} - |-9|$

$= 1.5 - 9$

$= -7.5$

**69.**

$f(x) = \dfrac{\sqrt{x^2 - 3} - 4}{|x-2|}$

(√ (X ^ 2 − 3) − 4) / ABS(X − 2)

**71.** $f(x) = -0.2x^2 + 3.6x + 8$

(a) The x-intercepts are (− 2, 0) and (20, 0). The y-intercept is (0, 8).

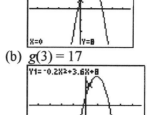

(b) $g(3) = 17$

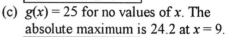

(c) $g(x) = 25$ for no values of x. The absolute maximum is 24.2 at $x = 9$.

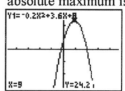

**73.** At the x-intercepts, $f(x) = 0$.

## Chapter 2 Looking Back

**1.** Evaluate $3 - \dfrac{3}{4}t$ for $t = 8$.

$3 - \dfrac{3}{4}(8) = 3 - \dfrac{24}{4} = 3 - 6 = -3$

**3.** 

The value of the expression is − 4 when $x = 2$.

**5.** $11x + 2(4 - 3x) = 11x + 8 - 6x$
$= 5x + 8$

**7.** Is 1 a solution of $2x + 1 = 4 - x$?
$2(1) + 1 = 4 - (1)$
$2 + 1 = 4 - 1$
$3 = 3$

Because the last statement is true, 1 is a solution of the equation.

**9.** (a) Let $L$ = length and $L - 9$ = width. The perimeter is $2L + 2(L - 9)$ or $2L + 2L - 18 = 4L - 18$.

(b) The value (in cents) of the coins is $5n + 10(9 - n) = 5n + 90 - 10n$
$= 90 - 5n$.

(c) The year-end value of an investment of $P$ dollars at 8% interest is
$P + Pr = P(1+r) = P(1 + 0.8)$
$= 1.08P$.

**11.** $|2x - 1|$

For $x = 3$:
$|2(3) - 1| = |6 - 1| = |5| = 5$

For $x = -2$:
$|2(-2) - 1| = |-4 - 1| = |-5| = 5$

# Chapter 2 Test

**1.** Evaluate $x^2 + 1$ for $x = 2$.
$x^2 + 1 = (2)^2 + 1 = 4 + 1 = 5$

**3.** For $3x - 4 = -10$ and the set $\{-2, 0, 2, 3\}$, the true statement $-10 = -10$ is the result when $x$ is replaced by $-2$. Replacing $x$ with any other member of the set results in a false statement. Thus, the solution is $x = -2$.

**5.** Terms are addends of an expression. Factors are quantities that are multiplied.

**7.** 
$$4 - 2[3x - (x - 1)] = 4 - 2(3x - x + 1)$$
$$= 4 - 2(2x + 1)$$
$$= 4 - 4x - 2$$
$$= -4x + 2$$

**9.** (a) $A(-3, 6)$
(b) $B(0, 3)$
(c) $C(3, 0)$
(d) $D(6, -3)$

**11.** $A(-2, 1)$, $B(3, -2)$, $C(2, 4)$
(a) Midpoint of line segment $\overline{BC}$:
$x_m = \dfrac{3+2}{2} = 2.5$, $y_m = \dfrac{4-2}{2} = 1$
The midpoint is $(2.5, 1)$.
(b) Distance between $A$ and $B$:
$d = \sqrt{(x_2 - x_1)^2 + (y_2 - y_1)^2}$
$d = \sqrt{(-2-3)^2 + (1+2)^2}$
$= \sqrt{(-5)^2 + (3)^2}$
$= \sqrt{25 + 9}$
$= \sqrt{34}$
$\approx 5.8$

**13.** When $x$ is replaced with $-4$, the value of the expression $3x - 2$ is
$3x - 2 = 3(-4) - 2 = -12 - 2 = -14$.
Thus the $y$-coordinate is $-14$.

**15.**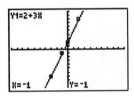

**17.** $g(4) = \sqrt{-3}$, which is not a real number.

**19.** (a) The domain of the given graph is $[-4, 4]$. The range is $[-5, 5]$.
(b) The domain of the relation $\{(1, 3), (2, 7), (3, -1), (5, 3)\}$ is $\{1, 2, 3, 5\}$. The range of the relation is $\{-1, 3, 7\}$.

**21.** $f(x) = \sqrt{2x - 1}$
$f(5) = \sqrt{2(5) - 1}$
$= \sqrt{10 - 1}$
$= \sqrt{9}$
$= 3$

**23.** $h(x) = \dfrac{1}{2}x - 8$
$h(6t) = \dfrac{1}{2}(6t) - 8$
$= 3t - 8$

**25.** $f(x) = x^2 - 9x - 10$
(a) The range is $[-30.25, \infty)$.
(b) The $x$-intercepts are $(-1, 0)$ and $(10, 0)$.

27. $f(x) = |x-2|$, $g(x) = -\frac{1}{2}x + 4$
   (a) The points of intersection are $(-4, 6)$ and $(4, 2)$
   
   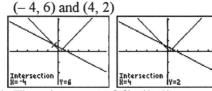
   
   (b) The $y$-intercept of $f$ is $(0, 2)$.
   
   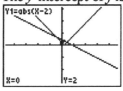

   (b) The first coordinate of the point whose second coordinate is 500 is 2.55 or 7.45. These are the ticket prices at which the profit is 500.
   
   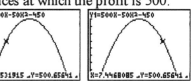

29. $P(a) = 500a - 50a^2 - 450$
   (a) The absolute maximum is 800. This is the maximum profit that can be earned.
   
   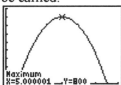

## Cumulative Test for Chapters 1-2

1. The set $\{x \mid -2 < x \leq 3\}$ can be written as $(-2, 3]$ in interval notation.

3. $-6 + [5 - (-4)] = -6 + (5+4)$
   $= -6 + 9$
   $= 3$

5. $2 - |-5| + (-7) = 2 - 5 - 7$
   $= -10$

7. $\frac{1}{2}(-1)(-4)(3)(-2) = -12$

9. $-1 \div 0$ is undefined because division by 0 is undefined.

11. $-5^2 = (-1)5^2 = (-1)25 = -25$

13. $8 - 2 \cdot 3 + 10 \div 5 = 8 - 6 + 2 = 4$

15. $\dfrac{2^2 + 4^2}{2 - 2(5-6)} = \dfrac{4 + 16}{2 - 2(-1)}$
    $= \dfrac{20}{2+2}$
    $= \dfrac{20}{4}$
    $= 5$

17. Using the Distributive Property,
    $9x + 12 = 3(3x) + 3(4) = \underline{3(3x+4)}$.

19. Using the Commutative Property of Multiplication, $x(2+5) = \underline{(2+5)x}$.

**21.** Evaluate $\dfrac{\sqrt{20-2x}}{x^2+4}$ for $x=2$.

$$\dfrac{\sqrt{20-2x}}{x^2+4} = \dfrac{\sqrt{20-2(2)}}{(2)^2+4}$$
$$= \dfrac{\sqrt{20-4}}{4+4}$$
$$= \dfrac{\sqrt{16}}{8}$$
$$= 0.5$$

**23.** For $\dfrac{x}{2} = -(3-x)$ and the set $\{-2, 0, 3, 6\}$, the true statement $3 = 3$ is the result when $x$ is replaced by 6. Replacing $x$ with any other member of the set results in a false statement. Thus, the solution is $x = 6$.

**25.** $-2(x-5) - (4-x) = -2x + 10 - 4 + x$
$\phantom{-2(x-5) - (4-x)} = -x + 6$

**27.** (a) Three more than the quotient of a number and 7 is represented by $\dfrac{x}{7} + 3$.
(b) If the difference of 6 and a number is multiplied by 2, the result is twice the number is represented by $2(6-x) = 2x$.

**29.** $P(-6, 3)$, $Q(8, -7)$
(a) $PQ$ represents the distance between the two points.
$$d = \sqrt{(x_2 - x_1)^2 + (y_2 - y_1)^2}$$
$$d = \sqrt{(-6-8)^2 + (3+7)^2}$$
$$= \sqrt{(-14)^2 + (10)^2}$$
$$= \sqrt{196 + 100}$$
$$= \sqrt{296}$$
$$\approx 17.20$$
(b) The midpoint of $\overline{PQ}$ is
$$x_m = \dfrac{8-6}{2} = 1,\ y_m = \dfrac{3-7}{2} = -2$$
The midpoint is $(1, -2)$.

**31.** (a) $g(x) = 4(3x-1) - (2-x)$
$g(-1) = 4[3(-1) - 1] - [2 - (-1)]$
$\phantom{g(-1)} = 4(-3-1) - (2+1)$
$\phantom{g(-1)} = 4(-4) - 3$
$\phantom{g(-1)} = -16 - 3$
$\phantom{g(-1)} = -19$
(b)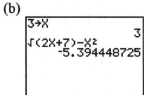

**33.** The relation in (i) is a function because each $x$-coordinate has exactly one $y$-coordinate associated with it. The relation in (ii) is not a function because there are numerous $y$-coordinates associated with the $x$-coordinate 3. The relation in (iii) is a function because for each real value of $x$, there is exactly one value of $y$.

**35.** $y = 0.5x^2 + 8x - 7$
The $x$-coordinates of the two points whose $y$-coordinates are both 11 are $-18$ and 2.
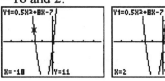

**37.**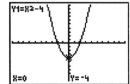
The domain is the set of real numbers, **R**. The range is $[-4, \infty)$.

# Chapter 3

# Linear Equations and Inequalities

## Section 3.1  Introduction to Linear Equations

1. An equation is a statement that two algebraic expressions have the same value.

3. $x - 3 + 2x = 5$ is an equation

5. $4(x-5) - 7(2x+3)$ is not an equation. It does not state that two algebraic expressions have the same value.

7. $0 = -7$ is an equation. It is not a true statement, but it is an equation.

9. $3x - 5 + 7x + 12$ is not an equation. It does not state that two algebraic expressions have the same value.

11. $x^2 = 5x + 6$ is an equation.

13. $\dfrac{4}{7}x + 3 = 0$ is a linear equation in one variable. It has the form $Ax + B = 0$, where $A \neq 0$.

15. $x^3 - 2x = 4$ is not a linear equation in one variable because $x^3$ is involved. It cannot be written in the form $Ax + B = 0$.

17. $2\pi^2 + 5x = 6$ is a linear equation in one variable. It can be written in the form $Ax + B = 0$, where $A \neq 0$, by writing $5x + 2\pi^2 - 6 = 0$.

19. $\sqrt{x} = 5$ is not a linear equation in one variable because $\sqrt{x}$ is involved. It cannot be written in the form $Ax + B = 0$.

21. The equation is true for some values of $x$ $(x \geq 0)$, but it is false for other values of $x$ $(x < 0)$.

23. Verify that $-3$ is a solution:
$$3x - 5 = 7x + 7$$
$$3(-3) - 5 = 7(-3) + 7$$
$$-9 - 5 = -21 + 7$$
$$-14 = -14$$

25. Verify that 5 is a solution:
$$4(x-2) + 5 = 5(3-x) + 3(x+4)$$
$$4(5-2) + 5 = 5(3-5) + 3(5+4)$$
$$4(3) + 5 = 5(-2) + 3(9)$$
$$12 + 5 = -10 + 27$$
$$17 = 17$$

27. Verify that 2 is a solution:
$$-2[2x - 3(1-x)] = -5(x+3) + 11$$
$$-2[2(2) - 3(1-2)] = -5(2+3) + 11$$
$$-2[4 - 3(-1)] = -5(5) + 11$$
$$-2(4+3) = -25 + 11$$
$$-2(7) = -14$$
$$-14 = -14$$

**29.** Verify that $-\dfrac{10}{3}$ is a solution:

$$\dfrac{3}{5}y = -2$$
$$\dfrac{3}{5}\left(-\dfrac{10}{3}\right) = -2$$
$$-\dfrac{30}{15} = -2$$
$$-2 = -2$$

**31.** Verify that $-7$ is a solution:

$$6x - (5x - 8) = 1$$
$$6(-7) - [5(-7) - 8] = 1$$
$$-42 - (-35 - 8) = 1$$
$$-42 - (-43) = 1$$
$$1 = 1$$

**33.** $x = 3$:

$$3x - 4 = 11$$
$$3(3) - 4 = 11$$
$$9 - 4 = 11$$
$$5 = 11 \quad \text{False}$$

$x = 3$ is not a solution.

$x = 5$:

$$3x - 4 = 11$$
$$3(5) - 4 = 11$$
$$15 - 4 = 11$$
$$11 = 11 \quad \text{True}$$

$x = 5$ is a solution.

$x = 7$:

$$3x - 4 = 11$$
$$3(7) - 4 = 11$$
$$21 - 4 = 11$$
$$17 = 11 \quad \text{False}$$

$x = 7$ is not a solution.

**35.** $x = -1$:

$$\dfrac{x}{2} + \dfrac{1}{3} = \dfrac{x}{3} - \dfrac{1}{2}$$
$$\dfrac{-1}{2} + \dfrac{1}{3} = \dfrac{-1}{3} - \dfrac{1}{2}$$
$$-\dfrac{3}{6} + \dfrac{2}{6} = -\dfrac{2}{6} - \dfrac{3}{6}$$
$$-\dfrac{1}{6} = -\dfrac{5}{6} \quad \text{False}$$

$x = -1$ is not a solution.

$x = -4$:

$$\dfrac{x}{2} + \dfrac{1}{3} = \dfrac{x}{3} - \dfrac{1}{2}$$
$$\dfrac{-4}{2} + \dfrac{1}{3} = \dfrac{-4}{3} - \dfrac{1}{2}$$
$$-\dfrac{12}{6} + \dfrac{2}{6} = -\dfrac{8}{6} - \dfrac{3}{6}$$
$$-\dfrac{10}{6} = -\dfrac{11}{6} \quad \text{False}$$

$x = -4$ is not a solution.

$x = -5$:

$$\dfrac{x}{2} + \dfrac{1}{3} = \dfrac{x}{3} - \dfrac{1}{2}$$
$$\dfrac{-5}{2} + \dfrac{1}{3} = \dfrac{-5}{3} - \dfrac{1}{2}$$
$$-\dfrac{15}{6} + \dfrac{2}{6} = -\dfrac{10}{6} - \dfrac{3}{6}$$
$$-\dfrac{13}{6} = -\dfrac{13}{6} \quad \text{True}$$

$x = -5$ is a solution.

**37.** $x = -3$:

$$3(x - 2) + 4 = 3x - 2$$
$$3(-3 - 2) + 4 = 3(-3) - 2$$
$$3(-5) + 4 = -9 - 2$$
$$-15 + 4 = -11$$
$$-11 = -11 \quad \text{True}$$

$x = -3$ is a solution.

$x = 0$:

$$3(x - 2) + 4 = 3x - 2$$
$$3(0 - 2) + 4 = 3(0) - 2$$
$$3(-2) + 4 = 0 - 2$$
$$-6 + 4 = -2$$
$$-2 = -2 \quad \text{True}$$

$x = 0$ is a solution.

$x = 2$:

$$3(x - 2) + 4 = 3x - 2$$
$$3(2 - 2) + 4 = 3(2) - 2$$
$$3(0) + 4 = 6 - 2$$
$$0 + 4 = 4$$
$$4 = 4 \quad \text{True}$$

$x = 2$ is a solution.

Section 3.1 Introduction to Linear Equations

**39.** $t = -11$:
$$5(t+2) = 3(t-4)$$
$$5(-11+2) = 3(-11-4)$$
$$5(-9) = 3(-15)$$
$$-45 = -45 \quad \text{True}$$
$t = -11$ is a solution.

$t = -1$:
$$5(t+2) = 3(t-4)$$
$$5(-1+2) = 3(-1-4)$$
$$5(1) = 3(-5)$$
$$5 = -15 \quad \text{False}$$
$t = -1$ is not a solution.

$t = 0$:
$$5(t+2) = 3(t-4)$$
$$5(0+2) = 3(0-4)$$
$$5(2) = 3(-4)$$
$$10 = -12 \quad \text{False}$$
$t = 0$ is not a solution.

**41.** The graphs of the left and right sides of an inconsistent linear equation are parallel lines. The solution set is $\varnothing$.

**43.** From the given graph, the estimated solution is $x = 3$.

**45.** From the given graph, the estimated solution is $x = 3$.

**47.** Graph $y_1 = 4x - 3$ and $y_2 = 5$. Trace to the point of intersection (2, 5). When $x = 2$, $y_1 = y_2$. The solution is $x = 2$. The screens show the solution and verification.

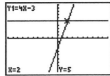

**49.** Graph $y_1 = 2 - 3x$ and $y_2 = -4$. Trace to the point of intersection (2, –4). When $x = 2$, $y_1 = y_2$. The solution is $x = 2$. The screens show the solution and verification.

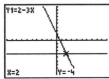 

**51.** Graph $y_1 = 2x + 3$ and $y_2 = 3 + 2x$. The graphs appear to coincide. The equation is an identity. The solution set is **R**.

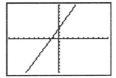

**53.** Graph $y_1 = 4 - x$ and $y_2 = 3x + 12$. Trace to the point of intersection (–2, 6). When $x = -2$, $y_1 = y_2$. The solution is $x = -2$. The screens show the solution and verification.

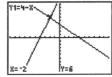 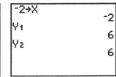

**55.** Graph $y_1 = 2(x+3)$ and $y_2 = 6 + 5x$. Trace to the point of intersection (0, 6). When $x = 0$, $y_1 = y_2$. The solution is $x = 0$. The screens show the solution and verification.

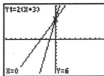 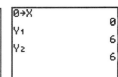

**57.** Graph $y_1 = 2 - 3x$ and $y_2 = -3(x+4)$. The graphs appear to be parallel. The equation is inconsistent. The solution set is $\varnothing$.

**59.** Graph $y_1 = 58 - (4 - 3x)$ and $y_2 = 0$. Trace to the point of intersection (–18, 0). When $x = -18$, $y_1 = y_2$. The solution is $x = -18$. The screens show the solution and verification.

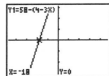 

**61.** Graph $y_1 = 1 - 2(x - 1)$ and $y_2 = -(2x - 3)$. The graphs appear to coincide. The equation is an identity. The solution set is **R**.

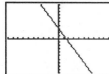

**63.** Graph $y_1 = \dfrac{1}{2}x + 2$ and $y_2 = -\dfrac{1}{6}x + 6$. Trace to the point of intersection (6, 5). When $x = 6$, $y_1 = y_2$. The solution is $x = 6$. The screens show the solution and verification.

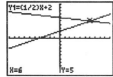 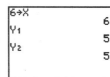

**65.** Graph $y_1 = \dfrac{x - 3}{3}$ and $y_2 = \dfrac{x}{6}$. Trace to the point of intersection (6, 1). When $x = 6$, $y_1 = y_2$. The solution is $x = 6$. The screens show the solution and verification.

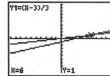 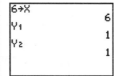

**67.** (a) The number $a$ is the solution of the equation.
(b) The number $b$ is the value of both the left and right sides when the variable is replaced with $a$.

**69.** Graph $y_1 = 2x + 3$ and $y_2 = 3x - 8$. Trace to the point of intersection (11, 25). When $x = 11$, $y_1 = y_2$. The solution is $x = 11$. The screens show the solution and verification.

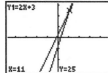 

**71.** Graph $y_1 = \dfrac{3}{2}(x + 7)$ and $y_2 = 3$. Trace to the point of intersection $(-5, 3)$. When $x = -5$, $y_1 = y_2$. The solution is $x = -5$. The screens show the solution and verification.

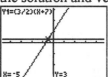 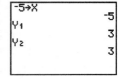

**73.** Graph $y_1 = 3.75 - 0.5x$ and $y_2 = -0.25x$. Trace to the point of intersection $(15, -3.75)$. When $x = 15$, $y_1 = y_2$. The solution is $x = 15$. The screens show the solution and verification.

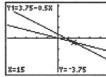

**75.** Graph $y_1 = \dfrac{5}{6}x + 3$ and $y_2 = -7$. Trace to the point of intersection $(-12, -7)$. When $x = -12$, $y_1 = y_2$. The solution is $x = -12$. The screens show the solution and verification.

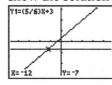 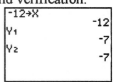

**77.** (a) The monthly cost of the first plan can be represented by the expression $20 + 2h$, where $h$ represents the number of on-line hours. The monthly cost of the second plan can be represented by the expression $10 + 2.25h$.
(b) Graph $y_1 = 20 + 2x$ and $y_2 = 10 + 2.25x$. The point of intersection is (40, 100). The solution is $h = 40$.

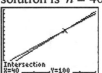

Section 3.2   Solving Linear Equations

(c) The estimated solution indicates that for 40 on-line hours, the two plans cost the same.

(d) The second coordinate of the solution plan indicates that for both plans, the cost is $100 for 40 on-line hours.

79.
$$E(x) - S(x) = 0.7$$
$$0.020x + 1.74 - (0.018x + 1.09) = 0.7$$
$$0.020x + 1.74 - 0.018x - 1.09 = 0.7$$
$$0.002x + 0.65 = 0.7$$

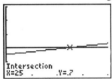

The solution is 25, which corresponds to 2015. In that year, there are projected to be 0.7 million more elementary teachers than secondary teachers.

81. An equation that estimates the age at which 40% of cardholders pay their bills monthly is $1.16a + 9.27 = 40$. By graphing $y_1 = 1.16x + 9.27$ and $y_2 = 40$, we find that the point of intersection is approximately (26, 40). The solution is $a \approx 26$.

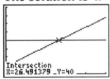

83. Graph $y_1 = 3x + 40$ and $y_2 = 3.5x - 20$. The point of intersection is (120, 400). The solution is $x = 120$.

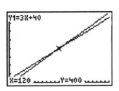

85. Graph $y_1 = \sqrt{3}\,x - 12$ and $y_2 = \pi x$. The point of intersection is approximately (−8.5, −26.7). The solution is $x \approx -8.5$.

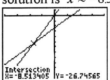

87. Graph $y_1 = |x - 12|$ and $y_2 = 18$. The points of intersection are (−6, 18) and (30, 18). Solutions: −6 and 30.

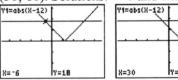

89. Graph $y_1 = 0.1x^2 - 4x + 20$ and $y_2 = -25$. Because the graphs do not intersect, the equation has no solution.

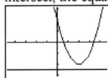

91. The equation $0 \cdot x + B = C$ is not a linear equation in one variable because it cannot be written as $Ax + B = 0$, where $A \neq 0$. The equation is an identity if $B = C$.

## Section 3.2   Solving Linear Equations

1. Equivalent equations are equations that have exactly the same solution sets.

3.
$$x + 3 = 7$$
$$x + 3 - 3 = 7 - 3 \quad \text{Add } -3 \text{ to both sides.}$$
$$x = 4$$

5.
$$z - 3 = -11$$
$$z - 3 + 3 = -11 + 3 \quad \text{Add 3 to both sides.}$$
$$z = -8$$

**7.**
$$9x = 8x$$
$$9x - 8x = 8x - 8x \quad \text{Add } -8x \text{ to both sides.}$$
$$x = 0$$

**9.**
$$7 - 2s = 10 - 3s$$
$$7 - 2s - 7 = 10 - 3s - 7$$
$$-2s = 3 - 3s$$
$$-2s + 3s = 3 - 3s + 3s$$
$$s = 3$$

**11.**
$$8 + 3z = 2z - 9$$
$$8 + 3z - 8 = 2z - 9 - 8$$
$$3z = 2z - 17$$
$$3z - 2z = 2z - 17 - 2z$$
$$z = -17$$

**13.**
$$4x - 3 + 3x = 7 + 6x - 4$$
$$7x - 3 = 6x + 3$$
$$7x - 3 + 3 = 6x + 3 + 3$$
$$7x = 6x + 6$$
$$7x - 6x = 6x + 6 - 6x$$
$$x = 6$$

**15.** For $x + 2 = 3$, 2 is added to $x$, and so we add $-2$ to both sides to isolate the variable. For $-2x = 6$, $x$ is multiplied by $-2$, and so we divide both sides by $-2$ to isolate the variable.

**17.**
$$7x = 0$$
$$\frac{7x}{7} = \frac{0}{7} \quad \text{Divide both sides by 7.}$$
$$x = 0$$

**19.**
$$-x = 9$$
$$(-1)(-x) = (-1)9 \quad \text{Multiply both sides by } -1.$$
$$x = -9$$

**21.**
$$-3x = \frac{2}{5}$$
$$\frac{-3x}{-3} = \frac{2}{5} \div (-3) \quad \text{Divide both sides by } -3.$$
$$x = \frac{2}{5} \cdot \left(-\frac{1}{3}\right)$$
$$x = -\frac{2}{15}$$

**23.**
$$\frac{y}{3} = 12$$
$$(3) \cdot \frac{y}{3} = (3) \cdot 12 \quad \text{Multiply both sides by 3.}$$
$$y = 36$$

**25.**
$$x + 3x + x = 30$$
$$5x = 30$$
$$\frac{5x}{5} = \frac{30}{5} \quad \text{Divide both sides by 5.}$$
$$x = 6$$

**27.**
$$5x + 8 = 23$$
$$5x + 8 - 8 = 23 - 8$$
$$5x = 15$$
$$\frac{5x}{5} = \frac{15}{5}$$
$$x = 3$$
Verification:
$$5x + 8 = 23$$
$$5(3) + 8 = 23$$
$$15 + 8 = 23$$
$$23 = 23$$

**29.**
$$7x - 5 = 8x + 7$$
$$7x - 5 - 8x = 8x + 7 - 8x$$
$$-x - 5 = 7$$
$$-x - 5 + 5 = 7 + 5$$
$$-x = 12$$
$$\frac{-x}{-1} = \frac{12}{-1}$$
$$x = -12$$

**31.**
$$3(3 - 2x) = 33 - 2x$$
$$9 - 6x = 33 - 2x$$
$$9 - 6x + 2x = 33 - 2x + 2x$$
$$9 - 4x = 33$$
$$9 - 4x - 9 = 33 - 9$$
$$-4x = 24$$
$$\frac{-4x}{-4} = \frac{24}{-4}$$
$$x = -6$$

Section 3.2 Solving Linear Equations

**33.**
$$7w - 5 = 11w - 5 - 4w$$
$$7w - 5 = 7w - 5$$
$$7w - 5 - 7w = 7w - 5 - 7w$$
$$-5 = -5$$
Because the resulting equation is true, the original equation is an identity, and the solution set is **R**.

**35.**
$$9x + 5 = 5x + 3(x - 1)$$
$$9x + 5 = 5x + 3x - 3$$
$$9x + 5 = 8x - 3$$
$$9x + 5 - 8x = 8x - 3 - 8x$$
$$x + 5 = -3$$
$$x + 5 - 5 = -3 - 5$$
$$x = -8$$

**37.**
$$16 + 7(6 - x) = 15 - 4(x + 2)$$
$$16 + 42 - 7x = 15 - 4x - 8$$
$$58 - 7x = 7 - 4x$$
$$58 - 7x + 4x = 7 - 4x + 4x$$
$$58 - 3x = 7$$
$$58 - 3x - 58 = 7 - 58$$
$$-3x = -51$$
$$\frac{-3x}{-3} = \frac{-51}{-3}$$
$$x = 17$$

**39.**
$$2(3a + 2) = 2(a + 1) + 4a$$
$$6a + 4 = 2a + 2 + 4a$$
$$6a + 4 = 6a + 2$$
$$6a + 4 - 6a = 6a + 2 - 6a$$
$$4 = 2$$
Because the resulting equation is false, the original equation is a contradiction, and the solution set is $\varnothing$.

**41.**
$$-2(x + 5) = 5(1 - x) + 3(7 - x)$$
$$-2x - 10 = 5 - 5x + 21 - 3x$$
$$-2x - 10 = 26 - 8x$$
$$-2x - 10 + 8x = 26 - 8x + 8x$$
$$6x - 10 = 26$$
$$6x - 10 + 10 = 26 + 10$$
$$6x = 36$$
$$\frac{6x}{6} = \frac{36}{6}$$
$$x = 6$$

**43.** The Multiplication Property of Equations permits us to multiply both *sides* by 15: $15\left(\frac{1}{3} - 2x\right) = 15\left(\frac{2}{5}\right)$. By the Distributive Property, we obtain $15 \cdot \frac{1}{3} - 15 \cdot 2x = 15 \cdot \frac{2}{5}$.

**45.** The LCD is 12. Multiply each term on both sides by the LCD.
$$2x - \frac{3}{4} = -\frac{5}{6}$$
$$12 \cdot 2x - 12 \cdot \frac{3}{4} = 12 \cdot \left(-\frac{5}{6}\right)$$
$$(12 \cdot 2)x - 12 \cdot \frac{3}{4} = 12 \cdot \left(-\frac{5}{6}\right)$$
$$24x - 9 = -10$$
$$24x - 9 + 9 = -10 + 9$$
$$24x = -1$$
$$\frac{24x}{24} = \frac{-1}{24}$$
$$x = -\frac{1}{24}$$

**47.** The LCD is 8.
$$\frac{y}{2} + \frac{y}{4} = \frac{7}{8} - \frac{y}{8}$$
$$8 \cdot \frac{y}{2} + 8 \cdot \frac{y}{4} = 8 \cdot \frac{7}{8} - 8 \cdot \frac{y}{8}$$
$$4y + 2y = 7 - y$$
$$6y = 7 - y$$
$$6y + y = 7 - y + y$$
$$7y = 7$$
$$\frac{7y}{7} = \frac{7}{7}$$
$$y = 1$$

**49.** The LCD is 4.
$$\frac{5}{4}(x+2) = \frac{x}{2}$$
$$4 \cdot \frac{5}{4}(x+2) = 4 \cdot \frac{x}{2}$$
$$5(x+2) = 2x$$
$$5x + 10 = 2x$$
$$5x + 10 - 2x = 2x - 2x$$
$$3x + 10 = 0$$
$$3x + 10 - 10 = 0 - 10$$
$$3x = -10$$
$$\frac{3x}{3} = \frac{-10}{3}$$
$$x = -\frac{10}{3}$$

**51.** The LCD is 9.
$$x + \frac{3x-1}{9} = 4 + \frac{3x+1}{3}$$
$$9 \cdot x + 9 \cdot \frac{3x-1}{9} = 9 \cdot 4 + 9 \cdot \frac{3x+1}{3}$$
$$9x + (3x - 1) = 36 + 3(3x + 1)$$
$$9x + 3x - 1 = 36 + 9x + 3$$
$$12x - 1 = 39 + 9x$$
$$12x - 1 - 9x = 39 + 9x - 9x$$
$$3x - 1 = 39$$
$$3x - 1 + 1 = 39 + 1$$
$$3x = 40$$
$$\frac{3x}{3} = \frac{40}{3}$$
$$x = \frac{40}{3}$$

**53.** The LCD is 12.
$$\frac{1}{4} + \frac{1}{6}(4a+5) = 2(a-3) + \frac{5}{12}$$
$$12 \cdot \frac{1}{4} + 12 \cdot \frac{1}{6}(4a+5) = 12 \cdot 2(a-3) + 12 \cdot \frac{5}{12}$$
$$3 + 2(4a+5) = 24(a-3) + 5$$
$$3 + 8a + 10 = 24a - 72 + 5$$
$$8a + 13 = 24a - 67$$
$$8a + 13 - 24a = 24a - 67 - 24a$$
$$-16a + 13 = -67$$
$$-16a + 13 - 13 = -67 - 13$$
$$-16a = -80$$
$$\frac{-16a}{-16} = \frac{-80}{-16}$$
$$a = 5$$

**55.**
$$2.6 = 0.4z + 1$$
$$2.6 - 0.4z = 0.4z + 1 - 0.4z$$
$$2.6 - 0.4z = 1$$
$$2.6 - 0.4z - 2.6 = 1 - 2.6$$
$$-0.4z = -1.6$$
$$\frac{-0.4z}{-0.4} = \frac{-1.6}{-0.4}$$
$$z = 4$$

**57.**
$$0.08x + 0.15(x + 200) = 7$$
$$100[0.08x + 0.15(x+200)] = 100(7)$$
$$8x + 15(x+200) = 700$$
$$8x + 15x + 3000 = 700$$
$$23x + 3000 = 700$$
$$23x + 3000 - 3000 = 700 - 3000$$
$$23x = -2300$$
$$\frac{23x}{23} = \frac{-2300}{23}$$
$$x = -100$$

**59.**
$$x + 4 = 7 \qquad 3x - 5 = 4$$
$$x + 4 - 4 = 7 - 4 \qquad 3x - 5 + 5 = 4 + 5$$
$$x = 3 \qquad 3x = 9$$
$$\qquad \frac{3x}{3} = \frac{9}{3}$$
$$\qquad x = 3$$
Because the equations have exactly the same solution set, they are equivalent.

Section 3.2  Solving Linear Equations

**61.**
$$x + 3 = 4 + x$$
$$x + 3 - x = 4 + x - x$$
$$3 = 4$$
Solution set: $\varnothing$
$$2x - 3 = 4 + 2x$$
$$2x - 3 - 2x = 4 + 2x - 2x$$
$$-3 = 4$$
Solution set: $\varnothing$
Because the equations have exactly the same solution set, they are equivalent.

**63.**
$$\frac{1}{4}x = -2$$
$$4 \cdot \frac{1}{4}x = 4 \cdot (-2)$$
$$x = -8$$

$$8 - x = 0$$
$$8 - x - 8 = 0 - 8$$
$$-x = -8$$
$$\frac{-x}{-1} = \frac{-8}{-1}$$
$$x = 8$$

Because the equations do not have exactly the same solution sets, they are not equivalent.

**65.**
$$x - 6 = 6 - x$$
$$x - 6 + x = 6 - x + x$$
$$2x - 6 = 6$$
$$2x - 6 + 6 = 6 + 6$$
$$2x = 12$$
$$\frac{2x}{2} = \frac{12}{2}$$
$$x = 6$$
This is a conditional equation.

**67.**
$$x + 6 = 6 + x$$
$$x + 6 - x = 6 + x - x$$
$$6 = 6$$
Because the resulting equation is true, the original equation is an identity, and the solution set is **R**.

**69.**
$$2x - 5 = 7 + 3x - (x - 3)$$
$$2x - 5 = 7 + 3x - x + 3$$
$$2x - 5 = 10 + 2x$$
$$2x - 5 - 2x = 10 + 2x - 2x$$
$$-5 = 10$$
Because the resulting equation is false, the original equation is an inconsistent equation, and the solution set is $\varnothing$.

**71.** The LCD is 15.
$$\frac{1}{5}x - 2 = \frac{1}{3}x - 4$$
$$15 \cdot \frac{1}{5}x - 15 \cdot 2 = 15 \cdot \frac{1}{3}x - 15 \cdot 4$$
$$3x - 30 = 5x - 60$$
$$3x - 30 - 5x = 5x - 60 - 5x$$
$$-2x - 30 = -60$$
$$-2x - 30 + 30 = -60 + 30$$
$$-2x = -30$$
$$\frac{-2x}{-2} = \frac{-30}{-2}$$
$$x = 15$$
This is a conditional equation.

**73.** First equation:
$$5 - 3x = c + 2x$$
$$5 - 3x - 2x = c + 2x - 2x$$
$$5 - 5x = c$$
$$5 - 5x - 5 = c - 5$$
$$-5x = c - 5$$
$$\frac{-5x}{-5} = \frac{c-5}{-5}$$
$$x = \frac{-c+5}{5}$$
Second equation:
$$2x - 3 = 5$$
$$2x - 3 + 3 = 5 + 3$$
$$2x = 8$$
$$\frac{2x}{2} = \frac{8}{2}$$
$$x = 4$$
In order for these equations to be equivalent, their solutions should be equal.
$$\frac{-c+5}{5} = 4$$
$$5 \cdot \frac{-c+5}{5} = 5 \cdot 4$$
$$-c + 5 = 20$$
$$-c + 5 - 5 = 20 - 5$$
$$-c = 15$$
$$\frac{-c}{-1} = \frac{15}{-1}$$
$$c = -15$$
So, for $c = -15$, the two given equations are equivalent.

**75.** First equation:
$$2x + c = 3x + 2$$
$$2x + c - 3x = 3x + 2 - 3x$$
$$-x + c = 2$$
$$-x + c - c = 2 - c$$
$$-x = 2 - c$$
$$\frac{-x}{-1} = \frac{2-c}{-1}$$
$$x = -2 + c$$

Second equation:
$$4x + 7 = 6x - 5$$
$$4x + 7 - 6x = 6x - 5 - 6x$$
$$-2x + 7 = -5$$
$$-2x + 7 - 7 = -5 - 7$$
$$-2x = -12$$
$$\frac{-2x}{-2} = \frac{-12}{-2}$$
$$x = 6$$

In order for these equations to be equivalent, their solutions should be equal.
$$-2 + c = 6$$
$$-2 + c + 2 = 6 + 2$$
$$c = 8$$
So, for $c = 8$, the two given equations are equivalent.

**77.**
$$3x - 2 = 4 + kx$$
$$3x - 2 - kx = 4 + kx - kx$$
$$3x - 2 - kx = 4$$
$$3x - 2 - kx + 2 = 4 + 2$$
$$3x - kx = 6$$
$$(3 - k)x = 6$$

The equation will be inconsistent if the last statement yields a false statement. The last statement will be false only if $k = 3$ because
$$(3 - 3)x = 6$$
$$0 = 6$$

**79.**
$$2x - 4 = x - 1 + kx$$
$$2x - 4 - x - kx = x - 1 + kx - x - kx$$
$$x - kx - 4 = -1$$
$$x - kx - 4 + 4 = -1 + 4$$
$$x - kx = 3$$
$$(1 - k)x = 3$$

The equation will be inconsistent if the last statement yields a false statement. The last statement will be false only if

The last statement will be false only if $k = 1$ because
$$(1 - 1)x = 3$$
$$0 = 3$$

**81.**
$$2x + 3 = c + 2x$$
$$2x + 3 - 2x = c + 2x - 2x$$
$$3 = c$$
The equation will be an identity if the last statement yields a true statement. The last statement will be true if $c = 3$.

**83.**
$$cx + 2 = 2 + 5x$$
$$cx + 2 - 5x = 2 + 5x - 5x$$
$$cx - 5x + 2 = 2$$
$$cx - 5x + 2 - 2 = 2 - 2$$
$$cx - 5x = 0$$
$$(c - 5)x = 0$$
The equation will be an identity if the last statement yields a true statement. The last statement will be true if $c = 5$.

**85.**
$$3x - 5 = m$$
$$3x - 5 + 5 = m + 5$$
$$3x = m + 5$$
$$\frac{3x}{3} = \frac{m+5}{3}$$
$$x = \frac{m+5}{3}$$

If the solution of the equation is 7, then
$$\frac{m+5}{3} = 7$$
$$3 \cdot \frac{m+5}{3} = 3 \cdot 7$$
$$m + 5 = 21$$
$$m + 5 - 5 = 21 - 5$$
$$m = 16$$

The value of $m$ must be 16 for the original equation to have 7 as its solution.

Section 3.2  Solving Linear Equations

**87.**
$$3x + m - 2 = 4 - x$$
$$3x + m - 2 + x = 4 - x + x$$
$$4x + m - 2 = 4$$
$$4x + m - 2 - m + 2 = 4 - m + 2$$
$$4x = 6 - m$$
$$\frac{4x}{4} = \frac{6-m}{4}$$
$$x = \frac{6-m}{4}$$

If the solution of the equation is 7, then
$$\frac{6-m}{4} = 7$$
$$4 \cdot \frac{6-m}{4} = 4 \cdot 7$$
$$6 - m = 28$$
$$6 - m - 6 = 28 - 6$$
$$-m = 22$$
$$\frac{-m}{-1} = \frac{22}{-1}$$
$$m = -22$$

The value of $m$ must be $-22$ for the original equation to have 7 as its solution.

**89.**
$$2x - 5 = 5 - 2x$$
$$2x - 5 + 2x = 5 - 2x + 2x$$
$$4x - 5 = 5$$
$$4x - 5 + 5 = 5 + 5$$
$$4x = 10$$
$$\frac{4x}{4} = \frac{10}{4}$$
$$x = \frac{5}{2}$$

**91.**
$$17 - 4z = 3z + 52$$
$$17 - 4z - 3z = 3z + 52 - 3z$$
$$17 - 7z = 52$$
$$17 - 7z - 17 = 52 - 17$$
$$-7z = 35$$
$$\frac{-7z}{-7} = \frac{35}{-7}$$
$$z = -5$$

**93.**
$$3(x+1) - 2 = 2(x+1) + x - 1$$
$$3x + 3 - 2 = 2x + 2 + x - 1$$
$$3x + 1 = 3x + 1$$
$$3x + 1 - 3x = 3x + 1 - 3x$$
$$1 = 1$$

Because the resulting equation is true, the original equation is an identity, and the solution set is **R**.

**95.**
$$5 + 3[t - (1 - 4t)] = 7$$
$$5 + 3(t - 1 + 4t) = 7$$
$$5 + 3(5t - 1) = 7$$
$$5 + 15t - 3 = 7$$
$$15t + 2 = 7$$
$$15t + 2 - 2 = 7 - 2$$
$$15t = 5$$
$$\frac{15t}{15} = \frac{5}{15}$$
$$t = \frac{1}{3}$$

**97.**
$$3(2x+1) = 8x - 2(x-2)$$
$$6x + 3 = 8x - 2x + 4$$
$$6x + 3 = 6x + 4$$
$$6x + 3 - 6x = 6x + 4 - 6x$$
$$3 = 4$$

Because the resulting equation is false, the original equation is a contradiction, and the solution set is $\varnothing$.

**99.** The LCD is 18.
$$\frac{4}{9}t - \frac{1}{6} = \frac{1}{3}t + 1$$
$$18 \cdot \frac{4}{9}t - 18 \cdot \frac{1}{6} = 18 \cdot \frac{1}{3}t + 18 \cdot 1$$
$$8t - 3 = 6t + 18$$
$$8t - 3 - 6t = 6t + 18 - 6t$$
$$2t - 3 = 18$$
$$2t - 3 + 3 = 18 + 3$$
$$2t = 21$$
$$\frac{2t}{2} = \frac{21}{2}$$
$$t = \frac{21}{2}$$

**101.** The LCD is 4.
$$\frac{3}{4}(x-3) - \frac{1}{2}(3x-5) = 2(3-x)$$
$$4 \cdot \frac{3}{4}(x-3) - 4 \cdot \frac{1}{2}(3x-5) = 4 \cdot 2(3-x)$$
$$3(x-3) - 2(3x-5) = 8(3-x)$$
$$3x - 9 - 6x + 10 = 24 - 8x$$
$$-3x + 1 = 24 - 8x$$
$$-3x + 1 + 8x = 24 - 8x + 8x$$
$$5x + 1 = 24$$
$$5x + 1 - 1 = 24 - 1$$
$$5x = 23$$
$$\frac{5x}{5} = \frac{23}{5}$$
$$x = \frac{23}{5}$$

**103.**
$$0.12 - 0.05y = 0.04y - 0.15$$
$$100(0.12 - 0.05y) = 100(0.04y - 0.15)$$
$$12 - 5y = 4y - 15$$
$$12 - 5y - 4y = 4y - 15 - 4y$$
$$12 - 9y = -15$$
$$12 - 9y - 12 = -15 - 12$$
$$-9y = -27$$
$$\frac{-9y}{-9} = \frac{-27}{-9}$$
$$y = 3$$

**105.**
$$x(x+6) = 5(x+1) + x^2$$
$$x^2 + 6x = 5x + 5 + x^2$$
$$x^2 + 6x - x^2 = 5x + 5 + x^2 - x^2$$
$$6x = 5x + 5$$
$$6x - 5x = 5x + 5 - 5x$$
$$x = 5$$

**107.** (a) An expression for the volume of the alarm for any setting $s$ is $30 + 8(s-1)$.

(b) 
$$30 + 8(s-1) = 62$$
$$30 + 8s - 8 = 62$$
$$22 + 8s = 62$$
$$22 + 8s - 22 = 62 - 22$$
$$8s = 40$$
$$\frac{8s}{8} = \frac{40}{8}$$
$$s = 5$$

The volume is 62 decibels when the setting is 5.

**109.**
$$1.475x + 13.4 = 50$$
$$1.475x + 13.4 - 13.4 = 50 - 13.4$$
$$1.475x = 36.6$$
$$\frac{1.475x}{1.475} = \frac{36.6}{1.475}$$
$$x \approx 25$$

Half of the United States will be enrolled in an HMO in approximately 2015 (that is, $1990 + 25$).

**111.** (a) Because the tax rate in 1945 was 2%, 10 times the tax rate in 1945 is 20%. An equation for estimating the year in which the tax rate will reach 20% is $0.28x - 1.07 = 20$.

(b)
$$0.28x - 1.07 = 20$$
$$0.28x - 1.07 + 1.07 = 20 + 1.07$$
$$0.28x = 21.07$$
$$\frac{0.28x}{0.28} = \frac{21.07}{0.28}$$
$$x \approx 75$$

According to the model, the tax rate will be 20% in 2010 (that is, $1935 + 75$).

Section 3.2  Solving Linear Equations

**113.** A function that models the total African-American and Hispanic population in any given year since 1990 is given by:
$h(t) = f(t) + g(t)$
$h(t) = 0.52t + 30.25 + 0.9t + 22.3$
$h(t) = 1.42t + 52.55$
Evaluate the function at $t = 15$ to estimate the total population in 2005.
$h(t) = 1.42t + 52.55$
$h(15) = 1.42(15) + 52.55$
$\quad\quad = 21.3 + 52.55$
$\quad\quad = 73.85$
In 2005 the combined population will be 73.85 million.

**115.** The equation describes the Hispanic population exceeding the African-American population by 2 million people.
$$g(t) - f(t) = 2$$
$0.9t + 22.3 - (0.52t + 30.25) = 2$
$0.9t + 22.3 - 0.52t - 30.25 = 2$
$\quad\quad\quad\quad 0.38t - 7.95 = 2$
$\quad 0.38t - 7.95 + 7.95 = 2 + 7.95$
$\quad\quad\quad\quad\quad 0.38t = 9.95$
$\quad\quad\quad\quad \dfrac{0.38t}{0.38} = \dfrac{9.95}{0.38}$
$\quad\quad\quad\quad\quad t \approx 26$
In the year 2016 (that is, 1990 + 26), the Hispanic population will be 2 million greater than the African-American population.

**117.** $\quad\quad b - ax = c$
$\quad b - ax - b = c - b$
$\quad\quad\quad -ax = c - b$
$\quad\quad \dfrac{-ax}{-a} = \dfrac{c-b}{-a}, \quad a \neq 0$
$\quad\quad\quad x = \dfrac{b-c}{a}, \quad a \neq 0$

**119.** $\quad\quad ax + b = cx + d$
$\quad ax + b - cx = cx + d - cx$
$\quad ax - cx + b = d$
$ax - cx + b - b = d - b$
$\quad\quad (a-c)x = d - b$
$\quad\quad \dfrac{(a-c)x}{(a-c)} = \dfrac{d-b}{(a-c)}, \quad a \neq c$
$\quad\quad\quad x = \dfrac{d-b}{a-c}, \quad a \neq c$

**121.** The LCD is 60.
$\dfrac{x}{2} + \dfrac{x+1}{3} + \dfrac{x+2}{4} + \dfrac{x+3}{5} = 1$
$60 \cdot \left(\dfrac{x}{2} + \dfrac{x+1}{3} + \dfrac{x+2}{4} + \dfrac{x+3}{5}\right) = 60 \cdot 1$
$30x + 20(x+1) + 15(x+2) + 12(x+3) = 60$
$30x + 20x + 20 + 15x + 30 + 12x + 36 = 60$
$\quad\quad\quad\quad\quad 77x + 86 = 60$
$\quad\quad 77x + 86 - 86 = 60 - 86$
$\quad\quad\quad\quad\quad 77x = -26$
$\quad\quad\quad\quad \dfrac{77x}{77} = \dfrac{-26}{77}$
$\quad\quad\quad\quad\quad x = -\dfrac{26}{77}$

**123.** The solution set for $h^2 = 25$ is $\{-5, 5\}$. The solution set for $h = 5$ is $\{5\}$. Because the solution sets are not exactly the same, no, these are not equivalent equations.

## Section 3.3 Formulas

1. To solve a formula for a given variable, treat all other variables as constants and use normal equation-solving procedures to isolate the given variable.

3. Solve $F = ma$ for $m$:
$$ma = F$$
$$\frac{ma}{a} = \frac{F}{a}$$
$$m = \frac{F}{a}$$

5. Solve $I = Prt$ for $r$:
$$Prt = I$$
$$\frac{Prt}{Pt} = \frac{I}{Pt}$$
$$r = \frac{I}{Pt}$$

7. Solve $v = \frac{s}{t}$ for $t$:
$$t \cdot v = t \cdot \frac{s}{t}$$
$$tv = s$$
$$\frac{tv}{v} = \frac{s}{v}$$
$$t = \frac{s}{v}$$

9. Solve $A = \frac{1}{2}bh$ for $b$:
$$\frac{1}{2}bh = A$$
$$2 \cdot \frac{1}{2}bh = 2 \cdot A$$
$$bh = 2A$$
$$\frac{bh}{h} = \frac{2A}{h}$$
$$b = \frac{2A}{h}$$

11. Solve $A = \frac{1}{2}h(a+b)$ for $a$:
$$2 \cdot \frac{1}{2}h(a+b) = 2 \cdot A$$
$$h(a+b) = 2A$$
$$ah + bh = 2A$$
$$ah + bh - bh = 2A - bh$$
$$ah = 2A - bh$$
$$\frac{ah}{h} = \frac{2A - bh}{h}$$
$$a = \frac{2A - bh}{h}$$

13. Solve $A = \pi r^2$ for $\pi$:
$$\pi r^2 = A$$
$$\frac{\pi r^2}{r^2} = \frac{A}{r^2}$$
$$\pi = \frac{A}{r^2}$$

15. Solve $E = mc^2$ for $m$:
$$mc^2 = E$$
$$\frac{mc^2}{c^2} = \frac{E}{c^2}$$
$$m = \frac{E}{c^2}$$

17. Solve $a = \frac{v-w}{t}$ for $v$:
$$\frac{v-w}{t} = a$$
$$t \cdot \frac{v-w}{t} = t \cdot a$$
$$v - w = at$$
$$v - w + w = at + w$$
$$v = at + w$$

19. Solve $A = P + Prt$ for $r$:
$$P + Prt = A$$
$$P + Prt - P = A - P$$
$$Prt = A - P$$
$$\frac{Prt}{Pt} = \frac{A - P}{Pt}$$
$$r = \frac{A - P}{Pt}$$

Section 3.3  Formulas

**21.** Solve $P = \dfrac{nRT}{V}$ for $R$:

$$\dfrac{nRT}{V} = P$$

$$V \cdot \dfrac{nRT}{V} = V \cdot P$$

$$nRT = PV$$

$$\dfrac{nRT}{nT} = \dfrac{PV}{nT}$$

$$R = \dfrac{PV}{nT}$$

**23.** Solve $R_T = \dfrac{R_1 + R_2}{2}$ for $R_1$:

$$\dfrac{R_1 + R_2}{2} = R_T$$

$$2 \cdot \dfrac{R_1 + R_2}{2} = 2 \cdot R_T$$

$$R_1 + R_2 = 2R_T$$

$$R_1 + R_2 - R_2 = 2R_T - R_2$$

$$R_1 = 2R_T - R_2$$

**25.** Solve $ax + by + c = 0$ for $y$:

$$ax + by + c - ax - c = 0 - ax - c$$

$$by = -(ax + c)$$

$$\dfrac{by}{b} = \dfrac{-(ax+c)}{b}$$

$$y = -\dfrac{(ax+c)}{b}$$

**27.** Solve $s = \dfrac{1}{2}gt^2 + vt$ for $v$:

$$\dfrac{1}{2}gt^2 + vt = s$$

$$2 \cdot \dfrac{1}{2}gt^2 + 2 \cdot vt = 2 \cdot s$$

$$gt^2 + 2vt = 2s$$

$$gt^2 + 2vt - gt^2 = 2s - gt^2$$

$$2vt = 2s - gt^2$$

$$\dfrac{2vt}{2t} = \dfrac{2s - gt^2}{2t}$$

$$v = \dfrac{2s - gt^2}{2t}$$

**29.** Solve $A = a + (n-1)d$ for $n$:

$$a + (n-1)d = A$$

$$a + dn - d = A$$

$$a + dn - d - a + d = A - a + d$$

$$dn = A - a + d$$

$$\dfrac{dn}{d} = \dfrac{A - a + d}{d}$$

$$n = \dfrac{A - a + d}{d}$$

**31.** Solve $S = \dfrac{n}{2}[2a + (n-1)d]$ for $a$:

$$\dfrac{n}{2}[2a + (n-1)d] = S$$

$$2 \cdot \dfrac{n}{2}[2a + (n-1)d] = 2 \cdot S$$

$$n[2a + (n-1)d] = 2S$$

$$2an + n(n-1)d = 2S$$

$$2an + n(n-1)d - n(n-1)d = 2S - n(n-1)d$$

$$2an = 2S - n(n-1)d$$

$$\dfrac{2an}{2n} = \dfrac{2S - n(n-1)d}{2n}$$

$$a = \dfrac{2S - n(n-1)d}{2n}$$

**33.** To isolate $y$, we can divide both sides by 3, or we can divide each term of both sides by 3.

**35.**
$$2x + 5y = 10$$
$$2x + 5y - 2x = 10 - 2x$$
$$5y = -2x + 10$$
$$\dfrac{5y}{5} = \dfrac{-2x}{5} + \dfrac{10}{5}$$
$$y = -\dfrac{2}{5}x + 2$$

**37.**
$$x - y + 5 = 0$$
$$x - y + 5 - x - 5 = 0 - x - 5$$
$$-y = -x - 5$$
$$(-1)(-y) = (-1)(-x) - (-1)(5)$$
$$y = x + 5$$

**39.**
$$2x + 3y - 9 = 0$$
$$2x + 3y - 9 - 2x + 9 = 0 - 2x + 9$$
$$3y = -2x + 9$$
$$\frac{3y}{3} = \frac{-2x}{3} + \frac{9}{3}$$
$$y = -\frac{2}{3}x + 3$$

**41.**
$$3x = 4y + 11$$
$$4y + 11 = 3x$$
$$4y + 11 - 11 = 3x - 11$$
$$4y = 3x - 11$$
$$\frac{4y}{4} = \frac{3x}{4} - \frac{11}{4}$$
$$y = \frac{3}{4}x - \frac{11}{4}$$

**43.**
$$3x - y = 5x + 3y$$
$$3x - y - 3y = 5x + 3y - 3y$$
$$3x - 4y = 5x$$
$$3x - 4y - 3x = 5x - 3x$$
$$-4y = 2x$$
$$\frac{-4y}{-4} = \frac{2x}{-4}$$
$$y = -\frac{1}{2}x$$

**45.**
$$0.4x - 0.3y + 12 = 0$$
$$0.4x - 0.3y + 12 - 0.4x = 0 - 0.4x$$
$$-0.3y + 12 = -0.4x$$
$$-0.3y + 12 - 12 = -0.4x - 12$$
$$-0.3y = -0.4x - 12$$
$$\frac{-0.3y}{-0.3} = \frac{-0.4x}{-0.3} - \frac{12}{-0.3}$$
$$y = \frac{4}{3}x + 40$$

**47.**
$$y - 4 = \frac{1}{2}(x + 6)$$
$$y - 4 = \frac{1}{2}x + 3$$
$$y - 4 + 4 = \frac{1}{2}x + 3 + 4$$
$$y = \frac{1}{2}x + 7$$

**49.**
$$\frac{4}{5}x - \frac{2}{3}y + 8 = 0$$
$$15 \cdot \frac{4}{5}x - 15 \cdot \frac{2}{3}y + 15 \cdot 8 = 15 \cdot 0$$
$$12x - 10y + 120 = 0$$
$$12x - 10y + 120 - 12x = 0 - 12x$$
$$-10y + 120 = -12x$$
$$-10y + 120 - 120 = -12x - 120$$
$$-10y = -12x - 120$$
$$\frac{-10y}{-10} = \frac{-12x}{-10} - \frac{120}{-10}$$
$$y = \frac{6}{5}x + 12$$

**51.**
$$\frac{y-2}{3} = \frac{x+3}{4}$$
$$12 \cdot \frac{y-2}{3} = 12 \cdot \frac{x+3}{4}$$
$$4(y - 2) = 3(x + 3)$$
$$4y - 8 = 3x + 9$$
$$4y - 8 + 8 = 3x + 9 + 8$$
$$4y = 3x + 17$$
$$\frac{4y}{4} = \frac{3x}{4} + \frac{17}{4}$$
$$y = \frac{3}{4}x + \frac{17}{4}$$

**53.** $A = 2\pi r^2 + 2\pi rh$
(a) If $r = 2.7$ and $h = 5.83$,
$$A = 2\pi(2.7)^2 + 2\pi(2.7)(5.83)$$
$$\approx 144.7 \text{ square inches}$$
(b)
$$2\pi r^2 + 2\pi rh = A$$
$$2\pi r^2 + 2\pi rh - 2\pi r^2 = A - 2\pi r^2$$
$$2\pi rh = A - 2\pi r^2$$
$$\frac{2\pi rh}{2\pi r} = \frac{A - 2\pi r^2}{2\pi r}$$
$$h = \frac{A - 2\pi r^2}{2\pi r}$$
(c) If $A = 2858.85$ and $r = 13$,
$$h = \frac{2858.85 - 2\pi(13)^2}{2\pi(13)} \approx 22 \text{ cm}$$

**55.**

$$b = \frac{2A}{h} \qquad\qquad b = 2A \cdot \frac{1}{h}$$
$$bh = 2A \qquad\qquad bh = 2A$$
$$\frac{1}{2}bh = A \qquad\qquad \frac{1}{2}bh = A$$

$$b = \frac{2}{h} \cdot A \qquad\qquad b = \frac{A}{\frac{1}{2}h}$$
$$bh = 2A \qquad\qquad \frac{1}{2}h \cdot b = A$$
$$\frac{1}{2}bh = A \qquad\qquad \frac{1}{2}bh = A$$

$$b = \frac{A}{2h}$$
$$2h \cdot b = A$$
$$2bh = A$$

All of the formulas, except for the last one given in (v), are equivalent to $\frac{1}{2}bh = A$.

**57.** The greatest distance across the track is the diameter of the track. If the runner jogged 4 miles in 10 laps, the circumference of the track is $4 \div 10 = 0.4$ miles or $0.4(5280) = 2112$ feet. The formula for circumference of a circle is $C = \pi d$. Solving for the diameter:
$$d = \frac{C}{\pi} = \frac{2112}{\pi} \approx 672.27 \text{ feet.}$$

**59.** We are given that the length is twice the width, so $L = 2W$. The perimeter is $P = 2(2W) + 2W = 6W$, and because the perimeter is 180 feet, the width is $W = P \div 6 = 180 \div 6 = 30$ feet. The length is 60 feet. The length of the diagonal is $c^2 = a^2 + b^2$:
$$c = \sqrt{a^2 + b^2}$$
$$c = \sqrt{30^2 + 60^2}$$
$$= \sqrt{4500}$$
$$\approx 67 \text{ feet}$$

**61.** The depth $H$ of the water in the second tank is calculated by
$$H = \frac{V}{LW} = \frac{15}{6(1.5)} = \frac{15}{9} \approx 1.7 \text{ feet. Thus}$$
the water in the second tank will be deeper.

**63.** The area of a trapezoid is given by $A = \frac{1}{2}h(a+b)$. With $A = 482{,}850$, $a = 1400$, and $b = 340$, we find $h$ as:
$$482{,}850 = \frac{1}{2}h(1400 + 340)$$
$$2 \cdot 482{,}500 = 2 \cdot \frac{1}{2}h(1740)$$
$$965{,}700 = 1740h$$
$$\frac{965{,}700}{1740} = \frac{1740h}{1740}$$
$$555 = h$$
The height of the dam is 555 feet.

**65.** The area of the front together with the area of the back is $2WH$. The area of the top and the bottom is $2WL$. The area of the two sides is $2LH$. The total area is the sum of these three expressions: $S = 2WH + 2LH + 2WL$.

**67.** $R = W + rW$
(a) $W = 78$ and $r = 0.32$
$R = 78 + (0.32)(78) = 102.96$
(b) 
$$W + rW = R$$
$$W + rW - W = R - W$$
$$rW = R - W$$
$$\frac{rW}{W} = \frac{R-W}{W}$$
$$r = \frac{R-W}{W}$$
If $R = 448.50$ and $W = 390$,
$$r = \frac{448.50 - 390}{390} = 0.15 \text{ or } 15\%.$$

**69.** 
$$2\pi r = C$$
$$\frac{2\pi r}{2\pi} = \frac{C}{2\pi}$$
$$r = \frac{C}{2\pi}$$

Circumference of Earth:
$$r = \frac{25{,}000}{2\pi} \approx 3978.87 \text{ mi}$$

(a) If the satellite travels in a circular orbit and the length of the orbit is 25,800 miles, then the satellite is

25,800 miles, then the satellite is $r = \dfrac{25{,}800}{2\pi} \approx 4106.20$ mi from the center of earth, or approximately 127 miles above the surface of earth.

(b) Similarly, if the orbit is 26,000 miles, the satellite is about 159 miles above the surface of earth.

(c) Similarly, if the orbit is 26,500 miles, the satellite is about 239 miles above the surface of earth.

**71.**
$$\tfrac{4}{3}\pi r^3 + \pi r^2 h = V$$
$$\tfrac{4}{3}\pi r^3 + \pi r^2 h - \tfrac{4}{3}\pi r^3 = V - \tfrac{4}{3}\pi r^3$$
$$\pi r^2 h = V - \tfrac{4}{3}\pi r^3$$
$$\dfrac{\pi r^2 h}{\pi r^2} = \dfrac{V - \tfrac{4}{3}\pi r^3}{\pi r^2}$$
$$h = \dfrac{V - \tfrac{4}{3}\pi r^3}{\pi r^2}$$

$V = 1450$ and $r = 4.2$:
$$h = \dfrac{1450 - \tfrac{4}{3}\pi(4.2)^3}{\pi(4.2)^2} \approx 20.56 \text{ meters}$$

**73.**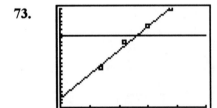

Based on a visual inspection of the graphs, function $S$ appears to be the better model.

**75. Model (i):**
1987: $C(7) = 1733$
1991: $C(11) = 1733$
1995: $C(15) = 1733$
1999: $C(19) = 1733$

$|M - A|$:
1987: $|1733 - 980| = 753$
1991: $|1733 - 1573| = 160$
1995: $|1733 - 1990| = 257$
1999: $|1733 - 2389| = 656$
Sum of the differences:
$753 + 160 + 257 + 656 = 1826$

**Model (ii):**
1987:
$S(7) = 116.1(7) + 223.7 = 1036.4$
1991:
$S(11) = 116.1(11) + 223.7 = 1500.8$
1995:
$S(15) = 116.1(15) + 223.7 = 1965.2$
1999:
$S(19) = 116.1(19) + 223.7 = 2429.6$

$M - A$:
1987: $|1036.4 - 980| = 56.4$
1991: $|1500.8 - 1573| = 72.2$
1995: $|1965.2 - 1990| = 24.8$
1999: $|2429.6 - 2389| = 40.6$
Sum of the differences:
$56.4 + 72.2 + 24.8 + 40.6 = 194$
From these results, function $S$ appears to be the better model.

**77.**
$$a(x + c) = b(x - c)$$
$$ax + ac = bx - bc$$
$$ax + ac - bx = bx - bc - bx$$
$$ax - bx + ac = -bc$$
$$ax - bx + ac - ac = -bc - ac$$
$$ax - bx = -ac - bc$$
$$(a - b)x = -(ac + bc)$$
$$\dfrac{(a-b)x}{(a-b)} = \dfrac{-(ac+bc)}{(a-b)}$$
$$x = \dfrac{ac + bc}{b - a}$$

**79.** Solve $S = \dfrac{a}{1-r}$ for $r$:
$$(1-r) \cdot S = (1-r) \cdot \dfrac{a}{1-r}$$
$$(1-r) \cdot S = a$$
$$S - rS = a$$
$$S - rS - S = a - S$$
$$-rS = a - S$$
$$\dfrac{-rS}{-S} = \dfrac{a-S}{-S}$$
$$r = \dfrac{S-a}{S}$$

**81.** Solve $\dfrac{P_1 V_1}{T_1} = \dfrac{P_2 V_2}{T_2}$ for $V_2$:

$$\dfrac{P_2 V_2}{T_2} = \dfrac{P_1 V_1}{T_1}$$

$$\dfrac{T_2}{P_2} \cdot \dfrac{P_2 V_2}{T_2} = \dfrac{T_2}{P_2} \cdot \dfrac{P_1 V_1}{T_1}$$

$$V_2 = \dfrac{P_1 V_1 T_2}{P_2 T_1}$$

**83.** Solve $A = 2LW + 2LH + 2WH$ for $L$:

$$2LW + 2LH + 2WH = A$$
$$2LW + 2LH + 2WH - 2WH = A - 2WH$$
$$2LW + 2LH = A - 2WH$$
$$L(2W + 2H) = A - 2WH$$
$$\dfrac{L(2W + 2H)}{(2W + 2H)} = \dfrac{A - 2WH}{(2W + 2H)}$$
$$L = \dfrac{A - 2WH}{2W + 2H}$$

## Section 3.4  Modeling and Problem Solving

**1.** The other number is $T - n$. The sum of the numbers is $n + (T - n) = T$.

**3.**  ① The goal is to determine the described number.
③ Let $x$ = the number.
④ $x + (x + 6) = 32$
⑤ $\quad 2x + 6 = 32$
$\quad 2x + 6 - 6 = 32 - 6$
$\quad\quad\quad 2x = 26$
$\quad\quad\quad \dfrac{2x}{2} = \dfrac{26}{2}$
$\quad\quad\quad\; x = 13$
⑥ The number is 13.
⑦ Check: Six more than the number is 19, and 19 + 13 is 32.

**5.**  ① The goal is to determine the described numbers.
③ Let $x$ = the smaller number and $x + 12$ = the larger number.
④ $\dfrac{1}{2}x = \dfrac{1}{3}(x + 12)$

⑤ $\quad 6 \cdot \dfrac{1}{2}x = 6 \cdot \dfrac{1}{3}(x + 12)$
$\quad\quad 3x = 2(x + 12)$
$\quad\quad 3x = 2x + 24$
$\quad 3x - 2x = 2x + 24 - 2x$
$\quad\quad\quad\; x = 24$
⑥ The smaller number is 24, and the larger number is 24 + 12 = 36.
⑦ Check: One-half of 24 is 12 and one-third of 36 is 12.

**7.** Both consecutive even integers and consecutive odd integers are 2 units apart on the number line. Whether the integers are even or odd depends on whether $x$ represents an even integer or an odd integer. The representations are the same in both cases.

9. ① The goal is to determine the two consecutive integers.
③ Let $x$ = the first consecutive integer and $x+1$ = the second consecutive integer.
④ $x+(x+1)=95$
⑤ $x+x+1=95$
$2x+1=95$
$2x+1-1=95-1$
$2x=94$
$\dfrac{2x}{2}=\dfrac{94}{2}$
$x=47$
⑥ The first number is 47, and the second number is $47+1=48$.
⑦ Check: The numbers are consecutive integers, and their sum is $47+48=95$.

11. ① The goal is to determine the two consecutive odd integers.
③ Let $x$ = the first consecutive odd integer and $x+2$ = the second consecutive odd integer.
④ $x+(x+2)=0$
⑤ $x+x+2=0$
$2x+2=0$
$2x+2-2=0-2$
$2x=-2$
$\dfrac{2x}{2}=\dfrac{-2}{2}$
$x=-1$
⑥ The first number is $-1$, and the second number is $-1+2=1$.
⑦ Check: The numbers are consecutive odd integers, and their sum is $-1+1=0$.

13. ① The goal is to determine the length of each piece.
②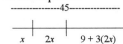
③ Because the second piece is described in terms of the first piece (and the third in terms of the second), let the variable represent the length of the first piece.
④ $x+(2x)+[9+3(2x)]=45$
⑤ $x+2x+9+6x=45$
$9x+9=45$
$9x+9-9=45-9$
$9x=36$
$\dfrac{9x}{9}=\dfrac{36}{9}$
$x=4$
⑥ The first piece is 4 feet, the second piece is $2(4)=8$ feet, and the third piece is $9+3(8)=33$ feet.
⑦ Check: The sum of the lengths is $4+8+33=45$ feet.

15. ① The goal is to determine the length of each piece.
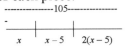
②
③ Because the second piece is described in terms of the first piece (and the third in terms of the second), let the variable represent the length of the first piece.
④ $x+(x-5)+2(x-5)=105$
⑤ $x+x-5+2x-10=105$
$4x-15=105$
$4x-15+15=105+15$
$4x=120$
$\dfrac{4x}{4}=\dfrac{120}{4}$
$x=30$
⑥ The first piece is 30 yd, the second piece is $30-5=25$ yd, and the third piece is $2(25)=50$ yd.
⑦ Check: The sum of the lengths is $30+25+50=105$ yards.

17. ① The goal is to determine the measures of all three angles.
②
③ Because the second and third angles are described in terms of the first angle, assign the variable to the measure of the first angle.
④ $x+(x+20)+(2x)=180$

Section 3.4   Modeling and Problem Solving

⑤  $x + x + 20 + 2x = 180$
$4x + 20 = 180$
$4x + 20 - 20 = 180 - 20$
$4x = 160$
$\dfrac{4x}{4} = \dfrac{160}{4}$
$x = 40$

⑥ The first angle is 40°, the second angle is $40 + 20 = 60°$, and the third angle is $2(40) = 80°$.

⑦ Check: The sum of the angle measures is $40 + 60 + 80 = 180°$.

**19.** ① The goal is to determine whether the triangle is a right triangle (i.e., if one of the angles measures 90°).

②

③ Because the second angle is described in terms of the first angle (and the third in terms of the second), assign the variable to the measure of the first angle.

④ $x + (2x) + \left[\dfrac{3}{2}(2x)\right] = 180$

⑤ $x + 2x + 3x = 180$
$6x = 180$
$\dfrac{6x}{6} = \dfrac{180}{6}$
$x = 30$

⑥ The first angle is 30°, the second angle is $2(30) = 60°$, and the third angle is $1.5(60) = 90°$. The triangle is a right triangle because one angle is a right angle (90°).

⑦ Check: The sum of the angle measures is $30 + 60 + 90 = 180°$.

**21.** ① The goal is to determine the dimensions of the rectangle.

② $L$

③ Let $L$ = length and $L - 8$ = width.
$2L + 2W = P$

④ $2L + 2(L - 8) = 64$

⑤ $2L + 2L - 16 = 64$
$4L - 16 = 64$
$4L - 16 + 16 = 64 + 16$
$4L = 80$
$\dfrac{4L}{4} = \dfrac{80}{4}$
$L = 20$

⑥ The length is 20 feet and the width is $20 - 8 = 12$ feet.

⑦ Check: The perimeter is $2(20) + 2(12) = 40 + 24 = 64$ feet.

**23.** ① The goal is to determine the dimensions of the rectangle.

② $W + 4$

Let $W$ = width and $W + 4$ = length.
$2L + 2W = P$

④ $2(W + 4) + 2W = 60$

⑤ $2W + 8 + 2W = 60$
$4W + 8 = 60$
$4W + 8 - 8 = 60 - 8$
$4W = 52$
$\dfrac{4W}{4} = \dfrac{52}{4}$
$W = 13$

⑥ The width is 13 yards and the length is $13 + 4 = 17$ yards.

⑦ Check: The perimeter is $2(17) + 2(13) = 34 + 26 = 60$ yards.

**25.** ① The goal is to determine the number of nickels, but to check the results we will also find the number of dimes.

②

|         | Number of coins | Value of coin | Value of coins in cents |
|---------|---------|---------|---------|
| Dimes   | $x$     | 10      | $10x$   |
| Nickels | $x + 9$ | 5       | $5(x + 9)$ |
| Total   |         |         | 225     |

③ The table shows the variable assignment.

④ $10x + 5(x + 9) = 225$

⑤ $10x + 5x + 45 = 225$
$15x + 45 = 225$
$15x + 45 - 45 = 225 - 45$
$15x = 180$
$\dfrac{15x}{15} = \dfrac{180}{15}$
$x = 12$

⑥ There are 12 dimes and $12 + 9 = 21$ nickels.

⑦ Check: The value of the coins is $10(12) + 5(21) = 120 + 105 = 225$.

**27.** ① The goal is to determine the number of dimes, but to check the results we will also find the number of quarters.

②

|  | Number of coins | Value of coin | Value of coins in cents |
|---|---|---|---|
| Quarters | $x$ | 25 | $25x$ |
| Dimes | $32 - x$ | 10 | $10(32-x)$ |
| Total |  |  | 515 |

③ The table shows the variable assignment.

④ $25x + 10(32 - x) = 515$

⑤ $25x + 320 - 10x = 515$
$15x + 320 = 515$
$15x + 320 - 320 = 515 - 320$
$15x = 195$
$\dfrac{15x}{15} = \dfrac{195}{15}$
$x = 13$

⑥ There are 13 quarters and $32 - 13 = 19$ dimes.

⑦ Check: The value of the coins is $25(13) + 10(19) = 515$.

**29.** Let $x$ = the first integer, $x + 1$ = the second integer, and $x + 2$ = the third integer. The average of the first and third is $\dfrac{x + (x+2)}{2} = \dfrac{2x+2}{2} = x + 1$, which is the expression for the second (middle) integer.

**31.** Because the sum of even integers is always even, the sum cannot be 19.

**33.** ① The goal is to determine the described number.

③ Let $x$ = the number.

④ $7(x + 5) = -5x - 1$

⑤ $7x + 35 = -5x - 1$
$7x + 35 + 5x = -5x - 1 + 5x$
$12x + 35 = -1$
$12x + 35 - 35 = -1 - 35$
$12x = -36$
$\dfrac{12x}{12} = \dfrac{-36}{12}$
$x = -3$

⑥ The number is $-3$.

⑦ Check: Increasing $-3$ by 5 is 2, which multiplied by 7 is 14. Multiplying $-3$ by $-5$ is 15, and subtracting 1 is 14.

**35.** ① The goal is to determine whether the triangle is a right triangle.

② (triangle with angles $2x - 10$, $x$, $x - 10$)

③ Because the first and third angles are described in terms of the second angle, assign the variable to the measure of the second angle.

④ $(x - 10) + x + (2x - 10) = 180$

⑤ $x - 10 + x + 2x - 10 = 180$
$4x - 20 = 180$
$4x - 20 + 20 = 180 + 20$
$4x = 200$
$\dfrac{4x}{4} = \dfrac{200}{4}$
$x = 50$

⑥ The second angle is 50°, the first angle is $50 - 10 = 40°$, and the third angle is $2(50) - 10 = 90°$. The triangle is a right triangle because the third angle is a right angle (90°).

⑦ Check: The sum of the angle measures is $50 + 40 + 90 = 180°$.

Section 3.4  Modeling and Problem Solving

**37.** ① The goal is to determine the measure of the angle between the guy wire and the ground.

② 
③ Let $x$ = the angle between the guy wire and the pole and $8x$ = the angle between the wire and the ground.
④ $x + 8x + 90 = 180$
⑤ $9x + 90 = 180$
$9x + 90 - 90 = 180 - 90$
$9x = 90$
$\dfrac{9x}{9} = \dfrac{90}{9}$
$x = 10$
⑥ The angle between the guy wire and the pole is 10°, and the angle between the guy wire and the ground is $8(10) = 80°$.
⑦ Check: The sum of the angle measures is $10 + 80 + 90 = 180°$.

**39.** ① The goal is to determine whether the book can be placed upright on a 12-inch high shelf. We need to find the height of the book.

② 
③ Let $W$ = width and $W + 4$ = height.
$2W + 2H = P$
④ $2W + 2(W + 4) = 48$
⑤ $2W + 2W + 8 = 48$
$4W + 8 = 48$
$4W + 8 - 8 = 48 - 8$
$4W = 40$
$\dfrac{4W}{4} = \dfrac{40}{4}$
$W = 10$
⑥ The width is 10 inches and the height is $10 + 4 = 14$ inches. No, the book cannot be placed upright on a 12-inch high shelf.
⑦ Check: The perimeter is $2(10) + 2(14) = 20 + 28 = 48$ inches.

**41.** ① The goal is to determine the length of the smaller pad. To do so, we must find the width of the smaller pad.

② Original rectangle: $W$, $W + 5$; Modified rectangle: $W + 8$, $2(W + 5)$
③ $2W + 2L = P$
④ Modified rectangle:
$2(W + 8) + 2[2(W + 5)] = 60$
⑤ $2W + 16 + 2(2W + 10) = 60$
$2W + 16 + 4W + 20 = 60$
$6W + 36 = 60$
$6W + 36 - 36 = 60 - 36$
$6W = 24$
$\dfrac{6W}{6} = \dfrac{24}{6}$
$W = 4$
⑥ The width of the original pad is 4 inches and the length is $4 + 5 = 9$ inches.
⑦ Check: The perimeter of the modified rectangle is $2(4 + 8) + 2[2(4 + 5)] = 24 + 36 = 60$ inches.

**43.** ① The goal is to determine whether the height of the window is no more than 40 inches.

② 
③ Let $W$ = width and $1.5W$ = height.
$2W + 2H = P$
④ $2W + 2(1.5W) = 150$
⑤ $2W + 3W = 150$
$5W = 150$
$\dfrac{5W}{5} = \dfrac{150}{5}$
$W = 30$
⑥ The width is 30 inches and the height is $1.5(30) = 45$ inches. No, the curtains will not cover the full height of the window.
⑦ Check: The perimeter is $2(30) + 2(45) = 60 + 90 = 150$ inches.

**45.** 
① The goal is to determine whether the triangle is an obtuse triangle (i.e., one angle is greater than 90°).
②
③ Because the first and third angles are described in terms of the second angle, assign the variable to the measure of the second angle.
④ $(4x) + x + (2x - 2) = 180$
⑤ $4x + x + 2x - 2 = 180$
$7x - 2 = 180$
$7x - 2 + 2 = 180 + 2$
$7x = 182$
$\dfrac{7x}{7} = \dfrac{182}{7}$
$x = 26$
⑥ The second angle is 26°, the first angle is $4(26) = 104°$, and the third angle is $2(26) - 2 = 50°$. The triangle is an obtuse triangle because the first angle is greater than 90°.
⑦ Check: The sum of the angle measures is $26 + 104 + 50 = 180°$.

**47.**
① The goal is to determine the measure of the smaller angle.
③ Let $x =$ the smaller angle and $2x + 10.8 =$ the larger angle.
④ $x + (2x + 10.8) = 180$
⑤ $x + 2x + 10.8 = 180$
$3x + 10.8 = 180$
$3x + 10.8 - 10.8 = 180 - 10.8$
$3x = 169.2$
$\dfrac{3x}{3} = \dfrac{169.2}{3}$
$x = 56.4$
⑥ The measure of the smaller angle is 56.4°.
⑦ Check: The sum of the two angles is $56.4 + [2(56.4) + 10.8] = 180°$.

**49.**
① The goal is to determine the measure of the angle between the ladder and the wall.
②
③ Let $x =$ the angle between the ladder and wall and $7x - 6 =$ the angle between the ladder and the ground.
④ $x + (7x - 6) + 90 = 180$
⑤ $x + 7x - 6 + 90 = 180$
$8x + 84 = 180$
$8x + 84 - 84 = 180 - 84$
$8x = 96$
$\dfrac{8x}{8} = \dfrac{96}{8}$
$x = 12$
⑥ The angle between the ladder and the wall is 12°.
⑦ Check: The sum of the angle measures is $12 + 7(12) - 6 + 90 = 12 + 84 - 6 + 90 = 180°$.

**51.**
① The goal is to determine the number of women in the class.
③ Let $x =$ the number of women and $\dfrac{1}{2}x =$ the number of men.
④ $x + \dfrac{1}{2}x = 39$
⑤ $1.5x = 39$
$\dfrac{1.5x}{1.5} = \dfrac{39}{1.5}$
$x = 26$
⑥ There are 26 women in the class.
⑦ Check: The number of students in the class is $26 + 0.5(26) = 39$.

**53.**
① The goal is to determine the amount of money each person had at the end of the game.
③ Let $x$ = the amount each player had at the beginning of the game.
④ The loser had $\left(\frac{1}{4}x+50\right)$ less than he started with, so he had $x-\left(\frac{1}{4}x+50\right)$ at the end of the game. The winner had $x+\left(\frac{1}{4}x+50\right)$. Because the winner had twice as much as the loser at the end of the game, the equation is
$$x+\left(\frac{1}{4}x+50\right) = 2\left[x-\left(\frac{1}{4}x+50\right)\right]$$
⑤ $x+0.25x+50 = 2(x-0.25x-50)$
$1.25x+50 = 2(0.75x-50)$
$1.25x+50 = 1.5x-100$
$1.25x+50-1.5x = 1.5x-100-1.5x$
$-0.25x+50 = -100$
$-0.25x+50-50 = -100-50$
$-0.25x = -150$
$\dfrac{-0.25x}{-0.25} = \dfrac{-150}{-0.25}$
$x = 600$
⑥ Each player started the game with $600. The loser ended the game with $600 - [0.25(600) + 50] = 600 - (150 + 50) = \$400$. The winner ended the game with $600 + [0.25(600) + 50] = 600 + (150 + 50) = \$800$.
⑦ Check: The winner had twice the amount of money the loser had: $2(\$400) = \$800$.

**55.**
① The goal is to determine the number of sofas received.
③ Let $x$ = the number of sofas ordered and $x+4$ = the number of chairs ordered.
④ $\dfrac{1}{2}x + \dfrac{1}{3}(x+4) = x$

⑤ $6 \cdot \dfrac{1}{2}x + 6 \cdot \dfrac{1}{3}(x+4) = 6 \cdot x$
$3x + 2(x+4) = 6x$
$3x + 2x + 8 = 6x$
$5x + 8 = 6x$
$5x + 8 - 5x = 6x - 5x$
$8 = x$
⑥ The number of sofas originally ordered is 8. The number of sofas received is $0.5(8) = 4$ sofas.
⑦ Check: The total items received was $\dfrac{1}{2}(8) + \dfrac{1}{3}(8+4) = 4 + 4 = 8$.

**57.**
① The goal is to determine how much the average had dropped at 10 A.M.
③ Let $x$ = the number of points the average had dropped by 10 A.M and $x+8$ = the number of points the average had dropped by noon.
④ $5(x+8) = 50$
⑤ $5x + 40 = 50$
$5x + 40 - 40 = 50 - 40$
$5x = 10$
$\dfrac{5x}{5} = \dfrac{10}{5}$
$x = 2$
⑥ The Dow average had dropped by 2 points at 10 A.M.
⑦ Check: The total drop in the average for the day was $5(2+8) = 5(10) = 50$ points.

**59.** You know that their lockers are not numbered consecutively because the sum of an odd integer and an even integer cannot be an even integer.

**61.**
① The goal is to determine the number of points scored by the forward.
③ Let $x$ = the number points scored by the guard, $x + 2$ = the number points scored by the center, and $(x + 2) + 1$ = the number points scored by the forward.
④ $x + (x+2) + [(x+2) + 1] = 143$

⑤ $x + x + 2 + x + 3 = 143$
$3x + 5 = 143$
$3x + 5 - 5 = 143 - 5$
$3x = 138$
$\dfrac{3x}{3} = \dfrac{138}{3}$
$x = 46$

⑥ The guard scored 46 points, the center scored 48 points, and the forward scored 49 points.

⑦ Check: The total points scored were $46 + 48 + 49 = 143$.

**63.** ① The goal is to determine the length of each piece.

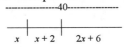

②
③ Because the second and third pieces are described in terms of the first piece, let the variable represent the length of the first piece.

④ $x + (x + 2) + (2x + 6) = 40$

⑤ $x + x + 2 + 2x + 6 = 40$
$4x + 8 = 40$
$4x + 8 - 8 = 40 - 8$
$4x = 32$
$\dfrac{4x}{4} = \dfrac{32}{4}$
$x = 8$

⑥ The first piece is 8 meters, the second piece is $8 + 2 = 10$ meters, and the third piece is $2(8) + 6 = 16 + 6 = 22$ meters.

⑦ Check: The sum of the lengths is $8 + 10 + 22 = 40$ meters.

**65.** ① The goal is to determine the number of members who attended the play.

②

|  | Number of tickets | Per-ticket cost | Cost of tickets |
|---|---|---|---|
| Adults | $x$ | 2.75 | $2.75x$ |
| Members | $x + 7$ | 1.25 | $1.25(x+7)$ |
| Total |  |  | 28.75 |

③ The table shows the variable assignment.

④ $2.75x + 1.25(x + 7) = 28.75$

⑤ $2.75x + 1.25x + 8.75 = 28.75$
$4x + 8.75 = 28.75$
$4x + 8.75 - 8.75 = 28.75 - 8.75$
$4x = 20$
$\dfrac{4x}{4} = \dfrac{20}{4}$
$x = 5$

⑥ Five adults attended the play, and $5 + 7 = 12$ members attended the play.

⑦ Check: The value of the tickets is $2.75(5) + 1.25(12) = 28.75$.

**67.** ① The goal is to determine the number of hot dogs sold.

②

|  | Number sold | Per-item price | Income |
|---|---|---|---|
| Hot dog | $x$ | 1.25 | $1.25x$ |
| Hamburg | $2x$ | 2 | $2(2x)$ |
| Total |  |  | 304.50 |

③ The table shows the variable assignment.

④ $1.25x + 2(2x) = 304.50$

⑤ $1.25x + 4x = 304.50$
$5.25x = 304.50$
$\dfrac{5.25x}{5.25} = \dfrac{304.50}{5.25}$
$x = 58$

⑥ There were 58 hot dogs sold.

⑦ Check: The income was $1.25(58) + 2[2(58)] = 72.5 + 2(116) = 304.50$.

**69.** ① The goal is to determine the amount of the inheritance.

③ Let $x$ = the amount of the inheritance, $\dfrac{1}{4}x$ = amount invested in bank stock, $\dfrac{1}{5}x$ = amount invested in bonds, and $\dfrac{1}{2}x$ = amount invested in mutual fund.

④ $\dfrac{1}{4}x + \dfrac{1}{5}x + \dfrac{1}{2}x = 38{,}000$

Section 3.4   Modeling and Problem Solving

⑤ $20 \cdot \left(\dfrac{1}{4}x + \dfrac{1}{5}x + \dfrac{1}{2}x\right) = 20 \cdot 38{,}000$

$5x + 4x + 10x = 760{,}000$

$19x = 760{,}000$

$\dfrac{19x}{19} = \dfrac{760{,}000}{19}$

$x = 40{,}000$

⑥ The person inherited $40,000.

⑦ Check: The total amount invested was $\dfrac{1}{4}(40{,}000) + \dfrac{1}{5}(40{,}000) + \dfrac{1}{2}(40{,}000) = 10{,}000 + 8000 + 20{,}000 = \$38{,}000$.

**71.**   ① The goal is to determine how many men, women, and children were at the picnic.

③ Let $x$ = the number of children, $4x$ = the number of men, and $2x$ = the number of women.

④ $x + 4x + 2x = 266$

⑤ $7x = 266$

$\dfrac{7x}{7} = \dfrac{266}{7}$

$x = 38$

⑥ There were 38 children, $4(38) = 152$ men, and $2(38) = 76$ women at the picnic.

⑦ Check: The total number of people at the picnic was $38 + 152 + 76 = 266$.

**73.**   She cannot know for sure. The solution of $x + (x+2) + (x+4) = 597$ is:

$x + x + 2 + x + 4 = 597$

$3x + 6 = 597$

$3x + 6 - 6 = 597 - 6$

$3x = 591$

$\dfrac{3x}{3} = \dfrac{591}{3}$

$x = 197$

Therefore, the highest score was $197 + 4 = 201$, but it was not necessarily the husband's score.

**75.**  (a) The variable $t$ represents the number of years after 1996 and $N$ is the number of college students aged 18-21. If the number of students increases by 0.175 million per year and there were 5.60 million in all in 1996, the model we need is $N(t) = 0.175t + 5.60$.

(b) The equation we need is:

$0.175t + 5.60 = 8.4$

$0.175t + 5.60 - 5.60 = 8.4 - 5.60$

$0.175t = 2.8$

$\dfrac{0.175t}{0.175} = \dfrac{2.8}{0.175}$

$t = 16$

The number of students in this age groups will reach 8.4 milion in 2012 (that is, 1996 + 16).

**77.**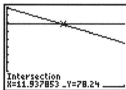

From the table, we see that the yearly decrease in emissions predicted by the function is 1.77 metric tons.

**79.** The equation needed to predict the year in which emissions are expected to be 20% less than the 1991 level is:

$-1.77x + 99.37 = 0.80(97.8)$

Emissions are expected to be 20% less than the 1991 level in about 2002 (found from 1990 + 12).

**81.** Let $x$ = the number chosen by the friend. Translating the number game into symbols, we have that "subtract 1 from the number and multiply by 3" is $3(x-1)$ and "twice the number increased by 1" is $2(x+1)$. Writing the entire expression and simplifying, we get:

$$3(x-1) - 2(x+1) + 5$$
$$= 3x - 3 - 2x - 2 + 5$$
$$= x - 5 + 5$$
$$= x$$

No matter what number is chosen, the result of the number game will be the number that was originally chosen.

## Section 3.5 Applications

**1.** In mathematics, *of* means *times*:
$0.15(30)$.

**3.** Let $x =$ total amount collected.
$$0.12x = 28$$
$$\frac{0.12x}{0.12} = \frac{28}{0.12}$$
$$x \approx 233.33$$
The boy collected $233.33 and kept $28. He turned in $233.33 − $28 = $205.33 to The Daily News, Inc.

**5.** Let $x =$ original price of the stock.
Value on Wed. $= x + 0.04x$
$$(x + 0.04x) - 0.04(x + 0.04x) = 349.44$$
$$x + 0.04x - 0.04x - 0.0016x = 349.44$$
$$0.9984x = 349.44$$
$$\frac{0.9984x}{0.9984} = \frac{349.44}{0.9984}$$
$$x = 350$$
The original price of the stock was $350. The investor lost money.

**7.** Let $x =$ monthly salary.
$7\% + 20\% + 6\% + 7\% = 40\%$
Take-home pay $= x - 0.40x$
$$x - 0.40x = 1200$$
$$0.60x = 1200$$
$$\frac{0.60x}{0.60} = \frac{1200}{0.60}$$
$$x = 2000$$
The worker's salary is $2000.

**9.** Let $x =$ the amount over $120,000
$7\%(120,000) = \$8400$
$7\%$ on first $120,000 + 9\%$ on amount over $120,000 =$ commission
$$8400 + 0.09x = 13,980$$
$$8400 - 8400 + 0.09x = 13,980 - 8400$$
$$0.09x = 5580$$
$$\frac{0.09x}{0.09} = \frac{5580}{0.09}$$
$$x = 62,000$$
The selling price of the house is $120,000 + $62,000 = $182,000.

**11.** Let $x =$ the original price.
Marked down price: $x - 0.30x$
$$x - 0.30x = 17.50$$
$$0.70x = 17.50$$
$$\frac{0.70x}{0.70} = \frac{17.50}{0.70}$$
$$x = 25$$
The original price was $25.

**13.** Let $x =$ the original price.
Marked down price: $x - 0.20x$
$$x - 0.20x = 239.20$$
$$0.80x = 239.20$$
$$\frac{0.80x}{0.80} = \frac{239.20}{0.80}$$
$$x = 299$$
The original price was $299.

**15.** Expenditures for materials (a), electric bill (c), and wages of hourly employees (d) are variable costs, and annual insurance (b) and the annual salary of the manager (e) are fixed costs.

Section 3.5 Applications

**17.** Let $h$ = number of hours.
Fixed cost = 100
Variable cost = $50h$
$$100 + 50h = 480$$
$$100 + 50h - 100 = 480 - 100$$
$$50h = 380$$
$$\frac{50h}{50} = \frac{380}{50}$$
$$h = 7.6$$
The bulldozer was used for 7.6 hours.

**19.** Let $c$ = number of copies.
Fixed cost = 300
Variable cost = $0.03c$
$$300 + 0.03c = 414$$
$$300 + 0.03c - 300 = 414 - 300$$
$$0.03c = 114$$
$$\frac{0.03c}{0.03} = \frac{114}{0.03}$$
$$c = 3800$$
The company made 3800 copies. The 3600-copy limit was exceeded by 200 copies.

**21.** The value is the principal $P$ plus 7% of the principal, which is represented by choice (iii).

**23.** Let $P$ = the amount invested.
Interest = $0.05P$
$$P + 0.05P = 600$$
$$1.05P = 600$$
$$\frac{1.05P}{1.05} = \frac{600}{1.05}$$
$$P \approx 571.43$$
Approximately $571.43 should be invested now at 5% interest to have $600 in one year.

**25.** Let $P$ = the amount invested.
Interest = $0.07P$
$$P + 0.07P = 1177$$
$$1.07P = 1177$$
$$\frac{1.07P}{1.07} = \frac{1177}{1.07}$$
$$P = 1100$$
The student's grant was $1100.

**27.** Let $P$ = the amount of the loan.
Interest = $0.08P$
$$P + 0.08P = 36,600 - 15,000$$
$$1.08P = 21,600$$
$$\frac{1.08P}{1.08} = \frac{21,600}{1.08}$$
$$P = 20,000$$
The amount of the loan was $20,000.

**29.** Let $x$ = the amount invested at 9%. The remaining money, $6500 - x$, is the amount invested at 6%.

|  | Amount invested | Interest rate | Interest earned |
|---|---|---|---|
| You | $x$ | 0.09 | $0.09x$ |
| Sister | $6500 - x$ | 0.06 | $0.06(6500-x)$ |
| Total |  |  | 465 |

$$0.09x + 0.06(6500 - x) = 465$$
$$0.09x + 390 - 0.06x = 465$$
$$0.03x + 390 = 465$$
$$0.03x + 390 - 390 = 465 - 390$$
$$0.03x = 75$$
$$\frac{0.03x}{0.03} = \frac{75}{0.03}$$
$$x = 2500$$
$2500 was invested at 9%, and $6500 - 2500 = \$4000$ was invested at 6%.

**31.** Let $x$ = the amount borrowed from the bank at 9%. The remaining money, $10,000 - x$, is the amount borrowed at 12%.

|  | Amount borrowed | Interest rate | Interest paid |
|---|---|---|---|
| Bank | $x$ | 0.09 | $0.09x$ |
| MIL | $10,000-x$ | 0.12 | $0.12(10,000-x)$ |
| Total |  |  | 1080 |

$$0.09x + 0.12(10,000 - x) = 1080$$
$$0.09x + 1200 - 0.12x = 1080$$
$$1200 - 0.03x = 1080$$
$$1200 - 0.03x - 1200 = 1080 - 1200$$
$$-0.03x = -120$$
$$\frac{-0.03x}{-0.03} = \frac{-120}{-0.03}$$
$$x = 4000$$
The buyer borrowed $4000 from the bank.

**33.** Let $r$ = the driving rate.

|  | Rate | Time | Distance |
|---|---|---|---|
| Bus | 27 | $\frac{2}{3}$ | 18 |
| Driving | $r$ | 0.4 | $0.4r$ |

The distances are equal.
$$0.4r = 18$$
$$\frac{0.4r}{0.4} = \frac{18}{0.4}$$
$$r = 45$$
The average speed for the drive home was 45 mph.

**35.** Let $r$ = the rate of the bus.

|  | Rate | Time | Distance |
|---|---|---|---|
| Bus | $r$ | 4 | $4r$ |
| Car | $r+6$ | 4 | $4(r+6)$ |
| Total |  |  | 472 |

The sum of the distances is 472.
$$4r + 4(r+6) = 472$$
$$4r + 4r + 24 = 472$$
$$8r + 24 = 472$$
$$8r + 24 - 24 = 472 - 24$$
$$8r = 448$$
$$\frac{8r}{8} = \frac{448}{8}$$
$$r = 56$$
The rate of the bus was 56 mph, and the rate of the car was $56 + 6 = 62$ mph.

**37.** Let $t$ = the number of hours each traveled.

|  | Rate | Time | Distance |
|---|---|---|---|
| Valdosta | 55 | $t$ | $55t$ |
| Atlanta | 70 | $t$ | $75t$ |

The sum of the distances is 300.
$$55t + 70t = 300$$
$$125t = 300$$
$$\frac{125t}{125} = \frac{300}{125}$$
$$t = 2.4$$
The drivers met 2.4 hours (or 2 hours and 24 minutes) after noon. The time was 2:24 P.M. The distance from Atlanta at that time is $70(2.4) = 168$ miles. The drivers meet 168 miles south of Atlanta.

**39.** Let $x$ = the number of pounds of cashews.

|  | Number of pounds | Price per pound | Total cost |
|---|---|---|---|
| Pecans | 20 | 1.80 | 36 |
| Cashews | $x$ | 2.40 | $2.40x$ |
| Mixture | $20 + x$ | 2.00 | $2(20 + x)$ |

$$36 + 2.4x = 2(20 + x)$$
$$36 + 2.4x = 40 + 2x$$
$$36 + 2.4x - 2x = 40 + 2x - 2x$$
$$36 + 0.4x = 40$$
$$36 + 0.4x - 36 = 40 - 36$$
$$0.4x = 4$$
$$\frac{0.4x}{0.4} = \frac{4}{0.4}$$
$$x = 10$$
Ten pounds of cashews should be used.

**41.** Let $x$ = the number of pounds of sunflower seed. Then $50 - x$ = the number of pounds of bird seed.

|  | Number of pounds | Price per pound | Total cost |
|---|---|---|---|
| Sunflwer | $x$ | 1.10 | $1.10x$ |
| Bird seed | $50 - x$ | 0.75 | $0.75(50-x)$ |
| Mixture | 50 | 0.90 | 45.00 |

$$1.10x + 0.75(50 - x) = 45$$
$$1.10x + 37.50 - 0.75x = 45$$
$$0.35x + 37.50 = 45$$
$$0.35x + 37.50 - 37.50 = 45 - 37.50$$
$$0.35x = 7.50$$
$$\frac{0.35x}{0.35} = \frac{7.50}{0.35}$$
$$x \approx 21.4$$
Approximately 21 pounds of sunflower seed should be used in the mixture.

**43.** Let $x$ = number of pounds of special lime.

|  | Number of pounds | Price per pound | Total cost |
|---|---|---|---|
| Special | $x$ | 90 | $90x$ |
| Standard | 5 | 50 | 250 |
| Mixture | $5 + x$ | 65 | $65(5 + x)$ |

Section 3.5   Applications

$$90x + 250 = 65(5+x)$$
$$90x + 250 = 325 + 65x$$
$$90x + 250 - 65x = 325 + 65x - 65x$$
$$25x + 250 = 325$$
$$25x + 250 - 250 = 325 - 250$$
$$25x = 75$$
$$\frac{25x}{25} = \frac{75}{25}$$
$$x = 3$$

The amount of special lime is 3 lb.

**45.** Let $x$ = the number of gallons of 3% milk. Then $150 - x$ = the number of gallons of 4.5% milk.

|        | Gallons of milk | % Butterfat | Gallons of Butterfat |
|--------|-----------------|-------------|----------------------|
| 3%     | $x$             | 0.03        | $0.03x$              |
| 4.5%   | $150 - x$       | 0.045       | $0.045(150-x)$       |
| Mixture| 150             | 0.04        | $0.04(150)$          |

$$0.03x + 0.045(150 - x) = 0.04(150)$$
$$0.03x + 6.75 - 0.045x = 6$$
$$6.75 - 0.015x = 6$$
$$6.75 - 0.015x - 6.75 = 6 - 6.75$$
$$-0.015x = -0.75$$
$$\frac{-0.015x}{-0.015} = \frac{-0.75}{-0.015}$$
$$x = 50$$

Fifty gallons of 3% butterfat milk and $150 - 50 = 100$ gallons of 4.5% butter fat milk were used in the mixture.

**47.** Let $x$ = the number of ounces of 0.8% HCl. Then $4 - x$ = the number of ounces of 1.4% HCl.

|        | Ounces solution | % HCl  | Ounces of HCl |
|--------|-----------------|--------|---------------|
| 0.8%   | $x$             | 0.008  | $0.008x$      |
| 1.4%   | $4 - x$         | 0.014  | $0.014(4-x)$  |
| Mixture| 4               | 0.011  | 0.044         |

$$0.008x + 0.014(4 - x) = 0.044$$
$$0.008x + 0.056 - 0.014x = 0.044$$
$$-0.006x + 0.056 = 0.044$$
$$-0.006x + 0.056 - 0.056 = 0.044 - 0.056$$
$$-0.006x = -0.012$$
$$\frac{-0.006x}{-0.006} = \frac{-0.012}{-0.006}$$
$$x = 2$$

The pharmacist should use 2 ounces of the 0.8% solution and 2 ounces of the 1.4% solution.

**49.** Let $x$ = the number of ounces of water.

|         | Ounces solution | % TSP | Ounces of TSP |
|---------|-----------------|-------|---------------|
| 20%     | 64              | 0.20  | $0.20(64)$    |
| Water   | $x$             | 0     | 0             |
| Mixture | $64 + x$        | 0.10  | $0.10(64+x)$  |

$$0.20(64) + 0 = 0.10(64 + x)$$
$$12.8 = 6.4 + 0.10x$$
$$12.8 - 6.4 = 6.4 + 0.10x - 6.4$$
$$\frac{6.4}{0.10} = \frac{0.10x}{0.10}$$
$$64 = x$$

The worker should add 64 ounces of water.

**51.** Let $P$ = the amount invested.
Interest $= 0.06P$
$$P + 0.06P = 1800 - 236.50$$
$$1.06P = 1563.5$$
$$\frac{1.06P}{1.06} = \frac{1563.5}{1.06}$$
$$P = 1475$$

The person originally invested $1475.

**53.** Let $x$ = price of English book,
$x + 4$ = price of biology book,
$x + 16$ = price of math book.
$$x + (x + 4) + (x + 16) = 209$$
$$x + x + 4 + x + 16 = 209$$
$$3x + 20 = 209$$
$$3x + 20 - 20 = 209 - 20$$
$$3x = 189$$
$$\frac{3x}{3} = \frac{189}{3}$$
$$x = 63$$

The English book was $63, the biology book was $63 + 4 = \$67$, and the math book was $63 + 16 = \$79$.

**55.** Let $x$ = pounds of plain confetti.

|           | Number of pounds | Price per pound | Total cost   |
|-----------|------------------|-----------------|--------------|
| Plain     | $x$              | 4.50            | $4.50x$      |
| Sparkling | 6                | 6.00            | 36.00        |
| Mixture   | $x + 6$          | 5.40            | $5.40(x+6)$  |

$$4.50x + 36 = 5.40(x + 6)$$
$$4.50x + 36 = 5.40x + 32.40$$
$$4.50x + 36 - 5.40x = 5.40x + 32.40 - 5.40x$$
$$36 - 0.9x = 32.40$$
$$36 - 0.9x - 36 = 32.40 - 36$$
$$-0.9x = -3.60$$
$$\frac{-0.9x}{-0.9} = \frac{-3.60}{-0.9}$$
$$x = 4$$

Four pounds of plain confetti should be used.

**57.** Let $x$ = the measure of the larger angle, and $\frac{1}{4}x - 20$ = measure of the smaller angle.

$$x + \left(\frac{1}{4}x - 20\right) = 180$$
$$x + \frac{1}{4}x - 20 = 180$$
$$\frac{5}{4}x - 20 = 180$$
$$\frac{5}{4}x - 20 + 20 = 180 + 20$$
$$\frac{5}{4}x = 200$$
$$\frac{4}{5} \cdot \frac{5}{4}x = \frac{4}{5} \cdot 200$$
$$x = 160$$

The measure of the larger angle is 160° and the measure of the smaller angle is $\frac{1}{4}(160) - 20 = 40 - 20 = 20°$.

**59.** Let $x$ = the number of cups of 12% vinegar. Then $2 - x$ = the number of cups of 25% vinegar.

|  | Ounces solution | % vinegar | Ounces of vinegar |
|---|---|---|---|
| 12% | $x$ | 0.12 | $0.12x$ |
| 25% | $2 - x$ | 0.25 | $0.25(2 - x)$ |
| Mixture | 2 | 0.15 | 0.3 |

$$0.12x + 0.25(2 - x) = 0.3$$
$$0.12x + 0.5 - 0.25x = 0.3$$
$$0.5 - 0.13x = 0.3$$
$$0.5 - 0.13x - 0.5 = 0.3 - 0.5$$
$$-0.13x = -0.2$$
$$\frac{-0.13x}{-0.13} = \frac{-0.2}{-0.13}$$
$$x \approx 1.54$$

The chef should use approximately 1.54 cups of 12% vinegar and 0.46 cup of 25% vinegar.

**61.** Because the cyclist rides 15 miles before stopping for a sundae, we assume that she rides an equal distance when she returns home, for a total distance of 30 miles. The clock is active for 1 hour and then for 50 minutes. The average speed displayed by the computer will be:

$$\frac{30 \text{ miles}}{1\frac{5}{6} \text{ hours}} \approx 16.4 \text{ mph}$$

**63.** Let $T$ = the total amount of loan.
$$80{,}000 + 0.025(80{,}000) = T$$
$$80{,}000 + 2000 = T$$
$$82{,}000 = T$$

The total amount of the loan is $82,000.

**65.** To find time, use the formula $t = \frac{d}{r}$.

(There are 1760 yards in a mile.)

Rabbit:
$$\frac{100 \text{ yd}}{\frac{35 \text{ mi}}{\text{hr}}} \times \frac{1 \text{ mi}}{1760 \text{ yd}} \times \frac{3600 \text{ sec}}{1 \text{ hour}} \approx 5.84 \text{ sec}$$

Cat:
$$\frac{100 \text{ yd}}{\frac{30 \text{ mi}}{\text{hr}}} \times \frac{1 \text{ mi}}{1760 \text{ yd}} \times \frac{3600 \text{ sec}}{1 \text{ hour}} \approx 6.82 \text{ sec}$$

Elephant:
$$\frac{100 \text{ yd}}{\frac{25 \text{ mi}}{\text{hr}}} \times \frac{1 \text{ mi}}{1760 \text{ yd}} \times \frac{3600 \text{ sec}}{1 \text{ hour}} \approx 8.18 \text{ sec}$$

Coyote:
$$\frac{100 \text{ yd}}{\frac{43 \text{ mi}}{\text{hr}}} \times \frac{1 \text{ mi}}{1760 \text{ yd}} \times \frac{3600 \text{ sec}}{1 \text{ hour}} \approx 4.76 \text{ sec}$$

Tortoise:
$$\frac{100 \text{ yd}}{\frac{0.17 \text{ mi}}{\text{hr}}} \times \frac{1 \text{ mi}}{1760 \text{ yd}} \times \frac{60 \text{ min}}{1 \text{ hour}} \approx 20.1 \text{ min}$$

Snail:
$$\frac{100 \text{ yd}}{\frac{0.03 \text{ mi}}{\text{hr}}} \times \frac{1 \text{ mi}}{1760 \text{ yd}} \approx 1.9 \text{ hours}$$

**67.** Settled before trial:
$80.44\% \times 163{,}511 \approx 131{,}528$
Settled after trial:
$12.40\% \times 163{,}511 \approx 20{,}275$
Dismissed:
$7.16\% \times 163{,}511 \approx 11{,}707$

**69.** From Exercise 68, we know that there were approximately 246,995 civil cases filed in 1988-89. There were $23.35\% \times 246{,}995 \approx 57{,}673$ cases dismissed because of a delay in prosecution in 1988-89. From Exercise 67, we know there were 11,707 cases dismissed in 1998-99. The decrease was $57{,}673 - 11{,}707 = 45{,}966$. The percent decrease was $\dfrac{45{,}966}{57{,}673} \times 100\% \approx 79.7\%$.

**71.** If $x =$ wholesale cost, then $1.25x =$ retail value. The buyer's offer is $0.75(1.25x) = 0.9375x$. Because this is less than the amount you paid for the inventory, you would lose money.

## Section 3.6  Linear Inequalities

**1.** The interval $[a, b]$ contains $a$, but the interval $(a, b]$ does not.

**3.** $x < 2$ can be represented by $(-\infty, 2)$ in interval notation, $\{x \mid x < 2\}$ in set notation, and by the following graph.

**5.** $x \geq 3$ can be represented by $[3, \infty)$ in interval notation, $\{x \mid x \geq 3\}$ in set notation, and by the following graph.

**7.** $-3 < x \leq 2$ can be represented by $(-3, 2]$ in interval notation, $\{x \mid -3 < x \leq 2\}$ in set notation, and by the following graph.

**9.** In both cases, we graph the left and right sides. For an equation, the solution is the $x$-coordinate of the point of intersection. For an inequality, the solution set is the set of $x$-values for which the graph of one side is above (or below) the graph of the other side.

**11.** (a) From the given graph, the solution set of $y_1 \leq y_2$ is $[-14, \infty)$.
(b) From the given graph, the solution set of $y_1 > y_2$ is $(-\infty, -14)$.

**13.** $3x - 16 < 8$
Enter $y_1 = 3x - 16$ and $y_2 = 8$.

Note that $y_1 < y_2$ for all values of $x$ that are less than 8. Thus the solution set of $y_1 < y_2$ is $(-\infty, 8)$.

15. $3x + 9 \geq -2x - 16$
Enter $y_1 = 3x + 9$ and $y_2 = -2x - 16$.

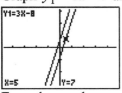

Note that $y_1 \geq y_2$ for all values of $x$ that are greater than or equal to $-5$. Thus the solution set of $y_1 \geq y_2$ is $[-5, \infty)$.

17. $3x - 8 \leq 3(2 + x)$
Graph $y_1 = 3x - 8$ and $y_2 = 3(2 + x)$.

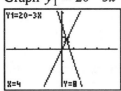

From the graph, we see that there is no point of intersection, but the graph of $y_1$ is always below $y_2$. Thus the solution set of $y_1 \leq y_2$ is **R**.

19. $20 - 3x \geq 2x$
Graph $y_1 = 20 - 3x$ and $y_2 = 2x$.

The estimate of the point of intersection is (4, 8). The graph of $y_1$ is above $y_2$ to the left of the point of intersection. Thus the solution set of $y_1 \geq y_2$ is $(-\infty, 4]$.

21. $2(x - 5) > 2x + 1$
Graph $y_1 = 2(x - 5)$ and $y_2 = 2x + 1$.

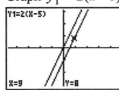

From the graph, we see that there is no point of intersection. The graph of $y_1$ is never above $y_2$. Thus the solution set of $y_1 > y_2$ is $\emptyset$.

23. $x + 3 < 8$
$x + 3 - 3 < 8 - 3$
$x < 5$
The solution set is $(-\infty, 5)$.

25. $-7 + x \leq -7$
$-7 + x + 7 \leq -7 + 7$
$x \leq 0$
The solution set is $(-\infty, 0]$.

27. $5x + 12 < 5 + 4x$
$5x + 12 - 4x < 5 + 4x - 4x$
$x + 12 < 5$
$x + 12 - 12 < 5 - 12$
$x < -7$
The solution set is $(-\infty, -7)$.

29. $3 - 6x \leq 2 - 7x$
$3 - 6x + 7x \leq 2 - 7x + 7x$
$3 + x \leq 2$
$3 + x - 3 \leq 2 - 3$
$x \leq -1$
The solution set is $(-\infty, -1]$.

31. (a) The graphs are parallel (or coincide for < or >).
(b) The graphs are parallel (or coincide for ≤ or ≥).

33. We are given that $a < b$. By the Multiplication Property of Inequalities, $2a \underline{\ <\ } 2b$.

35. We are given that $a < b$. Therefore, $b \underline{\ >\ } a$.

37. We are given that $a < b$. By the Multiplication Property of Inequalities, $-a \underline{\ >\ } -b$.

39. We are given that $a < b$. By the Addition Property of Inequalities, $-3 + a \underline{\ <\ } -3 + b$.

41. We are given that $a < b$. By the Multiplication Property of Inequalities, $ac > bc$ for $c < 0$.

Section 3.6  Linear Inequalities

**43.** We are given that $a < b$. By the Addition Property of Inequalities, $a + c < b + c$ for any real number $c$.

**45.** We are given that $a < b$. By the Multiplication Property of Inequalities, $\dfrac{a}{c} < \dfrac{b}{c}$ for $c > 0$.

**47.** The methods are the same, except that we must reverse the inequality symbol if we multiply or divide both sides of an inequality by a negative number.

**49.**
$$3x \leq 12$$
$$\frac{3x}{3} \leq \frac{12}{3}$$
$$x \leq 4$$
The solution set is $(-\infty, 4]$.

**51.**
$$-x < 5$$
$$\frac{-x}{-1} > \frac{5}{-1}$$
$$x > -5$$
The solution set is $(-5, \infty)$.

**53.**
$$-0.2y \geq 1.6$$
$$\frac{-0.2y}{-0.2} \leq \frac{1.6}{-0.2}$$
$$y \leq -8$$
The solution set is $(-\infty, -8]$.

**55.**
$$-\frac{2}{5}t > -6$$
$$-\frac{5}{2}\cdot\left(-\frac{2}{5}t\right) < -\frac{5}{2}\cdot(-6)$$
$$t < 15$$
The solution set is $(-\infty, 15)$.

**57.**
$$3x - 4 < 14$$
$$3x - 4 + 4 < 14 + 4$$
$$3x < 18$$
$$\frac{3x}{3} < \frac{18}{3}$$
$$x < 6$$
The solution set is $(-\infty, 6)$.

**59.**
$$7 - 3t \leq -8$$
$$7 - 3t - 7 \leq -8 - 7$$
$$-3t \leq -15$$
$$\frac{-3t}{-3} \geq \frac{-15}{-3}$$
$$t \geq 5$$
The solution set is $[5, \infty)$.

**61.**
$$5x + 3 < 7 + 5x$$
$$5x + 3 - 5x < 7 + 5x - 5x$$
$$3 < 7$$
The last inequality is true. The solution set is **R**.

**63.**
$$4x - 4 \leq 7x - 13$$
$$4x - 4 - 7x \leq 7x - 13 - 7x$$
$$-4 - 3x \leq -13$$
$$-4 - 3x + 4 \leq -13 + 4$$
$$-3x \leq -9$$
$$\frac{-3x}{-3} \geq \frac{-9}{-3}$$
$$x \geq 3$$
The solution set is $[3, \infty)$.

**65.**
$$8 - x > x - 8$$
$$8 - x - x > x - 8 - x$$
$$8 - 2x > -8$$
$$8 - 2x - 8 > -8 - 8$$
$$-2x > -16$$
$$\frac{-2x}{-2} < \frac{-16}{-2}$$
$$x < 8$$
The solution set is $(-\infty, 8)$.

**67.**
$$3 - x \geq 4x + 1$$
$$3 - x - 4x \geq 4x + 1 - 4x$$
$$3 - 5x \geq 1$$
$$3 - 5x - 3 \geq 1 - 3$$
$$-5x \geq -2$$
$$\frac{-5x}{-5} \leq \frac{-2}{-5}$$
$$x \leq \frac{2}{5}$$
The solution set is $\left(-\infty, \dfrac{2}{5}\right]$.

**69.** 
$3x - x < 2x$
$2x < 2x$
$2x - 2x < 2x - 2x$
$0 < 0$
The last inequality is false. The solution set is $\varnothing$.

**71.** 
$-(x-2) \leq 2(x+1)$
$-x + 2 \leq 2x + 2$
$-x + 2 - 2x \leq 2x + 2 - 2x$
$-3x + 2 \leq 2$
$-3x + 2 - 2 \leq 2 - 2$
$-3x \leq 0$
$\dfrac{-3x}{-3} \geq \dfrac{0}{-3}$
$x \geq 0$
The solution set is $[0, \infty)$.

**73.** 
$5 - 4x \geq 4(2 - x)$
$5 - 4x \geq 8 - 4x$
$5 - 4x + 4x \geq 8 - 4x + 4x$
$5 \geq 8$
The last inequality is false. The solution set is $\varnothing$.

**75.** 
$x - 4(x - 4) - 4x > x - 4$
$x - 4x + 16 - 4x > x - 4$
$-7x + 16 > x - 4$
$-7x + 16 - x > x - 4 - x$
$-8x + 16 > -4$
$-8x + 16 - 16 > -4 - 16$
$-8x > -20$
$\dfrac{-8x}{-8} < \dfrac{-20}{-8}$
$x < \dfrac{5}{2}$
The solution set is $\left(-\infty, \dfrac{5}{2}\right)$.

**77.** 
$5 + 4(x - 3) \geq 2x + 3(x - 1) + 4$
$5 + 4x - 12 \geq 2x + 3x - 3 + 4$
$4x - 7 \geq 5x + 1$
$4x - 7 - 5x \geq 5x + 1 - 5x$
$-7 - x \geq 1$
$-7 - x + 7 \geq 1 + 7$
$-x \geq 8$
$\dfrac{-x}{-1} \leq \dfrac{8}{-1}$
$x \leq -8$
The solution set is $(-\infty, -8]$.

**79.** 
$\dfrac{2}{3}x - 5 < \dfrac{5}{6}$
$\dfrac{2}{3}x - 5 + 5 < \dfrac{5}{6} + 5$
$\dfrac{2}{3}x < \dfrac{35}{6}$
$\dfrac{3}{2} \cdot \dfrac{2}{3}x < \dfrac{3}{2} \cdot \dfrac{35}{6}$
$x < \dfrac{35}{4}$
The solution set is $\left(-\infty, \dfrac{35}{4}\right)$.

**81.** 
$\dfrac{2x + 1}{-3} < 6$
$-3 \cdot \dfrac{2x + 1}{-3} > -3 \cdot 6$
$2x + 1 > -18$
$2x + 1 - 1 > -18 - 1$
$2x > -19$
$\dfrac{2x}{2} > \dfrac{-19}{2}$
$x > -\dfrac{19}{2}$
The solution set is $\left(-\dfrac{19}{2}, \infty\right)$.

Section 3.6   Linear Inequalities

**83.**
$$x - \frac{1}{3} + \frac{5}{6}x \le \frac{1}{2} - 3x$$
$$\frac{11}{6}x - \frac{1}{3} \le \frac{1}{2} - 3x$$
$$6 \cdot \frac{11}{6}x - 6 \cdot \frac{1}{3} \le 6 \cdot \frac{1}{2} - 6 \cdot 3x$$
$$11x - 2 \le 3 - 18x$$
$$11x - 2 + 18x \le 3 - 18x + 18x$$
$$29x - 2 \le 3$$
$$29x - 2 + 2 \le 3 + 2$$
$$29x \le 5$$
$$\frac{29x}{29} \le \frac{5}{29}$$
$$x \le \frac{5}{29}$$

The solution set is $\left(-\infty, \frac{5}{29}\right]$.

**85.**
$$\frac{3}{4}(x+3) + 2x < 1$$
$$\frac{3}{4}x + \frac{9}{4} + 2x < 1$$
$$\frac{11}{4}x + \frac{9}{4} < 1$$
$$4 \cdot \frac{11}{4}x + 4 \cdot \frac{9}{4} < 4 \cdot 1$$
$$11x + 9 < 4$$
$$11x + 9 - 9 < 4 - 9$$
$$11x < -5$$
$$\frac{11x}{11} < \frac{-5}{11}$$
$$x < -\frac{5}{11}$$

The solution set is $\left(-\infty, -\frac{5}{11}\right)$.

**87.**
$$0.2x + 1.7 \ge 2.94 - 5.3x$$
$$0.2x + 1.7 + 5.3x \ge 2.94 - 5.3x + 5.3x$$
$$5.5x + 1.7 \ge 2.94$$
$$5.5x + 1.7 - 1.7 \ge 2.94 - 1.7$$
$$5.5x \ge 1.24$$
$$\frac{5.5x}{5.5} \ge \frac{1.24}{5.5}$$
$$x \ge 0.22\overline{54}$$

The solution set is approximately $[0.23, \infty)$.

**89.**
$$2.8(1.3x - 0.9) \le 1.2 - 9.97x$$
$$3.64x - 2.52 \le 1.2 - 9.97x$$
$$3.64x - 2.52 + 9.97x \le 1.2 - 9.97x + 9.97x$$
$$13.61x - 2.52 \le 1.2$$
$$13.61x - 2.52 + 2.52 \le 1.2 + 2.52$$
$$13.61x \le 3.72$$
$$\frac{13.61x}{13.61} \le \frac{3.72}{13.61}$$
$$x \le 0.273328435$$

The solution set is approximately $(-\infty, 0.27]$.

**91.** The sum of the lengths of any two sides must be greater than the length of the third side: $a + b > c$, $a + c > b$, and $b + c > a$.

**93.** Let $x =$ the length of the first piece.
$$x + (2x) + (2x + 1) \ge 21$$
$$x + 2x + 2x + 1 \ge 21$$
$$5x + 1 \ge 21$$
$$5x + 1 - 1 \ge 21 - 1$$
$$5x \ge 20$$
$$\frac{5x}{5} \ge \frac{20}{5}$$
$$x \ge 4$$

The minimum length of the first piece is 4 feet.

**95.** Let $x =$ the points received in the last race. Based on the given finish positions, the total points for the first nine races are:
$10 + 3 + 10 + 7 + 7 + 10 + 7 + 7 + 5 = 66$
and the inequality is:
$$\frac{66 + x}{10} \ge 7$$
$$10 \cdot \frac{66 + x}{10} \ge 10 \cdot 7$$
$$66 + x \ge 70$$
$$66 + x - 66 \ge 70 - 66$$
$$x \ge 4$$

Because at least 4 points are needed, you must finish in third place or better.

**97.** Let $x$ = the number of miles driven.
$$26.20 + 0.22x \leq 200$$
$$26.20 + 0.22x - 26.20 \leq 200 - 26.20$$
$$0.22x \leq 173.80$$
$$\frac{0.22x}{0.22} \leq \frac{173.80}{0.22}$$
$$x \leq 790$$
The maximum number of miles that can be driven during a 1-day rental is 790.

**99.** Let $x$ = Calvin's number,
$x + 2$ = John's number, and
$x + 5$ = Carrie's number.
$$\frac{1}{7}x + \frac{1}{3}(x+2) + \frac{1}{6}(x+5) > 5$$
$$42 \cdot \left[\frac{1}{7}x + \frac{1}{3}(x+2) + \frac{1}{6}(x+5)\right] > 42 \cdot 5$$
$$6x + 14(x+2) + 7(x+5) > 210$$
$$6x + 14x + 28 + 7x + 35 > 210$$
$$27x + 63 > 210$$
$$27x + 63 - 63 > 210 - 63$$
$$27x > 147$$
$$\frac{27x}{27} > \frac{147}{27}$$
$$x > 5.\overline{4}$$
Calvin's number is 6. He inherits $6000.

**101.** (a) An inequality describing the years when an average of at least $7.2 million will be spent by the winning candidate is given by:
$$W(x) \geq 7.2$$
$$0.25x + 4.7 \geq 7.2$$
(b) Solve the inequality from part (a).
$$0.25x + 4.7 \geq 7.2$$
$$0.25x + 4.7 - 4.7 \geq 7.2 - 4.7$$
$$0.25x \geq 2.5$$
$$\frac{0.25x}{0.25} \geq \frac{2.5}{0.25}$$
$$x \geq 10$$
The solution set is $[10, \infty)$. At least $7.2 million will be spent by the winning candidate from 2006 onward.
(c) Visual evidence suggesting that $W(x) \geq L(x)$ will always be true is seen in the bar graph, where the bar of $W(x)$ starts higher than the bar for $L(x)$, and the values of $W(x)$ are increasing at a faster rate.

**103.** (a) The solution of the inequality follows:
$$N(x) \geq A(x)$$
$$0.17x + 9.04 \geq 0.16x + 9.80$$
$$0.17x + 9.04 - 9.04 \geq 0.16x + 9.80 - 9.04$$
$$0.17x \geq 0.16x + 0.76$$
$$0.17x - 0.16x \geq 0.16x - 0.16x + 0.76$$
$$0.01x \geq 0.76$$
$$\frac{0.01x}{0.01} \geq \frac{0.76}{0.01}$$
$$x \geq 76$$
The solution set is $[76, \infty)$.
(b) At the present rate, the average number of runs scored per game in the National League will be at least as many as in the American League in the year 2069 and beyond.

**105.**
$$5x + a \leq b$$
$$5x + a - a \leq b - a$$
$$5x \leq b - a$$
$$\frac{5x}{5} \leq \frac{b-a}{5}$$
$$x \leq \frac{b-a}{5}$$
The solution set is $\left(-\infty, \dfrac{b-a}{5}\right]$.

**107.**
$$a^2 x \geq 8, \quad a \neq 0$$
$$\frac{a^2 x}{a^2} \geq \frac{8}{a^2}, \quad a \neq 0$$
$$x \geq \frac{8}{a^2}$$
The solution set is $\left[\dfrac{8}{a^2}, \infty\right)$.

**109.** Let $x$ = the first even integer,
$x + 2$ = the second even integer, and
$x + 4$ = the third even integer.
The "sum of the first and third integers is no more than the second integer" is translated as:

$$x+(x+4) \leq x+2$$
$$x+x+4 \leq x+2$$
$$2x+4 \leq x+2$$
$$2x+4-x \leq x+2-x$$
$$x+4 \leq 2$$
$$x+4-4 \leq 2-4$$
$$x \leq -2$$

The first number is at most $-2$, (with the other numbers being at most 0 and 2). However, the conditions stipulate that at least two of the numbers are positive, and 0 is not considered a positive number. The best we can do is one positive integer. Your instructor will never have to pay.

## Section 3.7 Compound Inequalities

**1.** The solution set is the intersection of the solution sets of the individual inequalities.

**3.** $x = 2$:
$\quad 5x < 12 \qquad\qquad 3x > 3$
$\quad 5(2) < 12 \quad$ and $\quad 3(2) > 3$
$\quad 10 < 12 \quad$ True $\qquad 6 > 3 \quad$ True

Because $x = 2$ satisfies both inequalities, $x = 2$ is a solution of the compound inequality.

$x = 1$:
$\quad 5x < 12 \qquad\qquad 3x > 3$
$\quad 5(1) < 12 \quad$ and $\quad 3(1) > 3$
$\quad 5 < 12 \quad$ True $\qquad 3 > 3 \quad$ False

Because $x = 1$ does not satisfy both inequalities, $x = 1$ is not a solution of the compound inequality.

**5.** $x = 5$:
$\quad 2x+7 < -3 \qquad\qquad 3x+5 > 11$
$\quad 2(5)+7 < -3 \quad$ or $\quad 3(5)+5 > 11$
$\quad 17 < -3 \quad$ F $\qquad 20 > 11 \quad$ T

Because $x = 5$ satisfies at least one inequality, $x = 5$ is a solution of the compound inequality.

$x = -7$:
$\quad 2x+7 < -3 \qquad\qquad 3x+5 > 11$
$\quad 2(-7)+7 < -3 \quad$ or $\quad 3(-7)+5 > 11$
$\quad -7 < -3 \quad$ T $\qquad -16 > 11 \quad$ F

Because $x = -7$ satisfies at least one inequality, $x = -7$ is a solution of the compound inequality.

**7.** $x = 7$:
$\quad 0.5x > 4.5 \qquad\qquad 0.3x > 1.8$
$\quad 0.5(7) > 4.5 \quad$ or $\quad 0.3(7) > 1.8$
$\quad 3.5 > 4.5 \quad$ F $\qquad 2.1 > 1.8 \quad$ T

Because $x = 7$ satisfies at least one inequality, $x = 7$ is a solution of the compound inequality.

$x = 11$:
$\quad 0.5x > 4.5 \qquad\qquad 0.3x > 1.8$
$\quad 0.5(11) > 4.5 \quad$ or $\quad 0.3(11) > 1.8$
$\quad 5.5 > 4.5 \quad$ T $\qquad 3.3 > 1.8 \quad$ T

Because $x = 11$ satisfies at least one inequality, $x = 11$ is a solution of the compound inequality.

**9.** $x = 3.75$:
$\quad 0.5 \leq 2x-3 < 5$
$\quad 0.5 \leq 2(3.75)-3 < 5$
$\quad 0.5 \leq 7.5-3 < 5$
$\quad 0.5 \leq 4.5 < 5 \qquad$ True

Because $x = 3.75$ makes the double inequality true, $x = 3.75$ is a solution of the compound inequality.

$x = 5$:
$$0.5 \le 2x - 3 < 5$$
$$0.5 \le 2(5) - 3 < 5$$
$$0.5 \le 10 - 3 < 5$$
$$0.5 \le 7 < 5 \quad \text{False}$$

Because $x = 5$ makes the double inequality false, $x = 5$ is not a solution of the compound inequality.

**11.** A conjunction is a compound inequality in which the connective is *and*. A disjunction is a compound inequality in which the connective is *or*.

**13.** The solution set is the intersection of the individual sets.
$x \ge -5$:

$x < 3$:

$x \ge -5$ and $x < 3$:

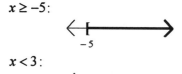

**15.** The solution set is the intersection of the individual sets.
$x < 5$:

$x \le 1$:

$x < 5$ and $x \le 1$:

**17.** The solution set is the intersection of the individual sets.
$x \le 4$:

$x > 6$:

The solution set for $x \le 4$ and $x > 6$ is $\varnothing$.

**19.** The solution set is the union of the individual sets.
$x < -5$:

$x > 4$:

$x < -5$ or $x > 4$:

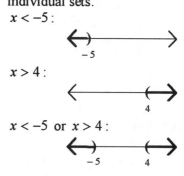

**21.** The solution set is the union of the individual sets.
$x \ge -3$:

$x > 4$:

$x \ge -3$ or $x > 4$:

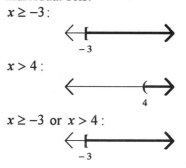

**23.** The solution set is the union of the individual sets.
$x \le 4$:

$x > 0$:

$x \le 4$ or $x > 0$:

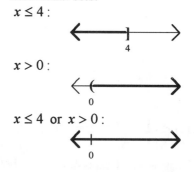

**25.** (a) The graph of $y_2$ intersects or is above the graph of $y_1$ at and to the right of the point $(-4, 8)$. The graph of $y_2$ intersects or is below the graph of $y_3$ at and to the left of the point $(8, 20)$. Points that satisfy both conditions have $x$-coordinates in the interval $[-4, 8]$.

(b) The graph of $y_2$ intersects or is below the graph of $y_1$ at and to the left of the point $(-4, 8)$. The graph of $y_2$ intersects or is above the

Section 3.7 Compound Inequalities

graph of $y_3$ at and to the right of the point (8, 20). Points that satisfy at least one condition have x-coordinates in the interval $(-\infty, 4] \cup [8, \infty)$.

**27.** $-12 \leq 3x - 2 \leq 8$. Graph $y_1 = -12$, $y_2 = 3x - 2$, and $y_3 = 8$.

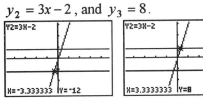

An estimate of the solution set of the compound inequality is $\left[-\dfrac{10}{3}, \dfrac{10}{3}\right]$.

**29.** $-2 < \dfrac{x-1}{-3} < 2$. Graph $y_1 = -2$, $y_2 = \dfrac{x-1}{-3}$, and $y_3 = 2$.

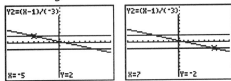

An estimate of the solution set of the compound inequality is $(-5, 7)$.

**31.** (a) The only solution to $x \leq c$ and $x \geq c$ is the intersection, the number $c$: $\{c\}$.
(b) No number can be less than $c$ and also greater than $c$. The solution to $x > c$ and $x < c$ is $\varnothing$.
(c) Every real number is either at most $c$ or at least $c$. The solution to $x \leq c$ or $x \geq c$ is **R**.
(d) The only number that does not satisfy at least one of the conditions $x > c$ or $x < c$ is the number $c$. The solution set is $\{x \mid x \neq c\}$.

**33.**
$$-4 < x + 3 < 7$$
$$-4 - 3 < x + 3 - 3 < 7 - 3$$
$$-7 < x < 4$$
The solution set is $(-7, 4)$

**35.**
$$2 \leq -3t \leq 18$$
$$\dfrac{2}{-3} \geq \dfrac{-3t}{-3} \geq \dfrac{18}{-3}$$
$$-\dfrac{2}{3} \geq t \geq -6 \quad \text{or} \quad -6 \leq t \leq -\dfrac{2}{3}$$
The solution set is $[-6, -2/3]$.

**37.**
$$9 \geq 2x + 1 \geq -2$$
$$9 - 1 \geq 2x + 1 - 1 \geq -2 - 1$$
$$8 \geq 2x \geq -3$$
$$\dfrac{8}{2} \geq \dfrac{2x}{2} \geq \dfrac{-3}{2}$$
$$4 \geq x \geq -1.5 \quad \text{or} \quad -1.5 \leq x \leq 4$$
The solution set is $[-1.5, 4]$.

**39.**
$$-5 \leq -2x - 7 < 2$$
$$-5 + 7 \leq -2x - 7 + 7 < 2 + 7$$
$$2 \leq -2x < 9$$
$$\dfrac{2}{-2} \geq \dfrac{-2x}{-2} > \dfrac{9}{-2}$$
$$-1 \geq x > -4.5 \quad \text{or} \quad -4.5 < x \leq -1$$
The solution set is $(-4.5, -1]$.

**41.**
$$-\dfrac{7}{4} \leq \dfrac{3}{4}x - 1 < 2$$
$$-\dfrac{7}{4} + 1 \leq \dfrac{3}{4}x - 1 + 1 < 2 + 1$$
$$-0.75 \leq 0.75x < 3$$
$$\dfrac{-0.75}{0.75} \leq \dfrac{0.75x}{0.75} < \dfrac{3}{0.75}$$
$$-1 \leq x < 4$$
The solution set is $[-1, 4)$.

**43.**
$$-\dfrac{17}{6} < \dfrac{2}{3}x - \dfrac{1}{6} \leq \dfrac{11}{3}$$
$$6 \cdot \left(-\dfrac{17}{6}\right) < 6 \cdot \left(\dfrac{2}{3}x - \dfrac{1}{6}\right) \leq 6 \cdot \left(\dfrac{11}{3}\right)$$
$$-17 < 4x - 1 \leq 22$$
$$-17 + 1 < 4x - 1 + 1 \leq 22 + 1$$
$$-16 < 4x \leq 23$$
$$\dfrac{-16}{4} < \dfrac{4x}{4} \leq \dfrac{23}{4}$$
$$-4 < x \leq 5.75$$
The solution set is $(-4, 5.75]$.

**45.** 
$$6.5 \geq 0.75x + 0.5 > -2.5$$
$$6.5 - 0.5 \geq 0.75x + 0.5 - 0.5 > -2.5 - 0.5$$
$$6 \geq 0.75x > -3$$
$$\frac{6}{0.75} \geq \frac{0.75x}{0.75} > \frac{-3}{0.75}$$
$$8 \geq x > -4 \quad \text{or} \quad -4 < x \leq 8$$
The solution set is $(-4, 8]$.

**47.** The solution set of a disjunction is the union of intervals, whereas the solution set of a conjunction is the intersection of intervals. Thus, for example, the union of $(-\infty, 1)$ and $[1, \infty)$ is **R**, but the intersection of those intervals is $\emptyset$.

**49.**
$$-x \leq 5 \qquad\qquad -x \geq -7$$
$$\frac{-x}{-1} \geq \frac{5}{-1} \quad \text{and} \quad \frac{-x}{-1} \leq \frac{-7}{-1}$$
$$x \geq -5 \qquad\qquad x \leq 7$$
The solution set is $[-5, 7]$.

**51.**
$$x + 1 \geq 5 \qquad\qquad x - 2 < 12$$
$$x + 1 - 1 \geq 5 - 1 \quad \text{and} \quad x - 2 + 2 < 12 + 2$$
$$x \geq 4 \qquad\qquad x < 14$$
The solution set is $[4, 14)$.

**53.**
$$2x - 5 \leq 7$$
$$2x - 5 + 5 \leq 7 + 5$$
$$2x \leq 12 \qquad \text{and}$$
$$\frac{2x}{2} \leq \frac{12}{2}$$
$$x \leq 6$$

$$x + 4 > -7$$
$$x + 4 - 4 > -7 - 4$$
$$x > -11$$
The solution set is $(-11, 6]$.

**55.**
$$-x \geq 4 \qquad\qquad -x \leq -3$$
$$\frac{-x}{-1} \leq \frac{4}{-1} \quad \text{or} \quad \frac{-x}{-1} \geq \frac{-3}{-1}$$
$$x \leq -4 \qquad\qquad x \geq 3$$
The solution set is $(-\infty, -4] \cup [3, \infty)$.

**57.**
$$x + 4 < -3 \qquad\qquad x + 4 > 3$$
$$x + 4 - 4 < -3 - 4 \quad \text{or} \quad x + 4 - 4 > 3 - 4$$
$$x < -7 \qquad\qquad x > -1$$
The solution set is $(-\infty, -7) \cup (-1, \infty)$.

**59.**
$$-4x - 5 > 0 \qquad\qquad 3x - 2 \geq 1$$
$$-4x - 5 + 5 > 0 + 5 \qquad 3x - 2 + 2 \geq 1 + 2$$
$$-4x > 5 \quad \text{or} \quad 3x \geq 3$$
$$\frac{-4x}{-4} < \frac{5}{-4} \qquad\qquad \frac{3x}{3} \geq \frac{3}{3}$$
$$x < -1.25 \qquad\qquad x \geq 1$$
The solution set is $(-\infty, -1.25) \cup [1, \infty)$.

**61.**
$$2x - 5 \leq -17$$
$$2x - 5 + 5 \leq -17 + 5 \qquad 3x > -27$$
$$2x \leq -12 \quad \text{and} \quad \frac{3x}{3} > \frac{-27}{3}$$
$$\frac{2x}{2} \leq \frac{-12}{2} \qquad\qquad x > -9$$
$$x \leq -6$$
The solution set is $(-9, -6]$.

**63.**
$$x - 2 > 3 \qquad\qquad x + 5 < 4$$
$$x - 2 + 2 > 3 + 2 \quad \text{or} \quad x + 5 - 5 < 4 - 5$$
$$x > 5 \qquad\qquad x < -1$$
The solution set is $(-\infty, -1) \cup (5, \infty)$.

**65.**
$$7 - x \geq 10 \qquad\qquad 3x - 5 > 7$$
$$7 - x - 7 \geq 10 - 7 \qquad 3x - 5 + 5 > 7 + 5$$
$$-x \geq 3 \quad \text{or} \quad 3x > 12$$
$$\frac{-x}{-1} \leq \frac{3}{-1} \qquad\qquad \frac{3x}{3} > \frac{12}{3}$$
$$x \leq -3 \qquad\qquad x > 4$$
The solution set is $(-\infty, -3] \cup (4, \infty)$.

**67.**
$$-4x - 5 \geq 0$$
$$-4x - 5 + 5 \geq 0 + 5$$
$$-4x \geq 5 \qquad \text{and}$$
$$\frac{-4x}{-4} \leq \frac{5}{-4}$$
$$x \leq -1.25$$

$$3x - 2 \geq 1$$
$$3x - 2 + 2 \geq 1 + 2$$
$$3x \geq 3$$
$$\frac{3x}{3} \geq \frac{3}{3}$$
$$x \geq 1$$

Because the connective is "and," the solution set is the intersection of the two intervals. The solution set is $\emptyset$.

Section 3.7   Compound Inequalities

**69.**
$$\frac{1}{2}x - 2 \geq 1 \qquad \frac{2}{3}x - 1 \geq \frac{5}{3}$$
$$\frac{1}{2}x - 2 + 2 \geq 1 + 2 \qquad \frac{2}{3}x - 1 + 1 \geq \frac{5}{3} + 1$$
$$\frac{1}{2}x \geq 3 \quad \text{or} \qquad \frac{2}{3}x \geq \frac{8}{3}$$
$$2 \cdot \frac{1}{2}x \geq 2 \cdot 3 \qquad \frac{3}{2} \cdot \frac{2}{3}x \geq \frac{3}{2} \cdot \frac{8}{3}$$
$$x \geq 6 \qquad x \geq 4$$

The solution set is $[4, \infty)$.

**71.**
$$\frac{2}{5}x + \frac{1}{2} \geq \frac{1}{2}$$
$$\frac{2}{5}x + \frac{1}{2} - \frac{1}{2} \geq \frac{1}{2} - \frac{1}{2}$$
$$\frac{2}{5}x \geq 0 \qquad \text{and}$$
$$\frac{5}{2} \cdot \frac{2}{5}x \geq \frac{5}{2} \cdot 0$$
$$x \geq 0$$

$$\frac{1}{4}x - \frac{1}{3} \leq \frac{1}{6}$$
$$\frac{1}{4}x - \frac{1}{3} + \frac{1}{3} \leq \frac{1}{6} + \frac{1}{3}$$
$$\frac{1}{4}x \leq \frac{1}{2}$$
$$4 \cdot \frac{1}{4}x \leq 4 \cdot \frac{1}{2}$$
$$x \leq 2$$

The solution set is $[0, 2]$.

**73.**
$$0.8 - 0.2x < -0.4$$
$$0.8 - 0.2x - 0.8 < -0.4 - 0.8$$
$$-0.2x < -1.2 \qquad \text{or}$$
$$\frac{-0.2x}{-0.2} > \frac{-1.2}{-0.2}$$
$$x > 6$$

$$0.75x - 0.25 < -0.25$$
$$0.75x - 0.25 + 0.25 < -0.25 + 0.25$$
$$0.75x < 0$$
$$\frac{0.75x}{0.75} < \frac{0}{0.75}$$
$$x < 0$$

The solution set is $(-\infty, 0) \cup (6, \infty)$.

**75.**
$$1.8x + 3.6 < 5.4$$
$$1.8x + 3.6 - 3.6 < 5.4 - 3.6$$
$$1.8x < 1.8 \qquad \text{and}$$
$$\frac{1.8x}{1.8} < \frac{1.8}{1.8}$$
$$x < 1$$

$$2.35x + 7.05 > -4.7$$
$$2.35x + 7.05 - 7.05 > -4.7 - 7.05$$
$$2.35x > -11.75$$
$$\frac{2.35x}{2.35} > \frac{-11.75}{2.35}$$
$$x > -5$$

The solution set is $(-5, 1)$.

**77.**
$$-2x > x - 6$$
$$-2x - x > x - 6 - x$$
$$-3x > -6 \qquad \text{and}$$
$$\frac{-3x}{-3} < \frac{-6}{-3}$$
$$x < 2$$

$$2x - 4 \geq 1 + x$$
$$2x - 4 - x \geq 1 + x - x$$
$$x - 4 \geq 1$$
$$x - 4 + 4 \geq 1 + 4$$
$$x \geq 5$$

The solution set is $\varnothing$.

**79.**
$$5x - 7 \geq 8 \qquad 2 - 3x < -4$$
$$5x - 7 + 7 \geq 8 + 7 \qquad 2 - 3x - 2 < -4 - 2$$
$$5x \geq 15 \quad \text{and} \qquad -3x < -6$$
$$\frac{5x}{5} \geq \frac{15}{5} \qquad \frac{-3x}{-3} > \frac{-6}{-3}$$
$$x \geq 3 \qquad x > 2$$

The solution set is $[3, \infty)$.

**81.**
$$5 - 2x < -9$$
$$5 - 2x - 5 < -9 - 5$$
$$-2x < -14 \quad \text{or}$$
$$\frac{-2x}{-2} > \frac{-14}{-2}$$
$$x > 7$$

$$6x - 2 > 5x$$
$$6x - 2 - 5x > 5x - 5x$$
$$x - 2 > 0$$
$$x - 2 + 2 > 0 + 2$$
$$x > 2$$

The solution set is $(2, \infty)$.

83. $\quad x - 3 \leq 1 \qquad\qquad x - 4 \geq 0$
$x - 3 + 3 \leq 1 + 3 \text{ and } x - 4 + 4 \geq 0 + 4$
$\qquad x \leq 4 \qquad\qquad\qquad x \geq 4$
The solution set is 4.

85. $\quad 7 - x < 0$
$7 - x - 7 < 0 - 7 \qquad\qquad x + 2 > 0$
$\quad -x < -7 \quad \text{or} \quad x + 2 - 2 > 0 - 2$
$\quad \dfrac{-x}{-1} > \dfrac{-7}{-1} \qquad\qquad x > -2$
$\quad\quad x > 7$
The solution set is $(-2, \infty)$.

87. $\quad -2x + 6 \geq -4$
$-2x + 6 - 6 \geq -4 - 6$
$\quad -2x \geq -10 \quad \text{and}$
$\quad \dfrac{-2x}{-2} \leq \dfrac{-10}{-2}$
$\quad\quad x \leq 5$

$\quad 2x - 1 > 9$
$2x - 1 + 1 > 9 + 1$
$\quad 2x > 10$
$\quad \dfrac{2x}{2} > \dfrac{10}{2}$
$\quad\quad x > 5$
The solution set is $\varnothing$.

89. $\quad -x \leq 20 - 3x$
$-x + 3x \leq 20 - 3x + 3x$
$\quad 2x \leq 20 \qquad\qquad \text{and}$
$\quad \dfrac{2x}{2} \leq \dfrac{20}{2}$
$\quad\quad x \leq 10$

$\quad 4 - x \leq -6$
$4 - x - 4 \leq -6 - 4$
$\quad -x \leq -10$
$\quad \dfrac{-x}{-1} \geq \dfrac{-10}{-1}$
$\quad\quad x \geq 10$
The solution set is 10.

91. $\quad 5 - x \leq 5$
$5 - x - 5 \leq 5 - 5$
$\quad -x \leq 0 \qquad \text{or}$
$\quad \dfrac{-x}{-1} \geq \dfrac{0}{-1}$
$\quad\quad x \geq 0$

$\quad 3(2x + 1) < 3 + 5x$
$\quad 6x + 3 < 3 + 5x$
$6x + 3 - 5x < 3 + 5x - 5x$
$\quad x + 3 < 3$
$\quad x + 3 - 3 < 3 - 3$
$\quad\quad x < 0$
The solution set is **R**.

93. Because $3 < x > 8$ is incorrect notation, it has no valid interpretation.

95. Let $x$ = the number of happy dolls and $\dfrac{2}{3}x$ = the number of grumpy dolls.

$75 \leq x + \dfrac{2}{3}x \leq 100$
$75 \leq \dfrac{5}{3}x \leq 100$
$\dfrac{3}{5} \cdot 75 \leq \dfrac{3}{5} \cdot \dfrac{5}{3}x \leq \dfrac{3}{5} \cdot 100$
$45 \leq x \leq 60$

The least number of happy dolls made is 45 and the maximum number of grumpy dolls made is $\dfrac{2}{3}(60) = 40$.

97. Let $h$ = the number of hours worked.
$130 \leq 32h + 50 \leq 170$
$130 - 50 \leq 32h + 50 - 50 \leq 170 - 50$
$\quad 80 \leq 32h \leq 120$
$\quad \dfrac{80}{32} \leq \dfrac{32h}{32} \leq \dfrac{120}{32}$
$\quad 2.5 \leq h \leq 3.75$

The mechanic's time is estimated at between 2.5 and 3.75 hours, inclusive.

99. Let $x$ = the number of toothbrushes sold and $2x$ = the number of toothpastes sold.
$206.00 \leq 1.30x + 3.00(2x) \leq 257.50$
$206.00 \leq 1.30x + 6.00x \leq 257.50$
$206.00 \leq 7.30x \leq 257.50$
$\dfrac{206.00}{7.30} \leq \dfrac{7.30x}{7.30} \leq \dfrac{257.50}{7.30}$
$28.21917808 \leq x \leq 35.2739726$

The number of toothbrushes sold is at least 29 and at most 35.

**101.** Let $x =$ the number of two-point goals and $\frac{2}{3}x =$ the number of free throws.

$$8 \le 2x + \frac{2}{3}x \le 16$$
$$8 \le \frac{8}{3}x \le 16$$
$$\frac{3}{8} \cdot 8 \le \frac{3}{8} \cdot \frac{8}{3}x \le \frac{3}{8} \cdot 16$$
$$3 \le x \le 6$$

The minimum number of two-point goals was 3 and the maximum number of two-point goals was 6.

**103.** Let $x =$ the number of delphinium seeds and $2x =$ the number of marigold seeds.

$$92 \le 0.50x + 0.90(2x) \le 115$$
$$92 \le 0.5x + 1.8x \le 115$$
$$92 \le 2.3x \le 115$$
$$\frac{92}{2.3} \le \frac{2.3x}{2.3} \le \frac{115}{2.3}$$
$$40 \le x \le 50$$

The number of marigold seeds planted is between $2(40) = 80$ and $2(50) = 100$, inclusive.

**105.** (a) From the table, we see that the first year in which the numer of road rage incidents was at least 2300 was 1998 (when $x = 3$).

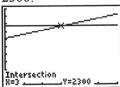

(b) From the graph, the point of intersection is (3, 2300). This means that in year 3 (1998), the number of road rage incidents was 2300.

(c) Because when $x > 3$, the graph of $N(x)$ is above the graph of $y_2$, the number of incidents was greater than 2300.

**107.** (a) Evaluate the model for $x = 4$:
$$V_1(4) = 3.15(4) + 4.97$$
$$= 12.6 + 4.97$$
$$= 17.57$$

The actual data for 1999 is $29.0 billion. This is 65% percent higher than that predicted by the model:
$$\frac{29.0 - 17.57}{17.57} \times 100\% \approx 65\%.$$

(b) From the calculator table, we see that $V_1$ and $V_2$ appear to be equally good models in 1996, when $x = 1$:

(c) According to the table, $V_2$ estimates the lower venture capital investment in 1995, when $x = 0$.

**109.**
$$240 \le T(x) \le 280$$
$$240 \le 20x + 200 \le 280$$
$$240 - 200 \le 20x + 200 - 200 \le 280 - 200$$
$$40 \le 20x \le 80$$
$$\frac{40}{20} \le \frac{20x}{20} \le \frac{80}{20}$$
$$2 \le x \le 4$$

The solution set is [2, 4]. This means that in the years 2000-2002, the number of ATM terminals will be between 240,000 and 280,000, inclusive.

**111.** A function $f(x)$ that represents the total number of transactions at all terminals for any year in the period 1999-2003 is:
$$f(x) = (12)(3491)(20x + 200).$$

**113.**
$$\begin{array}{cc} & k - x < 1 \\ x + k < 5 & k - x - k < 1 - k \\ x + k - k < 5 - k \text{ and } & -x < 1 - k \\ x < 5 - k & \frac{-x}{-1} > \frac{1-k}{-1} \\ & x > k - 1 \end{array}$$

If $k - 1$ is greater than or equal to $5 - k$, there will be no solution.

$$k - 1 \geq 5 - k$$
$$k - 1 + k \geq 5 - k + k$$
$$2k - 1 \geq 5$$
$$2k - 1 + 1 \geq 5 + 1$$
$$2k \geq 6$$
$$\frac{2k}{2} \geq \frac{6}{2}$$
$$k \geq 3$$

For any value of $k \geq 3$, the solution set is the empty set.

**115.** (a) $y_1 = 0.5x - 15$, $y_2 = x - 5$, $y_3 = -x + 21$

(b) The points of intersection are the points for which $y_1 = y_2$ and $y_2 = y_3$.

(c) The solution set is represented by that portion of the graph of $y_2$ that lies between the graphs of $y_1$ and $y_3$. As can be seen from the graphs, the solution set is $(-20, 13)$.

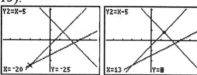

**117.** Write $3x + 2 \leq x + 8 \leq 2x + 15$ in the form of a conjunction and solve each inequality individually.

$3x + 2 \leq x + 8$ and $x + 8 \leq 2x + 15$

$$3x + 2 \leq x + 8$$
$$3x + 2 - x \leq x + 8 - x$$
$$2x + 2 \leq 8$$
$$2x + 2 - 2 \leq 8 - 2 \qquad \text{and}$$
$$2x \leq 6$$
$$\frac{2x}{2} \leq \frac{6}{2}$$
$$x \leq 3$$

$$x + 8 \leq 2x + 15$$
$$x + 8 - 2x \leq 2x + 15 - 2x$$
$$8 - x \leq 15$$
$$8 - x - 8 \leq 15 - 8$$
$$-x \leq 7$$
$$\frac{-x}{-1} \geq \frac{7}{-1}$$
$$x \geq -7$$

The solution set is $[-7, 3]$.

**119.** $$x \leq x + 1 \leq x + 2$$
$$x - x \leq x + 1 - x \leq x + 2 - x$$
$$0 \leq 1 \leq 2$$

The last inequality is true. Thus, the solution set is **R**.

## Section 3.8 Absolute Value: Equations and Inequalities

**1.** The number represented by $x$ is 5 units from 2.

**3.** Graph $y_1 = |x|$ and $y_2 = 12$.

The points of intersection are $(-12, 12)$ and $(12, 12)$. The solutions of the equation are $-12$ and $12$.

**5.** Graph $y_1 = |2x - 10| + 7$ and $y_2 = 7$.

The point of intersection is $(5, 7)$. The solution of the equation is 5.

7. Graph $y_1 = -|x+10|$ and $y_2 = -15$.

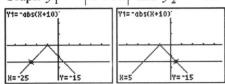

The points of intersection are $(-25, -15)$ and $(5, -15)$. The solutions of the equation are $-25$ and $5$.

9. $|2x-5| = 10$

$2x-5 = 10 \quad\quad 2x-5 = -10$
$\quad 2x = 15 \text{ or } \quad 2x = -5$
$\quad\quad x = 7.5 \quad\quad\quad\quad x = -2.5$

11. $|5x-2| = 0$

$5x - 2 = 0$
$5x = 2$
$x = 0.4$

13. $|x+4| - 3 = 7$

$|x+4| - 3 + 3 = 7 + 3$
$|x+4| = 10$
$x+4 = 10 \text{ or } x+4 = -10$
$x = 6 \quad\quad\quad\quad x = -14$

15. (a) From the graph, we see that
$y_1 < y_2$ for $(-\infty, -14) \cup (8, \infty)$.
(b) From the graph, we see that
$y_1 > y_2$ for $(-14, 8)$.
(c) From the graph, we see that
$y_1 = y_2$ for $\{-14, 8\}$.

17. Graph $y_1 = |x|$ and $y_2 = 12$.

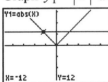

The points of intersection are $(-12, 12)$ and $(12, 12)$. The solution set of the inequality $y_1 < y_2$ is $(-12, 12)$.

19. Graph $y_1 = 15 - |x-6|$ and $y_2 = 6$.

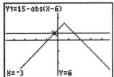 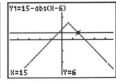

The points of intersection are $(-3, 6)$ and $(15, 6)$. The solution set of the inequality $y_1 > y_2$ is $(-3, 15)$.

21. Graph $y_1 = |2x-10| + 7$ and $y_2 = 7$.

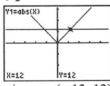

Because the graphs of the two equations intersect in only one point, the solution set of the inequality $y_1 \geq y_2$ is **R**.

23. The inequality $|x| \geq 5$ is equivalent to the disjunction $x \leq -5$ or $x \geq 5$ and has the solution set $(-\infty, -5] \cup [5, \infty)$. The inequality $|x| \leq 5$ is equivalent to the conjunction $-5 \leq x \leq 5$ and has the solution set $[-5, 5]$.

25. $|1 - 2t| < 9$

This inequality can be rewritten as:
$-9 < 1 - 2t < 9$
$-10 < -2t < 8$
$5 > t > -4 \text{ or } -4 < t < 5$

The solution set is $(-4, 5)$.

27. $|7+x| \leq 0$

$|7+x| < 0$ has no solutions. Because equality is permitted, the solutions to $|7+x| = 0$ are the only solutions of $|7+x| \leq 0$.

$7 + x = 0$
$x = -7$

The solution set is $\{-7\}$.

29. $|4-x| - 5 \leq 2$
$|4-x| - 5 + 5 \leq 2 + 5$
$|4-x| \leq 7$

This inequality can be rewritten as:
$-7 \leq 4 - x \leq 7$
$-11 \leq -x \leq 3$
$11 \geq x \geq -3 \text{ or } -3 \leq x \leq 11$

The solution set is $[-3, 11]$.

**31.** $|x+5| < -3$

Because the absolute value of an expression can never be less than a negative value, this inequality has no solution. The solution set is $\varnothing$.

**33.** $|3x+2| \geq 7$

This inequality can be rewritten as:
$$3x+2 \leq -7 \quad \quad 3x+2 \geq 7$$
$$3x \leq -9 \quad \text{or} \quad 3x \geq 5$$
$$x \leq -3 \quad \quad x \geq \frac{5}{3}$$

The solution set is $(-\infty, -3] \cup \left[\frac{5}{3}, \infty\right)$.

**35.** $|x-5| > 0$

This inequality can be rewritten as:
$$x-5 < 0 \quad \quad x-5 > 0$$
$$x < 5 \quad \text{or} \quad x > 5$$

The solution set is $(-\infty, 5) \cup (5, \infty)$.

**37.** $|3x-2| \geq -5$

Because the absolute value of an expression is always nonnegative, the value of this expression is always greater than $-5$. The solution set is **R**.

**39.** $|4t+3| + 2 > 10$
$|4t+3| + 2 - 2 > 10 - 2$
$|4t+3| > 8$

This inequality can be rewritten as:
$$4t+3 < -8 \quad \quad 4t+3 > 8$$
$$4t < -11 \quad \text{or} \quad 4t > 5$$
$$t < -\frac{11}{4} \quad \quad t > \frac{5}{4}$$

The solution set is $\left(-\infty, -\frac{11}{4}\right) \cup \left(\frac{5}{4}, \infty\right)$.

**41.** "The distance from 5 to $x$ is less than 7" can be represented as $|x-5| < 7$.

**43.** "The distance from $-7$ to $x$ is greater than or equal to 10" can be represented as $|x-(-7)| \geq 10$ or $|x+7| \geq 10$.

**45.** Because all the points in the given graph represent numbers that are less than 3 units from $-1$, an inequality in $x$ describing the graph is $|x+1| < 3$.

**47.** Because all the points in the given graph represent numbers that are 2 or more units from 0, an inequality in $x$ describing the graph is $|x| \geq 2$.

**49.** $|2k-3| = 8$
$$2k-3 = -8 \quad \quad 2k-3 = 8$$
$$2k = -5 \quad \text{or} \quad 2k = 11$$
$$k = -2.5 \quad \quad k = 5.5$$

**51.** $|y+1| - 3 = 7$
$|y+1| - 3 + 3 = 7 + 3$
$|y+1| = 10$
$$y+1 = -10 \quad \quad y+1 = 10$$
$$y = -11 \quad \text{or} \quad y = 9$$

**53.** $|t| + 8 = 5$
$|t| + 8 - 8 = 5 - 8$
$|t| = -3$

It is not possible for the absolute value of an expression to be negative. The solution set is $\varnothing$.

**55.** $5 - 3|x-5| = -13$
$5 - 3|x-5| - 5 = -13 - 5$
$-3|x-5| = -18$
$\dfrac{-3|x-5|}{-3} = \dfrac{-18}{-3}$
$|x-5| = 6$
$$x-5 = -6 \quad \quad x-5 = 6$$
$$x = -1 \quad \text{or} \quad x = 11$$

**57.** $|t+4| - 3 = -23$
$|t+4| - 3 + 3 = -23 + 3$
$|t+4| = -20$

It is not possible for the absolute value of an expression to be negative. The solution set is $\varnothing$.

**59.**
$$|3x-4|-4 = -4$$
$$|3x-4|-4+4 = -4+4$$
$$|3x-4| = 0$$
$$3x-4 = 0$$
$$3x = 4$$
$$x = \frac{4}{3}$$

**61.**
$$5+3|2x+1| = 20$$
$$5+3|2x+1|-5 = 20-5$$
$$3|2x+1| = 15$$
$$\frac{3|2x+1|}{3} = \frac{15}{3}$$
$$|2x+1| = 5$$

$2x+1 = -5 \qquad 2x+1 = 5$
$2x = -6 \quad$ or $\quad 2x = 4$
$x = -3 \qquad\qquad x = 2$

**63.**
$$\left|\frac{5-3x}{4}\right| = 3$$

$\dfrac{5-3x}{4} = -3 \qquad \dfrac{5-3x}{4} = 3$
$5-3x = -12 \quad$ or $\quad 5-3x = 12$
$-3x = -17 \qquad\qquad -3x = 7$
$x = \dfrac{17}{3} \qquad\qquad x = -\dfrac{7}{3}$

**65.** $|x-1| = |x-19|$

$x-1 = x-19 \qquad x-1 = -(x-19)$
$0 = -18 \quad$ or $\quad x-1 = -x+19$
False $\qquad\qquad 2x = 20$
$\qquad\qquad\qquad x = 10$

The only solution is 10.

**67.** $|2x-5| = |x|$

$\qquad\qquad\qquad 2x-5 = -x$
$2x-5 = x \qquad 3x-5 = 0$
$x-5 = 0 \quad$ or $\quad 3x = 5$
$x = 5 \qquad\qquad x = \dfrac{5}{3}$

**69.** $|x-2| = |2-x|$

$x-2 = 2-x \qquad x-2 = -(2-x)$
$2x-2 = 2 \quad$ or $\quad x-2 = -2+x$
$2x = 4 \qquad\qquad 0 = 0$
$x = 2 \qquad\qquad$ True

The solution set is **R**.

**71.** The first step in solving $-3|x-2| \geq 12$ is to divide both sides by $-3$ to obtain $|x-2| \leq -4$.

**73.**
$$-5|2x+3| < -35$$
$$\frac{-5|2x+3|}{-5} > \frac{-35}{-5}$$
$$|2x+3| > 7$$

This inequality can be rewritten as:
$2x+3 < -7 \qquad 2x+3 > 7$
$2x < -10 \quad$ or $\quad 2x > 4$
$x < -5 \qquad\qquad x > 2$

The solution set is $(-\infty, -5) \cup (2, \infty)$.

**75.**
$$5-|4-3x| > 2$$
$$5-|4-3x|-5 > 2-5$$
$$-|4-3x| > -3$$
$$\frac{-|4-3x|}{-1} < \frac{-3}{-1}$$
$$|4-3x| < 3$$

This inequality can be rewritten as:
$-3 < 4-3x < 3$
$-7 < -3x < -1$
$\dfrac{7}{3} > x > \dfrac{1}{3} \quad$ or $\quad \dfrac{1}{3} < x < \dfrac{7}{3}$

The solution set is $\left(\dfrac{1}{3}, \dfrac{7}{3}\right)$.

**77.**
$$|4x+3|+9 \geq 4$$
$$|4x+3|+9-9 \geq 4-9$$
$$|4x+3| \geq -5$$

Because the absolute value of an expression is always nonnegative, the value of this expression is always greater than $-5$. The solution set is **R**.

**79.**
$$|3y-5|+4 \geq 10$$
$$|3y-5|+4-4 \geq 10-4$$
$$|3y-5| \geq 6$$

This inequality can be rewritten as:
$$3y-5 \leq -6 \quad\quad 3y-5 \geq 6$$
$$3y \leq -1 \quad \text{or} \quad 3y \geq 11$$
$$y \leq -\frac{1}{3} \quad\quad y \geq \frac{11}{3}$$

The solution set is
$$\left(-\infty, -\frac{1}{3}\right] \cup \left[\frac{11}{3}, \infty\right).$$

**81.**
$$|3y-5|+6 \leq 6$$
$$|3y-5|+6-6 \leq 6-6$$
$$|3y-5| \leq 0$$

$|3y-5|<0$ has no solutions. Because equality is permitted, the solutions to $|3y-5|=0$ are the only solutions of $|3y-5| \leq 0$.
$$3y-5=0$$
$$3y=5$$
$$y=\frac{5}{3}$$

The solution set is $\left\{\frac{5}{3}\right\}$.

**83.**
$$|5-3x|+7 > 9$$
$$|5-3x|+7-7 > 9-7$$
$$|5-3x| > 2$$

This inequality can be rewritten as:
$$5-3x < -2 \quad\quad 5-3x > 2$$
$$-3x < -7 \quad \text{or} \quad -3x > -3$$
$$x > \frac{7}{3} \quad\quad x < 1$$

The solution set is $(-\infty, 1) \cup \left(\frac{7}{3}, \infty\right)$.

**85.**
$$5-|x-2| \geq 7$$
$$5-|x-2|-5 \geq 7-5$$
$$-|x-2| \geq 2$$
$$|x-2| \leq -2$$

It is not possible for the absolute value of an expression to be negative. The solution set is $\varnothing$.

**87.**
$$5-2|2x-1| > -1$$
$$5-2|2x-1|-5 > -1-5$$
$$-2|2x-1| > -6$$
$$|2x-1| < 3$$

This inequality can be rewritten as:
$$-3 < 2x-1 < 3$$
$$-2 < 2x < 4$$
$$-1 < x < 2$$

The solution set is $(-1, 2)$.

**89.**
$$\left|\frac{2x+1}{4}\right| > 1$$

This inequality can be rewritten as:
$$\frac{2x+1}{4} < -1 \quad\quad \frac{2x+1}{4} > 1$$
$$2x+1 < -4 \quad \text{or} \quad 2x+1 > 4$$
$$2x < -5 \quad\quad 2x > 3$$
$$x < -\frac{5}{2} \quad\quad x > \frac{3}{2}$$

The solution set is
$$\left(-\infty, -\frac{5}{2}\right) \cup \left(\frac{3}{2}, \infty\right).$$

**91.** Graph $y_1 = |x|$ and $y_2 = x+3$.

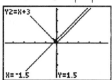

The point of intersection is $(-1.5, 1.5)$.
The solution set of the inequality $y_1 \geq y_2$ is $(-\infty, -1.5]$.

**93.** Graph $y_1 = |x-3|$ and $y_2 = 1-2x$.

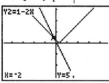

The point of intersection is $(-2, 5)$.
The solution set of the inequality $y_1 < y_2$ is $(-\infty, -2)$.

**95.** Graph $y_1 = |x+6|$ and $y_2 = |x-4|$.

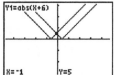

The point of intersection is (−1, 5).
The solution set of the inequality $y_1 < y_2$ is $(-\infty, -1)$.

**97.** Graph $y_1 = |t|$ and $y_2 = |15 - 2t|$.

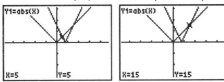

The points of intersection are (5, 5) and (15, 15). The solution set of the inequality $y_1 \geq y_2$ is $[5, 15]$.

**99.** By taking the absolute value, we guarantee that the distance is not negative.

**101.** The given conditions can be described by the absolute value inequality:
$|L - 12| \leq 0.02(12)$ or $|L - 12| \leq 0.24$.
The corresponding double inequality is $-0.24 \leq L - 12 \leq 0.24$.

**103.** The given conditions can be described by the absolute value inequality:
$|d - 3| \leq \dfrac{1}{16}$.
The corresponding double inequality is $-\dfrac{1}{16} \leq d - 3 \leq \dfrac{1}{16}$.

**105.** $|s - 50| \leq 6$
This inequality can be rewritten as:
$-6 \leq s - 50 \leq 6$
$44 \leq s \leq 56$
The range of actual scores is between 44 and 56, inclusive.

**107.** Let $t$ = the patient's body temperature.
$|t - 98.6| \leq 0.5$
This inequality can be rewritten as:
$-0.5 \leq t - 98.6 \leq 0.5$
$98.1 \leq t \leq 99.1$
The temperatures range between 98.1°F to 99.1°F, inclusive.

**109.**

$Y_1 \equiv 0.94X + 31.8$

The table gives the following numbers of uninsured persons for the given years:
1989 ($x = 2$): 33.7 million
1991 ($x = 4$): 35.6 million
1993 ($x = 6$): 37.4 million
1995 ($x = 8$): 39.3 million
1997 ($x = 10$): 41.2 million

**111.** From Exercise 110, the 16% of the population will be uninsured in 2001. The percent of uninsured persons in 1987 was $\dfrac{31.8}{241} \times 100\% \approx 13\%$. Thus, the percent of uninsured persons increased from 13% in 1987 to 16% in 2001.

**113.** (a) By the definition of absolute value, $|x - 2| = x - 2$ only if $x - 2 \geq 0$ or $x \geq 2$.
(b) By the definition of absolute value, $|x - 2| = -(x - 2)$ only if $x - 2 \leq 0$ or $x \leq 2$.
(c) The absolute values of a number and its opposite are equal. Therefore, the equation is true for all real numbers.

**115.** $|t - 3| = t + 3$

$t - 3 = t + 3$       $t - 3 = -(t + 3)$
$0 - 3 = 3$           $t - 3 = -t - 3$
$0 = 6$        or     $2t - 3 = -3$
False                 $2t = 0$
                      $t = 0$

The only solution to the equation is 0.

**117.** $|t + 5| = -t$

$t + 5 = -t$          $t + 5 = -(-t)$
$2t + 5 = 0$          $t + 5 = t$
$2t = -5$      or     $0 + 5 = 0$
$t = -2.5$            $0 = -5$
                      False

The only solution to the equation is −2.5.

**119.** $x + |x| = 0$ or $|x| = -x$
By the definition of absolute value,
$|x| = -x$ only if $x \leq 0$.

## Chapter 3 Project

1. 25.86 million people are members of families headed by full-time workers. If the average family size is 4, that means $25.86 \div 4 = 6.465$ million, or 6,465,000, uninsured families are headed by full-time workers.

3. 18-24: $0.26(32.4) = 8.424$ million, or 8,424,000
   25-44: $0.51(32.4) = 16.524$ million, or 16,524,000
   45-64: $0.23(32.4) = 7.452$ million, or 7,452,000

5. $5.4 + 3.3 + 3.3 = 12.0$ million uninsured workers have jobs with businesses that have fewer than 100 employees. That represents $\dfrac{12.0}{24.6} \times 100\% \approx 49\%$ of all uninsured workers.

## Chapter 3 Review Exercises

1. $3x^2 - 4x = 7$ is not a linear equation in one variable because $x^2$ is involved. It cannot be written in the form $Ax + B = 0$.

3. Verify that $-3$ is a solution:
   $3x - 7 = 2(x - 5)$
   $3(-3) - 7 = 2(-3 - 5)$
   $-9 - 7 = 2(-8)$
   $-16 = -16$

5. To estimate the solution, produce the graph of each side of the equation and trace to estimate the point of intersection. The x-coordinate of that point corresponds to the solution of the equation. To check the solution, store the value of the estimate solution in **X** and evaluate each expression.

7. Graph $y_1 = 4 - x$, $y_2 = -2$. Trace to the point of intersection $(6, -2)$. When $x = 6$, $y_1 = y_2$. The solution is $x = 6$. The screens show the solution and verification.

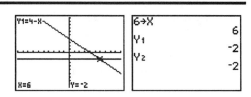

9. Graph $y_1 = 2(x+3)$, $y_2 = 2(x+1) + 4$. The graphs appear to coincide. The equation is an identity. Solution set: **R**.

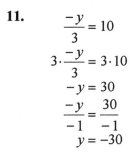

11. $\dfrac{-y}{3} = 10$

   $3 \cdot \dfrac{-y}{3} = 3 \cdot 10$

   $-y = 30$

   $\dfrac{-y}{-1} = \dfrac{30}{-1}$

   $y = -30$

**13.**
$$0.2n + 0.3 = 0.3n - 5.2$$
$$0.2n + 0.3 - 0.3n = 0.3n - 5.2 - 0.3n$$
$$0.3 - 0.1n = -5.2$$
$$0.3 - 0.1n - 0.3 = -5.2 - 0.3$$
$$-0.1n = -5.5$$
$$\frac{-0.1n}{-0.1} = \frac{-5.5}{-0.1}$$
$$n = 55$$

**15.**
$$7x - 5 = 12x - 5 - 4x$$
$$7x - 5 = 8x - 5$$
$$7x - 5 - 8x = 8x - 5 - 8x$$
$$-5 - x = -5$$
$$-5 - x + 5 = -5 + 5$$
$$-x = 0$$
$$\frac{-x}{-1} = \frac{0}{-1}$$
$$x = 0$$

**17.**
$$4(2s + 1) = s + 3(2s - 1)$$
$$8s + 4 = s + 6s - 3$$
$$8s + 4 = 7s - 3$$
$$8s + 4 - 7s = 7s - 3 - 7s$$
$$s + 4 = -3$$
$$s + 4 - 4 = -3 - 4$$
$$s = -7$$

**19.**
$$3x - 2(1 - 4x) = 6x + 5$$
$$3x - 2 + 8x = 6x + 5$$
$$11x - 2 = 6x + 5$$
$$11x - 2 - 6x = 6x + 5 - 6x$$
$$5x - 2 = 5$$
$$5x - 2 + 2 = 5 + 2$$
$$5x = 7$$
$$\frac{5x}{5} = \frac{7}{5}$$
$$x = 1.4$$
The equation is a conditional equation.

**21.**
$$2 - (2y - 3) = 5 - 2y$$
$$2 - 2y + 3 = 5 - 2y$$
$$5 - 2y = 5 - 2y$$
$$5 - 2y + 2y = 5 - 2y + 2y$$
$$5 = 5$$
Because the resulting equation is true, the original equation is an identity, and the solution set is **R**.

**23.**
$$3 - 2x = kx - 2 + 3x$$
$$3 - 2x - 3x - kx = kx - 2 + 3x - 3x - kx$$
$$3 - 5x - kx = -2$$
$$3 - 5x - kx - 3 = -2 - 3$$
$$-5x - kx = -5$$
$$(-5 - k)x = -5$$
The equation will be inconsistent if the last statement yields a false statement. The last statement will be false only if $k = -5$ because
$$[-5 - (-5)]x = -5$$
$$0 = -5$$

**25.** Solve $V = LWH$ for $H$:
$$LWH = V$$
$$\frac{LWH}{LW} = \frac{V}{LW}$$
$$H = \frac{V}{LW}$$

**27.** Solve $\frac{ac}{x} + \frac{cd}{y} = 6$ for $d$:
$$\frac{ac}{x} + \frac{cd}{y} - \frac{ac}{x} = 6 - \frac{ac}{x}$$
$$\frac{cd}{y} = 6 - \frac{ac}{x}$$
$$\frac{cd}{y} = \frac{6x - ac}{x}$$
$$y \cdot \frac{cd}{y} = y \cdot \frac{6x - ac}{x}$$
$$cd = \frac{6xy - acy}{x}$$
$$\frac{1}{c} \cdot cd = \frac{1}{c} \cdot \frac{6xy - acy}{x}$$
$$d = \frac{6xy - acy}{cx}$$

**29.** Solve $E = IR$ for $I$:
$$IR = E$$
$$\frac{IR}{R} = \frac{E}{R}$$
$$I = \frac{E}{R}$$

**31.**
$$3x - 4y - 25 = 0$$
$$3x - 4y - 25 - 3x = 0 - 3x$$
$$-4y - 25 = -3x$$
$$-4y - 25 + 25 = -3x + 25$$
$$-4y = -3x + 25$$
$$\frac{-4y}{-4} = \frac{-3x}{-4} + \frac{25}{-4}$$
$$y = \frac{3}{4}x - \frac{25}{4}$$

**33.**
$$P + Prt = A$$
$$P + Prt - P = A - P$$
$$Prt = A - P$$
$$\frac{Prt}{Pt} = \frac{A - P}{Pt}$$
$$r = \frac{A - P}{Pt}$$

For $P = \$2400$, $A = \$2556$, and $t = 1$:
$$r = \frac{2556 - 2400}{(2400)(1)} = 0.065$$
The rate is 6.5%.

**35.** We know that the formula for circumference is $C = 2\pi r$. Solving the formula for $r$, we have
$$\frac{2\pi r}{2\pi} = \frac{C}{2\pi}$$
$$r = \frac{C}{2\pi}$$
The width of the sidewalk is the difference of the lengths of the radii of the two circles.
$$\text{Width} = \frac{43.98}{2\pi} - \frac{31.42}{2\pi} \approx 2 \text{ feet}.$$

**37.** ① The goal is to determine the length of each piece.

② 

③ Because the second and third pieces are related to the first piece, let the variable represent the length of the first piece.

④ $x + (x + 1) + (x + 2) + 1 = 100$

⑤ 
$$x + x + 1 + x + 2 + 1 = 100$$
$$3x + 4 = 100$$
$$3x + 4 - 4 = 100 - 4$$
$$3x = 96$$
$$\frac{3x}{3} = \frac{96}{3}$$
$$x = 32$$

⑥ The first piece is 32 feet, the second piece is $32 + 1 = 33$ feet, and the third piece is $32 + 2 = 34$ feet.

⑦ Check: The sum of the lengths is $32 + 33 + 34 = 99$ feet. There is one foot left over.

**39.** ① The goal is to determine the number of nickels, but to check the results we will also find the number of dimes.

②

|  | Number of coins | Value of coin | Value of coins in cents |
|---|---|---|---|
| Dimes | $355 - x$ | 10 | $10(355-x)$ |
| Nickels | $x$ | 5 | $5x$ |
| Total |  |  | 2425 |

③ The table shows the variable assignment.

④ $10(355 - x) + 5x = 2425$

⑤ 
$$3550 - 10x + 5x = 2425$$
$$3550 - 5x = 2425$$
$$3550 - 5x - 3550 = 2425 - 3550$$
$$-5x = -1125$$
$$\frac{-5x}{-5} = \frac{-1125}{-5}$$
$$x = 225$$

⑥ There are 225 nickels and $355 - 225 = 130$ dimes.

⑦ Check: The value of the coins is $10(130) + 5(225) = 1300 + 1125 = 2425$.

**41.** Let $x$ = the original price.
Marked down price: $x - 0.40x$
$$x - 0.40x = 93$$
$$0.60x = 93$$
$$\frac{0.60x}{0.60} = \frac{93}{0.60}$$
$$x = 155$$
The original price was $155.

**43.** Let $h$ = number of hours.
Fixed cost = 275
Variable cost = $20h$
$$275 + 20h = 455$$
$$275 + 20h - 275 = 455 - 275$$
$$20h = 180$$
$$\frac{20h}{20} = \frac{180}{20}$$
$$h = 9$$
The operator worked for 9 hours.

**45.** Let $r$ = the rate of westbound truck.

|       | Rate  | Time | Distance   |
|-------|-------|------|------------|
| West  | $r$   | 3.5  | $3.5r$     |
| East  | $r-6$ | 3.5  | $3.5(r-6)$ |
| Total |       |      | 469        |

The sum of the distances is 469.
$$3.5r + 3.5(r-6) = 469$$
$$3.5r + 3.5r - 21 = 469$$
$$7r - 21 = 469$$
$$7r - 21 + 21 = 469 + 21$$
$$7r = 490$$
$$\frac{7r}{7} = \frac{490}{7}$$
$$r = 70$$
The average rate of westbound truck is 70 mph and the average rate of the eastbound truck is $70 - 6 = 64$ mph.

**47.** Let $x$ = the measure of the larger angle, and $90 - x$ = measure of the smaller angle.
$$\frac{1}{3}x - (90 - x) = 10$$
$$\frac{1}{3}x - 90 + x = 10$$
$$\frac{4}{3}x - 90 = 10$$
$$\frac{4}{3}x - 90 + 90 = 10 + 90$$
$$\frac{4}{3}x = 100$$
$$\frac{3}{4} \cdot \frac{4}{3}x = \frac{3}{4} \cdot 100$$
$$x = 75$$
The measure of the larger angle is 75° and the measure of the smaller angle is $90 - 75 = 15°$.

**49.** For the given graph, the set can be represented in interval notation as $[2, \infty)$.

**51.** For the given graph, the set can be represented in interval notation as $(-\infty, 4)$.

**53.** $[0, 3)$ can be represented on a number line as:

**55.** $2x - 1 > 7$
Graph $y_1 = 2x - 1$ and $y_2 = 7$.

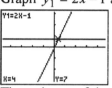

The estimate of the point of intersection is (4, 7). The graph of $y_1$ is above $y_2$ to the right of the point of intersection. Thus the solution set of $y_1 > y_2$ is $(4, \infty)$.

**57.** $5x + 7 - 3x \geq 7 + 2x$
Graph $y_1 = 5x + 7 - 3x$ and $y_2 = 7 + 2x$.

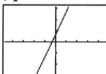

The graphs of $y_1$ and $y_2$ appear to coincide. Thus the solution set of $y_1 \geq y_2$ is **R**.

**59.**
$$3x - 4 < 14$$
$$3x - 4 + 4 < 14 + 4$$
$$3x < 18$$
$$\frac{3x}{3} < \frac{18}{3}$$
$$x < 6$$
The solution set is $(-\infty, 6)$.

**61.**
$$4x - 4 \leq 7x - 13$$
$$4x - 4 - 7x \leq 7x - 13 - 7x$$
$$-3x - 4 \leq -13$$
$$-3x - 4 + 4 \leq -13 + 4$$
$$-3x \leq -9$$
$$\frac{-3x}{-3} \geq \frac{-9}{-3}$$
$$x \geq 3$$
The solution set is $[3, \infty)$

**63.**
$$-17 < 4x - 7 \leq -8$$
$$-17 + 7 < 4x - 7 + 7 \leq -8 + 7$$
$$-10 < 4x \leq -1$$
$$\frac{-10}{4} < \frac{4x}{4} \leq \frac{-1}{4}$$
$$-2.5 < x \leq -0.25$$
The solution set is $(-2.5, -0.25]$.

**65.**
$$-\frac{2}{3}(x - 4) \leq \frac{2x - 3}{-4}$$
$$(-12) \cdot \left(-\frac{2}{3}\right)(x - 4) \geq (-12) \cdot \frac{2x - 3}{-4}$$
$$8(x - 4) \geq 3(2x - 3)$$
$$8x - 32 \geq 6x - 9$$
$$8x - 32 - 6x \geq 6x - 9 - 6x$$
$$2x - 32 \geq -9$$
$$2x - 32 + 32 \geq -9 + 32$$
$$2x \geq 23$$
$$\frac{2x}{2} \geq \frac{23}{2}$$
$$x \geq 11.5$$
The solution set is $[11.5, \infty)$.

**67.**
$$7 > x + 3 > -4$$
$$7 - 3 > x + 3 - 3 > -4 - 3$$
$$4 > x > -7 \quad \text{or} \quad -7 < x < 4$$
The solution set is $(-7, 4)$.

**69.**
$$-8 \leq -4x < 20$$
$$\frac{-8}{-4} \geq \frac{-4x}{-4} > \frac{20}{-4}$$
$$2 \geq x > -5 \quad \text{or} \quad -5 < x \leq 2$$
The solution set is $(-5, 2]$.

**71.** Let $L$ = the length of the rectangle, and $L - 7$ = the width of the rectangle. Perimeter is given by $P = 2L + 2W$.
$$2L + 2(L - 7) \leq 254$$
$$2L + 2L - 14 \leq 254$$
$$4L - 14 \leq 254$$
$$4L - 14 + 14 \leq 254 + 14$$
$$4L \leq 268$$
$$\frac{4L}{4} \leq \frac{268}{4}$$
$$L \leq 67$$
The maximum length is 67 meters, so the maximum width is $67 - 7 = 60$ meters.

**73.** Let $m$ = the number of miles driven. $20 + 0.10m$ represents the one-day cost of renting a car from Frugal Car Rental. $26 + 0.09m$ represents the one-day cost of renting a car from Big C Car Rental. Frugal is the better choice if the cost at Frugal is less than the cost at Big C.
$$20 + 0.10m < 26 + 0.09m$$
$$20 + 0.10m - 0.09m < 26 + 0.09m - 0.09m$$
$$20 + 0.01m < 26$$
$$20 + 0.01m - 20 < 26 - 20$$
$$0.01m < 6$$
$$\frac{0.01m}{0.01} < \frac{6}{0.01}$$
$$m < 600$$
Frugal is the better choice for 0 to 600 miles.

**75.**
$$-t \geq -2 \qquad \qquad \frac{t}{4} > -1$$
$$\frac{-t}{-1} \leq \frac{-2}{-1} \quad \text{or} \quad 4 \cdot \frac{t}{4} > 4 \cdot (-1)$$
$$t \leq 2 \qquad \qquad t > -4$$
The solution set is the union of these two intervals. The solution set is **R**.

## Chapter 3 Review Exercises

**77.**
$$\frac{x+2}{4} \leq 1$$
$$4 \cdot \frac{x+2}{4} \leq 4 \cdot 1$$
$$x+2 \leq 4 \quad \text{and}$$
$$x+2-2 \leq 4-2$$
$$x \leq 2$$

$$\frac{x+2}{4} \geq -1$$
$$4 \cdot \frac{x+2}{4} \geq 4 \cdot (-1)$$
$$x+2 \geq -4$$
$$x+2-2 \geq -4-2$$
$$x \geq -6$$

The solution set is $[-6, 2]$.

**79.**
$$\begin{array}{ll} x+1 > 0 & 3x-4 < 0 \\ x+1-1 > 0-1 \quad \text{or} & 3x-4+4 < 0+4 \\ x > -1 & 3x < 4 \\ & \dfrac{3x}{3} < \dfrac{4}{3} \\ & x < \dfrac{4}{3} \end{array}$$

The solution set is **R**.

**81.**
$$\begin{array}{ll} -5x+2 \geq 12 & 3x+4 \geq 25 \\ -5x+2-2 \geq 12-2 & 3x+4-4 \geq 25-4 \\ -5x \geq 10 \quad \text{or} & 3x \geq 21 \\ \dfrac{-5x}{-5} \leq \dfrac{10}{-5} & \dfrac{3x}{3} \geq \dfrac{21}{3} \\ x \leq -2 & x \geq 7 \end{array}$$

The solution set is $(-\infty, -2] \cup [7, \infty)$.

**83.**
$$\begin{array}{ll} 3x+4 < x & x+2 > 8-x \\ 3x+4-x < x-x & x+2+x > 8-x+x \\ 2x+4 < 0 & 2x+2 > 8 \\ 2x+4-4 < 0-4 \quad \text{or} & 2x+2-2 > 8-2 \\ 2x < -4 & 2x > 6 \\ \dfrac{2x}{2} < \dfrac{-4}{2} & \dfrac{2x}{2} > \dfrac{6}{2} \\ x < -2 & x > 3 \end{array}$$

The solution set is $(-\infty, -2) \cup (3, \infty)$.

**85.**
$$3x-4 > 5x-2$$
$$3x-4-5x > 5x-2-5x$$
$$-4-2x > -2$$
$$-4-2x+4 > -2+4 \quad \text{and}$$
$$-2x > 2$$
$$\frac{-2x}{-2} < \frac{2}{-2}$$
$$x < -1$$

$$3x-2 < 2x+3$$
$$3x-2-2x < 2x+3-2x$$
$$x-2 < 3$$
$$x-2+2 < 3+2$$
$$x < 5$$

The solution set is $(-\infty, -1)$.

**87.** $I = Prt$
$$150 \leq P(0.08)(1) \leq 200$$
$$150 \leq 0.08P \leq 200$$
$$\frac{150}{0.08} \leq \frac{0.08P}{0.08} \leq \frac{200}{0.08}$$
$$1875 \leq P \leq 2500$$

Between \$1875 and \$2500, inclusive, must be invested to earn between \$150 and \$200 annually.

**89.**
$$|-y| = 8$$
$$\begin{array}{ll} -y = 8 & -y = -8 \\ y = -8 \quad \text{or} & y = 8 \end{array}$$

**91.**
$$|2x-3|-7 = 5$$
$$|2x-3|-7+7 = 5+7$$
$$|2x-3| = 12$$
$$\begin{array}{ll} 2x-3 = 12 & 2x-3 = -12 \\ 2x = 15 \quad \text{or} & 2x = -9 \\ x = 7.5 & x = -4.5 \end{array}$$

**93.**
$$2+|x-2| = 3-|x-2|$$
$$2+|x-2|+|x-2| = 3-|x-2|+|x-2|$$
$$2+2|x-2| = 3$$
$$2+2|x-2|-2 = 3-2$$
$$2|x-2| = 1$$
$$\frac{2|x-2|}{2} = \frac{1}{2}$$
$$|x-2| = 0.5$$
$$\begin{array}{ll} x-2 = 0.5 & x-2 = -0.5 \\ x = 2.5 \quad \text{or} & x = 1.5 \end{array}$$

**95.** Graph $y_1 = |x-3|$ and $y_2 = \dfrac{1}{2}x - 1$.

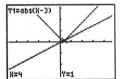 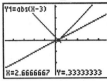

The points of intersection are $(4, 1)$ and $\left(\dfrac{8}{3}, \dfrac{1}{3}\right)$. The solutions are $4$ and $\dfrac{8}{3}$.

**97.** $|3x+5| < 14$

This inequality can be rewritten as:
$$-14 < 3x+5 < 14$$
$$-19 < 3x < 9$$
$$-\dfrac{19}{3} < x < 3$$

The solution set is $\left(-\dfrac{19}{3}, 3\right)$.

**99.**
$$7 - |c-5| \leq 4$$
$$7 - |c-5| - 7 \leq 4 - 7$$
$$-|c-5| \leq -3$$
$$\dfrac{-|c-5|}{-1} \geq \dfrac{-3}{-1}$$
$$|c-5| \geq 3$$

This inequality can be rewritten as:
$c - 5 \leq -3 \quad$ or $\quad c - 5 \geq 3$
$c \leq 2 \qquad\qquad\quad c \geq 8$

The solution set is $(-\infty, 2] \cup [8, \infty)$.

**101.** $|-t| \geq 7$

This inequality can be rewritten as:
$-t \leq -7 \quad$ or $\quad -t \geq 7$
$t \geq 7 \qquad\qquad t \leq -7$

The solution set is $(-\infty, -7] \cup [7, \infty)$.

**103.** $|x - 3| \leq 5$ is equivalent to
$-5 \leq x - 3 \leq 5$.

**105.** Graph $y_1 = |x+3|$ and $y_2 = 2x+1$.

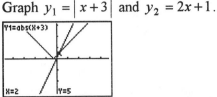

The point of intersection is $(2, 5)$. The solution set of the inequality $y_1 > y_2$ is $(-\infty, 2)$.

**107.** Graph $y_1 = |3 + x|$ and $y_2 = x + 4$.

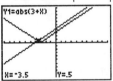

The point of intersection is $(-3.5, 0.5)$. The solution set of the inequality $y_1 < y_2$ is $(-3.5, \infty)$.

**109.** Let $x =$ the age of one brother.
$|x - 9| = 3$
$x - 9 = 3 \quad$ or $\quad x - 9 = -3$
$x = 12 \qquad\qquad\; x = 6$

He is either 6 or 12 years old.

**111.** Let $x =$ the original number.
Then $|x - 1| < x$, or $-x < x - 1 < x$.

$-x < x - 1 \qquad\quad x - 1 < x$
$-2x < -1 \quad$ and $\quad -1 < 0$
$x > \dfrac{1}{2} \qquad\qquad\quad$ True

The solution set is $\left(\dfrac{1}{2}, \infty\right)$. Thus, the original number cannot be negative.

## Chapter 3 Looking Back

**1.** (a) Evaluate $5x - 2y$ for $x = 2$, $y = \frac{3}{2}$

$$5x - 2y = 5(2) - 2\left(\frac{3}{2}\right)$$
$$= 10 - 3$$
$$= 7$$

(b) Evaluate $-x + 11$ for $x = -7$
$-x + 11 = -(-7) + 11 = 7 + 11 = 18$

**3.** $f(x) = \frac{1}{2}x - 10$

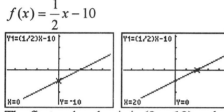

The first ordered pair is $(0, -10)$ and the second ordered pair is $(20, 0)$.

**5.**
$$x - y = 5$$
$$-y = -x + 5$$
$$y = x - 5$$

**7.** $x_1 = -2$, $y_1 = 4$, $x_2 = 5$, and $y_2 = 8$

$$\frac{y_2 - y_1}{x_2 - x_1} = \frac{8 - 4}{5 - (-2)} = \frac{4}{7}$$

**9.** (a) $1.5p + 2$

(b) $L = W - 4$

**11.** (a) Is 11 a solution of $2x - 15 = 7$?
$$2(11) - 15 = 7$$
$$22 - 15 = 7 \quad \text{Yes}$$
$$7 = 7$$

(b) Is $-2$ a solution of $1 - 3x = x - 9$?
$$1 - 3(-2) = -2 - 9$$
$$1 + 6 = -11 \quad \text{No}$$
$$7 = -11$$

## Chapter 3 Test

**1.**
$$2x - 3 = 7$$
$$2x - 3 + 3 = 7 + 3 \quad \text{(a) Addition Property of Equations}$$
$$2x = 10 \quad \text{(b) Property of Additive Inverses}$$
$$\frac{2x}{2} = \frac{10}{2} \quad \text{(c) Multiplication Property of Equations}$$
$$x = 5$$

**3.** $3x - 4 = 5x - 4 - 2x$
$3x - 4 = 3x - 4$
$-4 = -4$

Because the last statement is true, the solution set is **R**. The equation is an identity.

**5.** Solve $A = P + Prt$ for $t$.
$$P + Prt = A$$
$$Prt = A - P$$
$$t = \frac{A - P}{Pr}$$

**7.** $A = P + Prt = P(1 + rt)$
$2484 = P[1 + 0.08(1)]$
$P = 2300$

**9.** Let $x =$ the number of gallons of 50% antifreeze solution.

|  | Gallons solution | % Anti-freeze | Gallons of Antifreeze |
|---|---|---|---|
| 50% | $6 - x$ | 0.50 | $0.50(6 - x)$ |
| 100% | $x$ | 1.00 | $x$ |
| Mixture | 6 | 0.60 | 3.6 |

$$0.5(6 - x) + x = 3.6$$
$$3 - 0.5x + x = 3.6$$
$$3 + 0.5x = 3.6$$
$$0.5x = 0.6$$
$$x = 1.2$$

The radiator should be drained of 1.2 gal of 50% antifreeze and replaced with an equal amount of 100% antifreeze to yield 6 gallons of 60% antifreeze.

**11.** The graph shows all points greater than or equal to – 5 and less than 2. In interval notation, this is $[-5, 2)$.

**13.**
$$\begin{aligned} 4-2(x-3) &< 1-5x \\ 4-2x+6 &< 1-5x \\ 10-2x &< 1-5x \\ 3x+10 &< 1 \\ 3x &< -9 \\ x &< -3 \end{aligned}$$
The solution set is $(-\infty, -3)$.

**15.**
$$\begin{aligned} x+\frac{2}{3}(x-3) &< \frac{1}{2} \\ 6x+4(x-3) &< 3 \\ 6x+4x-12 &< 3 \\ 10x-12 &< 3 \\ 10x &< 15 \\ x &< 1.5 \end{aligned}$$
The solution set is $(-\infty, 1.5)$.

**17.**
$$\begin{aligned} 3-x &< 5 & & & 2x-1 &< 5 \\ -x &< 2 & &\text{and} & 2x &< 6 \\ x &> -2 & & & x &< 3 \end{aligned}$$
The solution set is $(-2, 3)$.

**19.**
$$\begin{aligned} 2x &\ge 0 & &\text{or} & x-3 &\le 1 \\ x &\ge 0 & & & x &\le 4 \end{aligned}$$
The solution set is **R**.

**21.**
$$\begin{aligned} 3-|x-1| &< 1 \\ -|x-1| &< -2 \\ |x-1| &> 2 \end{aligned}$$
This inequality can be rewritten as:
$$\begin{aligned} x-1 &< -2 & &\text{or} & x-1 &> 2 \\ x &< -1 & & & x &> 3 \end{aligned}$$
The solution set is $(-\infty, -1) \cup (3, \infty)$.

**23.** $|x-3| < -2$

It is not possible for the absolute value of an expression to be negative. The solution set is $\varnothing$.

**25.** Let $x =$ the snowfall on January 7.
$$\begin{aligned} 1.2+0+4.25+1.5+0+0.4+x &\le 8.7 \\ 7.35+x &\le 8.7 \\ x &\le 1.35 \end{aligned}$$
On January 7, at most 1.35 inches could fall without exceeding the average snowfall amount for the first week in January.

# Chapter 4

# Properties of Lines

## Section 4.1 Linear Equations in Two Variables

1. The equation $x = -5$ can be written as $1x + 0y = -5$, which is the standard form of a linear equation in two variables.

3. The equation $\dfrac{x}{2} + \dfrac{y}{3} - \dfrac{7}{8} = 0$ can be written in the form $Ax + By = C$ and, therefore, it is a linear equation in two variables.

5. The equation $y = x^2 - 3$ is not a linear equation in two variables. The exponents on both variables must be 1.

7. The equation $-6 = 2xy$ is not a linear equation in two variables. The variables cannot be part of the same term.

9. Is $(4, 5)$ a solution of $y = 2x - 3$?
$$5 = 2(4) - 3$$
$$5 = 8 - 3 \qquad \text{Yes}$$
$$5 = 5$$

11. Is $\left(2, \dfrac{3}{2}\right)$ a solution of $3x - 4y = 12$?
$$3(2) - 4\left(\dfrac{3}{2}\right) = 12$$
$$6 - 6 = 12 \qquad \text{No}$$
$$0 = 12$$

13. Is $(-2, -1)$ a solution of $x - 5 = 3$?
$$-2 - 5 = 3 \qquad \text{No}$$
$$-7 = 3$$

15. $(1, -1)$
$$2x - ky = 7$$
$$2(1) - k(-1) = 7$$
$$2 + k = 7$$
$$k = 5$$
For $(1, -1)$ to be a solution, $k$ must be 5.

17. $(12, -3)$
$$y = -\dfrac{2}{3}x + k$$
$$-3 = \left(-\dfrac{2}{3}\right)(12) + k$$
$$-3 = -8 + k$$
$$5 = k$$
For $(12, -3)$ to be a solution, $k$ must be 5.

19. $(k, k+1)$
$$y = -2x + k$$
$$k + 1 = -2k + k$$
$$k + 1 = -k$$
$$1 = -2k$$
$$-0.5 = k$$
For $(k, k+1)$ to be a solution, $k$ must be $-0.5$.

21. $y = -2x + 5$
For $(-3, a)$:
$$a = -2(-3) + 5$$
$$a = 6 + 5 = 11$$
For $(4, b)$:
$$b = -2(4) + 5$$
$$b = -8 + 5 = -3$$
For $(c, -7)$:
$$-7 = -2c + 5$$
$$-12 = -2c$$
$$6 = c$$

**23.** $x - 5 = -2$
For $(a, 5)$:
$$a - 5 = -2$$
$$a = 3$$
For $(b, 9)$:
$$b - 5 = -2$$
$$b = 3$$
For $(c, -6)$:
$$c - 5 = -2$$
$$c = 3$$

**25.** $x + 3y - 4 = 0$
For $(a, 3)$:
$$a + 3(3) - 4 = 0$$
$$a = -5$$
For $(7, b)$:
$$7 + 3(b) - 4 = 0$$
$$3b = -3$$
$$b = -1$$
For $(c, -4)$:
$$c + 3(-4) - 4 = 0$$
$$c = 16$$

**27.** For $|x| = 2$ there are two solutions, $-2$ and $2$. For $|x| < 2$ the solutions are all numbers in the open interval from $-2$ to $2$. For $y = x + 4$, one solution is the ordered pair $(-2, 2)$.

**29.** $y = 4 - x$

| $x$ | 10 | $-6$ | 7 |
|---|---|---|---|
| $y$ | $-6$ | 10 | $-3$ |

**31.** $f(x) = 3x + 1$

| $x$ | 0 | 3 | $-5$ |
|---|---|---|---|
| $f(x)$ | 1 | 10 | $-14$ |
| $(x, y)$ | $(0, 1)$ | $(3, 10)$ | $(-5, -14)$ |

**33.** (a) Because the points $(-3, -4)$ and $(6, 2)$ are on the graph, this is the graph of $2x - 3y - 6 = 0$.

(b) $(c, -3)$
$$2c - 3(-3) - 6 = 0$$
$$2c = -3$$
$$c = -1.5$$

(c) $(2, d)$
$$2(2) - 3d - 6 = 0$$
$$-3d = 2$$
$$d = -2/3$$

**35.** $y = 3x - 4$
Three solutions: $(0, -4), (2, 2), (3, 5)$

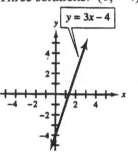

**37.** $y = \dfrac{2}{3}x - 2$
Three solutions: $(0, -2), (3, 0), (6, 2)$

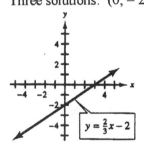

**39.** $3x - 5y = 15$
Three solutions: $(0, -3), (5, 0), (10, 3)$

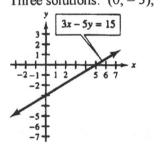

**41.** $y - 2 = 3$
Three solutions: $(0, 5), (3, 5), (7, 5)$

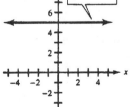

Section 4.1 Linear Equations in Two Variables

**43.** $x - 9 = -11$
Three solutions: $(-2, 0), (-2, 1), (-2, 5)$
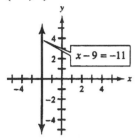

**45.** $y = x$
Three solutions: $(1, 1), (0, 0), (3, 3)$
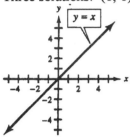

**47.** $3x = 21 - 7y$
Three solutions: $(7, 0), (0, 3), (14, -3)$
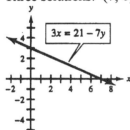

**49.**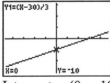
Intercepts: $(0, 12), (4, 0)$

**51.**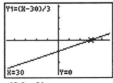
Intercepts: $(0, -10), (30, 0)$

**53.**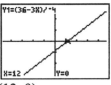
Intercepts: $(0, -9), (12, 0)$

**55.** $5x + 3y - 9 = 0$
$5(0) + 3(3) - 9 = 0$ True

$6x + 2y - 6 = 0$
$6(0) + 2(3) - 6 = 0$ True

$3x + 4y + 12 = 0$
$3(0) + 4(3) + 12 = 0$ False

**57.** $2x + 5y = 14$
$5y = -2x + 14$
$y = -\dfrac{2}{5}x + \dfrac{14}{5}$
$y$-intercept: $\left(0, \dfrac{14}{5}\right)$

**59.** $y - 2 = -4(x + 3)$
$y - 2 = -4x - 12$
$y = -4x - 10$
$y$-intercept: $(0, -10)$

**61.** $y$-intercept:
$y = -2x$
$y = -2(0) = 0$
$(0, 0)$

$x$-intercept:
$y = -2x$
$0 = -2x$
$0 = x$

**63.** $y$-intercept:
$3x - 4y = 12$
$3(0) - 4y = 12$
$y = -3$

$x$-intercept:
$3x - 4y = 12$
$3x - 4(0) = 12$
$x = 4$

$(0, -3)$ and $(4, 0)$

**65.** $y$-intercept:
$\dfrac{x}{3} + \dfrac{y}{4} = 1$
$\dfrac{0}{3} + \dfrac{y}{4} = 1$
$y = 4$

$x$-intercept:
$\dfrac{x}{3} + \dfrac{y}{4} = 1$
$\dfrac{x}{3} + \dfrac{0}{4} = 1$
$x = 3$

$(0, 4)$ and $(3, 0)$

**67.** $y$-intercept:
$y + 2 = 17$
$y = 15$
$(0, 15)$

$x$-intercept:
None

**69.** If the *x*- and *y*-intercepts are the same, the line contains the origin. Therefore, $b = 0$.

**71.** $y = x$

(a) A line shifted upward 4 units from the given line is $y = x + 4$.

(b) A line shifted downward 3 units from the given line is $y = x - 3$.

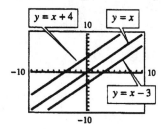

**73.** $(0, 2)$
$$ax + 4y = c$$
$$a(0) + 4(2) = c$$
$$8 = c$$
$(-4, 0)$
$$ax + 4y = c$$
$$a(-4) + 4(0) = 8$$
$$-4a = 8$$
$$a = -2$$

**75.** $(0, -5)$
$$ax + by = 15$$
$$a(0) + b(-5) = 15$$
$$-5b = 15$$
$$b = -3$$
$(3, 0)$
$$ax + by = 15$$
$$a(3) + b(0) = 15$$
$$3a = 15$$
$$a = 5$$

**77.** $3x = 12$
$x = 4$
Because the equation is of the form $x = $ a constant, the graph would be a vertical line.

**79.** $x - y = 0$
$x = y$
The graph of the equation is neither a vertical nor horizontal line.

**81.** $-2y = -8$
$y = 4$
Because the equation is of the form $y = $ a constant, the graph would be a horizontal line.

**83.** Two ordered pairs such that one number *y* is the opposite of another number *x* are $(5, -5)$ and $(-8, 8)$. The corresponding equation is $y = -x$ or $x = -y$. The verification of these two points is shown on the graphing calculator screens below.

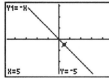 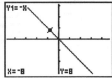

**85.** Two ordered pairs such that one number *y* is half of another number *x* are $(6, 3)$ and $(-8, -4)$. The corresponding equation is $y = \frac{1}{2}x$. The verification of these two points is shown on the screens below.

 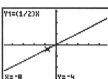

**87.** Two ordered pairs such that the sum of two numbers *x* and *y* is 5 are $(2, 3)$ and $(-5, 10)$. The corresponding equation is $x + y = 5$. The verification of these two points is shown on the screens below.

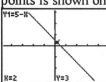 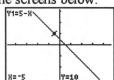

**89.** Let $x = $ the number of windows that have been washed. Then $y = -x + 22$ represents the number of windows that have not yet been washed.

(a) The intercepts are: $(0, 22), (22, 0)$.

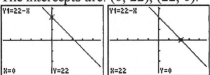

(b) The y-intercept corresponds to "no windows washed," and the x-intercept corresponds to "all windows washed."

**91.** Let $x$ = the number of tenths of miles. Then $y = 4 + 0.15x$ represents the total taxi fare. Fares for 2 and 2.1 miles are:

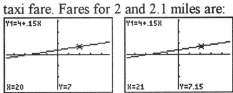

The fares are $7, $7.15, $7.30, $7.45, $7.60, and $7.75.

**93.** (a) The graph of a linear function is a line, and a line can be drawn relatively close to the tops of the bars.

(b) The graphs are shown below.

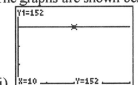

(i)

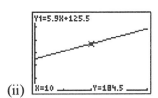

(ii)

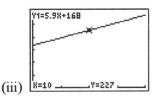

(iii)

The most reasonable model of the data is (ii).

(c) 50% more than in 1992 is:
$138 + 0.50(138) = 207$ billion.
Using model (ii):
$5.9x + 125.5 = E(x)$
$5.9x + 125.5 = 207$
$5.9x = 81.5$
$x \approx 14$

The first year in which expenditures will be at least 50% more than they were in 1992 is 2004 (1990 + 14).

**95.** (a) Find the y-intercept:
$L(x) = 0.15x + 73.7$
$L(0) = 0.15(0) + 73.7$
$= 0 + 73.7$
The y-intercept is (0, 73.7). This means that when $x = 0$ (in 1980), the life expectancy was 73.7 years.

(b) The average annual increase in our life expectancy is 0.15 year.

(c) Find the x-intercept:
$L(x) = 0.15x + 73.7$
$0 = 0.15x + 73.7$
$-0.15x = 73.7$
$x \approx -491$
The x-intercept is about (−491, 0). This means that when $x = -491$ (the year 1489), the life expectancy was 0, which is invalid.

**97.** If we let $x$ represent the number of years since 1996, then we would expect $b$ to be approximately equal to 1100, the production of E85 during 1996.

**99.** $k = \dfrac{1100 + 1300 + 1400}{3} = \dfrac{3800}{3} \approx 1267.\overline{6}$

The graph of $y = 1267.\overline{6}$ is a horizontal line, which implies that the annual production has and will remain constant.

**101.** In the equation $\dfrac{x}{g} + \dfrac{y}{h} = 1$, the x-intercept is $(g, 0)$ and the y-intercept is $(0, h)$.

**103.** $2x + by = 4b,\quad b < 0$
x-intercept: $(2b, 0)$
$2x + b(0) = 4b$
$2x = 4b$
$x = 2b$
y-intercept: $(0, 4)$
$2(0) + by = 4b$
$by = 4b$
$y = 4$

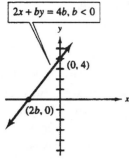

107. $y = x - 5,\ 10 \le x \le 15$

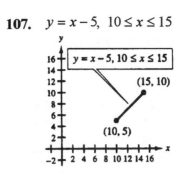

105. $y = 2x + 1,\ x \ge 0$

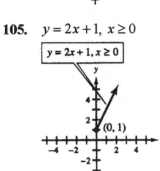

## Section 4.2  Slope of a Line

1. The geometric interpretation is the rise divided by the run.

3. Because the graph rises from left to right, the slope should be positive. Two points on the graph are (0, 3) and (1, 5).
   Slope $= \dfrac{y_2 - y_1}{x_2 - x_1} = \dfrac{5-3}{1-0} = \dfrac{2}{1} = 2$.

5. Because the graph falls from left to right, the slope should be negative. Two points on the graph are (– 2, 0) and (1, – 3).
   Slope $= \dfrac{y_2 - y_1}{x_2 - x_1} = \dfrac{-3-0}{1-(-2)} = \dfrac{-3}{3} = -1$.

7. Because the graph is a horizontal line, the slope is zero. Two points on the graph are (0, – 3) and (3, – 3).
   Slope $= \dfrac{y_2 - y_1}{x_2 - x_1} = \dfrac{-3-(-3)}{3-0} = \dfrac{0}{3} = 0$.

9. Because the graph is a vertical line, the slope is undefined.

11. (8, – 2) and (6, – 12)
    Slope $=$
    $\dfrac{y_2 - y_1}{x_2 - x_1} = \dfrac{-12-(-2)}{6-8} = \dfrac{-10}{-2} = 5$

13. (3, 4) and (5, 11)
    Slope $= \dfrac{y_2 - y_1}{x_2 - x_1} = \dfrac{11-4}{5-3} = \dfrac{7}{2}$

15. (5, – 6) and (5, 7)
    Slope $= \dfrac{y_2 - y_1}{x_2 - x_1} = \dfrac{7-(-6)}{5-5} = \dfrac{13}{0}$
    The slope is undefined.

17. (2.3, – 5.9) and (– 1.7, 5.4)
    Slope $= \dfrac{y_2 - y_1}{x_2 - x_1} = \dfrac{5.4-(-5.9)}{-1.7-2.3} = \dfrac{11.3}{-4}$
    $= -2.825$

Section 4.2   Slope of a Line

**19.** $(-2, 5)$ and $(8, -3)$
Slope $= \dfrac{y_2 - y_1}{x_2 - x_1} = \dfrac{-3 - 5}{8 - (-2)} = \dfrac{-8}{10} = -\dfrac{4}{5}$

**21.** $(-3, 7)$ and $(6, 7)$
Slope $= \dfrac{y_2 - y_1}{x_2 - x_1} = \dfrac{7 - 7}{6 - (-3)} = \dfrac{0}{9} = 0$

**23.** $\left(\dfrac{1}{2}, \dfrac{1}{3}\right)$ and $\left(\dfrac{2}{3}, \dfrac{3}{4}\right)$
Slope $= \dfrac{y_2 - y_1}{x_2 - x_1} = \dfrac{\frac{3}{4} - \frac{1}{3}}{\frac{2}{3} - \frac{1}{2}} = \dfrac{\frac{5}{12}}{\frac{1}{6}} = \dfrac{5}{2}$

**25.** Procedures (i) and (iii) are correct. Procedure (ii) refers to a rise and a run that are both negative. Thus the slope is $+3$.

**27.** $(-3, y), (2, -2), m = -1$
$\dfrac{y - (-2)}{-3 - 2} = -1$
$\dfrac{y + 2}{-5} = -1$
$y + 2 = 5$
$y = 3$

**29.** $(2, -1), (x, 0)$, slope is undefined.
$\dfrac{0 - (-1)}{x - 2} = $ undefined
$\dfrac{x - 2}{0 - (-1)} = 0$
$\dfrac{x - 2}{1} = 0$
$x - 2 = 0$
$x = 2$

**31.** $(-3, 4), (2, y), m = 0$
$\dfrac{y - 4}{2 - (-3)} = 0$
$\dfrac{y - 4}{5} = 0$
$y - 4 = 0$
$y = 4$

**33.** Method (ii) is the correct use of the slope formula. In method (i), the coordinates are not subtracted in the same order in the numerator and the denominator.

**35.** $3x + 5y = 15$
Two solutions are: $(0, 3)$ and $(5, 0)$
Slope $= \dfrac{y_2 - y_1}{x_2 - x_1} = \dfrac{0 - 3}{5 - 0} = -\dfrac{3}{5}$

**37.** $3 + y = 12$
Two solutions are: $(5, 9)$ and $(10, 9)$
Slope $= \dfrac{y_2 - y_1}{x_2 - x_1} = \dfrac{9 - 9}{10 - 5} = \dfrac{0}{5} = 0$

**39.** $y = \dfrac{3}{4}x - 3$
Two solutions are: $(0, -3)$ and $(4, 0)$
Slope $= \dfrac{y_2 - y_1}{x_2 - x_1} = \dfrac{0 - (-3)}{4 - 0} = \dfrac{3}{4}$

**41.** $y + 2x = 17$
$y = -2x + 17$
Slope: $m = -2$

**43.** $0 = 2x - y$
$y = 2x$    Slope: $m = 2$

**45.** $y + 2 = 17$
$y = 15$
Slope: $m = 0$

**47.** $x + 3y - 6 = 0$
$3y = -x + 6$
$y = -\dfrac{1}{3}x + 2$
Slope: $m = -\dfrac{1}{3}$

**49.** $4x - 3 = 13$
$4x = 16$
$x = 4$
Cannot be written in slope-intercept form. Slope is undefined.

**51.** $3y = 5x - 7$
$y = \dfrac{5}{3}x - \dfrac{7}{3}$
Slope: $m = \dfrac{5}{3}$

**53.** $\frac{3}{4}x - \frac{2}{3}y = 12$

$9x - 8y = 144$

$-8y = -9x + 144$

$y = \frac{9}{8}x - 18$

Slope: $m = \frac{9}{8}$

**55.** $Ax + By = C, \quad B \neq 0$

$By = -Ax + C, \quad B \neq 0$

$y = -\frac{A}{B}x + \frac{C}{B}, \quad B \neq 0$

The slope of the graph is $-A/B$.

**57.** (i) $6x + 3y - 7 = 0$

$3y = -6x + 7$

$y = -2x + \frac{7}{3}$

(ii) $4x - 2y + 3 = 0$

$-2y = -4x - 3$

$y = 2x + 1.5$

(iii) $x + \frac{1}{2}y - 1 = 0$

$\frac{1}{2}y = -x + 1$

$y = -2x + 2$

Lines (i) and (iii) have a slope of $-2$.

**59.** If the slope of a line is 0, the rise is 0.

**61.** $(0, 3), \; m = \frac{3}{4}$

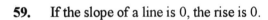

**63.** $(0, -1), \; m = -4$

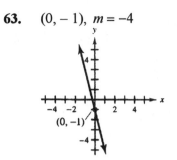

**65.** $(-5, 0), \; m$ is undefined

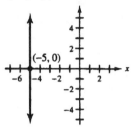

**67.** $(-2, -3), \; m = \frac{3}{4}$

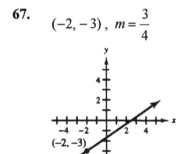

**69.** $(3, 7), \; m = 0$

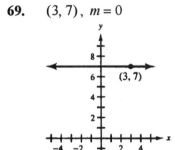

**71.** $(2, 6), \; m = -5$

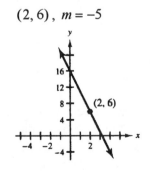

**73.** $y = \frac{3}{4}x + 5$; $m = \frac{3}{4}$; $y$-intercept: 5

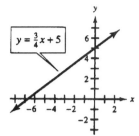

**75.** $y = -3x + 4$; $m = -3$; $y$-intercept: 4

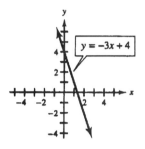

**77.** $y = 2$; $m = 0$; $y$-intercept: 2

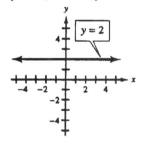

**79.** $y = x$; $m = 1$; $y$-intercept: 0

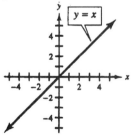

**81.** $y = -\frac{3}{4}x$; $m = -\frac{3}{4}$; $y$-intercept: 0

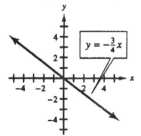

**83.** $x = 3$; $m$ is undefined; $y$-intercept: none

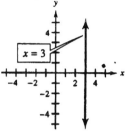

**85.** The graph retains the same slope, but it is shifted upward.

**87.** $2x + 5y = 10$
$5y = -2x + 10$
$y = -\frac{2}{5}x + 2$

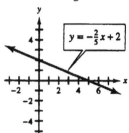

**89.** $x - 3 = 4$
$x = 7$

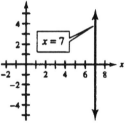

**91.** $2x - y - 6 = 2$
$-y = -2x + 8$
$y = 2x - 8$

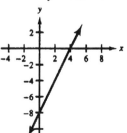

**93.** $y + 3 = 10$
$y = 7$

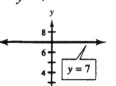

**95.** $y - 2x = 0$
$y = 2x$

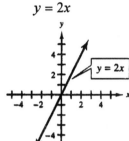

**97.** (a) In 8 years, the boy is expected to grow 16 inches. At age 14 he is expected to be 5'1".

(b) Slope $= \dfrac{16 \text{ inches}}{8 \text{ years}} = 2$

(c) The slope of the graph is the same as the average annual growth rate.

(d) Starting at (6, 3'9"), a rise of 8 units and a run of 4 units gives (10, 4'5"), or an age of 10 and a height of 4'5".

**99.** (a) The value of $m$ is the same as the average annual rate of change, or in this case, $m = -75$.

(b) The value of $b$ is the population when $x = 0$ (in 1990). In this case $b = 1200$.

(c) The run would be $1997 - 1990 = 7$ and the rise would be $7(-75) = -525$.

(d) The model is $y = -75x + 1200$. To find when the population is zero, substitute 0 for $y$ and solve for $x$:
$0 = -75x + 1200$
$75x = 1200$
$x = 16$
or in the year 2006.

**101.** $B(t) = 5.6t - 3.6$

(a) The slope of the graph is 5.6.

(b) (1, 4) and (10, 52)
$m = \dfrac{52 - 4}{10 - 1} = \dfrac{48}{9} = 5.\overline{3}$

**103.** (a) For semiprivate courses, the two points are (1995, 8142) and (1999, 9012). The slope is:
$m = \dfrac{9012 - 8142}{1999 - 1995} = \dfrac{870}{4} = 217.5$

For private courses, the two points are (1995, 4759) and (1999, 4708). The slope is:
$m = \dfrac{4708 - 4759}{1999 - 1995} = \dfrac{-51}{4} = -12.75$

(b) The positive slope indicates an increase in the number of semiprivate courses, whereas the negative slope indicates a decrease in the number of private courses.

**105.** Because the coefficient of $x$ is positive, the graph of the model function rises from left to right, which is consistent with the bar graph.

**107.** Because the percentage of women is increasing, the percentage of men is decreasing. Therefore, the coefficient of $x$ would be negative.

**109.** The range settings of X must be made small or the range settings for Y must be made large. The ratio of Y to X must be large.

**111.** $y = 90x + 200$
The following window settings give a good view of the intercepts:
xmin $= -5$, xmax $= 5$, ymin $= -100$, ymax $= 500$

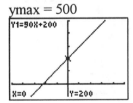

Section 4.3 Applications of Slope

**113.** $y = 35 - 0.02x$

The following window settings give a good view of the intercepts:
xmin = −1000, xmax = 3000,
ymin = −10, ymax = 50

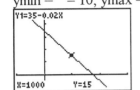

## Section 4.3 Applications of Slope

**1.** Perpendicular lines are lines that intersect to form a right angle.

**3.** $m_1 = 2$, $m_2 = -2$,
$m_1 \cdot m_2 = 2(-2) = -4$
Because the slopes are not equal, the lines are not parallel. Because the product of slopes is not $-1$, the lines are not perpendicular.

**5.** $m_1 = \dfrac{1}{2} = 0.5$, $m_2 = 0.5$,
Because the slopes are equal, the lines are parallel.

**7.** $m_1 = \dfrac{3}{5}$, $m_2 = -\dfrac{5}{3}$,
$m_1 \cdot m_2 = \dfrac{3}{5} \cdot \left(-\dfrac{5}{3}\right) = -1$
Because the slopes are not equal, the lines are not parallel. Because the product of slopes is $-1$, the lines are perpendicular.

**9.** (a) If $m_1 = -3$, $L_2$ is parallel to $L_1$ if $m_2 = -3$.
(b) If $m_1 = -3$, $L_2$ is perpendicular to $L_1$ if the product of their slopes is $-1$. Thus, $m_2 = \dfrac{1}{3}$.

**11.** (a) If $m_1 = -\dfrac{3}{4}$, $L_2$ is parallel to $L_1$ if $m_2 = -\dfrac{3}{4}$.
(b) If $m_1 = -\dfrac{3}{4}$, $L_2$ is perpendicular to $L_1$ if the product of their slopes is $-1$. Thus, $m_2 = \dfrac{4}{3}$.

**13.** (a) If $m_1 = 0$, $L_2$ is parallel to $L_1$ if $m_2 = 0$.
(b) If $m_1 = 0$, $L_1$ is a horizontal line. Thus, $L_2$ is perpendicular to $L_1$ if $L_2$ is a vertical line. Thus, $m_2$ is undefined.

**15.** $L_1$: $(3, 5)$ and $(-2, 1)$
$m_1 = \dfrac{y_2 - y_1}{x_2 - x_1} = \dfrac{1-5}{-2-3} = \dfrac{-4}{-5} = \dfrac{4}{5}$
$L_2$: $(-6, 9)$ and $(-1, 13)$
$m_2 = \dfrac{y_2 - y_1}{x_2 - x_1} = \dfrac{13-9}{-1-(-6)} = \dfrac{4}{5}$
Because the slopes are equal, the lines are parallel.

17. $L_1$: (0, 0) and (8, 8)

$$m_1 = \frac{y_2 - y_1}{x_2 - x_1} = \frac{8-0}{8-0} = \frac{8}{8} = 1$$

$L_2$: (−1, 4) and (3, −7)

$$m_2 = \frac{y_2 - y_1}{x_2 - x_1} = \frac{-7-4}{3-(-1)} = -\frac{11}{4}$$

Because the slopes are not equal, the lines are not parallel. Because the product of slopes is not −1, the lines are not perpendicular.

19. $L_1$: (−4, 2) and (−4, 7)

$$m_1 = \frac{y_2 - y_1}{x_2 - x_1} = \frac{7-2}{-4-(-4)} = \text{Undefined}$$

$L_2$: (0, 5) and (3, 5)

$$m_2 = \frac{y_2 - y_1}{x_2 - x_1} = \frac{5-5}{3-0} = 0$$

Because $L_1$ is vertical and $L_2$ is horizontal, the lines are perpendicular.

21. The slope of a vertical line is undefined.

23. $L_1$: $3y = 5 - 2x$

$$y = \frac{5}{3} - \frac{2}{3}x \qquad m_1 = -\frac{2}{3}$$

$L_2$: $y = -\frac{2}{3}x + 9 \qquad m_2 = -\frac{2}{3}$

Because the slopes are equal, the lines are parallel.

25. $L_1$: $x = -4$    Vertical line
    $L_2$: $y = -4$    Horizontal line
    The lines are perpendicular.

27. $L_1$: $y = 5x - 3 \qquad m_1 = 5$
    $L_2$: $y = -5x + 3 \qquad m_2 = -5$

Because the slopes are not equal, the lines are not parallel. Because the product of slopes is not −1, the lines are not perpendicular.

29. $L_1$: $x - y = 2$
    $-y = -x + 2$
    $y = x - 2 \qquad m_1 = 1$
    $L_2$: (5, 9) and (6, 8)

$$m_2 = \frac{y_2 - y_1}{x_2 - x_1} = \frac{8-9}{6-5} = \frac{-1}{1} = -1$$

Because the product of slopes is −1, the lines are perpendicular.

31. $L_1$: (−2, −5) and (0, 4)

$$m_1 = \frac{y_2 - y_1}{x_2 - x_1} = \frac{4-(-5)}{0-(-2)} = \frac{9}{2} = 4.5$$

$L_2$: $y = \frac{9}{2}x - 4 \qquad m_2 = 4.5$

Because the slopes are equal, the lines are parallel.

33. $L_1$: $y = 5 \qquad m_1 = 0$
    $L_2$: (2, 3) and (−6, 0)

$$m_2 = \frac{y_2 - y_1}{x_2 - x_1} = \frac{0-3}{-6-2} = \frac{-3}{-8} = \frac{3}{8}$$

Because the slopes are not equal, the lines are not parallel. Because the product of slopes is not −1, the lines are not perpendicular.

35. If $L$ is parallel to the $y$-axis, its slope is undefined.

37. If $L$ is perpendicular to the $y$-axis, its slope is 0.

39. $2x - y = 6$
    $-y = -2x + 6 \qquad m = 2$
    $y = 2x - 6$

(a) The slope of a line parallel to the given line is 2.

(b) Because $2\left(-\frac{1}{2}\right) = -1$, the slope of a line perpendicular to the given line is $-\frac{1}{2}$.

41. $4x + 3y - 6 = 0$
    $3y = -4x + 6 \qquad m = -\frac{4}{3}$
    $y = -\frac{4}{3}x + 2$

(a) The slope of a line parallel to the given line is $-\frac{4}{3}$.

Section 4.3  Applications of Slope

(b) Because $-\dfrac{4}{3} \cdot \left(\dfrac{3}{4}\right) = -1$, the slope of a line perpendicular to the given line is $\dfrac{3}{4}$.

**43.** $3y - 9 = 0$
$3y = 9 \quad m = 0$
$y = 3$

(a) The slope of a line parallel to the given line is 0.

(b) Because the given line is horizontal, a line perpendicular to it is vertical, which has an undefined slope.

**45.** Because the line is to be parallel to the line containing $(-2, 3)$ and $(4, -1)$, its slope should be:
$$m = \dfrac{y_2 - y_1}{x_2 - x_1} = \dfrac{-1 - 3}{4 - (-2)} = \dfrac{-4}{6} = -\dfrac{2}{3}$$
The required line contains $(-1, -4)$. The graph of this line follows.

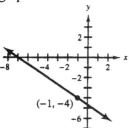

**47.** Because the line is to be parallel to the $y$-axis, its slope is undefined (the line is vertical). The required line contains $(3, 0)$. The graph of this line follows.

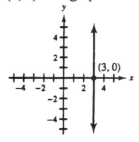

**49.** $L_1$: $(3, -4)$ and $(-1, 2)$
$$m_1 = \dfrac{y_2 - y_1}{x_2 - x_1} = \dfrac{2 - (-4)}{-1 - 3} = \dfrac{6}{-4} = -\dfrac{3}{2}$$

$L_2$: $(5, 0)$ and $(-2, k)$
$$m_2 = \dfrac{y_2 - y_1}{x_2 - x_1} = \dfrac{k - 0}{-2 - 5} = -\dfrac{k}{7}$$

(a) For the lines to be parallel, $m_1 = m_2$.
$-\dfrac{k}{7} = -\dfrac{3}{2}$
$2k = 21$
$k = 10.5$

(b) For the lines to be perpendicular, $m_1 \cdot m_2 = -1$.
$\left(-\dfrac{3}{2}\right)\left(-\dfrac{k}{7}\right) = -1$
$\dfrac{3k}{14} = -1$
$3k = -14$
$k = -\dfrac{14}{3}$

**51.** $(-4, 8)$ and $(-7, -1)$:
$$m = \dfrac{y_2 - y_1}{x_2 - x_1} = \dfrac{-1 - 8}{-7 - (-4)} = \dfrac{-9}{-3} = 3$$
$(-7, -1)$ and $(5, -5)$:
$$m = \dfrac{y_2 - y_1}{x_2 - x_1} = \dfrac{-5 - (-1)}{5 - (-7)} = \dfrac{-4}{12} = -\dfrac{1}{3}$$
Because the product of the slopes of the segment connecting $(-4, 8)$ and $(-7, -1)$ and the segment connecting $(-7, -1)$ and $(5, -5)$ is $-1$, the segments are perpendicular and the triangle is a right triangle.

**53.** $P(-5, 8)$, $Q(-1, 3)$, $R(7, -7)$
If the slope of the line segment joining $P$ and $Q$ is equal to the slope of the line segment joining $Q$ and $R$, then the points are collinear.

Slope of $PQ$: $m = \dfrac{3 - 8}{-1 - (-5)} = -\dfrac{5}{4}$

Slope of $QR$:
$$m = \dfrac{-7 - 3}{7 - (-1)} = -\dfrac{10}{8} = -\dfrac{5}{4}$$
The points are collinear.

**55.**

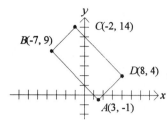

| Side | Points | Slope |
|---|---|---|
| $\overline{AB}$ | (3, −1), (−7, 9) | $\dfrac{9-(-1)}{-7-3} = -1$ |
| $\overline{CD}$ | (−2, 14), (8, 4) | $\dfrac{14-4}{-2-8} = -1$ |
| $\overline{AD}$ | (3, −1), (8, 4) | $\dfrac{-1-4}{3-8} = 1$ |
| $\overline{BC}$ | (−7, 9), (−2, 14) | $\dfrac{14-9}{-2-(-7)} = 1$ |

From the slopes we see that each pair of adjacent sides has slopes whose products are −1, verifying that each corner is a right angle. The figure is a rectangle.

**57.**

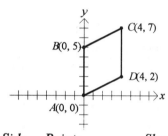

| Side | Points | Slope |
|---|---|---|
| $\overline{AB}$ | (0, 0), (0, 5) | Undefined |
| $\overline{CD}$ | (4, 7), (4, 2) | Undefined |
| $\overline{AD}$ | (0, 0), (4, 2) | $\dfrac{2-0}{4-0} = 0.5$ |
| $\overline{BC}$ | (0, 5), (4, 7) | $\dfrac{7-5}{4-0} = \dfrac{2}{4} = 0.5$ |

From the slopes we can verify that opposite sides of the quadrilateral are indeed parallel. Thus, the figure is a parallelogram.

**59.** The slope of a line represents the rate at which $y$ changes with respect to a change in $x$.

**61.** Because the slope of $y = \dfrac{1}{2}x + 3$ is $\dfrac{1}{2}$, the rate of change of $y$ with respect to $x$ is $\dfrac{1}{2}$.

**63.** Because the slope of $y = -4x + 9$ is $-4$, the rate of change of $y$ with respect to $x$ is $-4$.

**65.** (−1, −2) and (3, 4)

Slope $= \dfrac{-2-4}{-1-3} = \dfrac{-6}{-4} = \dfrac{3}{2}$

Because the slope of the line is $\dfrac{3}{2}$, the rate of change of $y$ with respect to $x$ is $\dfrac{3}{2}$.

**67.** (2, −2) and (−1, 5)

Slope $= \dfrac{-2-5}{2-(-1)} = \dfrac{-7}{3} = -\dfrac{7}{3}$

Because the slope of the line is $-\dfrac{7}{3}$, the rate of change of $y$ with respect to $x$ is $-\dfrac{7}{3}$.

**69.** Because the slope is 3, the rate of change of $y$ with respect to $x$ is 3. The change in $y$ corresponding to an increase of 2 in $x$ is an increase of 6 in $y$.

**71.** Because the slope is $-\dfrac{5}{4}$, the rate of change of $y$ with respect to $x$ is $-\dfrac{5}{4}$. The change in $x$ corresponding to an increase of 5 in $y$ is a decrease of 4 in $x$.

**73.** Because the slope is 5, the rate of change of $y$ with respect to $x$ is 5. The change in $y$ corresponding to a decrease of 2 in $x$ is a decrease of 10 in $y$.

**75.** Because the slope is $-\frac{3}{2}$, the rate of change of $y$ with respect to $x$ is $-\frac{3}{2}$.
The change in $x$ corresponding to a decrease of 6 in $y$ is an increase of 4 in $x$.

**77.** There exists a constant $k$ such that $A = kB$.

**79.** Because $y = 4x$ is written in the form $y = kx$, the equation represents a direct variation.

**81.** Because $\frac{y}{4} = x$ can be written in the form $y = kx$, the equation represents a direct variation.

**83.** Because $y = 5x - 3$ cannot be written in the form $y = kx$, the equation does not represent a direct variation.

**85.** Suppose that $y = kx$, $k > 0$. If $x$ increases, then $y$ <u>increases</u>, and if $y$ decreases, then $x$ <u>decreases</u>.

**87.** (a) Let $m$ = number of miles driven and $c$ = total cost. Then the equation $c = 0.10m + 29$ represents the total cost.
(b) The rate of change in total cost with respect to the number of miles driven is 0.1.

**89.** Using the points (1950, 5) and (2000, 55), the slope (annual rate of change) is $\frac{55 - 5}{2000 - 1950} = 1$ cent per year.

**91.** Let $y$ = the weight of bricks and $x$ = the number of bricks.
$$y = kx$$
$$1175 = k(500)$$
$$2.35 = k$$
The constant of variation is 2.35.
$$y = 2.35x$$
$$y = 2.35(1200)$$
$$y = 2820$$
A load of 1200 bricks will weigh 2820 pounds.

**93.** Let $y$ = the area of triangle and $x$ = the length of base.
$$y = kx$$
$$165 = k(12)$$
$$13.75 = k$$
The constant of variation is 13.75.
$$y = 13.75x$$
$$y = 13.75(18)$$
$$y = 247.5$$
A triangular sail with fixed height and base of 18 feet has an area of 247.5 square feet.

**95.** Let $y$ = the amount paid for typing and $x$ = the number of pages typed.
$$y = kx$$
$$100 = k(16)$$
$$6.25 = k$$
The constant of variation is 6.25.
$$y = 6.25x$$
$$y = 6.25(25)$$
$$y = 156.25$$
Having 25 pages typed will cost $156.25.

**97.** (a) The cost $C$ is $C(d) = 233d$, where $d$ is the number of days of vacation travel.
(b) The function is a direct variation because it is in the form $y = kx$.
(c) The constant of variation is 233.

**99.** (a) Because the function is $E(x) = 9.8x + 602.5$, we see that $m = 9.8$. This indicates that the average two-year increase in nuclear-based electricity generation for the period was 9.8 billion kilowatt-hours.
(b) The average rate of change is given by the slope, which is in this case 9.8.

**101.**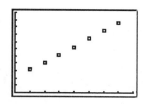

**103.** From Exercise 102, we saw that the slope of this line is 322.75. Using this slope and the point (2, 1875) for 1986, we know that the point for 1984 is $(0, 1875 - 2(322.75)) = (0, 1229.5)$. Thus, $y_1 = 322.75x + 1229.5$. The following table verifies the values of 1875 for 1986 ($x = 2$) and 5748 for 1998 ($x = 14$).

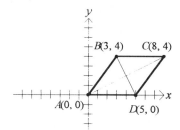

**105.** $kx + 3y = 10$
$3y = -kx + 10$
$y = -\dfrac{k}{3}x + \dfrac{10}{3}$

For this line to have a slope of $-2$, solve the following equation to find the necessary value of $k$.
$-\dfrac{k}{3} = -2$
$k = 6$

**107.** $3kx - 3y = 4$
$-3y = -3kx + 4$
$y = kx - \dfrac{4}{3}$

For this equation to be perpendicular to $x + y = 0$, or $y = -x$, which has a slope of $-1$, $k$ must be equal to 1.

**109.** $kx - 5y = 20$
$-5y = -kx + 20$
$y = \dfrac{k}{5}x - 4$

For this line to be parallel to the $x$-axis, its slope must be 0. Therefore, $k$ must be equal to 0.

**111.**

| Diagonal | Points | Slope |
|---|---|---|
| $\overline{AC}$ | (0, 0), (8, 4) | $\dfrac{4-0}{8-0} = 0.5$ |
| $\overline{BD}$ | (3, 4), (5, 0) | $\dfrac{0-4}{5-3} = \dfrac{-4}{2} = -2$ |

Because the product of the slopes of the two diagonals is $-1$, the diagonals are perpendicular.

## Section 4.4 Equations of Lines

**1.** To write an equation of a line with the slope-intercept form we must know the slope and the $y$-intercept.

**3.** $m = -4$, $b = -3$
$y = mx + b$
$y = -4x - 3$

**5.** $m = \dfrac{1}{2}$, $b = 3$
$y = mx + b$
$y = \dfrac{1}{2}x + 3$

**7.** $m = -5$, $y$-intercept: (0, 1), so, $b = 1$.
$y = mx + b$
$y = -5x + 1$

Section 4.4  Equations of Lines

**9.** $m = -\dfrac{3}{4}$, y-intercept: $(0, -5)$, so, $b = -5$.
$$y = mx + b$$
$$y = -\dfrac{3}{4}x - 5$$

**11.** $m = 7$, y-intercept: $\left(0, -\dfrac{2}{3}\right)$, so, $b = -\dfrac{2}{3}$.
$$y = mx + b$$
$$y = 7x - \dfrac{2}{3}$$

**13.** Two points on the graph: $(0, 2)$ and $(4, 5)$. The y-intercept is $b = 2$, and the slope is $\dfrac{5-2}{4-0} = \dfrac{3}{4}$. The equation is:
$$y = mx + b$$
$$y = \dfrac{3}{4}x + 2$$

**15.** The graph is a vertical line with x-intercept 4. The equation is $x = 4$.

**17.** Two points on the graph: $(0, -3)$ and $(-2, 2)$. The y-intercept is $b = -3$, and the slope is $\dfrac{2-(-3)}{-2-0} = \dfrac{5}{-2} = -\dfrac{5}{2}$. The equation is:
$$y = mx + b$$
$$y = -\dfrac{5}{2}x - 3$$

**19.** A horizontal line has a slope of 0, but a vertical line has no defined slope.

**21.** $m = 2$, $(1, 3)$
$$y - y_1 = m(x - x_1)$$
$$y - 3 = 2(x - 1)$$
$$y - 3 = 2x - 2$$
$$y = 2x + 1$$

**23.** $m = \dfrac{2}{5}$, $(10, -4)$
$$y - y_1 = m(x - x_1)$$
$$y - (-4) = \dfrac{2}{5}(x - 10)$$
$$y + 4 = \dfrac{2}{5}x - 4$$
$$y = \dfrac{2}{5}x - 8$$

**25.** $m = -\dfrac{2}{3}$, $(-2, -5)$
$$y - y_1 = m(x - x_1)$$
$$y - (-5) = -\dfrac{2}{3}[x - (-2)]$$
$$y + 5 = -\dfrac{2}{3}(x + 2)$$
$$y + 5 = -\dfrac{2}{3}x - \dfrac{4}{3}$$
$$y = -\dfrac{2}{3}x - \dfrac{19}{3}$$

**27.** $m = 0$, $(2, -5)$
$$y - y_1 = m(x - x_1)$$
$$y - (-5) = 0(x - 2)$$
$$y + 5 = 0$$
$$y = -5$$

**29.** $m$ is undefined, $(-3, 4)$. When $m$ is undefined, the line is vertical. Because the line has an x-intercept of $-3$, the equation is $x = -3$.

**31.** $m = 1$, $(7, 4)$
$$y - y_1 = m(x - x_1)$$
$$y - 4 = 1(x - 7)$$
$$y - 4 = x - 7$$
$$y = x - 3$$
$$y - x = -3$$
$$(-1)(y - x) = (-1)(-3)$$
$$x - y = 3$$

**33.** $m = \dfrac{5}{3}$, $(2, 0)$

$y - y_1 = m(x - x_1)$
$y - 0 = \dfrac{5}{3}(x - 2)$
$y = \dfrac{5}{3}x - \dfrac{10}{3}$
$3y = 5x - 10$
$3y - 5x = -10$
$(-1)(3y - 5x) = (-1)(-10)$
$5x - 3y = 10$

**35.** $m = -\dfrac{1}{4}$, $\left(\dfrac{1}{2}, -1\right)$

$y - y_1 = m(x - x_1)$
$y - (-1) = -\dfrac{1}{4}\left(x - \dfrac{1}{2}\right)$
$y + 1 = -\dfrac{1}{4}x + \dfrac{1}{8}$
$8y + 8 = -2x + 1$
$8y = -2x - 7$
$2x + 8y = -7$

**37.** Use the slope formula to determine the slope.

**39.** $(3, 4)$ and $(1, 8)$

Slope: $m = \dfrac{8 - 4}{1 - 3} = \dfrac{4}{-2} = -2$
$y - y_1 = m(x - x_1)$
$y - 4 = -2(x - 3)$
$y - 4 = -2x + 6$
$y = -2x + 10$

**41.** $(-2, 7)$ and $(6, -9)$

Slope: $m = \dfrac{-9 - 7}{6 - (-2)} = \dfrac{-16}{8} = -2$
$y - y_1 = m(x - x_1)$
$y - 7 = -2[x - (-2)]$
$y - 7 = -2(x + 2)$
$y - 7 = -2x - 4$
$y = -2x + 3$

**43.** $(2, 3)$ and $(-3, -2)$

Slope: $m = \dfrac{-2 - 3}{-3 - 2} = \dfrac{-5}{-5} = 1$
$y - y_1 = m(x - x_1)$
$y - 3 = 1(x - 2)$
$y - 3 = x - 2$
$y = x + 1$

**45.** $(2, -3)$ and $(-4, 1)$

Slope: $m = \dfrac{1 - (-3)}{-4 - 2} = \dfrac{4}{-6} = -\dfrac{2}{3}$
$y - y_1 = m(x - x_1)$
$y - (-3) = -\dfrac{2}{3}(x - 2)$
$y + 3 = -\dfrac{2}{3}x + \dfrac{4}{3}$
$y = -\dfrac{2}{3}x - \dfrac{5}{3}$

**47.** $(-4, 2)$ and $(-4, 5)$

Slope: $m = \dfrac{5 - 2}{-4 - (-4)} = \dfrac{3}{0} =$ undefined

A line with undefined slope is vertical. The equation of a vertical line through $(-4, 2)$ and $(-4, 5)$ is $x = -4$.

**49.** $(-3, -4)$ and $(1, -2)$

Slope: $m = \dfrac{-2 - (-4)}{1 - (-3)} = \dfrac{2}{4} = \dfrac{1}{2}$
$y - y_1 = m(x - x_1)$
$y - (-2) = \dfrac{1}{2}(x - 1)$
$y + 2 = \dfrac{1}{2}x - \dfrac{1}{2}$
$y = \dfrac{1}{2}x - \dfrac{5}{2}$

**51.** $(-3, 5)$ and $(4, 5)$

Slope: $m = \dfrac{5 - 5}{4 - (-3)} = \dfrac{0}{7} = 0$
$y - y_1 = m(x - x_1)$
$y - 5 = 0(x - 4)$
$y - 5 = 0$
$y = 5$

Section 4.4   Equations of Lines

**53.** (a) A line that is perpendicular to a vertical line is a horizontal line. The equation is in the form $y = b$.

(b) A line that is parallel to the $y$-axis is a vertical line. The equation is in the form $x = c$.

**55.** A horizontal line has the form $y = b$. Because the line contains $(3, -4)$, the constant is $-4$ and the equation is $y = -4$.

**57.** If the line has the same slope as the line with equation $y = 3$, the line is a horizontal line with the form $y = b$. Because the line contains $(-2, 5)$, the constant is 5 and the equation is $y = 5$.

**59.** $2x + 3y = 1$
$3y = -2x + 1$
$y = -\dfrac{2}{3}x + \dfrac{1}{3}$

The equation $2x + 3y = 1$ has slope $-\dfrac{2}{3}$.
The line containing $(-2, -3)$ and having slope $-\dfrac{2}{3}$ is:
$y - y_1 = m(x - x_1)$
$y - (-3) = -\dfrac{2}{3}[x - (-2)]$
$y + 3 = -\dfrac{2}{3}(x + 2)$
$y + 3 = -\dfrac{2}{3}x - \dfrac{4}{3}$
$y = -\dfrac{2}{3}x - \dfrac{13}{3}$

**61.** A vertical line has the form $x = c$. Because the line contains $(-2, 5)$, the constant is $-2$ and the equation is $x = -2$.

**63.** $2x + 3y = 6$
$3y = -2x + 6$
$y = -\dfrac{2}{3}x + 2$

The equation $2x + 3y = 6$ has a $y$-intercept of 2. The line containing $(0, 2)$ and $(1, 5)$ has slope
$m = \dfrac{5-2}{1-0} = \dfrac{3}{1} = 3$. The equation of the line is:
$y - y_1 = m(x - x_1)$
$y - 2 = 3(x - 0)$
$y - 2 = 3x$
$y = 3x + 2$

**65.** Because a line with undefined slope is a vertical line, the equation of the line has the form $x = c$. Because the line contains $(-2, -3)$, the constant is $-2$ and the equation is $x = -2$.

**67.** $L_2$: $y = 4x - 7$   $m_2 = 4$
Because $L_1$ is parallel to $L_2$, it has the same slope, so $m_1 = 4$. If the $y$-intercept of $L_1$ is $(0, 2)$, then $b = 2$. The equation of $L_1$ is:
$y = mx + b$
$y = 4x + 2$

**69.** $L_2$: $y = 4$, a horizontal line
Because $L_1$ is perpendicular to $L_2$, it is a vertical line having the form $x = c$. If $L_1$ contains the point $(-2, 7)$, the constant is $-2$ and the equation is $x = -2$.

**71.** From the figure: $m_2 = \dfrac{4-1}{6-2} = \dfrac{3}{4}$.
Using $(2, 1)$, an equation for $L_2$ is:
$y - y_1 = m(x - x_1)$
$y - 1 = \dfrac{3}{4}(x - 2)$
$y - 1 = \dfrac{3}{4}x - \dfrac{3}{2}$
$y = \dfrac{3}{4}x - \dfrac{1}{2}$

Because $L_1$ is perpendicular to $L_2$, the slope of $L_1$ is $m_1 = -\dfrac{4}{3}$. Using $(2, 1)$, an equation for $L_1$ is:

$$y - y_1 = m(x - x_1)$$
$$y - 1 = -\frac{4}{3}(x - 2)$$
$$y - 1 = -\frac{4}{3}x + \frac{8}{3}$$
$$y = -\frac{4}{3}x + \frac{11}{3}$$

73. $L_1$: $y = 3x - 7 \quad b = -7$
    $L_2$: $x + y = 3$
    $\quad y = -x + 3 \quad m = -1$
    
    If $L_3$ is parallel to $L_2$, it has the same slope as $L_2$. If $L_3$ also has the same $y$-intercept as $L_1$, then the equation is:
    $y = mx + b$
    $y = -1x - 7$
    $y = -x - 7$

75. $L_1$: $3x + 4y = 12$
    $\quad 4y = -3x + 12$
    $\quad y = -\frac{3}{4}x + 3 \quad m = -\frac{3}{4}, b = 3$
    $L_2$: $x - y = 3$
    $\quad -y = -x + 3$
    $\quad y = x - 3 \quad m = 1, b = -3$

    (a) If $L_3$ is perpendicular to $L_2$, its slope is $m = -1$. If $L_3$ has the same $y$-intercept as $L_1$, its $y$-intercept is $b = 3$. The equation of $L_3$ is
    $y = mx + b$
    $y = -1x + 3$
    $y = -x + 3$

    (b) If $L_4$ is parallel to $L_1$, its slope is $m = -\frac{3}{4}$. If $L_4$ has the same $y$-intercept as $L_2$, its $y$-intercept is $b = -3$. The equation of $L_4$ is
    $y = mx + b$
    $y = -\frac{3}{4}x - 3$

77. $kx + 3y = 10$
    $\quad 3y = -kx + 10$
    $\quad y = -\frac{k}{3}x + \frac{10}{3}$
    
    If $kx + 3y = 10$ is parallel to (has the same slope as) $y = 3 - 2x$ with slope $-2$, then the slopes are equal and
    $-\frac{k}{3} = -2$
    $k = 6$

79. $2x - ky = 2$
    $\quad -ky = -2x + 2$
    $\quad y = \frac{2}{k}x - \frac{2}{k}$
    
    If $2x - ky = 2$ is perpendicular to $y = \frac{3 - 4x}{2}$ with slope $-2$, then the product of the slopes is $-1$ and
    $\frac{2}{k} \cdot (-2) = -1$
    $-\frac{4}{k} = -1$
    $-4 = -k$
    $4 = k$

81. $x + 2y = 6k$
    $\quad 2y = -x + 6k$
    $\quad y = -\frac{1}{2}x + 3k$
    
    If $x + 2y = 6k$ has the same $y$-intercept as $y = \frac{1 - x}{-2}$ with intercept $-1/2$, then the intercepts are equal and
    $3k = -\frac{1}{2}$
    $k = -\frac{1}{6}$

83. To find the $x$-intercept of $x - y = 2k$, substitute 0 for $y$ and solve for $x$:
    $x - 0 = 2k$
    $x = 2k$
    To find the $x$-intercept of $y = 4x - 8$, substitute 0 for $y$ and solve for $x$:
    $0 = 4x - 8$
    $8 = 4x$
    $2 = x$
    Set these intercepts equal and solve for $k$:
    $2k = 2$
    $k = 1$

**85.**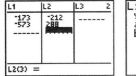

The equation of the line is
$y = 0.5x - 7.5$.

**87.**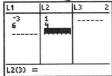

The equation of the line is
$y = -1.25x - 428.25$.

**89.** For $P$ and $Q$:

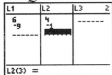

The equation of the line is $y = \dfrac{1}{3}x + 2$.

For $Q$ and $R$:

The equation of the line is $y = \dfrac{1}{3}x + 2$.

The points are collinear because the equations are the same.

**91.** (a) The computer originally cost \$1500. It depreciates \$1400 in 7 years, giving a rate of depreciation of \$200 per year. The equation for the value $v$ of the computer $t$ years since 2000 is $v = -200t + 1500$.

(b) The value of the computer in 2004 ($t = 4$) is:
$v = -200t + 1500$
$v = -200(4) + 1500$
$v = -800 + 1500 = 700$

**93.** Let $x$ represent the number of years since 1998 and $y$ represent the net loans (in millions). The data can be represented by two points: $(0, 92.1)$ and $(2, 105.7)$. The slope is: $m = \dfrac{105.7 - 92.1}{2 - 0} = \dfrac{13.6}{2} = 6.8$.

An equation for the line is:
$y - y_1 = m(x - x_1)$
$y - 92.1 = 6.8(x - 0)$
$y - 92.1 = 6.8x$
$\quad\quad y = 6.8x + 92.1$

The net loans in 2007 ($x = 9$) is:
$y = 6.8x + 92.1$
$y = 6.8(9) + 92.1$
$y = 61.2 + 92.1 = 153.3$

The net loans in 2007 is predicted to be \$153.3 million. The rate of change of the net loans is \$6.8 million per year.

**95.** (a) Using a calculator, the regression equation is $N(x) = 0.0133x + 1.166$

```
LinReg
y=ax+b
a=.0133
b=1.166
r²=.9853498217
r=.9926478841
```

(b) The graph of $N$ is a line whose slope is 0.0133.

(c) The graph of $N$ comes as close as possible to all the data points, but it does not necessarily contain any actual point.

**97.** (a) To have $(0, 87.2)$ as the $y$-intercept of the graph, we must define $x$ as the number of years since 1995.

(b) To write the function, we must find the slope between the two points $(0, 87.2)$ and $(3, 90.5)$:
$m = \dfrac{90.5 - 87.2}{3 - 0} = \dfrac{3.3}{3} = 1.1$
The function is $E(x) = 1.1x + 87.2$.

(c) Solve for $x$ when the function is set equal to 96:
$1.1x + 87.2 = E(x)$
$1.1x + 87.2 = 96$
$\quad\quad 1.1x = 8.8$
$\quad\quad\quad x = 8$
Energy use will reach 96 quadrillion BTUs in 2003 (that is, 1995 + 8).

**99.** Choose the points (0, 6.56) and (2, 6.68). The *y*-intercept is (0, 6.56). The slope is
$$m = \frac{6.68 - 6.56}{2 - 0} = \frac{0.12}{2} = 0.06$$
The model is $F = 0.06x + 6.56$.
The model is the same for any two data points because all the data points are points of the same line.

**101.** (0, 6.56) and (13, 7.19)
$$m = \frac{7.19 - 6.56}{13 - 0} \approx 0.048$$
The slope using these two points is approximately 0.048, which is less than the slope (coefficient of *x*) in the model from Exercise 99. Therefore, the coefficient of *x* in this new model would be smaller.

**103.** Given the points $(x_1, y_1)$ and $(x_2, y_2)$, the slope is $m = \dfrac{y_2 - y_1}{x_2 - x_1}$. Using the point-slope form and substituting for *m*, the equation of the line is:
$$y - y_1 = m(x - x_1)$$
$$y - y_1 = \frac{y_2 - y_1}{x_2 - x_1}(x - x_1)$$

**105.** $\dfrac{x}{3} + \dfrac{y}{2} = 1$

Using the results of Exercise 104, the intercepts are (3, 0) and (0, 2).

**107.** $\dfrac{2x}{5} - \dfrac{3y}{4} = 1$

$\dfrac{x}{5/2} + \dfrac{y}{-4/3} = 1$

Using the results of Exercise 104, the intercepts are $\left(\dfrac{5}{2}, 0\right)$ and $\left(0, -\dfrac{4}{3}\right)$.

**109.** $ax + by = c$
$by = -ax + c$
$y = -\dfrac{a}{b}x + \dfrac{c}{b}$

An equation of a line that is perpendicular to $ax + by = c$ has a slope of $\dfrac{b}{a}$. If the line contains the point $(a, b)$, the equation of the line is:
$$y - y_1 = m(x - x_1)$$
$$y - b = \frac{b}{a}(x - a)$$
$$ay - ab = b(x - a)$$
$$ay - ab = bx - ab$$
$$ay = bx$$
$$-bx + ay = 0$$
$$bx - ay = 0$$

## Section 4.5 Graphs of Linear Inequalities

**1.** The solution set of $x + 2y \leq 6$ contains ordered pairs that satisfy $x + 2y = 6$, but the solution set of $x + 2y < 6$ does not.

**3.** $y < 3x - 5$
(1, 0) is not a solution:
$0 < 3(1) - 5$
$0 < -2$   False
(0, −6) is a solution:
$-6 < 3(0) - 5$
$-6 < -3$   True
(2, 1) is not a solution:
$1 < 3(2) - 5$
$1 < 1$   False

**5.** $2x - 7y \geq 12$
(12, 1) is a solution:
$2(12) - 7(1) \geq 12$
$17 \geq 12$   True

Section 4.5    Graphs of Linear Inequalities

(6, 0) is a solution:
$2(6) - 7(0) \geq 12$
$12 \geq 12$    True

$(-4, -2)$ is not a solution:
$2(-4) - 7(-2) \geq 12$
$6 \geq 12$    False

7. $y \leq 4$

(1, 10) is not a solution:
$10 \leq 4$    False

(0, 4) is a solution:
$4 \leq 4$    True

(4, 5) is not a solution:
$5 \leq 4$    False

9. In the first case, we are assuming that $x + 3 > 7$ is an inequality in one variable. In the second case, we are assuming that $x + 3 > 7$ is an inequality in two variables.

11. To complete the graph, shade the half-plane below the boundary line.

13. To complete the graph, shade the half-plane above the boundary line.

15. To complete the graph, shade the half-plane below the boundary line.

17. $y < 7 - 3x$

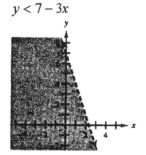

19. $-y \leq -2x$
$y \geq 2x$

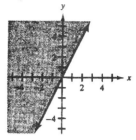

21. $3x + 2y \leq 6$
$2y \leq -3x + 6$
$y \leq -1.5 + 3$

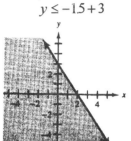

23. $-2y \geq 4x - 7$
$y \leq -2x + 3.5$

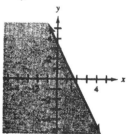

25. $-6 \leq x - 2y$
$2y \leq x + 6$
$y \leq 0.5x + 3$

27. $2x - 5y \geq 10$
$-5y \geq -2x + 10$
$y \leq 0.4x - 2$

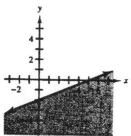

29. $y - 3 \leq -2(x - 3)$
$y - 3 \leq -2x + 6$
$y \leq -2x + 9$

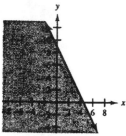

31. $\dfrac{x}{5} + \dfrac{y}{2} < 1$
$2x + 5y < 10$
$5y < -2x + 10$
$y < -0.4x + 2$

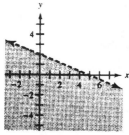

33. $y \geq 20$

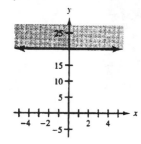

35. $x < 5$

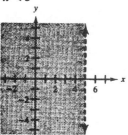

37. An equation of the boundary line is $y = 3$. Because the boundary line is solid and the half-plane below the boundary line is shaded, the inequality is $y \leq 3$.

39. Two points on the boundary line are $(-2, 0)$ and $(0, 1)$. The slope is $m = \dfrac{1 - 0}{0 - (-2)} = 0.5$ and the $y$-intercept is $b = 1$. An equation of the boundary line is $y = 0.5x + 1$. Because the boundary line is dashed and the half-plane below the boundary line is shaded, the inequality is $y < 0.5x + 1$.

41. An equation of the boundary line is $x = 2$. Because the boundary line is solid and the half-plane to the left of the boundary line is shaded, the inequality is $x \leq 2$.

43. An ordered pair solution that satisfies the given condition "one number $x$ is at least 1 more than another number $y$" is $(7, 5)$. The condition can be translated into the inequality $x \geq y + 1$.

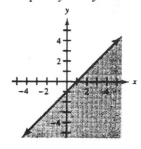

**45.** An ordered pair solution that satisfies the given condition "one number $y$ is not less than 4 less than 3 times another number $x$" is (2, 5). The condition can be translated into the inequality $y \geq 3x - 4$.

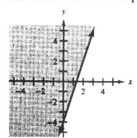

**47.** An ordered pair solution that satisfies the given condition "the sum of a number $x$ and twice another number $y$ exceeds 10" is (5, 3). The condition can be translated into the inequality $x + 2y > 10$.

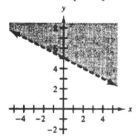

**49.** $x \leq 2$ and $x \geq 0$

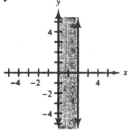

**51.** $x + y \geq 3$
$y \geq -x + 3$ and $x \geq 0$

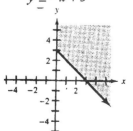

**53.** $y < \dfrac{1}{2}x$ and $y \geq 3x$

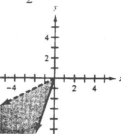

**55.** $y > -1$ or $y \leq -4$

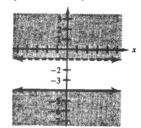

**57.** $y > -\dfrac{2}{3}x$ or $\begin{array}{l} y + 3x \leq 0 \\ y \leq -3x \end{array}$

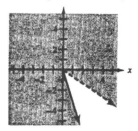

**59.** $y < x$ or $y \geq 0$

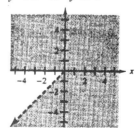

**61.** $y \leq -x + 3$ and $y \leq 2x + 1$

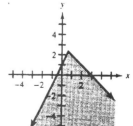

63. $3x + y < 5$      $x - y < 2$
    $y < -3x + 5$  or  $-y < -x + 2$
                       $y > x - 2$

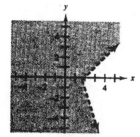

65. $2x - y - 5 < 0$        $4x - 2y > 0$
    $-y < -2x + 5$  and  $-2y > -4x$
    $y > 2x - 5$            $y < 2x$

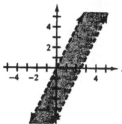

67. $5y + 3x < 30$        $5y + 3x < -20$
    $5y < -3x + 30$  and  $5y < -3x - 20$
    $y < -0.6x + 6$       $y < -0.6x - 4$

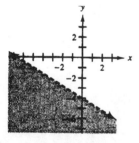

69. $x \geq y$ and $x \leq y$

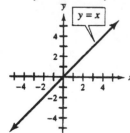

71. $x \geq 0$, $y \geq 0$, and $y \leq 10 - x$

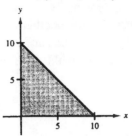

73. An equation of the upper boundary line is $y = 2$, and an equation of the lower boundary line is $y = -3$. Because the upper boundary line is dashed, the lower boundary is solid, and the region between the lines is shaded, the compound inequality is $y < 2$ and $y \geq -3$.

75. An equation of the left boundary line is $x = -2$, and an equation of the right boundary line is $x = 1$. Because the left boundary line is solid, the right boundary is dashed, and the region to the left and right of the boundaries is shaded, the compound inequality is $x \leq -2$ or $x > 1$.

77. Two points on the upper boundary line are $(-3, 0)$ and $(0, 3)$. The slope is $m = \dfrac{3 - 0}{0 - (-3)} = 1$ and the y-intercept is $b = 3$. An equation of the upper boundary line is $y = x + 3$. Two points on the lower boundary line are $(2, 0)$ and $(0, -2)$. The slope is $m = \dfrac{-2 - 0}{0 - 2} = 1$ and the y-intercept is $b = -2$. An equation of the lower boundary line is $y = x - 2$. Because both boundary lines are solid and the region above and below the boundary lines is shaded, the compound inequality is $y \geq x + 3$ or $y \leq x - 2$.

**79.** $|y| \leq 3$ is equivalent to $-3 \leq y \leq 3$

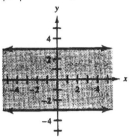

**81.** $|x - 2y| > 4$ is equivalent to

$$x - 2y < -4 \qquad x - 2y > 4$$
$$-2y < -x - 4 \quad \text{or} \quad -2y > -x + 4$$
$$y > 0.5x + 2 \qquad y < 0.5x - 2$$

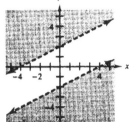

**83.** $|y - x| \geq 3$ is equivalent to

$$y - x \geq -3 \qquad y - x \leq 3$$
$$y \geq x - 3 \quad \text{or} \quad y \leq x + 3$$

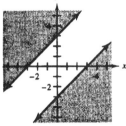

**85.** $|x| \leq -2$ has a solution set of $\varnothing$ because the absolute value of a number is never negative and, therefore, cannot be less than or equal to a negative number.

**87.** $|x - y| \leq 0$ is equivalent to $x - y = 0$ because the absolute value of an expression cannot be negative (less than 0).

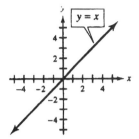

**89.** $|x + y| > 0$ is equivalent to

$$x + y < 0 \qquad x + y > 0$$
$$y < -x \quad \text{or} \quad y > -x$$

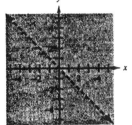

**91.** Let $x =$ the number of $4 \times 4$ pallets and $y =$ the number of $3 \times 5$ pallets. The area of a $4 \times 4$ pallet is 16 square feet and the area of a $3 \times 5$ pallet is 15 square feet. If the main hold is 6000 square feet, an inequality that describes the number of pallets that can be used is $16x + 15y \leq 6000$, $x \geq 0$, $y \geq 0$.

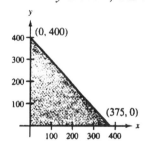

**93.** Let $x$ = the score of one team and $y$ = the score of the other team. An absolute value inequality that describes whether the teams are evenly matched is $|x-y| \leq 5$, $x \geq 0$, $y \geq 0$.

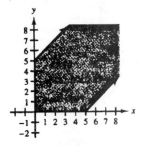

**95.** Let $c$ = the number of cats and $d$ = the number of dogs. If the maximum number of animals that can be kept is 100 but the shelter closes if the population falls below 20 can be described by the inequality $20 \leq c + d \leq 100$, $c \geq 0$, $d \geq 0$.

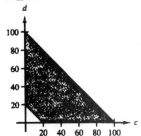

**97.** (a) Let $R$ = the number of votes for the Republican and $D$ = the number of votes for the Democrat. An inequality that describes the number of votes cast is $R + D \leq 0.536(190)$ or $R + D \leq 101.84$.
(b) The solution points of the graph of the inequality in part (a) represent the number of votes for the two candidates.

**99.** (a) Let $x$ = adjusted gross income and $y$ = tax. An inequality for the maximum tax that would be paid is $y \leq 6457.50 + 0.28(x - 43{,}050)$.
(b) The graph of the inequality in part (a) would be shaded below the boundary line.

**101.** For $W = 180$,
$$0.0875D - 0.00125W = 0.05$$
$$0.0875D - 0.00125(180) = 0.05$$
$$0.0875D - 0.225 = 0.05$$
$$0.0875D = 0.275$$
$$D \approx 3$$
The number of drinks that will produce a blood alcohol level of 0.05 in a 180-pound person is 3.

**103.** $0.0875D - 0.00125W > 0.05$
$-0.00125W > -0.0875D + 0.05$
$W < 70D - 40$
The graph of the inequality, with $W$ along the vertical axis and $D$ along the horizontal axis, would be shaded below the boundary line.

**105.** The graph of $|x| > 3$ is the region to the right of $x = 3$ or the region to the left of $x = -3$. The graph of $|y| \geq 2$ is the region on or above $y = 2$ or the region on or below $y = -2$. These two absolute value inequalities are connected by *and*, so the graph is the intersection of the two individual graphs.

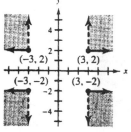

**107.** The graph of $|x + y| > -2$ is the entire plane. The graph of $|x| < -5$ is the empty set. These two absolute value inequalities are connected by *or*, so the graph is the union of the two individual graphs. The union is the entire plane.

**109.** An example of $y > b$, $b > 0$ is shown below.

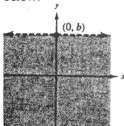

**111.** An example of $y < mx + b$, $m > 0$, $b < 0$ is shown below.

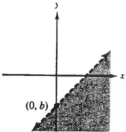

## Chapter 4 Project

**1.** The average yearly increase in the number of pairs of bald eagles is the same as the slope of the model giving the number of pairs of bald eagles, which is 326.27 pairs per year.

**3.** Set the model equal to 250,000 pairs and solve for $x$:
$$326.27x + 1164.86 = 250,000$$
$$326.27x = 248835.14$$
$$x \approx 763$$

**5.** $\$1,000,000 \div 5094 \approx \$196$ per pair of eagles

## Chapter 4 Review Exercises

**1.** The equation $3x = \dfrac{2}{y} + \dfrac{3}{2}$ cannot be written in the form $Ax + By = C$ and, therefore, it is not a linear equation in two variables.

**3.** The equation $x = \dfrac{y}{3} - \dfrac{4}{5}$ can be written in the form $Ax + By = C$ and, therefore, it is a linear equation in two variables.

**5.** Show that $(5, -11)$ is a solution of $y = -3x + 4$.
$$-11 = -3(5) + 4$$
$$-11 = -15 + 4$$
$$-11 = -11$$
Because the last statement is true, $(5, -11)$ is a solution of $y = -3x + 4$.

**7.** $y = 7 - x$
For $(-2, a)$:
$$a = 7 - (-2)$$
$$a = 7 + 2 = 9$$
For $(b, 14)$:
$$14 = 7 - b$$
$$7 = -b$$
$$-7 = b$$
For $(c, -5)$:
$$-5 = 7 - c$$
$$-12 = -c$$
$$12 = c$$

**9.** $2y - 3x = 9$

| $x$ | 3 | 7 | $-7$ |
|---|---|---|---|
| $y$ | 9 | 15 | $-6$ |

11. y-intercept:  x-intercept:
    $7x - 3y = 21$   $7x - 3y = 21$
    $7(0) - 3y = 21$  $7x - 3(0) = 21$
    $y = -7$         $x = 3$
    $(0, -7)$ and $(3, 0)$

13. y-intercept:  x-intercept:
    $y - 2 = 18$    None
    $y = 20$
    $(0, 20)$

15. $15 - 3x + 6y = 0$
    $6y = 3x - 15$
    $y = \frac{1}{2}x - \frac{5}{2}$
    The y-intercept is $\left(0, -\frac{5}{2}\right)$.

17. $2x - y = 8$
    $-y = -2x + 8$
    $y = 2x - 8$
    The y-intercept is $(0, -8)$.

19. $5x + 3y = 15$
    Three solutions: $(0, 5)$, $(3, 0)$, $\left(2, \frac{5}{3}\right)$

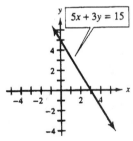

21. $y + 3 = 9$
    Three solutions: $(0, 6)$, $(3, 6)$, $(10, 6)$

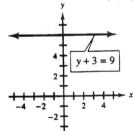

23. The graph contains the points $(-4, 0)$ and $(0, -3)$.
    (i) $3x + 4y = 12$
       $(-4, 0)$: $3(-4) + 4(0) = 12$
                  $-12 = 12$ False
       $(0, -3)$: $3(0) + 4(-3) = 12$
                  $-12 = 12$ False
    (ii) $3x + 4y = -12$
       $(-4, 0)$: $3(-4) + 4(0) = -12$
                  $-12 = -12$ True
       $(0, -3)$: $3(0) + 4(-3) = -12$
                  $-12 = -12$ True
    (iii) $3x - 4y = -12$
       $(-4, 0)$: $3(-4) - 4(0) = -12$
                  $-12 = -12$ True
       $(0, -3)$: $3(0) - 4(-3) = -12$
                  $12 = -12$ False

    The graph represents equation (ii) because it is the only equation that has both points from the graph as solutions.

25. $(-5, -8)$ and $(2, -3)$
    Slope $= \dfrac{y_2 - y_1}{x_2 - x_1} = \dfrac{-3 - (-8)}{2 - (-5)} = \dfrac{5}{7}$

27. $(2, -4)$ and $(-7, 6)$
    Slope $= \dfrac{y_2 - y_1}{x_2 - x_1} = \dfrac{6 - (-4)}{-7 - 2} = \dfrac{10}{-9} = -\dfrac{10}{9}$

29. $3x - 4y = 15$
    $-4y = -3x + 15$
    $y = \dfrac{3}{4}x - \dfrac{15}{4}$
    The slope is $m = \dfrac{3}{4}$.

31. $(0, -4)$, $m = -\dfrac{3}{5}$

**33.** $y = -x + 3$; $m = -1$; $y$-intercept: 3

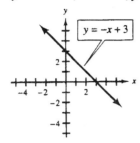

**35.** $y = 3$; $y$-intercept: 3; horizontal line

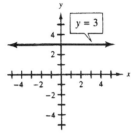

**37.** $m_1 = \dfrac{1}{4} = 0.25$, $m_2 = 0.25$,

Because the slopes are equal, the lines are parallel.

**39.** $m_1 = \dfrac{1}{3}$, $m_2 = -3$,

$m_1 \cdot m_2 = \dfrac{1}{3}(-3) = -1$

Because the slopes are not equal, the lines are not parallel. Because the product of slopes is $-1$, the lines are perpendicular.

**41.** $L_1$: $y = 0.5x - 3$  $m_1 = 0.5$
$L_2$: $y = -2x + 4$  $m_2 = -2$

Because $m_1 \cdot m_2 = 0.5(-2) = -1$, the lines are perpendicular.

**43.**

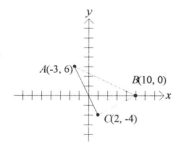

$\overline{AC}$: $m = \dfrac{-4-6}{2-(-3)} = \dfrac{-10}{5} = -2$

$\overline{BC}$: $m = \dfrac{-4-0}{2-10} = \dfrac{-4}{-8} = 0.5$

Because the product of the slopes of $\overline{AC}$ and $\overline{BC}$ is $-1$, the segments are perpendicular and the triangle is a right triangle.

**45.** Because the slope of $y = -3x + 5$ is $-3$, the rate of change of $y$ with respect to $x$ is $-3$.

**47.** Because the slope is $-5$, the rate of change of $y$ with respect to $x$ is $-5$. The change in $y$ corresponding to an increase of 3 in $x$ is a decrease of 15 in $y$.

**49.** Using the points (1988, 5) and (2000, 20), the slope (annual rate of change) is $\dfrac{20-5}{2000-1988} = \dfrac{15}{12} = 1.25$ dollars per year.

**51.** Because $y = 4x - 7$ cannot be written in the form $y = kx$, the equation does not represent a direct variation.

**53.** Because $y = 7x$ is written in the form $y = kx$, the equation represents a direct variation.

**55.** Let $n$ = the number of toothpicks produced and $h$ = the number of hours the machine is operating.

$n = kh$
$15{,}000 = k(6)$
$2500 = k$

The constant of variation is 2500.

$y = 2500x$
$y = 2500(45)$
$y = 112{,}500$

The machine will produce 112,500 toothpicks in 45 hours.

**57.** $m = -3$, $(-2, 5)$
$$y - y_1 = m(x - x_1)$$
$$y - 5 = -3[x - (-2)]$$
$$y - 5 = -3(x + 2)$$
$$y - 5 = -3x - 6$$
$$y = -3x - 1$$

**59.** $(2, -4)$ and $(-5, 3)$
Slope: $m = \dfrac{3 - (-4)}{-5 - 2} = \dfrac{7}{-7} = -1$
$$y - y_1 = m(x - x_1)$$
$$y - (-4) = -1(x - 2)$$
$$y + 4 = -x + 2$$
$$y = -x - 2$$

**61.** $(-3, -6)$ and $(1, -1)$
Slope: $m = \dfrac{-1 - (-6)}{1 - (-3)} = \dfrac{5}{4}$
$$y - y_1 = m(x - x_1)$$
$$y - (-1) = \dfrac{5}{4}(x - 1)$$
$$y + 1 = \dfrac{5}{4}x - \dfrac{5}{4}$$
$$y = \dfrac{5}{4}x - \dfrac{9}{4}$$

**63.** A horizontal line has the form $y = b$. Because the line contains $(-3, 6)$, the constant is 6 and the equation is $y = 6$.

**65.** $2x + 3y = 6$
$$3y = -2x + 6$$
$$y = -\dfrac{2}{3}x + 2$$
The equation $2x + 3y = 6$ has a slope of $-\dfrac{2}{3}$. The line containing $(1, 5)$ and having slope $-\dfrac{2}{3}$ (since the line is parallel to $2x + 3y = 6$) has the equation:

$$y - y_1 = m(x - x_1)$$
$$y - 5 = -\dfrac{2}{3}(x - 1)$$
$$y - 5 = -\dfrac{2}{3}x + \dfrac{2}{3}$$
$$y = -\dfrac{2}{3}x + \dfrac{17}{3}$$

**67.** The given line contains the points $(0, 2)$ and $(5, 0)$, so the $y$-intercept is $b = 2$ and slope is $m = \dfrac{0 - 2}{5 - 0} = -0.4$.
$$y = mx + b$$
$$y = -0.4x + 2$$

**69.** Let $x$ = the number of years since 1998 and $y$ = the annual sales. Using the points $(0, 150{,}000)$ and $(2, 400{,}000)$, the $y$-intercept is $b = 150{,}000$ and the slope is $m = \dfrac{400{,}000 - 150{,}000}{2 - 0} = 125{,}000$.
An equation is $y = 125{,}000x + 150{,}000$.
The sales will be 800,000 when:
$$800{,}000 = 125{,}000x + 150{,}000$$
$$675{,}000 = 125{,}000x$$
$$5 \approx x$$
Because $x$ represents the number of years since 1998, $x \approx 5$ corresponds to 2003.

**71.** $4x - y \geq -2$
$(3, 14)$ is a solution:
$$4(3) - (14) \geq -2$$
$$-2 \geq -2 \quad \text{True}$$
$(-2, 0)$ is not a solution:
$$4(-2) - (0) \geq -2$$
$$-8 \geq -2 \quad \text{False}$$
$(-4, -20)$ is a solution:
$$4(-4) - (-20) \geq -2$$
$$4 \geq -2 \quad \text{True}$$

**73.** $y \leq -3$

(1, 10) is not a solution:
$10 \leq -3$ False

(0, −4) is a solution:
$-4 \leq -3$ True

(4, 5) is not a solution:
$5 \leq -3$ False

**75.** $3x - 4y \leq 12$
$-4y \leq -3x + 12$
$y \geq 0.75x - 3$

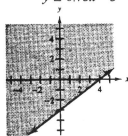

**77.** $y + 4 \leq 7$
$y \leq 3$

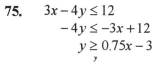

**79.** $5y - 2x > 7$
$5y > 2x + 7$ or
$y > 0.4x + 1.4$
$y + 3x < -2$
$y < -3x - 2$

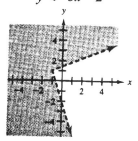

**81.** $y \leq 4$ and $\begin{array}{l} x + 2 \leq 0 \\ x \leq -2 \end{array}$

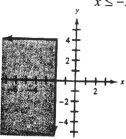

**83.** $|y| \geq 2$ is equivalent to $y \leq -2$ or $y \geq 2$

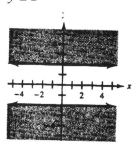

**85.** $|x - 2| - 2 < 5$
$|x - 2| < 7$
is equivalent to $\begin{array}{l} -7 < x - 2 < 7 \\ -5 < x < 9 \end{array}$.

**87.** Let $x$ = the number of grandstand tickets sold and $y$ = the number of bleacher tickets sold. An inequality that describes the number of each kind of ticket sold is $6x + 4y < 11{,}400$, $x \geq 0$, $y \geq 0$.

**89.** Let $x$ = the number of girls and $y$ = the number of boys. A compound inequality that describes the number of youngsters who can participate in the program is $x + y \leq 400$ and $x \geq 100$, $y \geq 0$.

## Chapter 4 Looking Back

1. (a) $2x - 3y = 6$

   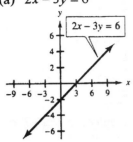

   (b) $2x - 3y \geq 6$

   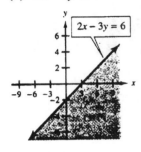

3. The graph of $x = 3$ is a vertical line containing $(3, 0)$. The graph of $y = 3$ is a horizontal line containing $(0, 3)$.

5. $f(x) = -5$, $g(x) = 3x - 2$

   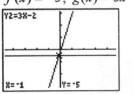

   The estimated point of intersection is $(-1, -5)$.

7.  $2y - (y - 5) + 17 = 0$
    $2y - y + 5 + 17 = 0$
    $y + 22 = 0$
    $y = -22$

9. $L_1: y = -x - 3 \quad m_1 = -1$
   $L_2: x + y = 4$
   $\phantom{L_2:} y = -x + 4 \quad m_2 = -1$
   Because $m_1 = m_2$, the lines are parallel.

11. Let $x = 4$, $y = \dfrac{1}{2}$, and $z = -3$.

    (a) $-3x + 4y + z = -3(4) + 4\left(\dfrac{1}{2}\right) + (-3)$
    $= -12 + 2 - 3 = -13$

    (b) $2x - 2y + 3z = 2(4) - 2\left(\dfrac{1}{2}\right) + 3(-3)$
    $= 8 - 1 - 9 = -2$

## Chapter 4 Test

1. $2x - y = -2$

   (a) Let $y = 0$:
   $2x - 0 = -2$
   $2x = -2 \quad (\underline{-1}, 0)$
   $x = -1$

   (b) Let $x = 0$:
   $2(0) - y = -2$
   $-y = -2 \quad (0, \underline{2})$
   $y = 2$

   (c) Let $x = -2$:
   $2(-2) - y = -2$
   $-4 - y = -2$
   $-y = 2 \quad (-2, \underline{-2})$
   $y = -2$

   (d) Let $y = 3$:
   $2x - 3 = -2$
   $2x = 1 \quad (\underline{0.5}, 3)$
   $x = 0.5$

3.  $y = -x - 2$, $m = -1$
    
    $x$-intercept: $(-2, 0)$
    $$0 = -x - 2$$
    $$x = -2$$
    
    $y$-intercept: $(0, -2)$
    $$y = -0 - 2$$
    $$y = -2$$

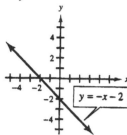

5.  $x - 2 = 3$
    $$x = 5$$
    slope: undefined; $x$-intercept: $(5, 0)$

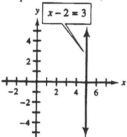

7.  $(2, -3)$ and $(-1, -5)$
    
    Slope: $m = \dfrac{-5 - (-3)}{-1 - 2} = \dfrac{-2}{-3} = \dfrac{2}{3}$
    
    $$y - y_1 = m(x - x_1)$$
    $$y - (-5) = \frac{2}{3}[x - (-1)]$$
    $$y + 5 = \frac{2}{3}(x + 1)$$
    $$y + 5 = \frac{2}{3}x + \frac{2}{3}$$
    $$y = \frac{2}{3}x - \frac{13}{3}$$

9.  A vertical line has the form $x = c$. Because the line contains $(2, 5)$, the constant is 2 and the equation is $x = 2$.

11. The slope of the line whose equation is $y = -\dfrac{1}{2}x + 3$ is $-\dfrac{1}{2}$. A line that is perpendicular to $y = -\dfrac{1}{2}x + 3$ has slope $m = 2$. If the line also contains the point $(1, 4)$, the equation of the line is
    $$y - y_1 = m(x - x_1)$$
    $$y - 4 = 2(x - 1)$$
    $$y - 4 = 2x - 2$$
    $$y = 2x + 2$$

13. $2x - y \leq 5$
    $$-y \leq -2x + 5$$
    $$y \geq 2x - 5$$

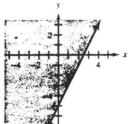

15. $y \geq 2x + 1$ and $\begin{array}{l} x + y \geq 3 \\ y \geq -x + 3 \end{array}$

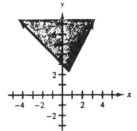

17. $|x + y| \leq 3$ is equivalent to $-3 \leq x + y \leq 3$

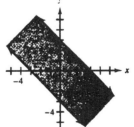

③ The table shows the variable assignment.
④ $10x + 5(x+9) = 225$
⑤ $10x + 5x + 45 = 225$
$15x + 45 = 225$
$15x + 45 - 45 = 225 - 45$
$15x = 180$
$\dfrac{15x}{15} = \dfrac{180}{15}$
$x = 12$
⑥ There are 12 dimes and $12 + 9 = 21$ nickels.
⑦ Check: The value of the coins is $10(12) + 5(21) = 120 + 105 = 225$.

**26.** ① The goal is to determine the number of quarters.
②

|  | Number of coins | Value of coin | Value of coins in cents |
|---|---|---|---|
| Dimes | $x$ | 10 | $10x$ |
| Quarters | $x + 4$ | 25 | $25(x+4)$ |
| Total |  |  | 520 |

③ The table shows the variable assignment.
④ $10x + 25(x+4) = 520$
⑤ $10x + 25x + 100 = 520$
$35x + 100 = 520$
$35x + 100 - 100 = 520 - 100$
$35x = 420$
$\dfrac{35x}{35} = \dfrac{420}{35}$
$x = 12$
⑥ There are 12 dimes and $12 + 4 = 16$ quarters.
⑦ Check: The value of the coins is $10(12) + 25(16) = 520$.

**27.** ① The goal is to determine the number of dimes, but to check the results we will also find the number of quarters.
②

|  | Number of coins | Value of coin | Value of coins in cents |
|---|---|---|---|
| Quarters | $x$ | 25 | $25x$ |
| Dimes | $32 - x$ | 10 | $10(32-x)$ |
| Total |  |  | 515 |

③ The table shows the variable assignment.
④ $25x + 10(32 - x) = 515$
⑤ $25x + 320 - 10x = 515$
$15x + 320 = 515$
$15x + 320 - 320 = 515 - 320$
$15x = 195$
$\dfrac{15x}{15} = \dfrac{195}{15}$
$x = 13$
⑥ There are 13 quarters and $32 - 13 = 19$ dimes.
⑦ Check: The value of the coins is $25(13) + 10(19) = 515$.

**28.** ① The goal is to determine the number of nickels.
②

|  | Number of coins | Value of coin | Value of coins in cents |
|---|---|---|---|
| Dimes | $x$ | 10 | $10x$ |
| Nickels | $x + 8$ | 5 | $5(x+8)$ |
| Total |  |  | 250 |

③ The table shows the variable assignment.
④ $10x + 5(x+8) = 250$
⑤ $10x + 5x + 40 = 250$
$15x + 40 = 250$
$15x + 40 - 40 = 250 - 40$
$15x = 210$
$\dfrac{15x}{15} = \dfrac{210}{15}$
$x = 14$
⑥ There are 14 dimes and $14 + 8 = 22$ nickels.
⑦ Check: The value of the coins is $10(14) + 5(22) = 250$.

# Chapter 5

# Systems of Linear Equations

## Section 5.1 Graphing and Substitution Methods

1. Substitute the numbers into each equation. If both equations are satisfied, the ordered pair is a solution.

3. $2x + y = 1$
   $3x - 2y = 12$
   $(0, 1)$:
   $\quad 3(0) - 2(1) = 12$
   $\quad\quad 0 - 2 = 12$
   $\quad\quad\quad -2 = 12 \quad$ False
   $(2, -3)$:
   $\quad 2(2) + (-3) = 1$
   $\quad\quad 4 - 3 = 1$
   $\quad\quad\quad 1 = 1 \quad$ True
   $\quad 3(2) - 2(-3) = 12$
   $\quad\quad 6 + 6 = 12$
   $\quad\quad\quad 12 = 12 \quad$ True
   $(4, 0)$:
   $\quad 2(4) + (0) = 1$
   $\quad\quad 8 + 0 = 1$
   $\quad\quad\quad 8 = 1 \quad$ False
   Because it makes both equations in the system true, only the point $(2, -3)$ is a solution of the system.

5. $2x + 7y = 4$
   $14y + 4x = 8$
   $(1, 1)$:
   $\quad 2(1) + 7(1) = 4$
   $\quad\quad 2 + 7 = 4$
   $\quad\quad\quad 9 = 4 \quad$ False

   $\left(-2, \dfrac{8}{7}\right)$:
   $\quad 2(-2) + 7\left(\dfrac{8}{7}\right) = 4$
   $\quad\quad -4 + 8 = 4$
   $\quad\quad\quad 4 = 4 \quad$ True
   $\quad 14\left(\dfrac{8}{7}\right) + 4(-2) = 8$
   $\quad\quad 16 - 8 = 8$
   $\quad\quad\quad 8 = 8 \quad$ True
   $(2, 0)$:
   $\quad 2(2) + 7(0) = 4$
   $\quad\quad 4 + 0 = 4$
   $\quad\quad\quad 4 = 4 \quad$ True
   $\quad 14(0) + 4(2) = 8$
   $\quad\quad 0 + 8 = 8$
   $\quad\quad\quad 8 = 8 \quad$ True
   Because they make both equations in the system true, the points $\left(-2, \dfrac{8}{7}\right)$ and $(2, 0)$ are solutions of the system.

7. To find the values of $a$ and $b$ for which $(3, -1)$ is a solution of the given system, substitute $(3, -1)$ into each equation and solve for $a$ and $b$.
   $\quad ax - 2y = 5 \quad\quad\quad x + by = 1$
   $\quad a(3) - 2(-1) = 5 \quad\quad 3 + b(-1) = 1$
   $\quad\quad 3a + 2 = 5 \quad\quad\quad\quad 3 - b = 1$
   $\quad\quad\quad 3a = 3 \quad\quad\quad\quad\quad -b = -2$
   $\quad\quad\quad\quad a = 1 \quad\quad\quad\quad\quad\quad b = 2$

9. To find the values of $a$ and $c$ for which $(3, -1)$ is a solution of the given system, substitute $(3, -1)$ into each equation and solve for $a$ and $c$.

$$ax + 3y = 9$$
$$a(3) + 3(-1) = 9$$
$$3a - 3 = 9$$
$$3a = 12$$
$$a = 4$$

$$2x - y = c$$
$$2(3) - (-1) = c$$
$$6 + 1 = c$$
$$7 = c$$

11. Two linear equations in two variables cannot have exactly two solutions because two lines cannot intersect at exactly two points.

13. $y = 3x - 4$
    $y = 3x + 5$

    Because the equations in this system have the same slope but different intercepts, the lines are parallel. Therefore, there are no solutions.

15. $y + 2x = 6$
    $y - 2x = 8$

    Because the equations in this system have different slopes, the lines will intersect. Therefore, there is one solution.

17. $y + x = 5$
    $y = -x + 5$

    Because the slopes and $y$-intercepts are equal, the lines coincide. There are infinitely many solutions.

19. (a) (i) The graphs do not coincide.
        (ii) The graphs coincide.
    (b) Yes, the lines could be parallel.

21. From the given graph, it appears that $(0, 6)$ is the solution.

23. From the given graph, it appears that $(0, 0)$ is the solution.

25. $y = 2x - 2$
    $y = -3x + 13$

    *Graphing Method:*
    Graph $y_1 = 2x - 2$ and $y_2 = -3x + 13$.

The estimated point of intersection is $(3, 4)$.

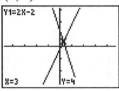

*Substitution Method:*
Because the first equation is already solved for $y$, substitute $2x - 2$ for $y$ in the second equation.

$$y = -3x + 13$$
$$2x - 2 = -3x + 13$$
$$5x - 2 = 13$$
$$5x = 15$$
$$x = 3$$

Now determine the value of $y$ by substituting 3 for $x$ in $y = 2x - 2$.

$$y = 2(3) - 2 = 4$$

The solution is $(3, 4)$.

27. $y = 0.6x - 10$
    $5y - 3x = 4$

    *Graphing Method:*
    Graph $y_1 = 0.6x - 10$ and $y_2 = \dfrac{3}{5}x + \dfrac{4}{5}$. The lines appear to be parallel. The system has no solution.

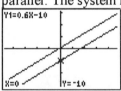

*Substitution Method:*
Because the first equation is already solved for $y$, substitute $0.6x - 10$ for $y$ in the second equation.

$$5y - 3x = 4$$
$$5(0.6x - 10) - 3x = 4$$
$$3x - 50 - 3x = 4$$
$$-50 = 4 \quad \text{False}$$

Because the resulting equation is false, it has no solution. The system has no solution.

29. $2x + y = 8$
$y + 2 = 0$

*Graphing Method:*
Graph $y_1 = 8 - 2x$ and $y_2 = -2$. The estimated point of intersection is $(5, -2)$.

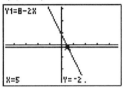

*Substitution Method:*
Solving the second equation for $y$, we obtain $y = -2$. Substitute $-2$ for $y$ in the first equation.
$$2x + y = 8$$
$$2x + (-2) = 8$$
$$2x = 10$$
$$x = 5$$
Now determine the value of $y$ by substituting 5 for $x$ in $2x + y = 8$.
$$2(5) + y = 8$$
$$10 + y = 8$$
$$y = -2$$
The solution is $(5, -2)$.

31. $x - y + 6 = 0$
$2x - y - 5 = 0$

*Graphing Method:*
Graph $y_1 = x + 6$ and $y_2 = 2x - 5$. The estimated point of intersection is $(11, 17)$.

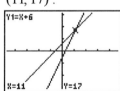

*Substitution Method:*
Solving the second equation for $y$, we obtain $y = 2x - 5$. Substitute $2x - 5$ for $y$ in the first equation.
$$x - y + 6 = 0$$
$$x - (2x - 5) + 6 = 0$$
$$x - 2x + 5 + 6 = 0$$
$$-x + 11 = 0$$
$$-x = -11$$
$$x = 11$$

Now determine the value of $y$ by substituting 11 for $x$ in $y = 2x - 5$.
$$y = 2(11) - 5 = 22 - 5 = 17$$
The solution is $(11, 17)$.

33. $y = -2(4 - x)$
$y + 8 = 2x$

*Graphing Method:*
Graph $y_1 = -2(4 - x)$ and $y_2 = 2x - 8$. The graphs appear to coincide.

*Substitution Method:*
Because the first equation is already solved for $y$, substitute $-2(4 - x)$ for $y$ in the second equation.
$$y + 8 = 2x$$
$$-2(4 - x) + 8 = 2x$$
$$-8 + 2x + 8 = 2x$$
$$2x = 2x$$
$$0 = 0 \quad \text{True}$$

The resulting equation is an identity that has infinitely many solutions. Thus, the solution set of this system contains the infinitely many solutions of the equation $y = 2x - 8$.

35. $\dfrac{1}{3}x + \dfrac{2}{9}y = 5$
$\dfrac{1}{9}x - \dfrac{2}{3}y = 15$

*Graphing Method:*
Graph $y_1 = -\dfrac{3}{2}x + \dfrac{45}{2}$ and $y_2 = \dfrac{1}{6}x - \dfrac{45}{2}$. The estimated point of intersection is $(27, -18)$.

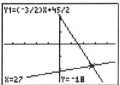

*Substitution Method:*
Solving the first equation for $x$, we obtain

$$\frac{1}{3}x + \frac{2}{9}y = 5$$
$$x + \frac{2}{3}y = 15$$
$$x = -\frac{2}{3}y + 15$$

Substitute $-\frac{2}{3}y + 15$ for $x$ in the second equation.
$$\frac{1}{9}x - \frac{2}{3}y = 15$$
$$\frac{1}{9}\left(-\frac{2}{3}y + 15\right) - \frac{2}{3}y = 15$$
$$\left(-\frac{2}{3}y + 15\right) - 6y = 135$$
$$-\frac{20}{3}y + 15 = 135$$
$$-\frac{20}{3}y = 120$$
$$y = -18$$

Now determine the value of $x$ by substituting $-18$ for $y$ in $x = -\frac{2}{3}y + 15$.
$$x = -\frac{2}{3}(-18) + 15 = 27$$

The solution is $(27, -18)$.

**37.**

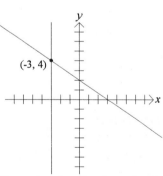

From the sketch of $x + 3 = 0$ and $2x + 3y = 6$, it appears that the solution is $(-3, 4)$.

**39.**

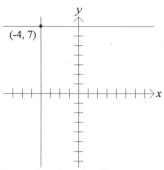

From the sketch of $x = -4$ and $y = 7$, it appears that the solution is $(-4, 7)$.

**41.** (a) Yes, either equation can be solved for either variable.
(b) Solving the second equation for $y$ is easier.

**43.** $x - y - 5 = 0$
$2x - y + 17 = 0$

Solving the first equation for $x$, we obtain $x = y + 5$. Substitute $y + 5$ for $x$ in the second equation.
$$2x - y + 17 = 0$$
$$2(y + 5) - y + 17 = 0$$
$$2y + 10 - y + 17 = 0$$
$$y + 27 = 0$$
$$y = -27$$

Now determine the value of $x$ by substituting $-27$ for $y$ in $x = y + 5$.
$$x = -27 + 5 = -22$$
The solution is $(-22, -27)$.

**45.** $y + 3x - 4 = 0$
$12 = 9x + 3y$

Solving the first equation for $y$, we obtain $y = -3x + 4$. Substitute $-3x + 4$ for $y$ in the second equation.
$$12 = 9x + 3y$$
$$12 = 9x + 3(-3x + 4)$$
$$12 = 9x - 9x + 12$$
$$12 = 12 \quad \text{True}$$

The resulting equation is an identity that has infinitely many solutions. Thus, the system of equations has infinitely many solutions. The equations are dependent.

**47.** $x = 3y + 9$
$3y - x = 6$

Because the first equation is already solved for $x$, substitute $3y + 9$ for $x$ in the second equation.
$$3y - x = 6$$
$$3y - (3y + 9) = 6$$
$$3y - 3y - 9 = 6$$
$$-9 = 6 \quad \text{False}$$

The resulting equation is false. Therefore, the system of equations has no solution and is inconsistent.

**49.** $y = 2x - 5$
$4x - 2y - 10 = 0$

Because the first equation is already solved for $y$, substitute $2x - 5$ for $y$ in the second equation.
$$4x - 2y - 10 = 0$$
$$4x - 2(2x - 5) - 10 = 0$$
$$4x - 4x + 10 - 10 = 0$$
$$0 = 0 \quad \text{True}$$

The resulting equation is an identity that has infinitely many solutions. Thus, the system of equations has infinitely many solutions. The equations are dependent.

**51.** $y = 2x - 7$
$3x + 2y = 0$

Because the first equation is already solved for $y$, substitute $2x - 7$ for $y$ in the second equation.
$$3x + 2y = 0$$
$$3x + 2(2x - 7) = 0$$
$$3x + 4x - 14 = 0$$
$$7x - 14 = 0$$
$$7x = 14$$
$$x = 2$$

Now determine the value of $y$ by substituting 2 for $x$ in $y = 2x - 7$.
$$y = 2(2) - 7 = -3$$
The solution is $(2, -3)$.

**53.** $6x = 5y$
$12x + 7 = 10y$

Solving the first equation for $x$, we obtain $x = \frac{5}{6}y$. Substitute $\frac{5}{6}y$ for $x$ in the second equation.
$$12x + 7 = 10y$$
$$12\left(\frac{5}{6}y\right) + 7 = 10y$$
$$10y + 7 = 10y$$
$$7 = 0 \quad \text{False}$$

The resulting equation is false. Therefore, the system of equations has no solution and is inconsistent.

**55.** $y = -2x + 5$
$21x + 10y = 50$

*Graphing Method:*
Graph $y_1 = -2x + 5$ and $y_2 = -\frac{21}{10}x + 5$. From the graph, the lines appear to coincide. If that is the case, there would be infinitely many solutions.

*Substitution Method:*
Because the first equation is already solved for $y$, substitute $-2x + 5$ for $y$ in the second equation.
$$21x + 10y = 50$$
$$21x + 10(-2x + 5) = 50$$
$$21x - 20x + 50 = 50$$
$$x + 50 = 50$$
$$x = 0$$

Now determine the value of $y$ by substituting 0 for $x$ in $y = -2x + 5$.
$$y = -2(0) + 5 = 5$$
The solution is $(0, 5)$.

**57.** $x = y$
$11x - 10y = 80$

*Graphing Method:*
Graph $y_1 = x$ and $y_2 = \frac{11}{10}x - 8$. There appears to be one point of intersection: at $(80, 80)$.

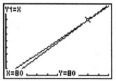

*Substitution Method:*

Because the first equation is already solved for $x$, substitute $y$ for $x$ in the second equation.

$$11x - 10y = 80$$
$$11y - 10y = 80$$
$$y = 80$$

Now determine the value of $x$ by substituting 80 for $y$ in $x = y$.

$$x = 80$$

The solution is $(80, 80)$.

**59.** $y = \dfrac{2}{3}x + 1 \quad m_1 = \dfrac{2}{3} \quad b_1 = 1$

$kx + 3y = 3$
$3y = -kx + 3$
$y = -\dfrac{k}{3}x + 1 \quad m_2 = -\dfrac{k}{3} \quad b_2 = 1$

For the equations to be dependent, the slopes must be equal and the $y$-intercepts must be equal.

$$-\dfrac{k}{3} = \dfrac{2}{3}$$
$$k = -2$$

The intercepts are already the same, so if $k = -2$, the equations are dependent.

**61.** $y = 3x + k \quad m_1 = 3 \quad b_1 = k$

$y + (2-k)x = 5 \quad m_2 = -(2-k) \quad b_2 = 5$

For the equations to be dependent, the slopes must be equal and the $y$-intercepts must be equal.

$$m_1 = m_2$$
$$3 = -(2-k)$$
$$3 = -2 + k$$
$$5 = k$$

With $k = 5$, the intercepts are the same. So if $k = 5$, the equations are dependent.

**63.** $kx + y = 2k \quad m_1 = -k \quad b_1 = 2k$

$y = 2x + 5 \quad m_2 = 2 \quad b_2 = 5$

For the system to be inconsistent, the slopes must be equal and the $y$-intercepts must not be the same.

$$m_1 = m_2$$
$$-k = 2$$
$$k = -2$$

With $k = -2$, the intercepts are not the same: $(0, -4)$ and $(0, 5)$. So, if $k = -2$, the system is inconsistent.

**65.** $kx + 3y = 12 \quad m_1 = -\dfrac{k}{3} \quad b_1 = 4$

$y = \dfrac{4}{3}x + (k+1) \quad m_2 = \dfrac{4}{3} \quad b_2 = k+1$

For the system to be inconsistent, the slopes must be equal and the $y$-intercepts must not be the same.

$$m_1 = m_2$$
$$-\dfrac{k}{3} = \dfrac{4}{3}$$
$$k = -4$$

With $k = -4$, the intercepts are not the same: $(0, 4)$ and $(0, -3)$. So, if $k = -4$, the system is inconsistent.

**67.** Let $x =$ one number and $y =$ the other number. The system is:

$$x + y = 16$$
$$x - y = 4$$

Solving the second equation for $x$, we obtain $x = y + 4$. Substitute $y + 4$ for $x$ in the first equation.

$$x + y = 16$$
$$(y + 4) + y = 16$$
$$2y + 4 = 16$$
$$2y = 12$$
$$y = 6$$

Now determine the value of $x$ by substituting 6 for $y$ in $x = y + 4$.

$$x = 6 + 4 = 10$$

The two numbers are 6 and 10.

**69.** Let $x =$ smaller number and $y =$ the larger number. The system is:

$$y = x + 8$$
$$\dfrac{3}{4}y = x + 3$$

Because the first equation is already solved for $y$, substitute $x + 8$ for $y$ in the second equation.

Section 5.1   Graphing and Substitution Methods

$$\frac{3}{4}y = x+3$$
$$\frac{3}{4}(x+8) = x+3$$
$$\frac{3}{4}x+6 = x+3$$
$$-\frac{1}{4}x+6 = 3$$
$$-\frac{1}{4}x = -3$$
$$x = 12$$

Now determine the value of $y$ by substituting 12 for $x$ in $y = x+8$.
$$y = 12+8 = 20$$
The two numbers are 12 and 20.

**71.** Let $x$ = the larger angle and $y$ = the smaller angle.
$$2x - 5y = 19$$
$$x + y = 90$$

Solving the second equation for $x$, we obtain $x = 90-y$. Substitute $90-y$ for $x$ in the first equation.
$$2x - 5y = 19$$
$$2(90-y) - 5y = 19$$
$$180 - 2y - 5y = 19$$
$$180 - 7y = 19$$
$$-7y = -161$$
$$y = 23$$

Now determine the value of $x$ by substituting 23 for $y$ in $x = 90-y$.
$$x = 90 - 23 = 67$$
The two angles are 23° and 67°.

**73.** Let $x$ = the larger angle and $y$ = the smaller angle.
$$x = 2y - 24$$
$$x + y = 180$$

Because the first equation is already solved for $x$, substitute $2y-24$ for $x$ in the second equation.
$$x + y = 180$$
$$2y - 24 + y = 180$$
$$3y - 24 = 180$$
$$3y = 204$$
$$y = 68$$

Now determine the value of $x$ by substituting 68 for $y$ in $x = 2y - 24$.
$$x = 2(68) - 24 = 112$$
The two angles are 68° and 112°.

**75.** Let $W$ = the width and $L$ = the length.
$$L = 2W + 7$$
$$2W + 2L = 74$$
Because the first equation is already solved for $L$, substitute $2W+7$ for $L$ in the second equation.
$$2W + 2L = 74$$
$$2W + 2(2W+7) = 74$$
$$2W + 4W + 14 = 74$$
$$6W + 14 = 74$$
$$6W = 60$$
$$W = 10$$
Now determine the value of $L$ by substituting 10 for $W$ in $L = 2W + 7$.
$$L = 2(10) + 7 = 27$$
The length is 27 feet and the width is 10 feet.

**77.** Let $W$ = the width and $L$ = the length.
$$W = \frac{1}{3}L$$
$$2W + 2L = 80$$
Because the first equation is already solved for $W$, substitute $\frac{1}{3}L$ for $W$ in the second equation.
$$2W + 2L = 80$$
$$2\left(\frac{1}{3}L\right) + 2L = 80$$
$$\frac{2}{3}L + 2L = 80$$
$$\frac{8}{3}L = 80$$
$$L = 30$$
Now determine the value of $W$ by substituting 30 for $L$ in $W = \frac{1}{3}L$.
$$W = \frac{1}{3}(30) = 10$$
The length is 30 inches and the width is 10 inches.

**79.** Let $x$ = the number of votes for Mills and $y$ = the number of votes for Lopez.
$$y = x + 200$$
$$y + x = 0.30(24,000)$$
Because the first equation is already solved for $y$, substitute $x + 200$ for $y$ in the second equation.
$$y + x = 0.30(24,000)$$
$$y + x = 7200$$
$$x + 200 + x = 7200$$
$$2x + 200 = 7200$$
$$2x = 7000$$
$$x = 3500$$
Mills received 3500 votes.

**81.** Let $x$ = the number of girls and $y$ = the number of boys.
$$x + y = 18$$
$$\frac{1}{2}x + \frac{1}{3}y = 7$$
Solving the first equation for $y$, we obtain $y = 18 - x$. Substituting for $y$ in the second equation, we obtain
$$\frac{1}{2}x + \frac{1}{3}y = 7$$
$$3x + 2y = 42$$
$$3x + 2(18 - x) = 42$$
$$3x + 36 - 2x = 42$$
$$x + 36 = 42$$
$$x = 6$$
There are 6 girls on the team.

**83.** $y = -0.57x + 38.29$
$y = -0.01x + 9.10$

(a) From the graph, it appears that the solution is approximately (52, 8.6).

(b) Because the first equation is already solved for $y$, use it to substitute for $y$ in the second equation.
$$-0.57x + 38.29 = -0.01x + 9.10$$
$$-0.56x + 38.29 = 9.10$$
$$-0.56x = -29.19$$
$$x = 52.125$$

Substitute $x = 52.125$ in the second equation to find the value of $y$:
$$y = -0.01(52.125) + 9.10 = 8.57875$$
The solution is (52.125, 8.57875).

(c) In 2002 ($x \approx 52$), the percent of men and women over age 65 in the work force will be equal (about 8.6%).

**85.** Using pairs of the form (number of years since 1965, percent of Ph.D.s awarded to women), two points are (2, 13) and (32, 41). The slope is
$$m = \frac{41 - 13}{32 - 2} = \frac{28}{30} = 0.9\overline{3}.$$
The linear equation is:
$$y - y_1 = m(x - x_1)$$
$$y - 13 = 0.9\overline{3}(x - 2)$$
$$y - 13 = 0.9\overline{3}x - 1.8\overline{6}$$
$$y = 0.9\overline{3}x + 11.1\overline{3}$$
The linear function that models the percent of Ph.D.s awarded to women is $W(x) = 0.9\overline{3}x + 11.1\overline{3}$.

**87.** At the point of intersection, the percent of Ph.D.s awarded to men and women is the same. Therefore, the $y$-coordinate would be 50 (50%).

**89.** $x^2 + y = 9$
$2x^2 - y = 3$

Graph $y_1 = 9 - x^2$ and $y_2 = 2x^2 - 3$. Using the features of the graphing calculator, we find that the points of intersection are (−2, 5) and (2, 5).

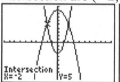 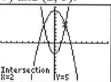

**91.** $x = y + 12$
$y - x = 9$
$2x + y = 12$
$y + 6 = -2x$

Graph $y_1 = x - 12$, $y_2 = x + 9$, $y_3 = -2x + 12$, and $y_4 = -2x - 6$.

Section 5.2   The Addition Method

Using a graphing calculator, we can identify the vertices as $(2, -10)$, $(8, -4)$, $(-5, 4)$, and $(1, 10)$. Because the opposite sides are parallel, the figure is a parallelogram.

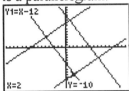

**93.**  $y - x = 5$
$3y = x + 9$
$x + y + 1 = 0$

Graph $y_1 = x + 5$, $y_2 = \dfrac{1}{3}x + 3$, and $y_3 = -x - 1$. As can be seen from the graph, the lines intersect at the point $(-3, 2)$.

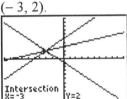

**95.**  $x - y = k$
$3x - 3y = 2$

Solving the first equation for $x$, we obtain $x = y + k$. Using this to substitute for $x$ in the second equation, we obtain

$$3x - 3y = 2$$
$$3(y + k) - 3y = 2$$
$$3y + 3k - 3y = 2$$
$$3k = 2$$

Because it is impossible to solve this equation in terms of one of the variables $x$ or $y$, the equation does not have exactly one solution for any value of $k$. There are either infinitely many solutions or no solution.

## Section 5.2  The Addition Method

**1.**  (a) Multiply the second equation by $-3$ and add the equations.
  (b) Multiply the first equation by 4 and add the equations.

**3.**  Eliminate $y$ by adding the equations.
$$\begin{array}{rcr} x + y &=& 10 \\ x - y &=& 2 \\ \hline 2x &=& 12 \\ x &=& 6 \end{array}$$

Replace $x$ with 6 in the first equation to find $y$.
$x + y = 10$
$6 + y = 10$
$y = 4$
The solution is $(6, 4)$.

**5.**  Eliminate $x$ by adding the equations.

**7.**  $3x + 2y = 4$  $\rightarrow$  $3x + 2y = 4$
$4y = 3x + 26$  $\rightarrow$  $-3x + 4y = 26$

Add the equations to eliminate $x$.
$$\begin{array}{rcr} 3x + 2y &=& 4 \\ -3x + 4y &=& 26 \\ \hline 6y &=& 30 \\ y &=& 5 \end{array}$$

Replace $y$ with 5 in the first equation to find $x$.
$$3x + 2y = 4$$
$$3x + 2(5) = 4$$
$$3x + 10 = 4$$
$$3x = -6$$
$$x = -2$$
The solution is $(-2, 5)$.

9.  $4x = 3y + 6 \quad \rightarrow \quad 4x - 3y = 6$
    $12 = 5x + 3y \quad \rightarrow \quad 5x + 3y = 12$
    Add to eliminate $y$.
    $$\begin{array}{rcl} 4x - 3y &=& 6 \\ 5x + 3y &=& 12 \\ \hline 9x &=& 18 \\ x &=& 2 \end{array}$$
    Replace $x$ with 2 in the first equation to find $y$.
    $$4x = 3y + 6$$
    $$4(2) = 3y + 6$$
    $$8 = 3y + 6$$
    $$2 = 3y$$
    $$\frac{2}{3} = y$$
    The solution is $\left(2, \frac{2}{3}\right)$.

11. No, the result of $0 = 0$ means that there are infinitely many solutions.

13. $2x + 3y + 1 = 0 \quad \rightarrow \quad 2x + 3y = -1$
    $5x + 3y = 29 \quad \rightarrow \quad 5x + 3y = 29$
    Multiply the first equation by $-1$ and then add to eliminate $y$.
    $$\begin{array}{rcl} -2x - 3y &=& 1 \\ 5x + 3y &=& 29 \\ \hline 3x &=& 30 \\ x &=& 10 \end{array}$$
    Replace $x$ with 10 in the second equation to find $y$.
    $$5x + 3y = 29$$
    $$5(10) + 3y = 29$$
    $$50 + 3y = 29$$
    $$3y = -21$$
    $$y = -7$$
    The solution is $(10, -7)$.

15. $3x - 2y = 21$
    $5x + 4y = 13$
    Multiply the first equation by 2 and then add to eliminate $y$.
    $$\begin{array}{rcl} 6x - 4y &=& 42 \\ 5x + 4y &=& 13 \\ \hline 11x &=& 55 \\ x &=& 5 \end{array}$$
    Replace $x$ with 5 in the first equation to find $y$.
    $$3x - 2y = 21$$
    $$3(5) - 2y = 21$$
    $$15 - 2y = 21$$
    $$-2y = 6$$
    $$y = -3$$
    The solution is $(5, -3)$.

17. $2x + 3y = 3 \quad \rightarrow \quad 2x + 3y = 3$
    $4x = 6y - 2 \quad \rightarrow \quad 4x - 6y = -2$
    Multiply the first equation by $-2$ and then add to eliminate $x$.
    $$\begin{array}{rcl} -4x - 6y &=& -6 \\ 4x - 6y &=& -2 \\ \hline -12y &=& -8 \\ y &=& \frac{2}{3} \end{array}$$
    Replace $y$ with $\frac{2}{3}$ in the first equation to find $x$.
    $$2x + 3y = 3$$
    $$2x + 3\left(\frac{2}{3}\right) = 3$$
    $$2x + 2 = 3$$
    $$2x = 1$$
    $$x = \frac{1}{2}$$
    The solution is $\left(\frac{1}{2}, \frac{2}{3}\right)$.

19. $6x = 5y + 10 \quad \rightarrow \quad 6x - 5y = 10$
    $3x + 2y = 23 \quad \rightarrow \quad 3x + 2y = 23$
    Multiply the second equation by $-2$ and then add to eliminate $x$.
    $$\begin{array}{rcl} 6x - 5y &=& 10 \\ -6x - 4y &=& -46 \\ \hline -9y &=& -36 \\ y &=& 4 \end{array}$$

Replace $y$ with 4 in the first equation to find $x$.
$$6x = 5y + 10$$
$$6x = 5(4) + 10$$
$$6x = 20 + 10$$
$$6x = 30$$
$$x = 5$$
The solution is $(5, 4)$.

21. The lines intersect at $(3, 2)$. Many other pairs of lines also intersect at this point.

23. $6x + 11y = 17$
$4x - 5y = -1$

Multiply the first equation by 2 and the second equation by $-3$. Then add to eliminate $x$.
$$\begin{aligned} 12x + 22y &= 34 \\ -12x + 15y &= 3 \\ \hline 37y &= 37 \\ y &= 1 \end{aligned}$$

Replace $y$ with 1 in $4x - 5y = -1$ to find $x$.
$$4x - 5(1) = -1$$
$$4x - 5 = -1$$
$$4x = 4$$
$$x = 1$$
The solution is $(1, 1)$.

25. $-6x + 5y = 10$
$5x + 4y = 8$

Multiply the first equation by 5 and the second equation by 6. Then add to eliminate $x$.
$$\begin{aligned} -30x + 25y &= 50 \\ 30x + 24y &= 48 \\ \hline 49y &= 98 \\ y &= 2 \end{aligned}$$

Replace $y$ with 2 in $5x + 4y = 8$ to find $x$.
$$5x + 4(2) = 8$$
$$5x + 8 = 8$$
$$5x = 0$$
$$x = 0$$
The solution is $(0, 2)$.

27. $2x - 7y + 11 = 0 \quad \rightarrow \quad 2x - 7y = -11$
$7x + 4y + 10 = 0 \quad \rightarrow \quad 7x + 4y = -10$

Multiply the first equation by 4 and the second equation by 7. Then add to eliminate $y$.
$$\begin{aligned} 8x - 28y &= -44 \\ 49x + 28y &= -70 \\ \hline 57x &= -114 \\ x &= -2 \end{aligned}$$

Replace $x$ with $-2$ in $2x - 7y = -11$ to find $y$.
$$2(-2) - 7y = -11$$
$$-4 - 7y = -11$$
$$-7y = -7$$
$$y = 1$$
The solution is $(-2, 1)$.

29. $5y - 2x = 5 \quad \rightarrow \quad -2x + 5y = 5$
$5x + 2y = 2 \quad \rightarrow \quad 5x + 2y = 2$

Multiply the first equation by 5 and the second equation by 2. Then add to eliminate $x$.
$$\begin{aligned} -10x + 25y &= 25 \\ 10x + 4y &= 4 \\ \hline 29y &= 29 \\ y &= 1 \end{aligned}$$

Replace $y$ with 1 in $5x + 2y = 2$ to find $x$.
$$5x + 2(1) = 2$$
$$5x + 2 = 2$$
$$5x = 0$$
$$x = 0$$
The solution is $(0, 1)$.

31. You can tell when the equations of a system are dependent because both variables are eliminated and the resulting equation is true.

33. $2x + 3y + 18 = 0 \quad \rightarrow \quad 2x + 3y = -18$
$5y = 6x - 2 \quad \rightarrow \quad -6x + 5y = -2$

Multiply the first equation by 3 and then add to eliminate $x$.
$$\begin{aligned} 6x + 9y &= -54 \\ -6x + 5y &= -2 \\ \hline 14y &= -56 \\ y &= -4 \end{aligned}$$

Replace $y$ with $-4$ in $2x + 3y = -18$ to find $x$.

$$2x + 3(-4) = -18$$
$$2x - 12 = -18$$
$$2x = -6$$
$$x = -3$$
The solution is $(-3, -4)$.

**35.**  $3x - 2y = 8 \quad \rightarrow \quad 3x - 2y = 8$
$6x = 4y + 17 \quad \rightarrow \quad 6x - 4y = 17$

Multiply the first equation by $-2$ and then add to eliminate $x$.
$$-6x + 4y = -16$$
$$\underline{\phantom{-}6x - 4y = \phantom{-}17}$$
$$0 = 1 \quad \text{False}$$
The system has no solution. The system is inconsistent.

**37.**  $3x - y = 7 \quad \rightarrow \quad 3x - y = 7$
$2y = 6x - 14 \quad \rightarrow \quad -6x + 2y = -14$

Multiply the first equation by 2 and then add to eliminate $x$.
$$6x - 2y = 14$$
$$\underline{-6x + 2y = -14}$$
$$0 = 0 \quad \text{True}$$
The system has infinitely many solutions. The equations are dependent.

**39.**  $2 + 7y = 5x \quad \rightarrow \quad -5x + 7y = -2$
$15x = 21y - 6 \quad \rightarrow \quad 15x - 21y = -6$

Multiply the first equation by 3 and then add to eliminate $x$.
$$-15x + 21y = -6$$
$$\underline{\phantom{-}15x - 21y = -6}$$
$$0 = -12 \quad \text{False}$$
The system has no solution. The system is inconsistent.

**41.**  $6y + 7x = 16 \quad \rightarrow \quad 7x + 6y = 16$
$3x = 2y + 16 \quad \rightarrow \quad 3x - 2y = 16$

Multiply the second equation by 3 and then add to eliminate $y$.
$$7x + 6y = 16$$
$$\underline{9x - 6y = 48}$$
$$16x \phantom{- 6y} = 64$$
$$x \phantom{- 6y} = 4$$
Replace $x$ with 4 in $3x = 2y + 16$ to find $y$.

$$3(4) = 2y + 16$$
$$12 = 2y + 16$$
$$-4 = 2y$$
$$-2 = y$$
The solution is $(4, -2)$.

**43.**  $2y = 3x - 5 \quad \rightarrow \quad -3x + 2y = -5$
$-12x = -20 - 8y \quad \rightarrow \quad -12x + 8y = -20$

Multiply the first equation by $-4$ and then add to eliminate $x$.
$$12x - 8y = 20$$
$$\underline{-12x + 8y = -20}$$
$$0 = 0 \quad \text{True}$$
The system has infinitely many solutions. The equations are dependent.

**45.**  Simplify first:
$$4(x - 2) - 5(y - 3) = 14$$
$$3(x + 4) - 2(y + 5) = 13$$

$$4x - 8 - 5y + 15 = 14$$
$$3x + 12 - 2y - 10 = 13$$

$$4x - 5y = 7$$
$$3x - 2y = 11$$

Multiply the first equation by $-3$ and the second equation by 4. Then add to eliminate $x$.
$$-12x + 15y = -21$$
$$\underline{\phantom{-}12x - \phantom{1}8y = \phantom{-}44}$$
$$7y = 23$$
$$y = \frac{23}{7}$$

Replace $y$ with $\frac{23}{7}$ in $3x - 2y = 11$ to find $x$.
$$3x - 2\left(\frac{23}{7}\right) = 11$$
$$3x - \frac{46}{7} = 11$$
$$3x = \frac{123}{7}$$
$$x = \frac{41}{7}$$
The solution is $\left(\frac{41}{7}, \frac{23}{7}\right)$.

Section 5.2  The Addition Method

**47.** Begin by multiplying the first equation by its LCD of 10 and the second equation by its LCD of 3.
$$\frac{1}{2}x + \frac{3}{10}y = \frac{1}{2} \rightarrow 5x + 3y = 5$$
$$-\frac{5}{3}x - y = \frac{4}{3} \rightarrow -5x - 3y = 4$$

Now add the equations to eliminate $x$.
$$\begin{aligned} 5x + 3y &= 5 \\ -5x - 3y &= 4 \\ \hline 0 &= 9 \quad \text{False} \end{aligned}$$

The system has no solution. The system is inconsistent.

**49.**
$$x - 1.5y = 0.25 \rightarrow$$
$$0.9y - 0.6x = -0.15 \rightarrow$$
$$x - 1.5y = 0.25$$
$$-0.6x + 0.9y = -0.15$$

Multiply the first equation by 0.6 and then add to eliminate $x$.
$$\begin{aligned} 0.6x - 0.9y &= 0.15 \\ -0.6x + 0.9y &= -0.15 \\ \hline 0 &= 0 \quad \text{True} \end{aligned}$$

The system has infinitely many solutions. The equations are dependent.

**51.**
$$0.1x - 0.25y = 1.05$$
$$0.625x + 0.4y = 0.675$$

Begin by multiplying the first equation by 100 and the second equation by 1000 to eliminate decimals.
$$10x - 25y = 105$$
$$625x + 400y = 675$$

Multiply the first equation by 16 and then add to eliminate $y$.
$$\begin{aligned} 160x - 400y &= 1680 \\ 625x + 400y &= 675 \\ \hline 785x &= 2355 \\ x &= 3 \end{aligned}$$

Replace $x$ with 3 in $10x - 25y = 105$ to find $y$.
$$\begin{aligned} 10(3) - 25y &= 105 \\ -25y &= 75 \\ y &= -3 \end{aligned}$$

The solution is $(3, -3)$.

**53.**
$$\frac{x+2}{9} + \frac{y+2}{6} = 1$$
$$\frac{x-1}{4} - \frac{y+1}{3} = -1$$

Begin by multiplying the first equation by its LCD of 18 and the second equation by its LCD of 12.
$$2(x+2) + 3(y+2) = 18$$
$$3(x-1) - 4(y+1) = -12$$

Now simplify each equation.
$$2x + 4 + 3y + 6 = 18$$
$$3x - 3 - 4y - 4 = -12$$

$$2x + 3y = 8$$
$$3x - 4y = -5$$

Multiply the first equation by 4 and the second equation by 3. Then add to eliminate $y$.
$$\begin{aligned} 8x + 12y &= 32 \\ 9x - 12y &= -15 \\ \hline 17x &= 17 \\ x &= 1 \end{aligned}$$

Replace $x$ with 1 in $2x + 3y = 8$ to find $y$.
$$\begin{aligned} 2(1) + 3y &= 8 \\ 3y &= 6 \\ y &= 2 \end{aligned}$$

The solution is $(1, 2)$.

**55.**
$$\frac{x-y}{6} + \frac{x+y}{3} = 1$$
$$y = 6 - 3x$$

Begin by multiplying the first equation by its LCD of 6.
$$\frac{x-y}{6} + \frac{x+y}{3} = 1$$
$$x - y + 2(x + y) = 6$$
$$x - y + 2x + 2y = 6$$
$$3x + y = 6$$

Rewriting the $2^{nd}$ equation, the system is
$$3x + y = 6$$
$$3x + y = 6$$

Because these equations are the same, the system is dependent and has infinitely many solutions.

**57.** $x + by = -2a$
$ax - 4y = b$  Solution: $(-2, -2)$

Replace $x$ with $-2$ and $y$ with $-2$.
$-2 + b(-2) = -2a$
$a(-2) - 4(-2) = b$

$2a - 2b = 2$
$-2a - b = -8$

Add the equations to eliminate $a$.
$\phantom{-}2a - 2b = \phantom{-}2$
$-2a - \phantom{2}b = -8$
$\overline{\phantom{-2a} - 3b = -6}$
$\phantom{-2a - 3}b = \phantom{-}2$

Substitute 2 for $b$ in $-2a - b = -8$ to find $a$.
$-2a - 2 = -8$
$-2a = -6$
$a = 3$

To have a solution of $(-2, -2)$, $a = 3$ and $b = 2$.

**59.** $ax - by = -17$
$bx + ay = -1$  Solution: $(-2, 5)$

Replace $x$ with $-2$ and $y$ with 5.
$a(-2) - b(5) = -17$
$b(-2) + a(5) = -1$

$-2a - 5b = -17$
$5a - 2b = -1$

Multiply the first equation by 5 and the second equation by 2. Then add the equations to eliminate $a$.
$-10a - 25b = -85$
$\phantom{-}10a - \phantom{2}4b = \phantom{-}-2$
$\overline{\phantom{-10a} - 29b = -87}$
$\phantom{-10a - 29}b = \phantom{-}3$

Substitute 3 for $b$ in $-2a - 5b = -17$ to find $a$.
$-2a - 5(3) = -17$
$-2a = -2$
$a = 1$

To have a solution of $(-2, 5)$, $a = 1$ and $b = 3$.

**61.** $x + 2y = 5 \rightarrow y = -\dfrac{1}{2}x + \dfrac{5}{2}$

$2x - ky = 7 \rightarrow y = \dfrac{2}{k}x - \dfrac{7}{k}$

For the system to be inconsistent, the slopes of the two lines must be equal and the $y$-intercepts must not be equal.

$-\dfrac{1}{2} = \dfrac{2}{k}$
$k = -4$

If $k = -4$, the $y$-intercepts are $\dfrac{5}{2}$ and $\dfrac{7}{4}$. Thus, if $k = -4$, the system is inconsistent.

**63.** $y = -2x + 3$

$kx - 3y = 4 \rightarrow y = \dfrac{k}{3}x - \dfrac{4}{3}$

For the system to be inconsistent, the slopes of the two lines must be equal and the $y$-intercepts must not be equal.

$-2 = \dfrac{k}{3}$
$-6 = k$

The $y$-intercepts are already different. If $k = -6$, the system is inconsistent.

**65.** $x + 2y = c \rightarrow y = -\dfrac{1}{2}x + \dfrac{c}{2}$

$3x + 6y = 12 \rightarrow y = -\dfrac{1}{2}x + 2$

For the equations to be dependent, the slopes of the two lines must be equal and the $y$-intercepts must be equal. The slopes are already equal.

$\dfrac{c}{2} = 2$
$c = 4$

So, if $c = 4$, the equations are dependent.

Section 5.2    The Addition Method                                                                                  **169**

**67.**
$$ax + by = 3 \rightarrow y = -\frac{a}{b}x + \frac{3}{b}$$
$$x - y = 1 \rightarrow y = x - 1$$

For the equations to be dependent, the slopes of the two lines must be equal and the $y$-intercepts must be equal.

$$-1 = \frac{3}{b}$$
$$b = -3$$

Use $b = -3$ when equating the slopes:

$$-\frac{a}{b} = 1$$
$$-\frac{a}{-3} = 1$$
$$-a = -3$$
$$a = 3$$

So, if $a = 3$ and $b = -3$, the equations are dependent.

**69.**
$$\begin{aligned} x + y &= b \\ x - y &= c \\ \hline 2x &= b + c \\ x &= \frac{b+c}{2} \end{aligned}$$

Substitute for $x$ in $x + y = b$ to find $y$.

$$\frac{b+c}{2} + y = b$$
$$y = b - \frac{b+c}{2}$$
$$y = \frac{b-c}{2}$$

The solution is $\left(\frac{b+c}{2}, \frac{b-c}{2}\right)$.

**71.**
$$\begin{aligned} ax + y &= 3 \\ x - y &= c \\ \hline (a+1)x &= c + 3 \\ x &= \frac{c+3}{a+1}, \quad a \neq -1 \end{aligned}$$

Substitute for $x$ in $x - y = c$ to find $y$.

$$\frac{c+3}{a+1} - y = c$$
$$\frac{c+3}{a+1} - c = y$$
$$\frac{c+3 - c(a+1)}{a+1} = y$$
$$\frac{3 - ac}{a+1} = y$$

The solution is $\left(\frac{c+3}{a+1}, \frac{3-ac}{a+1}\right)$, $a \neq -1$.

**73.**
$$2ax + y = 3$$
$$-ax + 3y = 2$$

Multiply the second equation by 2 and then add to eliminate $x$.

$$\begin{aligned} 2ax + y &= 3 \\ -2ax + 6y &= 4 \\ \hline 7y &= 7 \\ y &= 1 \end{aligned}$$

Substitute for $y$ in $-ax + 3y = 2$ to find $x$.

$$-ax + 3(1) = 2$$
$$-ax = -1$$
$$x = \frac{1}{a}, \quad a \neq 0$$

The solution is $\left(\frac{1}{a}, 1\right)$, $a \neq 0$.

**75.**
$$ax + by = a$$
$$x - y = 1$$

Multiply the second equation by $b$ and then add to eliminate $y$.

$$\begin{aligned} ax + by &= a \\ bx - by &= b \\ \hline (a+b)x &= a + b \\ x &= 1, \quad a \neq -b \end{aligned}$$

Substitute for $x$ in $x - y = 1$ to find $y$.

$$1 - y = 1$$
$$y = 0$$

The solution is $(1, 0)$, $a \neq -b$.

**77.** Let $x =$ the price of a hamburger and $y =$ the price of a milk shake.

$$8x + 4y = 20$$
$$3x + 2y = 8.10$$

Multiply the second equation by $-2$.

$$\begin{aligned} 8x + 4y &= 20 \\ -6x - 4y &= -16.2 \\ \hline 2x &= 3.8 \\ x &= 1.9 \end{aligned}$$

Substituting $x = 1.9$ into $8x + 4y = 20$,

$$8(1.9) + 4y = 20$$
$$15.2 + 4y = 20$$
$$4y = 4.8$$
$$y = 1.2$$

A hamburger costs $1.90 and a milk shake costs $1.20.

**79.** Let $x$ = the number of dimes and $y$ = the number of nickels.
$$x + y = 50$$
$$10x + 5y = 385$$
Multiply the first equation by $-5$.
$$-5x - 5y = -250$$
$$10x + 5y = 385$$
$$5x = 135$$
$$x = 27$$
There are 27 dimes in the meter.

**81.** Let $x$ = the number of dimes and $y$ = the number of nickels.
$$x + y = 64$$
$$10x + 5y = 425$$
Multiply the first equation by $-5$.
$$-5x - 5y = -320$$
$$10x + 5y = 425$$
$$5x = 105$$
$$x = 21$$
Substituting $x = 21$ into $x + y = 64$,
$$21 + y = 64$$
$$y = 43$$
There are 21 dimes and 43 nickels in the cash register.

**83.** Let $x$ = the number of carpenters and $y$ = the number of unskilled laborers.
$$x + y = 16$$
$$22x + 8y = 212$$
Multiply the first equation by $-8$.
$$-8x - 8y = -128$$
$$22x + 8y = 212$$
$$14x = 84$$
$$x = 6$$
The subcontractor employs 6 carpenters.

**85.** Let $x$ = the number of pages with graphics and $y$ = the number of pages without graphics.
$$x + y = 60$$
$$5x + 2y = 147$$
Multiply the first equation by $-2$.

$$-2x - 2y = -120$$
$$5x + 2y = 147$$
$$3x = 27$$
$$x = 9$$
The report has 9 out of 60 pages with graphics, or 15% of the pages have graphics.

**87.** (a) To solve the system, multiply the second equation (for Internet) by $-1$ and add the equations to eliminate $y$.
$$y = 0.6x + 93$$
$$-y = -3.5x - 24.9$$
$$0 = -2.9x + 68.1$$
$$2.9x = 68.1$$
$$x \approx 23.5$$
Substitute 23.5 for $x$ in the first equation.
$$y = 0.6x + 93$$
$$y = 0.6(23.5) + 93$$
$$y \approx 107$$
The solution is about (23.5, 107).
(b) In 2018, adults will spend 107 hours per year reading and 107 hours per year on the Internet.

**89.** Email: $17x - 3y = 2$
$$-3y = -17x + 2$$
$$y = \frac{17}{3}x - \frac{2}{3}$$
Paper: $19x + 6y - 416 = 0$
$$6y = -19x + 416$$
$$y = -\frac{19}{6}x + \frac{416}{6}$$

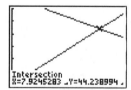

The estimated solution is $(7.9, 44.2)$.

**91.** The percentages for E-mail and paper do not total 100% because there are other forms of communication, such as telephone, voice mail, and personal conversations.

Section 5.2   The Addition Method

**93.**
$$\frac{3}{x}+\frac{2}{y}=2 \rightarrow 3u+2v=2$$
$$\frac{1}{x}-\frac{6}{y}=-1 \rightarrow u-6v=-1$$

Multiply the second equation by $-3$.
Then add equations to eliminate $u$.

$$\begin{aligned} 3u + 2v &= 2 \\ -3u + 18v &= 3 \\ \hline 20v &= 5 \\ v &= \frac{1}{4} \end{aligned}$$

Use this to substitute in $u-6v=-1$ to find $u$.

$$u - 6\left(\frac{1}{4}\right) = -1$$
$$u - \frac{3}{2} = -1$$
$$u = \frac{1}{2}$$

Now solve for $x$ and $y$:

$$u = \frac{1}{x} \qquad v = \frac{1}{y}$$
$$\frac{1}{2} = \frac{1}{x} \quad \text{and} \quad \frac{1}{4} = \frac{1}{y}$$
$$x = 2 \qquad y = 4$$

The solution is $(2, 4)$.

**95.**
$$\frac{5}{x}+\frac{10}{y}=3 \rightarrow 5u+10v=3$$
$$\frac{1}{x}+\frac{3}{y}=1 \rightarrow u+3v=1$$

Multiply the second equation by $-5$.
Then add equations to eliminate $u$.

$$\begin{aligned} 5u + 10v &= 3 \\ -5u - 15v &= -5 \\ \hline -5v &= -2 \\ v &= \frac{2}{5} \end{aligned}$$

Use this to substitute in $5u+10v=3$ to find $u$.

$$5u + 10\left(\frac{2}{5}\right) = 3$$
$$5u + 4 = 3$$
$$5u = -1$$
$$u = -\frac{1}{5}$$

Now solve for $x$ and $y$:

$$u = \frac{1}{x} \qquad v = \frac{1}{y}$$
$$-\frac{1}{5} = \frac{1}{x} \quad \text{and} \quad \frac{2}{5} = \frac{1}{y}$$
$$x = -5 \qquad 2y = 5$$
$$\qquad \qquad y = 2.5$$

The solution is $(-5, 2.5)$.

**97.**
$$\frac{3}{x}+\frac{5}{y}=10 \rightarrow 3u+5v=10$$
$$\frac{2}{x}+\frac{1}{y}=2 \rightarrow 2u+v=2$$

Multiply the second equation by $-5$.
Then add equations to eliminate $v$.

$$\begin{aligned} 3u + 5v &= 10 \\ -10u - 5v &= -10 \\ \hline -7u &= 0 \\ u &= 0 \end{aligned}$$

Use this to substitute in $2u+v=2$ to find $v$.

$$2(0) + v = 2$$
$$v = 2$$

Now solve for $x$ and $y$:

$$u = \frac{1}{x} \qquad v = \frac{1}{y}$$
$$0 = \frac{1}{x} \quad \text{and} \quad 2 = \frac{1}{y}$$
No solution $\qquad y = 0.5$

The system has no solution because $1/x = 0$ has no solution.

**99.**
$$-3x + 2y = 5$$
$$2y = 3x + 5$$
$$y = 1.5x + 2.5 \quad m_1 = 1.5 \quad b_1 = 2.5$$
$$6x + by = 7$$
$$by = -6x + 7$$
$$y = -\frac{6}{b}x + \frac{7}{b} \quad m_2 = -\frac{6}{b} \quad b_2 = \frac{7}{b}$$

The system will be consistent if $m_1 \neq m_2$ or if $m_1 = m_2$ and $b_1 = b_2$.

$$m_1 = m_2$$
$$1.5 = -\frac{6}{b}$$
$$b = -4$$

With $b = -4$, $b_1 = 2.5$ and $b_2 = -1.75$. Because $b_1 \neq b_2$ when $m_1 = m_2$, the

**101.** $4x - 8y = 20$
$-8y = -4x + 20$
$y = 0.5x - 2.5 \quad m_1 = 0.5 \quad b_1 = -2.5$
$x - 2y = c$
$-2y = -x + c$
$y = 0.5x - \dfrac{c}{2} \quad m_2 = 0.5 \quad b_2 = -\dfrac{c}{2}$

The system will be consistent if $m_1 \ne m_2$ or if $m_1 = m_2$ and $b_1 = b_2$. Because we already have $m_1 = m_2$, we need to find the value of $c$ for which $b_1 = b_2$.
$b_1 = b_2$
$-2.5 = -\dfrac{c}{2}$
$-5 = -c$
$5 = c$

Thus, for $c = 5$, the system is consistent.

**103.** $ax + 3y = 2$
$3y = -ax + 2$
$y = -\dfrac{a}{3}x + \dfrac{2}{3} \quad m_1 = -\dfrac{a}{3} \quad b_1 = \dfrac{2}{3}$
$2x + 6y = 4$
$6y = -2x + 4$
$y = -\dfrac{1}{3}x + \dfrac{2}{3} \quad m_2 = -\dfrac{1}{3} \quad b_2 = \dfrac{2}{3}$

Because $b_1 = b_2$, the equations are independent only if $m_1 \ne m_2$.

system is consistent only when $m_1 \ne m_2$. Therefore, the system is consistent for any value of $b \ne -4$.

$m_1 = m_2$
$-\dfrac{a}{3} = -\dfrac{1}{3}$
$a = 1$

The equations are independent for any value of $a \ne 1$.

**105.** $x + y = 5$
$y = -x + 5 \quad m_1 = -1 \quad b_1 = 5$
$2x + 2y = c$
$2y = -2x + c$
$y = -1x + \dfrac{c}{2} \quad m_2 = -1 \quad b_2 = \dfrac{c}{2}$

Because $m_1 = m_2$, the equations are independent only if $b_1 \ne b_2$.
$b_1 = b_2$
$5 = \dfrac{c}{2}$
$10 = c$

The equations are independent for any value of $c \ne 10$.

## Section 5.3 Systems of Equations in Three Variables

**1.** No, not all coefficients of a linear equation can be zero.

**3.** $5x + 7y - 2z = -1$
$x - 2y + z = 8$
$3x - y + 3z = 14$
$(3, -2, 1)$:
$5x + 7y - 2z = -1$
$5(3) + 7(-2) - 2(1) = -1$
$15 - 14 - 2 = -1$
$-1 = -1 \quad$ True

$x - 2y + z = 8$
$3 - 2(-2) + 1 = 8$
$8 = 8$  True

$3x - y + 3z = 14$
$3(3) - (-2) + 3(1) = 14$
$9 + 2 + 3 = 14$
$14 = 14$  True

Yes, $(3, -2, 1)$ is a solution.

$(2, -1, 2):$

$5x + 7y - 2z = -1$
$5(2) + 7(-1) - 2(2) = -1$
$10 - 7 - 4 = -1$
$-1 = -1$  True

$x - 2y + z = 8$
$2 - 2(-1) + 2 = 8$
$6 = 8$  False

Because $(2, -1, 2)$ does not satisfy all three equations in the system, it is not a solution.

**5.**  $3x + 4y - z = -12$
$-x + 2y + 3z = 8$
$2x + 6y + z = -8$

$(-3, 1, 7):$

$3x + 4y - z = -12$
$3(-3) + 4(1) - 7 = -12$
$-9 + 4 - 7 = -12$
$-12 = -12$  True

$-x + 2y + 3z = 8$
$-(-3) + 2(1) + 3(7) = 8$
$3 + 2 + 21 = 8$
$26 = 8$  False

Because $(-3, 1, 7)$ does not satisfy all three equations in the system, it is not a solution.

$(0, -2, 4):$

$3x + 4y - z = -12$
$3(0) + 4(-2) - 4 = -12$
$-12 = -12$  True

$-x + 2y + 3z = 8$
$-(0) + 2(-2) + 3(4) = 8$
$8 = 8$  True

$2x + 6y + z = -8$
$2(0) + 6(-2) + 4 = -8$
$-8 = -8$  True

Yes, $(0, -2, 4)$ is a solution.

**7.**  $ax - 2y + z = 7$
$x - y + cz = -2$
$2x + by - z = 0$

Solution: $(1, -1, 2)$

Replace $x$ with 1, $y$ with $-1$, and $z$ with 2.

$a(1) - 2(-1) + (2) = 7$
$(1) - (-1) + c(2) = -2$
$2(1) + b(-1) - (2) = 0$

$a + 2 + 2 = 7$
$1 + 1 + 2c = -2$
$2 - b - 2 = 0$

$a + 4 = 7$         $a = 3$
$2 + 2c = -2$  →   $c = -2$
$-b = 0$             $b = 0$

If $a = 3$, $b = 0$, and $c = -2$, the point $(1, -1, 2)$ is a solution of the system.

**9.**  $2x + y + z = d$
$x + by - z = 7$
$-3x + 2y + cz = 4$

Solution: $(0, 2, -1)$

Replace $x$ with 0, $y$ with 2, and $z$ with $-1$.

$2(0) + (2) + (-1) = d$
$(0) + b(2) - (-1) = 7$
$-3(0) + 2(2) + c(-1) = 4$

$2 - 1 = d$
$2b + 1 = 7$
$4 - c = 4$

$1 = d$
$b = 3$
$c = 0$

If $b = 3$, $c = 0$, and $d = 1$, the point $(0, 2, -1)$ is a solution of the system.

**11.** The graph of a linear equation in two variables is a line in a plane. The graph of a linear equation in three variables is a plane in three-dimensional space.

**13.** (a) There is no solution if at least two planes are parallel or a plane is parallel to the line of intersection of the other two planes.

(b) There are infinitely many solutions if the three planes coincide or intersect in a common line.

15. $\quad x + y + z = 0 \quad (1)$
    $\quad 3x + y \phantom{+ z} = 0 \quad (2)$
    $\quad \phantom{3x +} y - 2z = 7 \quad (3)$

Because the $x$ term is already missing from equation (3), eliminate $x$ from equations (1) and (2). Multiply equation (1) by $-3$ and add the equations.
$$-3x - 3y - 3z = 0 \quad (1)$$
$$\underline{\phantom{-}3x + y \phantom{- 3z} = 0} \quad (2)$$
$$\phantom{-3x} -2y - 3z = 0 \quad (4)$$

Now we have two equations, (3) and (4), in two variables.
$$y - 2z = 7 \quad (3)$$
$$-2y - 3z = 0 \quad (4)$$

Multiply equation (3) by 2 and add the equations to eliminate $y$.
$$2y - 4z = 14$$
$$\underline{-2y - 3z = \phantom{1}0}$$
$$\phantom{-2y} -7z = 14$$
$$z = -2$$

Substitute $z = -2$ in equation (3) to find the value of $y$.
$$y - 2z = 7 \quad (3)$$
$$y - 2(-2) = 7$$
$$y + 4 = 7$$
$$y = 3$$

To find $x$, substitute $z = -2$ and $y = 3$ in equation (1).
$$x + y + z = 0 \quad (1)$$
$$x + 3 - 2 = 0$$
$$x + 1 = 0$$
$$x = -1$$

The solution is $(-1, 3, -2)$.

17. $\quad x + 2y - 3z = 5 \quad (1)$
    $\quad x \phantom{+ 2y} + 2z = 15 \quad (2)$
    $\quad \phantom{x +} 2y - z = 6 \quad (3)$

Because the $x$ term is already missing from equation (3), eliminate $x$ from equations (1) and (2). Multiply equation (2) by $-1$ and add the equations.
$$x + 2y - 3z = 5 \quad (1)$$
$$\underline{-x \phantom{+ 2y} - 2z = -15} \quad (2)$$
$$\phantom{-x +} 2y - 5z = -10 \quad (4)$$

Now we have two equations, (3) and (4), in two variables.
$$2y - z = 6 \quad (3)$$
$$2y - 5z = -10 \quad (4)$$

Multiply equation (4) by $-1$ and add the equations to eliminate $y$.
$$2y - z = 6$$
$$\underline{-2y + 5z = 10}$$
$$4z = 16$$
$$z = 4$$

Substitute $z = 4$ in equation (4) to find $y$.
$$2y - 5z = -10 \quad (4)$$
$$2y - 5(4) = -10$$
$$2y - 20 = -10$$
$$2y = 10$$
$$y = 5$$

To find $x$, substitute $z = 4$ and $y = 5$ in equation (2).
$$x + 2z = 15 \quad (2)$$
$$x + 2(4) = 15$$
$$x + 8 = 15$$
$$x = 7$$

The solution is $(7, 5, 4)$.

19. $\quad 2x - y + z = 9 \quad (1)$
    $\quad 3x + 2y - z = 4 \quad (2)$
    $\quad 4x + 3y + 2z = 8 \quad (3)$

Eliminate $z$ from equations (1) and (2).
$$2x - y + z = 9 \quad (1)$$
$$\underline{3x + 2y - z = 4} \quad (2)$$
$$5x + y \phantom{- z} = 13 \quad (4)$$

Eliminate $z$ from equations (2) and (3) by multiplying equation (2) by 2 and then adding the equations.
$$6x + 4y - 2z = 8 \quad (2)$$
$$\underline{4x + 3y + 2z = 8} \quad (3)$$
$$10x + 7y \phantom{+ 2z} = 16 \quad (5)$$

Now we have two equations in two variables.
$$5x + y = 13 \quad (4)$$
$$10x + 7y = 16 \quad (5)$$

Multiply equation (4) by $-2$ and add the equations to eliminate $x$.
$$-10x - 2y = -26$$
$$\underline{\phantom{-}10x + 7y = \phantom{-}16}$$
$$5y = -10$$
$$y = -2$$

Substitute $y = -2$ in equation (4) to find $x$.

$$5x + y = 13 \quad (4)$$
$$5x + (-2) = 13$$
$$5x = 15$$
$$x = 3$$

To find $z$, substitute $x = 3$ and $y = -2$ in equation (1).
$$2x - y + z = 9 \quad (1)$$
$$2(3) - (-2) + z = 9$$
$$8 + z = 9$$
$$z = 1$$

The solution is $(3, -2, 1)$.

**21.**
$$x + 2y - 3z = 1 \quad (1)$$
$$x + y + 2z = -1 \quad (2)$$
$$3x + 3y - z = 4 \quad (3)$$

Eliminate $y$ from equations (1) and (2) by multiplying equation (2) by $-2$ and then adding the equations.
$$x + 2y - 3z = 1 \quad (1)$$
$$-2x - 2y - 4z = 2 \quad (2)$$
$$\overline{-x \quad\quad - 7z = 3} \quad (4)$$

Eliminate $y$ from equations (2) and (3) by multiplying equation (2) by $-3$ and then adding the equations.
$$-3x - 3y - 6z = 3 \quad (2)$$
$$3x + 3y - z = 4 \quad (3)$$
$$\overline{\quad\quad\quad -7z = 7}$$
$$z = -1$$

Because both $x$ and $y$ were eliminated, we solved for $z$, getting $z = -1$. Now substitute for $z$ in equation (4) to find $x$.
$$-x - 7z = 3 \quad (4)$$
$$-x - 7(-1) = 3$$
$$-x + 7 = 3$$
$$-x = -4$$
$$x = 4$$

To find $y$, substitute $x = 4$ and $z = -1$ in equation (2).
$$x + y + 2z = -1 \quad (2)$$
$$4 + y + 2(-1) = -1$$
$$2 + y = -1$$
$$y = -3$$

The solution is $(4, -3, -1)$.

**23.**
$$4x + 5y - 2z = 23 \quad (1)$$
$$-6x + 2y + 7z = -14 \quad (2)$$
$$8x + 3y + 3z = 11 \quad (3)$$

Eliminate $x$ from equations (1) and (2) by multiplying equation (1) by 3, multiplying equation (2) by 2, and then adding the equations.
$$12x + 15y - 6z = 69 \quad (1)$$
$$-12x + 4y + 14z = -28 \quad (2)$$
$$\overline{\quad\quad 19y + 8z = 41} \quad (4)$$

Eliminate $x$ from equations (1) and (3) by multiplying equation (1) by $-2$, and then adding the equations.
$$-8x - 10y + 4z = -46 \quad (1)$$
$$8x + 3y + 3z = 11 \quad (3)$$
$$\overline{\quad\quad -7y + 7z = -35} \quad (5)$$

We can divide through equation (5) by 7. Now we have two equations in two variables.
$$19y + 8z = 41 \quad (4)$$
$$-y + z = -5 \quad (5)$$

Multiply equation (5) by $-8$ and add the equations to eliminate $z$.
$$19y + 8z = 41 \quad (4)$$
$$8y - 8z = 40 \quad (5)$$
$$\overline{27y \quad\quad = 81}$$
$$y = 3$$

Substitute $y = 3$ in equation (5) to find $z$.
$$-y + z = -5 \quad (5)$$
$$-3 + z = -5$$
$$z = -2$$

To find $x$, substitute $y = 3$ and $z = -2$ in equation (1).
$$4x + 5y - 2z = 23 \quad (1)$$
$$4x + 5(3) - 2(-2) = 23$$
$$4x + 19 = 23$$
$$4x = 4$$
$$x = 1$$

The solution is $(1, 3, -2)$.

**25.**
$$7x - 2y + 3z = 19 \quad (1)$$
$$x + 8y - 6z = 9 \quad (2)$$
$$2x + 4y + 9z = 5 \quad (3)$$

Eliminate $x$ from equations (1) and (2) by multiplying equation (2) by $-7$ and then adding the equations.
$$7x - 2y + 3z = 19 \quad (1)$$
$$-7x - 56y + 42z = -63 \quad (2)$$
$$\overline{\quad -58y + 45z = -44} \quad (4)$$

Eliminate $x$ from equations (2) and (3) by multiplying equation (2) by $-2$ and then adding the equations.

$$-2x - 16y + 12z = -18 \quad (2)$$
$$2x + 4y + 9z = 5 \quad (3)$$
$$\overline{\phantom{-2x} - 12y + 21z = -13} \quad (5)$$

Now we have two equations in two variables.
$$-58y + 45z = -44 \quad (4)$$
$$-12y + 21z = -13 \quad (5)$$

Multiply equation (4) by $-7$, multiply equation (5) by 15, and add the equations to eliminate $z$.
$$406y - 315z = 308 \quad (4)$$
$$-180y + 315z = -195 \quad (5)$$
$$\overline{226y \phantom{-315z} = 113}$$
$$y = \frac{1}{2}$$

Substitute $y = \frac{1}{2}$ in equation (5) to find $z$.
$$-12y + 21z = -13 \quad (5)$$
$$-12\left(\frac{1}{2}\right) + 21z = -13$$
$$-6 + 21z = -13$$
$$21z = -7$$
$$z = -\frac{1}{3}$$

To find $x$, substitute $y = \frac{1}{2}$ and $z = -\frac{1}{3}$ in equation (2).
$$x + 8y - 6z = 9 \quad (2)$$
$$x + 8\left(\frac{1}{2}\right) - 6\left(-\frac{1}{3}\right) = 9$$
$$x + 4 + 2 = 9$$
$$x = 3$$

The solution is $\left(3, \frac{1}{2}, -\frac{1}{3}\right)$.

**27.**
$$2x + 3y + 2z = 2 \quad (1)$$
$$-3x - 6y - 4z = -3 \quad (2)$$
$$-\frac{1}{6}x + y + z = 0 \quad (3)$$

Eliminate $y$ from equations (1) and (2) by multiplying equation (1) by 2 and then adding the equations.
$$4x + 6y + 4z = 4 \quad (1)$$
$$-3x - 6y - 4z = -3 \quad (2)$$
$$\overline{x \phantom{+ 6y + 4z} = 1}$$

Eliminate $y$ from equations (2) and (3) by multiplying equation (3) by 6 and then adding the equations.
$$-3x - 6y - 4z = -3 \quad (2)$$
$$-x + 6y + 6z = 0 \quad (3)$$
$$\overline{-4x \phantom{+ 6y} + 2z = -3} \quad (4)$$

Substitute $x = 1$ in equation (4) to find $z$.
$$-4x + 2z = -3 \quad (4)$$
$$-4(1) + 2z = -3$$
$$2z = 1$$
$$z = \frac{1}{2}$$

Substitute $x = 1$ and $z = \frac{1}{2}$ in equation (1) to find $y$.
$$2x + 3y + 2z = 2 \quad (1)$$
$$2(1) + 3y + 2\left(\frac{1}{2}\right) = 2$$
$$3y + 3 = 2$$
$$3y = -1$$
$$y = -\frac{1}{3}$$

The solution is $\left(1, -\frac{1}{3}, \frac{1}{2}\right)$.

**29.**
$$4x + 3y - 3z = 5 \quad (1)$$
$$2x - 6y + 9z = -7 \quad (2)$$
$$6x + 6y - 3z = 7 \quad (3)$$

Eliminate $y$ from equations (1) and (2) by multiplying equation (1) by 2 and then adding the equations.
$$8x + 6y - 6z = 10 \quad (1)$$
$$2x - 6y + 9z = -7 \quad (2)$$
$$\overline{10x \phantom{+ 6y} + 3z = 3} \quad (4)$$

Eliminate $y$ from equations (2) and (3) by adding the equations.
$$2x - 6y + 9z = -7 \quad (2)$$
$$6x + 6y - 3z = 7 \quad (3)$$
$$\overline{8x \phantom{+ 6y} + 6z = 0} \quad (5)$$

Now we have two equations in two variables.
$$10x + 3z = 3 \quad (4)$$
$$8x + 6z = 0 \quad (5)$$

Multiply equation (4) by $-2$ and add the equations to eliminate $z$.

Section 5.3  Systems of Equations in Three Variables

$$\begin{aligned}-20x - 6z &= -6 \quad (4)\\ 8x + 6z &= 0 \quad (5)\\ \hline -12x &= -6\\ x &= \frac{1}{2}\end{aligned}$$

Substitute $x = \frac{1}{2}$ in equation (5) to find $z$.

$$\begin{aligned}8x + 6z &= 0 \quad (5)\\ 8\left(\frac{1}{2}\right) + 6z &= 0\\ 4 + 6z &= 0\\ 6z &= -4\\ z &= -\frac{2}{3}\end{aligned}$$

Substitute $x = \frac{1}{2}$ and $z = -\frac{2}{3}$ in equation (1) to find $y$.

$$\begin{aligned}4x + 3y - 3z &= 5 \quad (1)\\ 4\left(\frac{1}{2}\right) + 3y - 3\left(-\frac{2}{3}\right) &= 5\\ 3y + 4 &= 5\\ 3y &= 1\\ y &= \frac{1}{3}\end{aligned}$$

The solution is $\left(\frac{1}{2}, \frac{1}{3}, -\frac{2}{3}\right)$.

**31.** (a) There is no solution if after eliminating one variable, the resulting system of two equations in two variables is inconsistent.

(b) There are infinitely many solutions if after eliminating one variable, the two equations in two variables in the resulting system are dependent.

**33.**
$$\begin{aligned}x + 2y - z &= 5 \quad (1)\\ x - 2y + z &= 2 \quad (2)\\ 2x + 4y - 2z &= 7 \quad (3)\end{aligned}$$

Eliminate $y$ from equations (1) and (2) by adding the equations.

$$\begin{aligned}x + 2y - z &= 5 \quad (1)\\ x - 2y + z &= 2 \quad (2)\\ \hline 2x &= 7\\ x &= 3.5\end{aligned}$$

Eliminate $y$ from equations (2) and (3) by multiplying equation (2) by 2 and then adding the equations.

$$\begin{aligned}2x - 4y + 2z &= 4 \quad (2)\\ 2x + 4y - 2z &= 14 \quad (3)\\ \hline 4x &= 18\\ x &= 4.5\end{aligned}$$

Because the resulting equations cannot both be true, the resulting system of two equations in two variables is inconsistent. So, the original system is inconsistent, and the solution set is $\varnothing$.

**35.**
$$\begin{aligned}x + 2y - z &= 3 \quad (1)\\ -4x - 8y + 4z &= -12 \quad (2)\\ 3x + 6y - 3z &= 9 \quad (3)\end{aligned}$$

Eliminate $x$ from equations (1) and (2) by multiplying equation (1) by 4 and then adding the equations.

$$\begin{aligned}4x + 8y - 4z &= 12 \quad (1)\\ -4x - 8y + 4z &= -12 \quad (2)\\ \hline 0 &= 0\end{aligned}$$

Eliminate $x$ from equations (1) and (3) by multiplying equation (1) by $-3$ and then adding the equations.

$$\begin{aligned}-3x - 6y + 3z &= -9 \quad (1)\\ 3x + 6y - 3z &= 9 \quad (3)\\ \hline 0 &= 0\end{aligned}$$

The resulting equations are dependent. Equation (2) is equation (1) multiplied by $-4$. Equation (3) is equation (1) multiplied by 3. The solution set is the set of all solutions to $x + 2y - z = 3$.

**37.**
$$\begin{aligned}x + y &= 1 \quad (1)\\ 2x + 3y - z &= -1 \quad (2)\\ 3x - y - 4z &= 7 \quad (3)\end{aligned}$$

Eliminate $z$ from equations (2) and (3) by multiplying equation (2) by $-4$ and then adding the equations.

$$\begin{aligned}-8x - 12y + 4z &= 4 \quad (2)\\ 3x - y - 4z &= 7 \quad (3)\\ \hline -5x - 13y &= 11 \quad (4)\end{aligned}$$

Now we have two equations in two variables.

$$\begin{aligned}x + y &= 1 \quad (1)\\ -5x - 13y &= 11 \quad (4)\end{aligned}$$

Multiply equation (1) by 5 and add the equations to eliminate $x$.

$$\begin{aligned} 5x + 5y &= 5 \quad (1)\\ -5x - 13y &= 11 \quad (4)\\ \hline -8y &= 16\\ y &= -2 \end{aligned}$$

Substitute $y = -2$ in equation (1) to find $x$.
$$\begin{aligned} x + y &= 1 \quad (1)\\ x + (-2) &= 1\\ x &= 3 \end{aligned}$$

Substitute $x = 3$ and $y = -2$ in equation (2) to find $z$.
$$\begin{aligned} 2x + 3y - z &= -1 \quad (2)\\ 2(3) + 3(-2) - z &= -1\\ 6 - 6 - z &= -1\\ -z &= -1\\ z &= 1 \end{aligned}$$

The solution is $(3, -2, 1)$.

**39.**
$$\begin{aligned} x - y + z &= 2 \quad (1)\\ 2x + y - 2z &= -2 \quad (2)\\ 3x - 2y + z &= 2 \quad (3) \end{aligned}$$

Eliminate $y$ from equations (1) and (2) by adding the equations.
$$\begin{aligned} x - y + z &= 2 \quad (1)\\ 2x + y - 2z &= -2 \quad (2)\\ \hline 3x - z &= 0 \quad (4) \end{aligned}$$

Eliminate $y$ from equations (2) and (3) multiplying equation (2) by 2 and adding the equations.
$$\begin{aligned} 4x + 2y - 4z &= -4 \quad (2)\\ 3x - 2y + z &= 2 \quad (3)\\ \hline 7x - 3z &= -2 \quad (5) \end{aligned}$$

Now we have two equations in two variables.
$$\begin{aligned} 3x - z &= 0 \quad (4)\\ 7x - 3z &= -2 \quad (5) \end{aligned}$$

Multiply equation (4) by $-3$ and add the equations to eliminate $z$.
$$\begin{aligned} -9x + 3z &= 0 \quad (4)\\ 7x - 3z &= -2 \quad (5)\\ \hline -2x &= -2\\ x &= 1 \end{aligned}$$

Substitute $x = 1$ in equation (4) to find $z$.
$$\begin{aligned} 3x - z &= 0 \quad (4)\\ 3(1) - z &= 0\\ 3 - z &= 0\\ 3 &= z \end{aligned}$$

Substitute $x = 1$ and $z = 3$ in equation (1) to find $y$.
$$\begin{aligned} x - y + z &= 2 \quad (1)\\ 1 - y + 3 &= 2\\ 4 - y &= 2\\ -y &= -2\\ y &= 2 \end{aligned}$$

The solution is $(1, 2, 3)$.

**41.**
$$\begin{aligned} x + y + z &= 0 \quad (1)\\ 2x + 3y &= -2 \quad (2)\\ y - 4z &= 0 \quad (3) \end{aligned}$$

Eliminate $x$ from equations (1) and (2) by multiplying equation (1) by $-2$ and then adding the equations.
$$\begin{aligned} -2x - 2y - 2z &= 0 \quad (1)\\ 2x + 3y &= -2 \quad (2)\\ \hline y - 2z &= -2 \quad (4) \end{aligned}$$

Now we have two equations in two variables.
$$\begin{aligned} y - 4z &= 0 \quad (3)\\ y - 2z &= -2 \quad (4) \end{aligned}$$

Multiply equation (3) by $-1$ and add the equations to eliminate $y$.
$$\begin{aligned} -y + 4z &= 0 \quad (3)\\ y - 2z &= -2 \quad (4)\\ \hline 2z &= -2\\ z &= -1 \end{aligned}$$

Substitute $z = -1$ in equation (3) to find $y$.
$$\begin{aligned} y - 4z &= 0 \quad (3)\\ y - 4(-1) &= 0\\ y + 4 &= 0\\ y &= -4 \end{aligned}$$

Substitute $z = -1$ and $y = -4$ in equation (2) to find $x$.
$$\begin{aligned} 2x + 3y &= -2 \quad (2)\\ 2x + 3(-4) &= -2\\ 2x - 12 &= -2\\ 2x &= 10\\ x &= 5 \end{aligned}$$

The solution is $(5, -4, -1)$.

Section 5.3   Systems of Equations in Three Variables

**43.**
$$x + y = 1 \quad (1)$$
$$y - 2z = 2 \quad (2)$$
$$x - 3z = 14 \quad (3)$$

Eliminate $x$ from equations (1) and (3) by multiplying equation (1) by $-1$ and then adding the equations.
$$-x - y = -1 \quad (1)$$
$$x - 3z = 14 \quad (3)$$
$$-y - 3z = 13 \quad (4)$$

Now we have two equations in two variables.
$$y - 2z = 2 \quad (2)$$
$$-y - 3z = 13 \quad (4)$$

Eliminate $y$ by adding the equations.
$$y - 2z = 2 \quad (2)$$
$$-y - 3z = 13 \quad (4)$$
$$-5z = 15$$
$$z = -3$$

Substitute $z = -3$ in equation (2) to find $y$.
$$y - 2z = 2 \quad (2)$$
$$y - 2(-3) = 2$$
$$y + 6 = 2$$
$$y = -4$$

Substitute $y = -4$ in equation (1) to find $x$.
$$x + y = 1 \quad (1)$$
$$x + (-4) = 1$$
$$x = 5$$

The solution is $(5, -4, -3)$.

**45.**
$$x - y + z = 0 \quad (1)$$
$$\frac{1}{2}y + \frac{3}{4}z = \frac{1}{4} \quad (2)$$
$$\frac{1}{2}x + z = -\frac{1}{2} \quad (3)$$

Eliminate $x$ from equations (1) and (3) by multiplying equation (3) by $-2$ and then adding the equations.
$$x - y + z = 0 \quad (1)$$
$$-x - 2z = 1 \quad (3)$$
$$-y - z = 1 \quad (4)$$

Now we have two equations in two variables.
$$\frac{1}{2}y + \frac{3}{4}z = \frac{1}{4} \quad (2)$$
$$-y - z = 1 \quad (4)$$

Multiply equation (2) by 4, multiply equation (4) by 2, and add the equations to eliminate $y$.
$$2y + 3z = 1 \quad (2)$$
$$-2y - 2z = 2 \quad (4)$$
$$z = 3$$

Substitute $z = 3$ in equation (4) to find $y$.
$$-y - z = 1 \quad (4)$$
$$-y - 3 = 1$$
$$-y = 4$$
$$y = -4$$

Substitute $y = -4$ and $z = 3$ in equation (1) to find $x$.
$$x - y + z = 0 \quad (1)$$
$$x - (-4) + 3 = 0$$
$$x + 7 = 0$$
$$x = -7$$

The solution is $(-7, -4, 3)$.

**47.** First simplify the given equations.
$$3x - y + 2z = 3 \quad (1)$$
$$2x + 2y + z = 2 \quad (2)$$
$$-x + 3y - z = -4 \quad (3)$$

Eliminate $z$ from equations (1) and (3) by multiplying equation (3) by 2 and then adding the equations.
$$3x - y + 2z = 3 \quad (1)$$
$$-2x + 6y - 2z = -8 \quad (3)$$
$$x + 5y = -5 \quad (4)$$

Eliminate $z$ from equations (2) and (3) by adding the equations.
$$2x + 2y + z = 2 \quad (2)$$
$$-x + 3y - z = -4 \quad (3)$$
$$x + 5y = -2 \quad (5)$$

The result is a system of two equations in two variables that is inconsistent (notice that the $x$- and $y$-variable coefficients in equations (4) and (5) are the same but the constant is different). Thus the original system is inconsistent and there are no solutions. The solution set is $\varnothing$.

**49.** First rewrite the given equations.
$$4x - 3y + 6z = 12 \quad (1)$$
$$-2x + 1.5y - 3z = -6 \quad (2)$$
$$0.5x - 0.375y + 0.75z = 1.5 \quad (3)$$

Notice that equation (1) is $-2$ times equation (2). Also, equation (2) is $-4$ times equation (3). The equations are dependent. The solution set is the set of all solutions to $4x - 3y + 6z = 12$.

**51.** First rewrite the given equations.
$$4x + 15y - 6z = -126 \quad (1)$$
$$8x - 4y + z = 4 \quad (2)$$
$$-2x + y + 2z = 8 \quad (3)$$
Eliminate $x$ from equations (1) and (2) by multiplying equation (1) by $-2$ and then adding the equations.
$$-8x - 30y + 12z = 252 \quad (1)$$
$$8x - 4y + z = 4 \quad (2)$$
$$\overline{\phantom{-8x}-34y + 13z = 256} \quad (4)$$
Eliminate $x$ from equations (1) and (3) by multiplying equation (3) by 2 and then adding the equations.
$$4x + 15y - 6z = -126 \quad (1)$$
$$-4x + 2y + 4z = 16 \quad (3)$$
$$\overline{\phantom{4x+}17y - 2z = -110} \quad (5)$$
Now we have two equations in two variables.
$$-34y + 13z = 256 \quad (4)$$
$$17y - 2z = -110 \quad (5)$$
Multiply equation (5) by 2 and then add the equations to eliminate $y$.
$$-34y + 13z = 256 \quad (4)$$
$$34y - 4z = -220 \quad (5)$$
$$\overline{\phantom{-34y+}9z = 36}$$
$$z = 4$$
Substitute $z = 4$ in equation (5).
$$17y - 2z = -110 \quad (5)$$
$$17y - 2(4) = -110$$
$$17y - 8 = -110$$
$$17y = -102$$
$$y = -6$$
Substitute $z = 4$ and $y = -6$ in equation (3) to find $x$.
$$-2x + y + 2z = 8 \quad (3)$$
$$-2x + (-6) + 2(4) = 8$$
$$-2x - 6 + 8 = 8$$
$$-2x + 2 = 8$$
$$-2x = 6$$
$$x = -3$$
The solution is $(-3, -6, 4)$.

**53.** 
$$2x + y = -4 \quad (1)$$
$$x + z = 2 \quad (2)$$
$$2y - z = -1 \quad (3)$$

Solve equation (1) for $y$:
$$2x + y = -4$$
$$y = -2x - 4$$
Solve equation (2) for $z$:
$$x + z = 2$$
$$z = -x + 2$$
Substitute for $y$ and $z$ in equation (3):
$$2y - z = -1$$
$$2(-2x - 4) - (-x + 2) = -1$$
$$-4x - 8 + x - 2 = -1$$
$$-3x - 10 = -1$$
$$-3x = 9$$
$$x = -3$$
Use $x = -3$ in equation (1) to find $y$.
$$y = -2x - 4$$
$$y = -2(-3) - 4 = 2$$
Use $x = -3$ in equation (2) to find $z$.
$$z = -x + 2$$
$$z = -(-3) + 2 = 5$$
The solution is $(-3, 2, 5)$.

**55.**
$$x + y + z = -2 \quad (1)$$
$$x - 3y = 8 \quad (2)$$
$$-2y + z = 8 \quad (3)$$
Solve equation (2) for $x$:
$$x - 3y = 8$$
$$x = 3y + 8$$
Solve equation (3) for $z$:
$$-2y + z = 8$$
$$z = 2y + 8$$
Substitute for $x$ and $z$ in equation (1):
$$x + y + z = -2$$
$$(3y + 8) + y + (2y + 8) = -2$$
$$6y + 16 = -2$$
$$6y = -18$$
$$y = -3$$
Use $y = -3$ in equation (2) to find $x$.
$$x = 3y + 8$$
$$x = 3(-3) + 8 = -1$$
Use $y = -3$ in equation (3) to find $z$.
$$z = 2y + 8$$
$$z = 2(-3) + 8 = 2$$
The solution is $(-1, -3, 2)$.

**57.** Let $a =$ the measure of angle $A$, $b =$ the measure of angle $B$, and $c =$ the measure of angle $C$. The equations are $b = 3a - 5$,

## Section 5.3  Systems of Equations in Three Variables

$c = a + b + 6$, $a + b + c = 180$. Rewriting these as a system, we obtain:
$$-3a + b \phantom{+c} = -5 \quad (1)$$
$$-a - b + c = 6 \quad (2)$$
$$a + b + c = 180 \quad (3)$$

Eliminate $a$ and $b$ by adding equations (2) and (3).
$$-a - b + c = 6 \quad (2)$$
$$\underline{a + b + c = 180} \quad (3)$$
$$2c = 186$$
$$c = 93$$

Eliminate $a$ in equations (1) and (3) by multiplying equation (3) by 3 and then adding the equations.
$$-3a + b \phantom{+3c} = -5 \quad (1)$$
$$\underline{3a + 3b + 3c = 540} \quad (3)$$
$$4b + 3c = 535 \quad (4)$$

Substitute $c = 93$ in equation (4) to find $b$.
$$4b + 3c = 535 \quad (4)$$
$$4b + 3(93) = 535$$
$$4b + 279 = 535$$
$$4b = 256$$
$$b = 64$$

Substitute for $b$ and $c$ in equation (3) to find $a$.
$$a + b + c = 180$$
$$a + 64 + 93 = 180$$
$$a = 23$$

Angle $A$ measures $23°$, angle $B$ measures $64°$, and angle $C$ measures $93°$.

**59.** Let $x =$ the length of the first piece, $y =$ the length of the second piece, and $z =$ the length of the third piece. Three equations are $y = x + z + 4$ (note that *12 feet* longer is the same as *4 yards* longer), $z = x - 6$, and $x + y + z = 100$. Rewriting these as a system, we obtain:
$$-x + y - z = 4 \quad (1)$$
$$-x \phantom{+y} + z = -6 \quad (2)$$
$$x + y + z = 100 \quad (3)$$

Eliminate $x$ and $z$ by adding equations (1) and (3).
$$-x + y - z = 4 \quad (1)$$
$$\underline{x + y + z = 100} \quad (3)$$
$$2y = 104$$
$$y = 52$$

Eliminate $z$ in equations (1) and (2) by adding the equations.
$$-x + y - z = 4 \quad (1)$$
$$\underline{-x \phantom{+y} + z = -6} \quad (2)$$
$$-2x + y \phantom{+z} = -2 \quad (4)$$

Substitute $y = 52$ in equation (4) to find $x$.
$$-2x + y = -2 \quad (4)$$
$$-2x + 52 = -2$$
$$-2x = -54$$
$$x = 27$$

Substitute for $x$ and $y$ in equation (3) to find $z$.
$$x + y + z = 100 \quad (3)$$
$$27 + 52 + z = 100$$
$$79 + z = 100$$
$$z = 21$$

The three pieces are 27 yards, 52 yards and 21 yards.

**61.** Let $n =$ the number of nickels, $d =$ the number of dimes, and $q =$ the number of quarters. The equations are $d = 2n + 2$, $5n + 10d + 25q = 2495$, and $n + d + q = 155$. Rewriting these as a system of equations, we obtain:
$$-2n + d \phantom{+25q} = 2 \quad (1)$$
$$5n + 10d + 25q = 2495 \quad (2)$$
$$n + d + q = 155 \quad (3)$$

Eliminate $q$ in equations (2) and (3) by multiplying equation (3) by $-25$ and then adding the equations.
$$5n + 10d + 25q = 2495 \quad (2)$$
$$\underline{-25n - 25d - 25q = -3875} \quad (3)$$
$$-20n - 15d \phantom{+25q} = -1380 \quad (4)$$

Now we have two equations in two variables.
$$-2n + d = 2 \quad (1)$$
$$-20n - 15d = -1380 \quad (4)$$

Eliminate $d$ by multiplying equation (1) by 15 and then adding the equations.
$$-30n + 15d = 30 \quad (1)$$
$$\underline{-20n - 15d = -1380} \quad (4)$$
$$-50n \phantom{+15d} = -1350$$
$$n = 27$$

Use this in equation (1) to find $d$.

$$-2n + d = 2 \quad (1)$$
$$-2(27) + d = 2$$
$$-54 + d = 2$$
$$d = 56$$

Substitute $n = 27$ and $d = 56$ in equation (3) to find $q$.
$$n + d + q = 155 \quad (3)$$
$$27 + 56 + q = 155$$
$$q = 72$$

There are 27 nickels, 56 dimes, and 72 quarters.

**63.** (a) A system of equations for the given conditions is
$$R + D = 0.08(1200) \quad (1)$$
$$R + D + I = 1200 \quad (2)$$
$$R = D + 4 \quad (3)$$

(b) We can use substitution on this system very easily. Use equation (3) to substitute for $R$ in equation (1) and solve for $D$.
$$R + D = 0.08(1200)$$
$$R + D = 96$$
$$(D + 4) + D = 96$$
$$2D + 4 = 96$$
$$2D = 92$$
$$D = 46$$

If $D = 46$, then from equation (3), $R = 46 + 4 = 50$. Substituting for $R$ and $D$ in equation (2), we can find $I$.
$$R + D + I = 1200$$
$$50 + 46 + I = 1200$$
$$I = 1104$$

There are 50 Republicans, 46 Democrats, and 1104 Independents.

**65.** (a) A system of equations for the given conditions is
$$b + w + p = 100 \quad (1)$$
$$b = 2w \quad (2)$$
$$w = 3p \quad (3)$$

(b) We can use substitution very easily with this system. Use equations (2) and (3) to substitute into equation (1).

$$b + w + p = 100 \quad (1)$$
$$2w + 3p + p = 100$$
$$2(3p) + 3p + p = 100$$
$$6p + 3p + p = 100$$
$$10p = 100$$
$$p = 10$$

Now substitute for $p$ in equation (3).
$$w = 3p \quad (3)$$
$$w = 3(10)$$
$$w = 30$$

Substitute for $w$ in equation (2).
$$b = 2w \quad (2)$$
$$b = 2(30)$$
$$b = 60$$

There are 60 brass, 30 woodwind, and 10 percussion players in the band.

**67.** Ordered pairs of the form $(t, p)$ are
1930: (30, 11.6)
1970: (70, 4.8)
2000: (100, 9.5)

**69.**
$$11.6 = 900a + 30b + c \quad (1)$$
$$4.8 = 4900a + 70b + c \quad (2)$$
$$9.5 = 10{,}000a + 100b + c \quad (3)$$

Eliminate $c$ in equations (1) and (2) by multiplying equation (2) by $-1$ and adding the equations.
$$11.6 = 900a + 30b + c \quad (1)$$
$$-4.8 = -4900a - 70b - c \quad (2)$$
$$\overline{6.8 = -4000a - 40b \quad (4)}$$

Eliminate $c$ in equations (2) and (3) by multiplying equation (2) by $-1$ and adding the equations.

$$-4.8 = -4900a - 70b - c \quad (2)$$
$$9.5 = 10{,}000a + 100b + c \quad (3)$$
$$\overline{4.7 = 5100a + 30b \quad (5)}$$

Now we have two equations in two variables.
$$6.8 = -4000a - 40b \quad (4)$$
$$4.7 = 5100a + 30b \quad (5)$$

To eliminate $b$, multiply equation (4) by 3 and equation (5) by 4 and add the equations.

$$20.4 = -12{,}000a - 120b \quad (4)$$
$$18.8 = 20{,}400a + 120b \quad (5)$$
$$\overline{39.2 = 8400a}$$
$$0.005 \approx a$$

Substitute for $a$ in equation (5) to find $b$.
$$4.7 = 5100a + 30b$$
$$4.7 = 5100(39.2/8400) + 30b$$
$$4.7 = 23.8 + 30b$$
$$-0.637 \approx b$$

Finally, substituting for $a$ and $b$ in any of the original equations, we can find a value of $c$ of 26.5. The model is
$$p(t) = 0.005t^2 - 0.637t + 26.5.$$

**71.**
$$x + y + z - w = -1 \quad (1)$$
$$2x - y + 2z + w = 1 \quad (2)$$
$$x - y - z + 2w = 6 \quad (3)$$
$$x + 2y + 3z - 2w = -4 \quad (4)$$

Eliminate $w$ from equations (1) and (2) by adding the equations.
$$x + y + z - w = -1 \quad (1)$$
$$2x - y + 2z + w = 1 \quad (2)$$
$$\overline{3x + 3z = 0} \quad (5)$$

Eliminate $w$ from equations (1) and (3) by multiplying equation (1) by 2 and then adding the equations.
$$2x + 2y + 2z - 2w = -2 \quad (1)$$
$$x - y - z + 2w = 6 \quad (3)$$
$$\overline{3x + y + z = 4} \quad (6)$$

Eliminate $w$ from equations (3) and (4) by adding the equations.
$$x - y - z + 2w = 6 \quad (3)$$
$$x + 2y + 3z - 2w = -4 \quad (4)$$
$$\overline{2x + y + 2z = 2} \quad (7)$$

Now we have three equations in three variables.
$$x + z = 0 \quad (5)$$
$$3x + y + z = 4 \quad (6)$$
$$2x + y + 2z = 2 \quad (7)$$

To eliminate $y$, multiply equation (6) by $-1$ and add the result to equation (7).
$$-3x - y - z = -4 \quad (6)$$
$$2x + y + 2z = 2 \quad (7)$$
$$\overline{-x + z = -2} \quad (8)$$

To eliminate $x$, add equations (5) & (8).

$$x + z = 0 \quad (5)$$
$$-x + z = -2 \quad (8)$$
$$\overline{2z = -2}$$
$$z = -1$$

To find $x$, substitute for $z$ in equation (5).
$$x + z = 0 \quad (5)$$
$$x - 1 = 0$$
$$x = 1$$

To find $y$, substitute for $x$ and $z$ in equation (6).
$$3x + y + z = 4 \quad (6)$$
$$3(1) + y - 1 = 4$$
$$y + 2 = 4$$
$$y = 2$$

Finally, to find $w$, substitute for $x$, $y$, and $z$ in equation (2).
$$2x - y + 2z + w = 1 \quad (2)$$
$$2(1) - 2 + 2(-1) + w = 1$$
$$w - 2 = 1$$
$$w = 3$$

The solution is $(1, 2, -1, 3)$.

**73.** Letting $u = \dfrac{1}{x}$, $v = \dfrac{1}{y}$, and $w = \dfrac{1}{z}$, the given system can be rewritten as:
$$3u + 2v - w = -1 \quad (1)$$
$$u + v + w = 2 \quad (2)$$
$$u - v + 2w = 6 \quad (3)$$

Eliminate $w$ from equations (1) and (2) by adding the equations.
$$3u + 2v - w = -1 \quad (1)$$
$$u + v + w = 2 \quad (2)$$
$$\overline{4u + 3v = 1} \quad (4)$$

Eliminate $w$ from equations (2) and (3) by multiplying equation (2) by $-2$ and adding the equations.
$$-2u - 2v - 2w = -4 \quad (2)$$
$$u - v + 2w = 6 \quad (3)$$
$$\overline{-u - 3v = 2} \quad (5)$$

Now we have two equations in two variables.
$$4u + 3v = 1 \quad (4)$$
$$-u - 3v = 2 \quad (5)$$

To eliminate $v$, add the equations.

$$\begin{aligned} 4u + 3v &= 1 \quad (4) \\ -u - 3v &= 2 \quad (5) \\ \hline 3u &= 3 \\ u &= 1 \end{aligned}$$

Substitute for $u$ in equation (4) to find $v$.
$$4u + 3v = 1 \quad (4)$$
$$4(1) + 3v = 1$$
$$3v = -3$$
$$v = -1$$

Substitute for $u$ and $v$ in equation (2) to find $w$.
$$u + v + w + = 2 \quad (2)$$
$$1 - 1 + w = 2$$
$$w = 2$$

If $u = 1$, $v = -1$, and $w = 2$, then

$$u = \frac{1}{x} \qquad v = \frac{1}{y} \qquad w = \frac{1}{z}$$
$$1 = \frac{1}{x} \qquad -1 = \frac{1}{y} \qquad 2 = \frac{1}{z}$$
$$x = 1 \qquad y = -1 \qquad 2z = 1$$
$$z = 0.5$$

The solution to the original system is $(1, -1, 0.5)$.

**75.**
$$\begin{aligned} x + y &= a \quad (1) \\ y + z &= b \quad (2) \\ x + z &= c \quad (3) \end{aligned}$$

Eliminate $z$ in equations (2) and (3) by multiplying equation (2) by $-1$ and adding the equations.

$$\begin{aligned} -y - z &= -b \quad (2) \\ x + z &= c \quad (3) \\ \hline x - y &= c - b \quad (4) \end{aligned}$$

Now we have two equations in two variables.
$$\begin{aligned} x + y &= a \quad (1) \\ x - y &= c - b \quad (4) \end{aligned}$$

Eliminate $y$ by adding the two equations.
$$\begin{aligned} x + y &= a \quad (1) \\ x - y &= c - b \quad (4) \\ \hline 2x &= a - b + c \\ x &= \frac{a - b + c}{2} \end{aligned}$$

Substitute for $x$ in equation (1) to find $y$.
$$x + y = a \quad (1)$$
$$\frac{a - b + c}{2} + y = a$$
$$y = a - \frac{a - b + c}{2}$$

Substitute for $x$ in equation (3) to find $z$.
$$x + z = c \quad (3)$$
$$\frac{a - b + c}{2} + z = c$$
$$z = c - \frac{a - b + c}{2}$$

The solution is
$$\left( \frac{a - b + c}{2}, a - \frac{a - b + c}{2}, c - \frac{a - b + c}{2} \right)$$

## Section 5.4 Matrices

1. A $2 \times 4$ matrix is an array of numbers with two rows and four columns.

3. Because there are two rows and four columns, this is a $2 \times 4$ matrix.

5. Because there are two rows and two columns, this is a $2 \times 2$ matrix.

7. Because there are three rows and four columns, this is a $3 \times 4$ matrix.

9. (a) A coefficient matrix is a matrix whose elements are the coefficients of the variables of the equations in a system of equations.
   (b) A constant matrix is a matrix whose elements are the constants of the equations in a system of equations.
   (c) An augmented matrix is a matrix whose elements are the coefficients and the constants of the equations in a system of equations.

11. $3x + 2y = 6$
    $x - 4y = 9$
    The augmented matrix for this system is
    $$\begin{bmatrix} 3 & 2 & | & 6 \\ 1 & -4 & | & 9 \end{bmatrix}$$

13. $x + 2y - 3z = 5$
    $x + 2z = 15$
    $2y - z = 6$
    The augmented matrix for this system is
    $$\begin{bmatrix} 1 & 2 & -3 & | & 5 \\ 1 & 0 & 2 & | & 15 \\ 0 & 2 & -1 & | & 6 \end{bmatrix}$$

15. $$\begin{bmatrix} 2 & 1 & | & 1 \\ 3 & -2 & | & 12 \end{bmatrix}$$
    A system of linear equations for this matrix is
    $2x + y = 1$
    $3x - 2y = 12$

17. $$\begin{bmatrix} 1 & 1 & -1 & | & 2 \\ 2 & -3 & 1 & | & 5 \\ 3 & 2 & -4 & | & 3 \end{bmatrix}$$
    A system of equations for this matrix is
    $x + y - z = 2$
    $2x - 3y + z = 5$
    $3x + 2y - 4z = 3$

19. The goal is to have 1's along the main diagonal of the matrix and 0's elsewhere.

21. *Matrix*     *Row Operation*

    $\begin{bmatrix} 3 & 6 & | & 12 \\ 2 & -3 & | & 1 \end{bmatrix}$   Augmented matrix

    $\begin{bmatrix} 1 & 2 & | & 4 \\ 2 & -3 & | & 1 \end{bmatrix}$   (a) Row 1 was divided by 3: $R_1 \div 3$.

    $\begin{bmatrix} 1 & 2 & | & 4 \\ 0 & -7 & | & -7 \end{bmatrix}$   (b) Row 1 was multiplied by $-2$ and then added to Row 2: $-2R_1 + R_2 \rightarrow R_2$.

    $\begin{bmatrix} 1 & 2 & | & 4 \\ 0 & 1 & | & 1 \end{bmatrix}$   (c) Row 2 was divided by $-7$: $R_2 \div (-7)$.

    $\begin{bmatrix} 1 & 0 & | & 2 \\ 0 & 1 & | & 1 \end{bmatrix}$   (d) Row 2 was multiplied by $-2$ and then added to Row 1: $-2R_2 + R_1 \rightarrow R_1$

    (e) The solution is $(2, 1)$.

23. *Matrix*     *Row Operation*

    $\begin{bmatrix} 1 & 1 & -1 & | & 4 \\ 2 & -1 & 1 & | & -1 \\ 1 & 1 & -2 & | & 5 \end{bmatrix}$ Augmented matrix

    $\begin{bmatrix} 1 & 1 & -1 & | & 4 \\ 0 & -3 & 3 & | & -9 \\ 0 & 0 & -1 & | & 1 \end{bmatrix}$ (a) Row 1 was multiplied by $-2$ and then added to Row 2: $-2R_1 + R_2 \rightarrow R_2$

    (b) Row 1 was multiplied by $-1$ and then added to Row 3: $-1R_1 + R_3 \rightarrow R_3$

$\begin{bmatrix} 1 & 1 & -1 & | & 4 \\ 0 & 1 & -1 & | & 3 \\ 0 & 0 & 1 & | & -1 \end{bmatrix}$ (c) Row 2 was divided by $-3$: $R_2 \div (-3)$.

(d) Row 3 was multiplied by $-1$: $-1R_3$

$\begin{bmatrix} 1 & 1 & 0 & | & 3 \\ 0 & 1 & 0 & | & 2 \\ 0 & 0 & 1 & | & -1 \end{bmatrix}$ (e) Row 3 was added to Row 1: $R_3 + R_1 \to R_1$.

(f) Row 3 was added to Row 2: $R_3 + R_2 \to R_2$.

$\begin{bmatrix} 1 & 0 & 0 & | & 1 \\ 0 & 1 & 0 & | & 2 \\ 0 & 0 & 1 & | & -1 \end{bmatrix}$ (g) Row 2 was multiplied by $-1$ and then added to Row 1: $-1R_2 + R_1 \to R_1$

(h) The solution is $(1, 2, -1)$.

**25.** *Matrix*                *Result*

$\begin{bmatrix} 4 & 8 & | & 52 \\ 3 & -1 & | & -17 \end{bmatrix}$

$R_1 \div 4 \to \begin{bmatrix} 1 & (a) & | & 13 \\ 3 & -1 & | & -17 \end{bmatrix}$ (a) 2

$-3R_1 + R_2 \to \begin{bmatrix} 1 & (b) & | & 13 \\ 0 & -7 & | & (c) \end{bmatrix}$ (b) 2   (c) $-56$

$R_2 \div (-7) \to \begin{bmatrix} 1 & (d) & | & 13 \\ 0 & (e) & | & 8 \end{bmatrix}$ (d) 2   (e) 1

$-2R_2 + R_1 \to \begin{bmatrix} 1 & 0 & | & (f) \\ 0 & (g) & | & 8 \end{bmatrix}$ (f) $-3$   (g) 1

(h) Solution: $(-3, 8)$

**27.** *Matrix*                *Result*

$\begin{bmatrix} 2 & -3 & 1 & | & 0 \\ 1 & 1 & -1 & | & 3 \\ 3 & -2 & -2 & | & 7 \end{bmatrix}$

$R_2 \leftrightarrow R_1 \begin{bmatrix} 1 & 1 & -1 & | & (a) \\ 2 & (b) & 1 & | & 0 \\ 3 & -2 & -2 & | & 7 \end{bmatrix}$ (a) 3   (b) $-3$

$-2R_1 + R_2 \to \begin{bmatrix} 1 & 1 & -1 & | & 3 \\ 0 & -5 & (c) & | & -6 \\ 0 & -5 & 1 & | & (d) \end{bmatrix}$ (c) 3   (d) $-2$
$-3R_1 + R_3 \to$

$\begin{bmatrix} 1 & (e) & -1 & | & 3 \\ 0 & -5 & 3 & | & -6 \\ 0 & 0 & (f) & | & 4 \end{bmatrix}$ (e) 1   (f) $-2$
$-R_2 + R_3 \to$

$R_3 \div (-2) \to \begin{bmatrix} 1 & 1 & -1 & | & 3 \\ 0 & -5 & 3 & | & (g) \\ 0 & 0 & 1 & | & (h) \end{bmatrix}$ (g) $-6$   (h) $-2$

$R_3 + R_1 \to \begin{bmatrix} 1 & 1 & 0 & | & (i) \\ 0 & -5 & (j) & | & 0 \\ 0 & 0 & 1 & | & -2 \end{bmatrix}$ (i) 1   (j) 0
$-3R_3 + R_2 \to$

$R_2 \div (-5) \to \begin{bmatrix} 1 & (k) & 0 & | & 1 \\ 0 & 1 & 0 & | & (l) \\ 0 & 0 & 1 & | & -2 \end{bmatrix}$ (k) 1   (l) 0

$-R_2 + R_1 \to \begin{bmatrix} 1 & 0 & (m) & | & 1 \\ (n) & 1 & 0 & | & 0 \\ 0 & 0 & 1 & | & -2 \end{bmatrix}$ (m) 0   (n) 0

(o) Solution: $(1, 0, -2)$

**29.** The equations must be written in standard form before the coefficient matrix and constant matrix can be written.

**31.** First, write the augmented matrix for the given system. Then perform row operations.

$\begin{bmatrix} 2 & -3 & | & 18 \\ 1 & 2 & | & -5 \end{bmatrix}$

$R_1 \leftrightarrow R_2 \begin{bmatrix} 1 & 2 & | & -5 \\ 2 & -3 & | & 18 \end{bmatrix}$

$-2R_1 + R_2 \to \begin{bmatrix} 1 & 2 & | & -5 \\ 0 & -7 & | & 28 \end{bmatrix}$

$R_2 \div (-7) \to \begin{bmatrix} 1 & 2 & | & -5 \\ 0 & 1 & | & -4 \end{bmatrix}$
$-2R_2 + R_1 \to \begin{bmatrix} 1 & 0 & | & 3 \\ 0 & 1 & | & -4 \end{bmatrix}$

The solution is $(3, -4)$.

Section 5.4   Matrices

**33.** First, write the augmented matrix for the given system. Then perform row operations.

$$\begin{bmatrix} 4 & 8 & | & 0 \\ 1 & -2 & | & 1 \end{bmatrix}$$

$R_1 \leftrightarrow R_2 \begin{bmatrix} 1 & -2 & | & 1 \\ 4 & 8 & | & 0 \end{bmatrix}$

$-4R_1 + R_2 \to \begin{bmatrix} 1 & -2 & | & 1 \\ 0 & 16 & | & -4 \end{bmatrix}$

$R_2 \div 16 \to \begin{bmatrix} 1 & -2 & | & 1 \\ 0 & 1 & | & -0.25 \end{bmatrix}$

$2R_2 + R_1 \to \begin{bmatrix} 1 & 0 & | & 0.5 \\ 0 & 1 & | & -0.25 \end{bmatrix}$

The solution is $(0.5, -0.25)$.

**35.** $2x - 4y = 7 \to 2x - 4y = 7$
$2y - x = 4 \to -x + 2y = 4$

First, write the augmented matrix for the system written in standard form. Then perform row operations.

$$\begin{bmatrix} 2 & -4 & | & 7 \\ -1 & 2 & | & 4 \end{bmatrix}$$

$R_1 \leftrightarrow R_2 \begin{bmatrix} -1 & 2 & | & 4 \\ 2 & -4 & | & 7 \end{bmatrix}$

$2R_1 + R_2 \to \begin{bmatrix} -1 & 2 & | & 4 \\ 0 & 0 & | & 15 \end{bmatrix}$

The second row corresponds to the equation $0x + 0y = 15$ or $0 = 15$. This false equation indicates that the system is inconsistent (has no solution).

**37.** First, write the given system in standard form.

$3x - 2y = 6$
$-9x + 6y = -18$

Write the augmented matrix for this system. Then perform row operations.

$$\begin{bmatrix} 3 & -2 & | & 6 \\ -9 & 6 & | & -18 \end{bmatrix}$$

$3R_1 + R_2 \to \begin{bmatrix} 3 & -2 & | & 6 \\ 0 & 0 & | & 0 \end{bmatrix}$

The second row corresponds to the equation $0 = 0$, which is always true. The equations are dependent.

**39.**  $x \quad\quad + z = 1$
$\quad\quad y - z = 5$
$x - y \quad\quad = 2$

Write the augmented matrix for this system. Then perform row operations.

$$\begin{bmatrix} 1 & 0 & 1 & | & 1 \\ 0 & 1 & -1 & | & 5 \\ 1 & -1 & 0 & | & 2 \end{bmatrix}$$

$-R_1 + R_3 \to \begin{bmatrix} 1 & 0 & 1 & | & 1 \\ 0 & 1 & -1 & | & 5 \\ 0 & -1 & -1 & | & 1 \end{bmatrix}$

$R_2 + R_3 \to \begin{bmatrix} 1 & 0 & 1 & | & 1 \\ 0 & 1 & -1 & | & 5 \\ 0 & 0 & -2 & | & 6 \end{bmatrix}$

$R_3 \div (-2) \to \begin{bmatrix} 1 & 0 & 1 & | & 1 \\ 0 & 1 & -1 & | & 5 \\ 0 & 0 & 1 & | & -3 \end{bmatrix}$

$R_3 + R_2 \to \begin{bmatrix} 1 & 0 & 1 & | & 1 \\ 0 & 1 & 0 & | & 2 \\ 0 & 0 & 1 & | & -3 \end{bmatrix}$

$-1R_3 + R_1 \to \begin{bmatrix} 1 & 0 & 0 & | & 4 \\ 0 & 1 & 0 & | & 2 \\ 0 & 0 & 1 & | & -3 \end{bmatrix}$

The solution is $(4, 2, -3)$.

**41.**  $x + y - z = -4$
$-x - 2y + z = 7$
$2x + 2y + z = -2$

Write the augmented matrix for this system. Then perform row operations.

$$\begin{bmatrix} 1 & 1 & -1 & | & -4 \\ -1 & -2 & 1 & | & 7 \\ 2 & 2 & 1 & | & -2 \end{bmatrix}$$

$R_1 + R_2 \to \begin{bmatrix} 1 & 1 & -1 & | & -4 \\ 0 & -1 & 0 & | & 3 \\ 0 & 0 & 3 & | & 6 \end{bmatrix}$
$-2R_1 + R_3 \to$

$R_2 \div (-1) \to \begin{bmatrix} 1 & 1 & -1 & | & -4 \\ 0 & 1 & 0 & | & -3 \\ 0 & 0 & 1 & | & 2 \end{bmatrix}$
$R_3 \div 3 \to$

$$R_3 + R_1 \rightarrow \begin{bmatrix} 1 & 1 & 0 & | & -2 \\ 0 & 1 & 0 & | & -3 \\ 0 & 0 & 1 & | & 2 \end{bmatrix}$$

$$-R_2 + R_1 \rightarrow \begin{bmatrix} 1 & 0 & 0 & | & 1 \\ 0 & 1 & 0 & | & -3 \\ 0 & 0 & 1 & | & 2 \end{bmatrix}$$

The solution is $(1, -3, 2)$.

**43.** First, rewrite the given system in standard form.
$$\begin{aligned} 2x - 5y &= -5 \\ 5x + 2y &= 2 \end{aligned}$$
The augmented matrix is:
$$\begin{bmatrix} 2 & -5 & | & -5 \\ 5 & 2 & | & 2 \end{bmatrix}$$

**45.** First, rewrite the given system in standard form.
$$\begin{aligned} 3x - 4y + z &= 2 \\ x + y + 3z &= 1 \\ 2x + 3y - z &= 0 \end{aligned}$$
The augmented matrix is:
$$\begin{bmatrix} 3 & -4 & 1 & | & 2 \\ 1 & 1 & 3 & | & 1 \\ 2 & 3 & -1 & | & 0 \end{bmatrix}$$

**47.** First, rewrite the given system in standard form.
$$\begin{aligned} 3x + 4y &= -5 \\ x &= -3 \end{aligned}$$
Let $A$ = the augmented matrix.

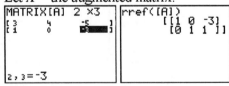

The solution is $(-3, 1)$.

**49.** First, rewrite the given system in standard form.
$$\begin{aligned} 3x + y + z &= 5 \\ 2x - y + z &= 6 \\ y - 2z &= 2 \end{aligned}$$
Let $A$ = the augmented matrix.

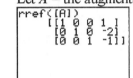

The solution is $(3, -2, -2)$.

**51.** Let $A$ = the augmented matrix.

The solution is $(0, -1, 2)$.

**53.** Let $A$ = the augmented matrix.

The solution is $(1, -2, -1)$.

**55.** (a) The data written as a $3 \times 2$ matrix is
$$\begin{bmatrix} 428 & 453 \\ 408 & 430 \\ 483 & 544 \end{bmatrix}.$$
The entry in row 3, column 2 represents average number of minutes in a woman's work day in developing countries.

(b) The data written as a $2 \times 3$ matrix is
$$\begin{bmatrix} 428 & 408 & 483 \\ 453 & 430 & 544 \end{bmatrix}.$$
To find the difference between men's and women's times in the United States, look in the first column.

**57.** (a) Because the variables in the system of equations are written from left to right as $z$, $x$, and $y$, the columns of the given coefficient matrix correspond to $z$, $x$, and $y$ from left to right. Therefore, the given calculator result corresponds to a solution of $z = -3$, $x = 4$, and $y = -5$.

(b) Rewriting the given system in standard form, we have

Section 5.5 Applications

$$x - 2y + z = 11$$
$$-2x + y + 3z = -22$$
$$4x - y - z = 24$$

Let $A$ = the augmented matrix. Using a calculator to solve the system, we obtain

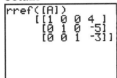

The solution is $x = 4$, $y = -5$, and $z = -3$. This is the same solution as that obtained in part (a).

**59.** To solve the given system, let $A$ = the augmented matrix.

The solution is $(-2, -3, 1, 2)$.

**61.** First, write the given system in standard form.

$$x + y \phantom{+ 2z + 3w} = 0$$
$$\phantom{x +} y + 2z \phantom{+ 3w} = -3$$
$$\phantom{x + y +} 2z + 3w = 4$$
$$4x \phantom{+ y + 2z} + 3w = 10$$

Let $A$ = the augmented matrix.

The solution is $(1, -1, -1, 2)$.

**63.** First, write the augmented matrix for the given system. Then perform row operations.

$$\begin{bmatrix} 1 & 3 & | & 2 \\ a & 1 & | & 1 \end{bmatrix}$$

$$-aR_1 + R_2 \to \begin{bmatrix} 1 & 3 & | & 2 \\ 0 & 1-3a & | & 1-2a \end{bmatrix}$$

We know that if the last row yields a false equation (such as 0 = a constant), the system has no solution. Thus, if $1 - 3a = 0$, or equivalently $a = \dfrac{1}{3}$, the equation has no solution. Otherwise, if $a \neq \dfrac{1}{3}$, the solution is unique.

## Section 5.5 Applications

**1.** Let $x$ = pounds of assorted creams and $y$ = pounds of chocolate-covered almonds.

|         | Number of pounds | Cost per pound | Total cost |
|---------|------------------|----------------|------------|
| Creams  | $x$              | 2.95           | $2.95x$    |
| Almonds | $y$              | 3.70           | $3.70y$    |
| Mixture | 20               | 3.25           | 65         |

The system of equations is:
$$x + y = 20 \quad (1)$$
$$2.95x + 3.70y = 65 \quad (2)$$

Multiply equation (1) by $-3.70$ and then add the equations to eliminate $y$.

$$-3.70x - 3.70y = -74 \quad (1)$$
$$2.95x + 3.70y = 65 \quad (2)$$
$$\overline{-0.75x \phantom{+ 3.70y} = -9}$$
$$x = 12$$

There are 12 pounds of assorted creams in the 20-pound mixture.

3. Let $x$ = pounds of apples and $y$ = pounds grapes.

|        | Number of pounds | Cost per pound | Total cost |
|--------|------------------|----------------|------------|
| Apples | $x$              | 0.45           | $0.45x$    |
| Grapes | $y$              | 1.29           | $1.29y$    |
| Total  | 8                |                | 6.12       |

The system of equations is:
$$x + y = 8 \quad (1)$$
$$0.45x + 1.29y = 6.12 \quad (2)$$

Multiply equation (1) by $-1.29$ and then add the equations to eliminate $y$.
$$-1.29x - 1.29y = -10.32 \quad (1)$$
$$0.45x + 1.29y = 6.12 \quad (2)$$
$$\overline{-0.84x \qquad\qquad = -4.20}$$
$$x = 5$$

The shopper bought 5 pounds of apples.

5. Let $x$ = number of cakes, $y$ = number of pies, and $z$ = number of dozens of cookies.

|         | Number of items sold | Cost per item | Total cost |
|---------|----------------------|---------------|------------|
| Cakes   | $x$                  | 3.50          | $3.5x$     |
| Pies    | $0.75y$              | 4.00          | $3y$       |
| Cookies | $z$                  | 1.40          | $1.4z$     |
| Total   | 128                  |               | 361        |

The system of equations is:
$$x + 0.75y + z = 128 \quad (1)$$
$$3.5x + 3y + 1.4z = 361 \quad (2)$$
$$-x + y = 6 \quad (3)$$

Let $A$ = the augmented matrix.

There were 42 cakes, 48 pies, and 50 dozen cookies donated for the sale.

7. Let $x$ = number of patron's tickets $y$ = number of other tickets.

|        | Number of tickets | Cost of ticket | Total cost |
|--------|-------------------|----------------|------------|
| Patron | $x$               | 2              | $2x$       |
| Other  | $y$               | 3              | $3y$       |
| Total  | 456               |                | 1131       |

The system of equations is:
$$x + y = 456 \quad (1)$$
$$2x + 3y = 1131 \quad (2)$$

Multiply equation (1) by $-2$ and then add the equations to eliminate $x$.
$$-2x - 2y = -912 \quad (1)$$
$$2x + 3y = 1131 \quad (2)$$
$$\overline{\qquad\qquad y = 219}$$

Substituting $y = 219$ in equation (1), we get $x = 237$. There were 237 patron's tickets sold and 219 other tickets sold for the play.

9. Let $x$ = number of box seats $y$ = number of reserved seats.

|          | Number of seats | Cost of ticket | Total cost |
|----------|-----------------|----------------|------------|
| Box      | $x$             | 12             | $12x$      |
| Reserved | $y$             | 8              | $8y$       |
| Total    | 42,245          |                | 395,004    |

The system of equations is:
$$x + y = 42{,}245 \quad (1)$$
$$12x + 8y = 395{,}004 \quad (2)$$

Multiply equation (1) by $-8$ and then add the equations to eliminate $y$.
$$-8x - 8y = -337{,}960 \quad (1)$$
$$12x + 8y = 395{,}004 \quad (2)$$
$$\overline{4x \qquad\qquad = 57{,}044}$$
$$x = 14{,}261$$

There were 14,261 box seats sold.

11. Let $x$ = number of orchestra seats, $y$ = number of loge seats, and $z$ = number of balcony seats.

|           | Number of seats | Cost of ticket | Total cost |
|-----------|-----------------|----------------|------------|
| Orchestra | $x$             | 65             | $65x$      |
| Loge      | $y$             | 50             | $50y$      |
| Balcony   | $z$             | 30             | $30z$      |
| Total     | 5500            |                | 246,750    |

Section 5.5 Applications

The system of equations is:
$$x + y + z = 5500 \quad (1)$$
$$65x + 50y + 30z = 246{,}750 \quad (2)$$
$$-x + y = 100 \quad (3)$$

Let $A$ = the augmented matrix.

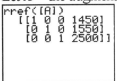

There were 1450 orchestra seats, 1550 loge seats, and 2500 balcony seats sold.

**13.** Let $x$ = the amount invested at 8.5% and $y$ = the amount invested at 6.9%.

|  | Amount invested | Interest rate | Interest earned |
|---|---|---|---|
| 8.5% | $x$ | 0.085 | $0.085x$ |
| 6.9% | $y$ | 0.069 | $0.069y$ |
| Total | 10,000 |  | 745.84 |

The system of equations is:
$$x + y = 10{,}000 \quad (1)$$
$$0.085x + 0.069y = 745.84 \quad (2)$$

Multiply equation (1) by $-0.069$ and add the equations to eliminate $y$.

$$-0.069x - 0.069y = -690 \quad (1)$$
$$\underline{0.085x + 0.069y = 745.84} \quad (2)$$
$$0.016x = 55.84$$
$$x = 3490$$

If \$3490 was invested at 8.5% and a total of \$10,000 was invested, \$6510 was invested at 6.9%.

**15.** Let $x$ = the amount borrowed at 6% and $y$ = the amount borrowed at 7%.

|  | Amount borrowed | Interest rate | Interest owed |
|---|---|---|---|
| 6% | $x$ | 0.06 | $0.06x$ |
| 7% | $y$ | 0.07 | $0.07y$ |
| Total | 7000 |  | 452 |

The system of equations is:
$$x + y = 7000 \quad (1)$$
$$0.06x + 0.07y = 452 \quad (2)$$

Multiply equation (1) by $-0.06$ and add the equations to eliminate $x$.

$$-0.06x - 0.06y = -420 \quad (1)$$
$$\underline{0.06x + 0.07y = 452} \quad (2)$$
$$0.01y = 32$$
$$y = 3200$$

If \$3200 was borrowed at 7% and a total of \$7000 was borrowed, \$3800 was borrowed at 6%.

**17.** Let $x$ = the amount invested in the stock fund, $y$ = the amount invested in the bond fund, and $z$ = the amount invested in the mutual fund.

|  | Amount invested | Interest rate | Interest earned |
|---|---|---|---|
| Stock | $x$ | 0.06 | $0.06x$ |
| Bond | $y$ | 0.075 | $0.075y$ |
| Mutual | $z$ | 0.08 | $0.08z$ |
| Total | 16,500 |  | 1191.50 |

The system of equations is:
$$x + y + z = 16{,}500$$
$$0.06x + 0.075y + 0.08z = 1191.50$$
$$2y - z = 1200$$

Let $A$ = the augmented matrix.

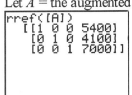

There was \$5400 invested in the stock fund, \$4100 invested in the bond fund, and \$7000 invested in the mutual fund.

**19.** The number of ounces of acid is found by multiplying the total volume of solution (20 ounces) by the concentration (0.10).

**21.** Let $x$ = liters of 34% muriatic acid solution and $y$ = liters of 55% muriatic acid solution

|  | Liters of solution | % acid | Liters of acid |
|---|---|---|---|
| Weak | $x$ | 0.34 | $0.34x$ |
| Strong | $y$ | 0.55 | $0.55y$ |
| Mixture | 70 | 0.40 | 28 |

The system of equations is:
$$x + y = 70 \quad (1)$$
$$0.34x + 0.55y = 28 \quad (2)$$
Multiply equation (1) by $-0.34$ and add the equations to eliminate $x$.
$$-0.34x - 0.34y = -23.8 \quad (1)$$
$$\underline{0.34x + 0.55y = \phantom{-}28 \quad (2)}$$
$$0.21y = 4.2$$
$$y = 20$$
If 20 liters of the 55% solution were used to make 70 liters of 40% mixture, then 50 liters of the 34% solution were used.

**23.** Let $x$ = gallons of 2% chlorine solution and $y$ = gallons of 3.5% chlorine solution

|  | Gallons of solution | % chlorine | Gallons of chlorine |
|---|---|---|---|
| Weak | $x$ | 0.02 | $0.02x$ |
| Strong | $y$ | 0.035 | $0.035y$ |
| Mixture | 600 | 0.025 | 15 |

The system of equations is:
$$x + y = 600 \quad (1)$$
$$0.02x + 0.035y = 15 \quad (2)$$
Multiply equation (1) by $-0.02$ and add the equations to eliminate $x$.
$$-0.02x - 0.02y = -12 \quad (1)$$
$$\underline{0.02x + 0.035y = 15 \quad (2)}$$
$$0.015y = 3$$
$$y = 200$$
To obtain the desired result, 200 gallons of 3.5% chlorine solution were used.

**25.** Let $x$ = liters of 10% alcohol solution, $y$ = liters of 20% alcohol solution, and $z$ = liters of 50% alcohol solution.

|  | Liters of solution | % alcohol | Liters of alcohol |
|---|---|---|---|
| 10% | $x$ | 0.1 | $0.1x$ |
| 20% | $y$ | 0.2 | $0.2y$ |
| 50% | $z$ | 0.5 | $0.5z$ |
| Mixture | 18 | 0.3 | 5.4 |

The system of equations is:
$$x + y + z = 18$$
$$0.1x + 0.2y + 0.5z = 5.4$$
$$2y - z = 0$$

Let $A$ = the augmented matrix.

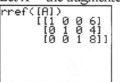

The mixture contained 6 liters of 10% solution, 4 liters of 20% solution, and 8 liters of 50% solution.

**27.** Because the boat is pushed by the current, the effective rate is the sum of the rates of the boat and the current.

**29.** Let $b$ = the speed of the boat in still water and $c$ = the speed of the current.

|  | Rate (mph) | Time (hours) | Distance (miles) |
|---|---|---|---|
| Upstream | $b - c$ | 7 | $7(b-c)$ |
| Downstream | $b + c$ | 3 | $3(b+c)$ |

Because the boat traveled 21 miles upstream and 21 miles downstream, we get the following system of equations.
$$7(b-c) = 21 \;\rightarrow\;$$
$$3(b+c) = 21 \;\rightarrow\;$$
$$b - c = 3 \quad (1)$$
$$b + c = 7 \quad (2)$$
Add the two equations to eliminate $c$.
$$b - c = 3$$
$$\underline{b + c = 7}$$
$$2b = 10$$
$$b = 5$$
The speed of the boat is 5 mph. Substituting in equation (2), we find that the speed of the current is 2 mph.

**31.** Let $a$ = the speed of the airplane in still air and $w$ = the speed of the wind.

|  | Rate (mph) | Time (hours) | Distance (miles) |
|---|---|---|---|
| Against | $a - w$ | 8 | $8(a-w)$ |
| With | $a + w$ | 7 | $7(a+w)$ |

Because the airplane traveled 3360 miles in each direction, we get the following system of equations.

Section 5.5   Applications

$8(a-w) = 3360 \rightarrow a - w = 420$  (1)
$7(a+w) = 3360 \rightarrow a + w = 480$  (2)
Add the two equations to eliminate $w$

$$\begin{aligned} a - w &= 420 \\ a + w &= 480 \\ \hline 2a &= 900 \\ a &= 450 \end{aligned}$$

The speed of the airplane is 450 mph. Substituting in equation (2), we find that the speed of the wind is 30 mph.

**33.** Let $x$ = the speed of the boat in still water, $y$ = the speed of the current, and $z$ = the distance paddled the first day.

|  | Rate (mph) | Time (hours) | Distance (miles) |
|---|---|---|---|
| Upstream | $x - y$ | 6 | $z$ |
| Downstream | $x + y$ | 2 | $z$ |
| Drifting | $y$ | 7 | $z + 2$ |

The system of equations is:
$6(x - y) = z \rightarrow 6x - 6y - z = 0$
$2(x + y) = z \rightarrow 2x + 2y - z = 0$
$7y = z + 2 \rightarrow 7y - z = 2$

Let $A$ = the augmented matrix.

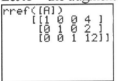

The speed of the boat in still water is 4 mph and the speed of the current is 2 mph.

**35.** Substitute each ordered pair into the equation $y = ax + b$.
$(3, -2): -2 = a(3) + b$
$(-1, 6): 6 = a(-1) + b$
The system of equations is:
$3a + b = -2$
$-a + b = 6$
Let $A$ = the augmented matrix.

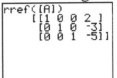

The solution matrix shows that $a = -2$ and $b = 4$. The equation is $y = -2x + 4$.

**37.** Substitute each ordered pair into the equation $y = ax^2 + bx + c$.
$(-3, 22): 22 = a(-3)^2 + b(-3) + c$
$(4, 15): 15 = a(4)^2 + b(4) + c$
$(1, -6): -6 = a(1)^2 + b(1) + c$
The system of equations is:
$9a - 3b + c = 22$
$16a + 4b + c = 15$
$a + b + c = -6$

Let $A$ = the augmented matrix.

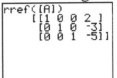

The solution matrix shows that $a = 2$, $b = -3$, and $c = -5$. The equation is $y = 2x^2 - 3x - 5$.

**39.** Substitute each ordered pair into the equation $y = ax^2 + bx + c$.
$(-1, 2): 2 = a(-1)^2 + b(-1) + c$
$(2, -7): -7 = a(2)^2 + b(2) + c$
$(3, -26): -26 = a(3)^2 + b(3) + c$
The system of equations is:
$a - b + c = 2$
$4a + 2b + c = -7$
$9a + 3b + c = -26$

Let $A$ = the augmented matrix.

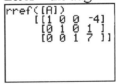

The solution matrix shows that $a = -4$, $b = 1$, and $c = 7$. The equation is $y = -4x^2 + x + 7$.

**41.** Let $x$ = the amount the father pays and $y$ = the amount the son pays.
The system of equations is:
$x + y = 51$  (1)
$x = 2y$  (2)

Using equation (2) to substitute for $x$ in equation (1), we obtain
$$x + y = 51$$
$$2y + y = 51$$
$$3y = 51$$
$$y = 17$$
Substitute for $y$ in equation (2) to find $x$.
$$x = 2y$$
$$x = 2(17)$$
$$x = 34$$
The father pays $34 and the son pays $17.

**43.** Let $x$ = the number of five-dollar bills and $y$ = the number of ten-dollar bills.
The system of equations is:
$$x + y = 70 \quad (1)$$
$$5x + 10y = 505 \quad (2)$$
Solving equation (1) for $y$, we obtain $y = 70 - x$. Substitute into equation (2).
$$5x + 10y = 505$$
$$5x + 10(70 - x) = 505$$
$$5x + 700 - 10x = 505$$
$$700 - 5x = 505$$
$$-5x = -195$$
$$x = 39$$
Substitute for $x$ in $y = 70 - x$ to find $y$.
$$y = 70 - 39 = 31$$
There were 39 five-dollar bills and 31 ten-dollar bills.

**45.** Let $x$ = the number of heavy-duty drills and $y$ = the number of homeowner's drills.
The system of equations is:
$$x + y = 31 \quad (1)$$
$$46.50x + 18.95y = 918.05 \quad (2)$$
Solving equation (1) for $y$, we obtain $y = 31 - x$. Substitute into equation (2).
$$46.50x + 18.95y = 918.05$$
$$46.50x + 18.95(31 - x) = 918.05$$
$$46.50x + 587.45 - 18.95x = 918.05$$
$$27.55x + 587.45 = 918.05$$
$$27.55x = 330.60$$
$$x = 12$$
Substitute for $x$ in $y = 31 - x$ to find $y$:
$y = 31 - 12 = 19$. There were 12 heavy-duty drills and 19 homeowner's drills.

**47.** Let $x$ = the number of cans of high-gloss and $y$ = the number of cans of flat latex.
The system of equations is:
$$x + y = 75 \quad (1)$$
$$19.95x + 10.95y = 1109.25 \quad (2)$$
Solving equation (1) for $y$, we obtain $y = 75 - x$. Substitute into equation (2).
$$19.95x + 10.95y = 1109.25$$
$$19.95x + 10.95(75 - x) = 1109.25$$
$$19.95x + 821.25 - 10.95x = 1109.25$$
$$9x + 821.25 = 1109.25$$
$$9x = 288$$
$$x = 32$$
Substitute for $x$ in $y = 75 - x$ to find $y$.
$$y = 75 - 32 = 43$$
There were 32 cans of high gloss enamel and 43 cans of flat latex.

**49.** Let $x$ = the number of orchid corsages and $y$ = the number of carnation corsages.
The system of equations is:
$$x + y = 95 \quad (1)$$
$$23x + 14y = 1663 \quad (2)$$
Solving equation (1) for $y$, we obtain $y = 95 - x$. Substitute into equation (2).
$$23x + 14y = 1663$$
$$23x + 14(95 - x) = 1663$$
$$23x + 1330 - 14x = 1663$$
$$9x + 1330 = 1663$$
$$9x = 333$$
$$x = 37$$
Substitute for $x$ in $y = 95 - x$ to find $y$.
$$y = 95 - 37 = 58$$
There were 37 orchid corsages and 58 carnation corsages sold.

**51.** Let $x$ = the number of Democrats who voted and $y$ = the number of Republicans who voted.
The system of equations is:
$$x + y = 700 \quad (1)$$
$$0.88x + 0.72y = 536 \quad (2)$$
Solving equation (1) for $x$, we obtain $x = 700 - y$. Substitute into equation (2).

## Section 5.5  Applications

$$0.88x + 0.72y = 536$$
$$0.88(700 - y) + 0.72y = 536$$
$$616 - 0.88y + 0.72y = 536$$
$$616 - 0.16y = 536$$
$$-0.16y = -80$$
$$y = 500$$

There were 500 Republicans who voted in the election. If 72% of Republicans voted for the referendum, then 28% voted against it. There were $0.28(500) = 140$ Republicans who voted against it.

**53.** Let $x$ = the number climbing the 8-mile trail, $y$ = the number climbing the 6-mile trail, and $z$ = the number climbing the 5.5-mile trail.

The system of equations is:
$$x + y + z = 40$$
$$y = z + 2$$
$$8x + 6y + 5.5z = 280$$

Writing these equations in standard form, we obtain
$$x + y + z = 40$$
$$y - z = 2$$
$$8x + 6y + 5.5z = 280$$

Let $A$ = the augmented matrix.

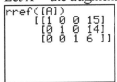

The solution matrix shows that 22 members climbed the 8-mile trail, 10 members climbed the 6-mile trail, and 8 members climbed the 5.5-mile trail.

**55.** Let $x$ = the number of grade A credit hours, $y$ = the number of grade B credit hours, and $z$ = the number of grade C credit hours.

The system of equations is:
$$x + y + z = 17$$
$$y = 2z$$
$$4x + 3y + 2z = 52$$

Writing these equations in standard form, we obtain
$$x + y + z = 17$$
$$y - 2z = 0$$
$$4x + 3y + 2z = 52$$

Let $A$ = the augmented matrix.

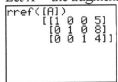

The solution matrix shows that the student got an A for 5 credit hours of classes, a B for 8 credit hours of classes and a C for 4 credit hours of classes.

**57.** Let $x$ = the number of 1-inch volumes in the first set, $y$ = the number of 1.5-inch volumes in the second set, and $z$ = the number of 2-inch volumes in the third set.

The system of equations is:
$$x + y + z = 35$$
$$x = y + 1$$
$$x + 1.5y + 2z = 48$$

Writing these equations in standard form, we obtain
$$x + y + z = 35$$
$$x - y = 1$$
$$x + 1.5y + 2z = 48$$

Let $A$ = the augmented matrix.

```
rref([A])
    [[1 0 0 15]
     [0 1 0 14]
     [0 0 1 6 ]]
```

The solution matrix shows that there were 15 volumes in the first set, 14 volumes in the second set, and 6 volumes in the third set.

**59.** Let $x$ = the price of a hamburger, $y$ = the price of an order of fries, and $z$ = the price of a shake.

The system of equations is:
$$2x + 3y + 2z = 9.75$$
$$4x + 2z = 10.80$$
$$5y + 3z = 9.25$$

Writing these equations in standard form, we obtain
$$2x + 3y + 2z = 9.75$$
$$4x + 2z = 10.80$$
$$5y + 3z = 9.25$$

Let $A$ = the augmented matrix.

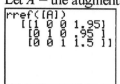

The solution matrix shows that the price of a hamburger is $1.95, the price of an order of fries is $0.95, and the price of a shake is $1.50.

**61.** (a) Using linear regression on a calculator, we find that a linear equation that models the percent of crime stories is $y = -6x + 28$.

Using linear regression on a calculator, we find that a linear equation that models the percent of human interest stories is approximately $y = 6x + 5$.

(b) From the graph, the solution of the system is approximately (1.9, 16.5).

(c) Solve the system by substituting $6x + 5$ for $y$ in the first equation.
$$6x + 5 = -6x + 28$$
$$12x + 5 = 28$$
$$12x = 23$$
$$x = \frac{23}{12} \approx 1.9$$

Now substitute for $x$ in the second equation to find $y$.
$$y = 6x + 5$$
$$y = 6\left(\frac{23}{12}\right) + 5$$
$$y = 16.5$$

The solution is also approximately (1.9, 16.5). This means that in 2000 ($x \approx 2$), the percentages of news stories involving crime and human interest was approximately equal (about 16.5%).

**63.** Substitute each ordered pair into the equation $y = ax^2 + bx + c$.

$(2, 343)$: $343 = a(2)^2 + b(2) + c$

$(6, 1033)$: $1033 = a(6)^2 + b(6) + c$

$(10, 2124)$: $2124 = a(10)^2 + b(10) + c$

The system of equations is:
$$4a + 2b + c = 343$$
$$36a + 6b + c = 1033$$
$$100a + 10b + c = 2124$$

**65.**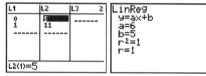

The model is
$y = 12.53x^2 + 72.25x + 148.38$. This is the same model as was obtained in Exercise 64.

**67.** Let $x$ = the number of ounces of oil added to correct the problem and $y$ = the number of ounces of oil in the original mixture.

The man had $6(128) = 768$ ounces of 2% mixture. The amount of oil in his original mixture was $0.02(768) = 15.36$ ounces. The system of equations is:
$$x + y = 0.025(768 + x)$$
$$y = 15.36$$

This system can be rewritten as
$$0.975x + y = 19.2 \quad (1)$$
$$y = 15.36 \quad (2)$$

Use equation (2) to substitute for $y$ in equation (1) and solve for $x$.
$$0.975x + y = 19.2$$
$$0.975x + 15.36 = 19.2$$
$$0.975x = 3.84$$
$$x \approx 3.94$$

He should add approximately 3.94 ounces of oil to the mixture to obtain a 2.5% mixture.

Section 5.6 Systems of Linear Inequalities

**69.** Let $x$ = the rate for the supervisor and $y$ = the rate for the inspector.
The supervisor walked 0.12 miles in the time it took the train to go $0.12 + 0.6 = 0.72$ miles.

Train: $\dfrac{d}{r} = t \;\to\; \dfrac{0.72}{30} = 0.024$ hour

Supervisor: $rt = d \;\to\; \begin{array}{l} x(0.024) = 0.12 \\ x = 5 \text{ mph} \end{array}$

It took the supervisor 3 minutes, or 1/20 hour, to get to the end of the tunnel. The distance he traveled was $d = rt = 5(1/20) = 0.25$ mile. The distance the inspector traveled in the 3 minutes was $0.6 - 0.25 = 0.35$ mile. The rate for the inspector was $y = \dfrac{d}{t} = \dfrac{0.35}{1/20} = 7$ mph.

So, the supervisor walked at a rate of 5 mph and the inspector walked at a rate of 7 mph.

## Section 5.6  Systems of Linear Inequalities

**1.** The graph of each linear inequality is a half-plane. The graph of the solution set of the system is the intersection of the half-planes.

**3.** $x \geq 3$
$y \geq 2$

**5.** $16 \geq 3x - 4y \;\;\to\;\; y \geq 0.75x - 4$
$y < 6 \;\;\to\;\; y < 6$

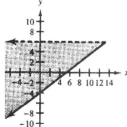

**7.** $3y - 4x < 15 \;\;\to\;\; y < \dfrac{4}{3}x + 5$
$y > 0 \;\;\to\;\; y > 0$

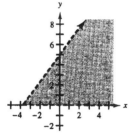

**9.** $y \geq \dfrac{1}{2}x$
$y \leq 3x$

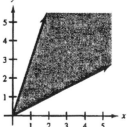

**11.** $6 \leq 2x + 3y \quad \rightarrow \quad y \geq -\dfrac{2}{3}x + 2$
$2y - 5x \geq 14 \quad \rightarrow \quad y \geq 2.5x + 7$

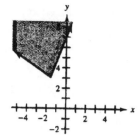

**13.** $y - 2x > 7 \quad \rightarrow \quad y > 2x + 7$
$y + 3 < 2x \quad \rightarrow \quad y < 2x - 3$

Notice that the boundary lines have the same slope but different intercepts. Therefore, the lines do not intersect. Because the solution of the first inequality is the half-plane above the boundary and the solution of the second inequality is the half-plane below the boundary, the solution is $\varnothing$.

**15.** From the given figure, the system of inequalities is $y \leq 4$, $x \leq 5$, $y \geq 0$, and $x \geq 0$.

**17.** From the given figure, using slopes and intercepts as a guide, the system of inequalities is $y \leq -8x + 8$ and $y > -\dfrac{1}{3}x - 2$.

**19.** The graph of the given system of inequalities is a square, centered at the origin, and its interior. The sides are parallel to the axes, and the length of each side is 10.

**21.** $y + 3x \leq 3 \quad \rightarrow \quad y \leq -3x + 3$
$y + x + 1 \geq 0 \quad \rightarrow \quad y \geq -x - 1$
$y < x + 7 \quad \rightarrow \quad y < x + 7$

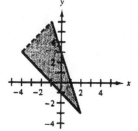

**23.** $x + 5 \geq 0 \quad \rightarrow \quad x \geq -5$
$y - x > 1 \quad \rightarrow \quad y > x + 1$
$2x + y < 6 \quad \rightarrow \quad y < -2x + 6$

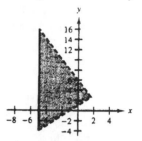

**25.** $y - 2x < 1 \quad \rightarrow \quad y < 2x + 1$
$y + x + 2 \geq 0 \quad \rightarrow \quad y \geq -x - 2$
$y + 7 \geq 0 \quad \rightarrow \quad y \geq -7$

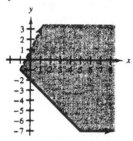

**27.** The number of people cannot be negative: $y \geq 0$ and $x \geq 0$.

**29.** $2x + 3y \leq 18 \quad \rightarrow \quad y \leq -\dfrac{2}{3}x + 6$
$x \geq 0 \quad \rightarrow \quad x \geq 0$
$y \geq 0 \quad \rightarrow \quad y \geq 0$

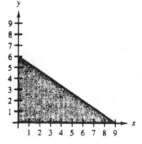

**31.** $y \geq x \qquad x \geq 0$
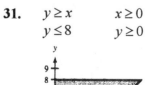
$y \leq 8 \qquad y \geq 0$

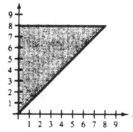

Section 5.6  Systems of Linear Inequalities

33. $-2x + y \leq 7$  →  $y \leq 2x + 7$
    $x \leq 5$  →  $x \leq 5$
    $x \geq 0$  →  $x \geq 0$
    $y \geq 0$  →  $y \geq 0$

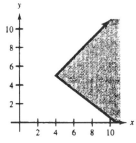

35. $-x + y \leq 1$  →  $y \leq x + 1$
    $3x + 4y \geq 32$  →  $y \geq -0.75x + 8$
    $x \geq 0$  →  $x \geq 0$
    $y \geq 0$  →  $y \geq 0$

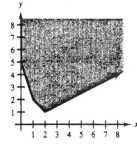

37. $3x + y \geq 5$  →  $y \geq -3x + 5$
    $x + y \geq 3$  →  $y \geq -x + 3$
    $y \geq \frac{1}{2}x$  →  $y \geq \frac{1}{2}x$
    $x \geq 0$  →  $x \geq 0$
    $y \geq 0$  →  $y \geq 0$

39. $-x + y \geq 3$  →  $y \geq x + 3$
    $x + y \leq -2$  →  $y \leq -x - 2$
    $x \geq 0$  →  $x \geq 0$
    $y \geq 0$  →  $y \geq 0$

The last two inequalities indicate that the solution set is in the first quadrant. However, the intersection of the first two inequalities is ∅ in the first quadrant. Therefore, the solution of the system is ∅.

41. Let $x$ = the number of millions of dollars in short-term notes and $y$ = the number of millions of dollars in mortgages. The system of inequalities is
    $x + y \leq 10$  →  $y \leq -x + 10$
    $y \leq 7$  →  $y \leq 7$
    $x \geq 0$  →  $x \geq 0$
    $y \geq 0$  →  $y \geq 0$

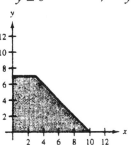

The most that can be invested in mortgages is $7 million. If $6 million is invested in short-term notes, the most that can be invested in mortgages is $4 million.

43. Let $x$ = the number of tuna fruit salads and $y$ = the number of chicken Caesar salads. The system of inequalities is
    $x \leq \frac{2}{3}y$  →  $y \geq \frac{3}{2}x$
    $x + y \geq 50$  →  $y \geq -x + 50$
    $x \geq 0$  →  $x \geq 0$
    $y \geq 0$  →  $y \geq 0$

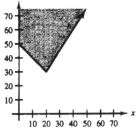

If $y = 60$, then $x$ ranges from 0 to 40. If 60 chicken Caesar salads are prepared, then between 0 and 40 tuna fruit salads should be prepared.

**45.** (a) Let $x =$ the number of barns and $y =$ the number of tool buildings. The system of inequalities is

$$15x + 5y \le 90 \rightarrow y \le -3x + 18$$
$$7x + 6y \le 84 \rightarrow y \le -\frac{7}{6}x + 14$$
$$x \ge 0 \rightarrow x \ge 0$$
$$y \ge 0 \rightarrow y \ge 0$$

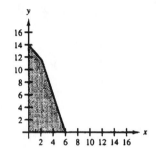

(b) Yes, the builder can build 4 barns and 5 tool buildings because $x = 4$ and $y = 5$ satisfies all the inequalities.

(c) No, the builder cannot construct 5 barns and 4 tool buildings because $x = 5$ and $y = 4$ does not satisfy the inequality $15x + 5y \le 90$.

**47.** Because mortgages yield the higher interest rate, the most amount of money possible should be invested in mortgages. The maximum amount possible is $7 million. That leaves $3 million for short-term notes. So, in order to maximize the return to the bank, it should invest $7 million in mortgages and $3 million in short-term notes.

## Chapter 5 Project

1. For 1994, the percentage of office communications conducted by either E-mail or by paper was $20 + 57 = 77\%$. For 1997, the percentage conducted by either E-mail or by paper was $40 + 47 = 87\%$.

3. To find the year in which half (50%) of all communications will be conducted by E-mail, substitute $y = 50$ into the given equation and solve for $x$.
$$17x - 3y = 2$$
$$17x - 3(50) = 2$$
$$17x = 152$$
$$x \approx 9$$
The model predicts that in 1999 half of all communications will be conducted by E-mail.

5. The percentages reflect the consumption of paper only for office communications. Paper is consumed for many other purposes, such as printing E-mail.

## Chapter 5 Review Exercises

1. The point represents a solution of both equations.

3. $3x - 5y = 1$
$y = 2x - 3$

$(2, 1)$:
$$3x - 5y = 1$$
$$3(2) - 5(1) = 1$$
$$6 - 5 = 1$$
$$1 = 1 \quad \text{True}$$

$$y = 2x - 3$$
$$1 = 2(2) - 3$$
$$1 = 4 - 3$$
$$1 = 1 \quad \text{True}$$

Because it makes both equations in the system true, the point $(2, 1)$ is a solution of the system.

**5.** $y = 2x - 5$
$y = -x + 4$

Graph $y_1 = 2x - 5$ and $y_2 = -x + 4$.
The estimated point of intersection is $(3, 1)$.

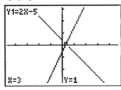

**7.** $2x - 5y = 10$
$10y - 4x + 20 = 0$

Graph $y_1 = 0.4x - 2$ and $y_2 = 0.4x - 2$.
As we can see from the graph, the two equations coincide. The system has infinitely many solutions.

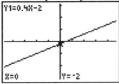

**9.** $y = -x + 4$
$y = x - 3$

Because the equations in this system have different slopes, the lines will intersect. Therefore, there is one solution.

**11.** $y = 5 \quad \rightarrow \quad y = 5$
$x - 2 = 4 \quad \rightarrow \quad x = 6$

Because the graph of the first equation is a horizontal line and the graph of the second equation is a vertical line, the lines will intersect in one point. Therefore, there is one solution.

**13.** $y = x + 1$
$2x + 3y = 6$

Because the first equation is already solved for $y$, substitute $x + 1$ for $y$ in the second equation.
$$2x + 3y = 6$$
$$2x + 3(x + 1) = 6$$
$$2x + 3x + 3 = 6$$
$$5x + 3 = 6$$
$$5x = 3$$
$$x = 0.6$$

Now determine the value of $y$ by substituting $0.6$ for $x$ in $y = x + 1$.
$$y = 0.6 + 1 = 1.6$$

The solution is $(0.6, 1.6)$.

**15.** $y = x + 3$
$2x - 2y + 6 = 0$

Because the first equation is already solved for $y$, substitute $x + 3$ for $y$ in the second equation.
$$2x - 2y + 6 = 0$$
$$2x - 2(x + 3) + 6 = 0$$
$$2x - 2x - 6 + 6 = 0$$
$$0 = 0 \quad \text{True}$$

The resulting equation is an identity that has infinitely many solutions. Thus, the solution set of this system is the infinitely many solutions of $y = x + 3$.

**17.** $3x + y = 2$
$4x + 8y = 1$

Solving the first equation for $y$, we obtain $y = 2 - 3x$. Substitute $2 - 3x$ for $y$ in the second equation.
$$4x + 8y = 1$$
$$4x + 8(2 - 3x) = 1$$
$$4x + 16 - 24x = 1$$
$$16 - 20x = 1$$
$$-20x = -15$$
$$x = 0.75$$

Now determine the value of $y$ by substituting $0.75$ for $x$ in $y = 2 - 3x$.
$$y = 2 - 3(0.75) = -0.25$$

The solution is $(0.75, -0.25)$.

**19.** Return to the original system and use the addition method again, but this time to eliminate $x$ instead of $y$.

**21.** $3x - 4y = 22$
$2x + 5y = -16$

Multiply the first equation by 5 and the second equation by 4. Then add to eliminate $y$.

$$\begin{array}{rcl} 15x - 20y &=& 110 \\ 8x + 20y &=& -64 \\ \hline 23x &=& 46 \\ x &=& 2 \end{array}$$

Replace $x$ with 2 in $2x + 5y = -16$ to find $y$.

$2x + 5y = -16$
$2(2) + 5y = -16$
$4 + 5y = -16$
$5y = -20$
$y = -4$

The solution is $(2, -4)$.

**23.** $y - 3x = 4 \quad \rightarrow \quad -3x + y = 4$
$6x - 2y + 8 = 0 \quad \rightarrow \quad 6x - 2y = -8$

Multiply the first equation by 2. Then add to eliminate $y$.

$$\begin{array}{rcl} -6x + 2y &=& 8 \\ 6x - 2y &=& -8 \\ \hline 0 &=& 0 \quad \text{True} \end{array}$$

The system has infinitely many solutions. The equations are dependent.

**25.** $\dfrac{3}{4}x + \dfrac{1}{3}y - 4 = 0 \quad \rightarrow \quad 9x + 4y = 48$
$\dfrac{1}{2}x = \dfrac{2}{3}y + 8 \quad \rightarrow \quad 3x - 4y = 48$

Add the equations to eliminate $y$.

$$\begin{array}{rcl} 9x + 4y &=& 48 \\ 3x - 4y &=& 48 \\ \hline 12x &=& 96 \\ x &=& 8 \end{array}$$

Replace $x$ with 8 in $9x + 4y = 48$ to find $y$.

$9(8) + 4y = 48$
$72 + 4y = 48$
$4y = -24$
$y = -6$

The solution is $(8, -6)$.

**27.** $\dfrac{1}{5}x - \dfrac{2}{3}y = 2 \quad \rightarrow \quad 3x - 10y = 30$
$3x = 10y + 19 \quad \rightarrow \quad 3x - 10y = 19$

Multiply the second equation by $-1$. Then add to eliminate $x$.

$$\begin{array}{rcl} 3x - 10y &=& 30 \\ -3x + 10y &=& -19 \\ \hline 0 &=& 11 \quad \text{False} \end{array}$$

The system has no solution. The system is inconsistent.

**29.** $x + 3y = 6 \quad \rightarrow \quad y = -\dfrac{1}{3}x + 2$
$4x - cy = 8 \quad \rightarrow \quad y = \dfrac{4}{c}x - \dfrac{8}{c}$

For the system to be inconsistent, the slopes of the two lines must be equal and the $y$-intercepts must not be equal.

$-\dfrac{1}{3} = \dfrac{4}{c}$
$c = -12$

If $c = -12$, the $y$-intercepts are 2 and $\dfrac{2}{3}$.

Thus, if $c = -12$, the system is inconsistent.

**31.** $x - 2y = 5 \quad \rightarrow \quad y = \dfrac{1}{2}x - \dfrac{5}{2}$
$cx + 6y + 15 = 0 \quad \rightarrow \quad y = -\dfrac{c}{6}x - \dfrac{5}{2}$

For the system to be dependent, the slopes of the two lines must be equal and the $y$-intercepts must be equal. The $y$-intercepts are already equal.

$\dfrac{1}{2} = -\dfrac{c}{6}$
$2c = -6$
$c = -3$

So, if $c = -3$, the equations are dependent.

**33.** The solution set is represented by a plane drawn in a three-dimensional coordinate space.

**35.**
$$x + y + z = 2$$
$$2x - 3y + 3z = 2$$
$$x - 2y + z = 1$$
$(-2, 1, 3)$:
$$x + y + z = 2$$
$$-2 + 1 + 3 = 2$$
$$2 = 2 \quad \text{True}$$

$$2x - 3y + 3z = 2$$
$$2(-2) - 3(1) + 3(9) = 2$$
$$-4 - 3 + 27 = 2$$
$$20 = 2 \quad \text{False}$$

Because $(-2, 1, 3)$ does not satisfy all three equations in the system, it is not a solution.

**37.**
$$x + 2y + z = 5 \quad (1)$$
$$2x + y - 2z = -3 \quad (2)$$
$$4x + 3y - z = 0 \quad (3)$$

Eliminate $z$ from equations (1) and (3).
$$x + 2y + z = 5 \quad (1)$$
$$4x + 3y - z = 0 \quad (3)$$
$$\overline{5x + 5y \phantom{-z} = 5}$$
$$x + y = 1 \quad (4)$$

Eliminate $z$ from equations (1) and (2) by multiplying equation (1) by 2 and then adding the equations.
$$2x + 4y + 2z = 10 \quad (1)$$
$$2x + y - 2z = -3 \quad (2)$$
$$\overline{4x + 5y \phantom{- 2z} = 7} \quad (5)$$

Now we have two equations in two variables.
$$x + y = 1 \quad (4)$$
$$4x + 5y = 7 \quad (5)$$

Multiply equation (4) by $-4$ and add the equations to eliminate $x$.
$$-4x - 4y = -4 \quad (4)$$
$$4x + 5y = 7 \quad (5)$$
$$\overline{\phantom{-4x} y = 3}$$

Substitute $y = 3$ in equation (4) to find $x$.
$$x + y = 1 \quad (4)$$
$$x + 3 = 1$$
$$x = -2$$

To find $z$, substitute $x = -2$ and $y = 3$ in equation (1).

$$x + 2y + z = 5 \quad (1)$$
$$-2 + 2(3) + z = 5$$
$$4 + z = 5$$
$$z = 1$$
The solution is $(-2, 3, 1)$.

**39.**
$$5x - 2y + 7z = -19 \quad (1)$$
$$x + 3y - 2z = 3 \quad (2)$$
$$4x - y - z = 16 \quad (3)$$

Eliminate $y$ from equations (1) and (3) by multiplying equation (3) by $-2$ and then adding the equations.
$$5x - 2y + 7z = -19 \quad (1)$$
$$-8x + 2y + 2z = -32 \quad (3)$$
$$\overline{-3x \phantom{+ 2y} + 9z = -51}$$
$$-x + 3z = -17 \quad (4)$$

Eliminate $y$ from equations (2) and (3) by multiplying equation (3) by 3 and then adding the equations.
$$x + 3y - 2z = 3 \quad (2)$$
$$12x - 3y - 3z = 48 \quad (3)$$
$$\overline{13x \phantom{- 3y} - 5z = 51} \quad (5)$$

Now we have two equations in two variables.
$$-x + 3z = -17 \quad (4)$$
$$13x - 5z = 51 \quad (5)$$

Multiply equation (4) by 13 and then add the equations to eliminate $x$.
$$-13x + 39z = -221 \quad (4)$$
$$13x - 5z = 51 \quad (5)$$
$$\overline{\phantom{-13x +} 34z = -170}$$
$$z = -5$$

Substitute for $z$ in equation (4) to find $x$.
$$-x + 3z = -17 \quad (4)$$
$$-x + 3(-5) = -17$$
$$-x = -2$$
$$x = 2$$

To find $y$, substitute $x = 2$ and $z = -5$ in equation (3).
$$4x - y - z = 16 \quad (3)$$
$$4(2) - y - (-5) = 16$$
$$13 - y = 16$$
$$-y = 3$$
$$y = -3$$
The solution is $(2, -3, -5)$.

**41.**
$$x - 2y + z = 3 \quad (1)$$
$$7x - 3y + 5z = 1 \quad (2)$$
$$2x - 4y + 2z = -5 \quad (3)$$

Eliminate $z$ from equations (1) and (3) by multiplying equation (1) by $-2$ and then adding the equations.
$$-2x + 4y - 2z = -6 \quad (1)$$
$$\underline{\phantom{-}2x - 4y + 2z = -5} \quad (3)$$
$$0 = -11 \quad \text{False}$$

The system has no solution. The system is inconsistent.

**43.**
$$2x - 2y + 10z = 12 \quad (1)$$
$$x - 2y + 3z = -1 \quad (2)$$
$$4x - 3y + z = -32 \quad (3)$$

Eliminate $x$ from equations (1) and (2) by multiplying equation (2) by $-2$ and then adding the equations.
$$2x - 2y + 10z = 12 \quad (1)$$
$$\underline{-2x + 4y - 6z = 2} \quad (2)$$
$$2y + 4z = 14$$
$$y + 2z = 7 \quad (4)$$

Eliminate $x$ from equations (1) and (3) by multiplying equation (1) by $-2$ and then adding the equations.
$$-4x + 4y - 20z = -24 \quad (1)$$
$$\underline{\phantom{-}4x - 3y + z = -32} \quad (3)$$
$$y - 19z = -56 \quad (5)$$

We have two equations in two variables.
$$y + 2z = 7 \quad (4)$$
$$y - 19z = -56 \quad (5)$$

Multiply equation (5) by $-1$ and then add the equations to eliminate $y$.
$$y + 2z = 7 \quad (4)$$
$$\underline{-y + 19z = 56} \quad (5)$$
$$21z = 63$$
$$z = 3$$

Substitute for $z$ in equation (4) to find $y$.
$$y + 2z = 7 \quad (4)$$
$$y + 2(3) = 7$$
$$y = 1$$

To find $x$, substitute $y = 1$ and $z = 3$ in equation (2).
$$x - 2y + 3z = -1 \quad (2)$$
$$x - 2(1) + 3(3) = -1$$
$$x + 7 = -1$$
$$x = -8$$

The solution is $(-8, 1, 3)$.

**45.**
$$x + y - z = 8 \quad (1)$$
$$y + 2z = -8 \quad (2)$$
$$x - 2y = -3 \quad (3)$$

To eliminate $x$, multiply equation (3) by $-1$ and add the result to equation (1).
$$x + y - z = 8 \quad (1)$$
$$\underline{-x + 2y = 3} \quad (3)$$
$$3y - z = 11 \quad (4)$$

Now we have two equations in two variables.
$$y + 2z = -8 \quad (2)$$
$$3y - z = 11 \quad (4)$$

To eliminate $z$, multiply equation (4) by 2 and then add the equations.
$$y + 2z = -8 \quad (2)$$
$$\underline{6y - 2z = 22} \quad (4)$$
$$7y = 14$$
$$y = 2$$

Substitute for $y$ in equation (4) to find $z$.
$$3y - z = 11 \quad (4)$$
$$3(2) - z = 11$$
$$6 - z = 11$$
$$-z = 5$$
$$z = -5$$

Substitute for $y$ and $z$ in equation (1) to find $x$.
$$x + y - z = 8 \quad (1)$$
$$x + 2 - (-5) = 8$$
$$x + 7 = 8$$
$$x = 1$$

The solution is $(1, 2, -5)$.

**47.** A coefficient matrix consists of the coefficients of the variables of a system of equations. An augmented matrix is the combination of a coefficient matrix and a constant matrix.

**49.**
$$4x - 3y = 12$$
$$x - 2y = -3$$

The augmented matrix for this system is
$$\begin{bmatrix} 4 & -3 & | & 12 \\ 1 & -2 & | & -3 \end{bmatrix}$$

**51.** $\begin{bmatrix} 1 & 2 & -3 & | & 4 \\ 3 & 1 & -2 & | & 5 \\ 1 & 1 & 0 & | & 0 \end{bmatrix}$

A system of linear equations for this matrix is

$$\begin{aligned} x + 2y - 3z &= 4 \\ 3x + y - 2z &= 5 \\ x + y &= 0 \end{aligned}$$

**53.** $\begin{aligned} 2x - y &= 5 \\ x - 2y &= 4 \end{aligned}$

First, write the augmented matrix for the given system. Then perform row operations.

$\begin{bmatrix} 2 & -1 & | & 5 \\ 1 & -2 & | & 4 \end{bmatrix}$

$R_1 \leftrightarrow R_2 \begin{bmatrix} 1 & -2 & | & 4 \\ 2 & -1 & | & 5 \end{bmatrix}$

$-2R_1 + R_2 \to \begin{bmatrix} 1 & -2 & | & 4 \\ 0 & 3 & | & -3 \end{bmatrix}$

$R_2 \div (3) \to \begin{bmatrix} 1 & -2 & | & 4 \\ 0 & 1 & | & -1 \end{bmatrix}$

$2R_2 + R_1 \to \begin{bmatrix} 1 & 0 & | & 2 \\ 0 & 1 & | & -1 \end{bmatrix}$

The solution is $(2, -1)$.

**55.** $\begin{aligned} x - y - 5 &= 0 \\ 2x - y + 17 &= 0 \end{aligned} \to \begin{aligned} x - y &= 5 \\ 2x - y &= -17 \end{aligned}$

First, write the augmented matrix for the given system. Then perform row operations.

$\begin{bmatrix} 1 & -1 & | & 5 \\ 2 & -1 & | & -17 \end{bmatrix}$

$-2R_1 + R_2 \to \begin{bmatrix} 1 & -1 & | & 5 \\ 0 & 1 & | & -27 \end{bmatrix}$

$R_2 + R_1 \to \begin{bmatrix} 1 & 0 & | & -22 \\ 0 & 1 & | & -27 \end{bmatrix}$

The solution is $(-22, -27)$.

**57.** $\begin{aligned} x - y - 2z &= -2 \\ x + y + 6z &= 6 \\ y - z &= -1 \end{aligned}$

Write the augmented matrix for this system. Then perform row operations.

$\begin{bmatrix} 1 & -1 & -2 & | & -2 \\ 1 & 1 & 6 & | & 6 \\ 0 & 1 & -1 & | & -1 \end{bmatrix}$

$-R_1 + R_2 \to \begin{bmatrix} 1 & -1 & -2 & | & -2 \\ 0 & 2 & 8 & | & 8 \\ 0 & 1 & -1 & | & -1 \end{bmatrix}$

$R_2 \leftrightarrow R_3 \begin{bmatrix} 1 & -1 & -2 & | & -2 \\ 0 & 1 & -1 & | & -1 \\ 0 & 2 & 8 & | & 8 \end{bmatrix}$

$-2R_2 + R_3 \to \begin{bmatrix} 1 & -1 & -2 & | & -2 \\ 0 & 1 & -1 & | & -1 \\ 0 & 0 & 10 & | & 10 \end{bmatrix}$

$R_3 \div (10) \to \begin{bmatrix} 1 & -1 & -2 & | & -2 \\ 0 & 1 & -1 & | & -1 \\ 0 & 0 & 1 & | & 1 \end{bmatrix}$

$R_3 + R_2 \to \begin{bmatrix} 1 & -1 & -2 & | & -2 \\ 0 & 1 & 0 & | & 0 \\ 0 & 0 & 1 & | & 1 \end{bmatrix}$

$2R_3 + R_1 \to \begin{bmatrix} 1 & -1 & 0 & | & 0 \\ 0 & 1 & 0 & | & 0 \\ 0 & 0 & 1 & | & 1 \end{bmatrix}$

$R_2 + R_1 \to \begin{bmatrix} 1 & 0 & 0 & | & 0 \\ 0 & 1 & 0 & | & 0 \\ 0 & 0 & 1 & | & 1 \end{bmatrix}$

The solution is $(0, 0, 1)$.

**59.** First, rewrite the given system in standard form.

$$\begin{aligned} x &= -10 \\ 0.3x + y &= 6 \end{aligned}$$

Let $A$ = the augmented matrix.

```
rref([A])
    [[1 0 -10]
     [0 1  9 ]]
```

The solution is $(-10, 9)$.

**61.** Let $A$ = the augmented matrix.

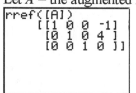

The solution is $(-1, 4, 0)$.

**63.** Let $x$ = the number of five-dollar bills and $y$ = the number of twenty-dollar bills.
The system of equations is:
$$x + y = 60 \quad (1)$$
$$5x + 20y = 705 \quad (2)$$

Solving equation (1) for $x$, we obtain $x = 60 - y$. Substitute into equation (2).
$$5(60 - y) + 20y = 705$$
$$300 - 5y + 20y = 705$$
$$300 + 15y = 705$$
$$15y = 405$$
$$y = 27$$

If there are 27 twenty-dollar bills in the cash drawer, there are $60 - 27 = 33$ five-dollar bills in drawer.

**65.** Let $x$ = the number of sledge hammers and $y$ = the number of claw hammers.
The system of equations is:
$$x + y = 42 \quad (1)$$
$$22.95x + 10.95y = 639.90 \quad (2)$$

Let $A$ = the augmented matrix.

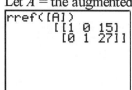

There are 15 sledge hammers and 27 claw hammers in stock.

**67.** Let $x$ = the first digit, $y$ = the second digit, and $z$ = the third digit.
The system of equations is:
$$x + y + z = 12 \quad (1)$$
$$x + y = 7 \quad (2)$$
$$y + z = 8 \quad (3)$$

Let $A$ = the augmented matrix.

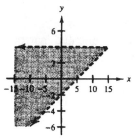

The license plate number is 435.

**69.** 
$$y < 4 \quad \rightarrow \quad y < 4$$
$$2x - 5y < 10 \quad \rightarrow \quad y > \frac{2}{5}x - 2$$

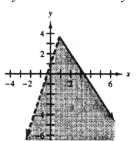

**71.** 
$$y < 3x + 1 \quad \rightarrow \quad y < 3x + 1$$
$$2y + 3x \leq 10 \quad \rightarrow \quad y \leq -1.5x + 5$$

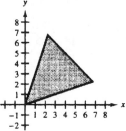

**73.** 
$$y \leq 3x \quad\quad\quad y \leq 3x$$
$$3y \geq x \quad \rightarrow \quad y \geq \frac{1}{3}x$$
$$x + y \leq 9 \quad\quad\quad y \leq -x + 9$$

**75.** $y \geq 2$
$y \geq -3$
$x \geq -5$

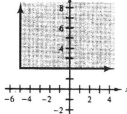

**77.** $x + y \leq 10 \quad \rightarrow \quad y \leq -x + 10$
$x + y \geq 2 \quad \rightarrow \quad y \geq -x + 2$
$x \geq 0 \quad \rightarrow \quad x \geq 0$
$y \geq 0 \quad \rightarrow \quad y \geq 0$

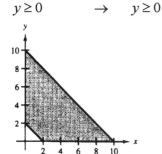

**79.** Let $x =$ the number of bags of mortar and $y =$ the number of bags of lime. The system of inequalities is:
$x + y \leq 20 \quad \rightarrow \quad y \leq -x + 20$
$80x + 40y \leq 1000 \quad \rightarrow \quad y \leq -2x + 25$
$x \geq 0 \quad \rightarrow \quad x \geq 0$
$y \geq 0 \quad \rightarrow \quad y \geq 0$

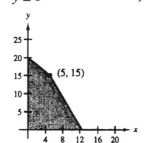

## Chapter 5 Looking Back

**1.** (a) $9^2 = 9 \cdot 9 = 81$
(b) $(-2)^5 = (-2)(-2)(-2)(-2)(-2) = -32$
(c) $10^4 = 10 \cdot 10 \cdot 10 \cdot 10 = 10{,}000$

**3.** Let $x = -2$ and $y = 3$.
$4y^2 + \dfrac{1}{2}x^3 = 4(3)^2 + \dfrac{1}{2}(-2)^3$
$= 4(9) + \dfrac{1}{2}(-8)$
$= 36 - 4$
$= 32$

**5.** $5(2x - 1) = 10x - 5$

**7.** $3x + 1 - x + 4 = 2x + 5$

**9.** $3(t - 2) - t = 3t - 6 - t$
$= 2t - 6$

**11.** (a) $x \cdot x \cdot x \cdot x \cdot x = x^5$
(b) $(a-1)(a-1) = (a-1)^2$

## Chapter 5 Test

1. $y = -2x + 3 \rightarrow y_1 = -2x + 3$
   $x - y = 3 \rightarrow y_2 = x - 3$

   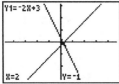

   The estimated solution is $(2, -1)$.
   Check:
   Substitute $-2x + 3$ for $y$ in the second equation.
   $$x - y = 3$$
   $$x - (-2x + 3) = 3$$
   $$x + 2x - 3 = 3$$
   $$3x - 3 = 3$$
   $$3x = 6$$
   $$x = 2$$
   Substitute 2 for $x$ in $y = -2x + 3$.
   $$y = -2(2) + 3 = -1$$
   The solution is $(2, -1)$.

3. $2x + y = 7$
   $x - y = 8$

   Solving the first equation for $y$, we obtain $y = 7 - 2x$. Substituting for $y$ in the second equation, we obtain
   $$x - y = 8$$
   $$x - (7 - 2x) = 8$$
   $$x - 7 + 2x = 8$$
   $$3x - 7 = 8$$
   $$3x = 15$$
   $$x = 5$$
   Substitute for $x$ in $y = 7 - 2x$ to find $y$.
   $$y = 7 - 2(5) = -3$$
   The solution is $(5, -3)$.

5. $x - \frac{3}{2}y = 13$
   $\frac{3}{2}x - y = 17$
   $\rightarrow$ $2x - 3y = 26$
   $3x - 2y = 34$

   Multiply the first equation by $-3$ and the second equation by 2. Then add the equations to eliminate $x$.

   $$-6x + 9y = -78$$
   $$6x - 4y = 68$$
   $$\overline{\phantom{-6x+}5y = -10}$$
   $$y = -2$$

   Substitute for $y$ in the first equation in the original system to find $x$.
   $$x - \frac{3}{2}y = 13$$
   $$x - \frac{3}{2}(-2) = 13$$
   $$x + 3 = 13$$
   $$x = 10$$
   The solution is $(10, -2)$.

7. $3x - 6y = 24$
   $5x + 4y = 12$

   First, write the augmented matrix for the given system. Then perform row operations.

   $$\begin{bmatrix} 3 & -6 & | & 24 \\ 5 & 4 & | & 12 \end{bmatrix}$$
   $R_1 \div 3 \rightarrow \begin{bmatrix} 1 & -2 & | & 8 \\ 5 & 4 & | & 12 \end{bmatrix}$
   $-5R_1 + R_2 \rightarrow \begin{bmatrix} 1 & -2 & | & 8 \\ 0 & 14 & | & -28 \end{bmatrix}$
   $R_2 \div 14 \rightarrow \begin{bmatrix} 1 & -2 & | & 8 \\ 0 & 1 & | & -2 \end{bmatrix}$
   $2R_2 + R_1 \rightarrow \begin{bmatrix} 1 & 0 & | & 4 \\ 0 & 1 & | & -2 \end{bmatrix}$

   The solution is $(4, -2)$.

9. Let $A$ = the augmented matrix.

   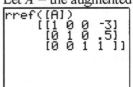

   The solution is $(-3, 0.5, 1)$.

11. $2x + y = 2$
    $8x + 4y = 7$

    Multiply the first equation by $-4$. Then add the equations to eliminate $y$.
    $-8x - 4y = -8$
    $\underline{8x + 4y = 7}$
    $0 = -1$    False

    The system is inconsistent.

13. (a) If the equations are dependent, the graphs of the equations coincide.
    (b) If the system has a unique solution, the graphs of the equations intersect at one point.
    (c) If the system is inconsistent, the graphs of the equations are parallel.

15. $2x + y \leq 12 \quad \rightarrow \quad y \leq -2x + 12$
    $y \leq 8 \quad \rightarrow \quad y \leq 8$
    $x \geq 0 \quad \rightarrow \quad x \geq 0$
    $y \geq 0 \quad \rightarrow \quad y \geq 0$

    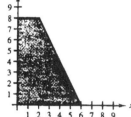

17. Let $x$ = the measure of one angle and $y$ = the measure of the other angle.
    The system of equations is
    $x + y = 90 \quad \rightarrow \quad x + y = 90$
    $x = 2y - 5 \quad \rightarrow \quad x - 2y = -5$

    Multiply the second equation by $-1$ and then add the equations to eliminate $x$.

    $x + y = 90$
    $\underline{-x + 2y = 5}$
    $3y = 95$
    $y = 31\dfrac{2}{3}$

    Substitute for $y$ in $x = 2y - 5$ to find $x$.
    $x = 2\left(31\dfrac{2}{3}\right) - 5 = 58\dfrac{1}{3}$

    The two angles are $31\dfrac{2}{3}°$ and $58\dfrac{1}{3}°$.

19. Let $x$ = the number of pounds of shrimp, $y$ = the number of pounds of crabs, and $z$ = the number of pounds of lobster.
    The system of equations is:
    $x = \dfrac{1}{2}(y + z)$
    $y = z - 5$
    $5x + 7y + 9z = 960.50$

    These can be rewritten in standard form as:
    $2x - y - z = 0$
    $y - z = -5$
    $5x + 7y + 9z = 960.50$

    Let $A$ = the augmented matrix.

    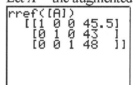

    The caterer can purchase 45.5 pounds of shrimp, 43 pounds of crab, and 48 pounds of lobster.

# Cumulative Test for Chapters 3-5

**1.** The solution is 4. When $x$ is replaced with 4, the value of each side of the equation is 17.

**3.** Solve for $b$:
$$A = \frac{1}{2}bh$$
$$2A = bh$$
$$\frac{2A}{h} = b$$

**5.** Let $x$ = the time in transit.

|  | Rate (mph) | Time (hours) | Distance (miles) |
|---|---|---|---|
| Bicycle | 12 | $x$ | $12x$ |
| Car | 40 | $x$ | $40x$ |

The sum of the distances traveled by the bicycle and the car is 39 miles.
$$12x + 40x = 39$$
$$52x = 39$$
$$x = 0.75$$
The bicycle and car meet after 0.75 hour or 45 minutes. They will meet at 1:45 P.M.

**7.** (a) $x - 2(x-2) < 5x + 8$
$$x - 2x + 4 < 5x + 8$$
$$-x + 4 < 5x + 8$$
$$-6x + 4 < 8$$
$$-6x < 4$$
$$x > -\frac{2}{3}$$
The solution set is $\left(-\frac{2}{3}, \infty\right)$.

(b) $\frac{3-x}{-4} \geq x$
$$3 - x \leq -4x$$
$$3 + 3x \leq 0$$
$$3x \leq -3$$
$$x \leq -1$$
The solution set is $(-\infty, -1]$.

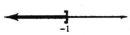

**9.** (a) $|x-3| < 8$ is equivalent to
$$-8 < x - 3 < 8$$
$$-5 < x < 11$$
The solution set is $(-5, 11)$.

(b) $2|x+1| - 3 = 5$
$$2|x+1| = 8$$
$$|x+1| = 4$$
This can be written as two equations:
$x + 1 = 4$ or $x + 1 = -4$
$x = 3$  $\quad x = -5$

(c) $-2|3+x| \leq -4$
$$|3+x| \geq 2$$
This can be rewritten as:
$3 + x \leq -2$ or $3 + x \geq 2$
$x \leq -5$ $\quad$ $x \geq -1$
The solution set is $(-\infty, -5] \cup [-1, \infty)$.

**11.** $(3, a)$:
$$2x + 3y = 7$$
$$2(3) + 3a = 7$$
$$6 + 3a = 7$$
$$3a = 1$$
$$a = \frac{1}{3}$$
If $a = \frac{1}{3}$, then $(3, a)$ is a solution.

$(b, 1)$:
$$2x + 3y = 7$$
$$2b + 3(1) = 7$$
$$2b = 4$$
$$b = 2$$
If $b = 2$, then $(b, 1)$ is a solution.

**13.** (a) $x + 2y = 3$
$$2y = -x + 3$$
$$y = -\frac{1}{2}x + \frac{3}{2}$$
The slope is negative.

(b) $x + 2 = 3$ is the same as $x = 1$. Because this is the equation of a vertical line, the slope is undefined.

Cumulative Test for Chapters 3-5

(c) $0 \cdot x + 2y = 3$. Because the coefficient of $x$ is 0, this is the equation of a horizontal line. The slope is zero.

(d) $x = 2y + 3$
$x - 3 = 2y$
$\frac{1}{2}x - \frac{3}{2} = y$

The slope is positive.

15. The rate of change of $y$ with respect to $x$ is the slope of the line: $\frac{2}{3}$.

17. $A(-2, -5)$ and $B(6, 3)$

Slope: $m = \frac{3 - (-5)}{6 - (-2)} = \frac{8}{8} = 1$

$y - y_1 = m(x - x_1)$
$y - 3 = 1(x - 6)$
$y - 3 = x - 6$
$y = x - 3$
$-x + y = -3$
$x - y = 3$

19. (a) The boundary line is the dashed line $y = x - 7$.
    (b) Either test a point or write the inequality as $y > x - 7$ and shade above the line.

21. (a) Because the lines do not intersect, the solution set is empty.
    (b) Because the lines intersect, the solution set contains exactly one ordered pair.
    (c) Because the lines coincide, the solution set contains infinitely many ordered pairs.

23. (a) $2x - 3y = -21$

$y = \frac{2}{3}x + 7 \quad \rightarrow \quad -2x + 3y = 21$

When these equations are added, the result is $0 = 0$. The equations are dependent.

(b) $4x - y = -9 \quad \rightarrow \quad 4x - y = -9$
$2y - x = 11 \quad \rightarrow \quad -x + 2y = 11$

Multiply the second equation by 4 and add the equations to eliminate $x$.

$\begin{aligned} 4x - y &= -9 \\ -4x + 8y &= 44 \\ \hline 7y &= 35 \\ y &= 5 \end{aligned}$

Substitute for $y$ in the second equation to find $x$.
$2(5) - x = 11$
$10 - x = 11$
$-x = 1$
$x = -1$

The solution set is $(-1, 5)$.

(c) $y = 5x + 1 \quad \rightarrow \quad -5x + y = 1$

$x - \frac{y}{5} = 0.2 \quad \rightarrow \quad 5x - y = 1$

When these equations are added the result is $0 = 2$, which is false. This is an inconsistent system.

25. $2x + 3y + 5z = 18$ (1)
$3x - 2y + 4z = 13$ (2)
$4x - 3y + 3z = 8$ (3)

Eliminate $x$ from equations (1) and (2) by multiplying equation (1) by $-3$, multiplying equation (2) by 2, and then adding the equations.

$\begin{aligned} -6x - 9y - 15z &= -54 \quad (1) \\ 6x - 4y + 8z &= 26 \quad (2) \\ \hline -13y - 7z &= -28 \quad (4) \end{aligned}$

Eliminate $x$ from equations (1) and (3) by multiplying equation (1) by $-2$ and then adding the equations.

$\begin{aligned} -4x - 6y - 10z &= -36 \quad (1) \\ 4x - 3y + 3z &= 8 \quad (3) \\ \hline -9y - 7z &= -28 \quad (5) \end{aligned}$

Now we have two equations in two variables.

$-13y - 7z = -28$ (4)
$-9y - 7z = -28$ (5)

Multiply equation (4) by $-1$ and add the equations to eliminate $z$.

$\begin{aligned} 13y + 7z &= +28 \quad (4) \\ -9y - 7z &= -28 \quad (5) \\ \hline 4y &= 0 \\ y &= 0 \end{aligned}$

Substitute for $y$ in equation (5) to find $z$.

$$-9y - 7z = -28$$
$$-9(0) - 7z = -28$$
$$-7z = -28$$
$$z = 4$$

Substitute for $y$ and $z$ in equation (1) to find $x$.
$$2x + 3y + 5z = 18$$
$$2x + 3(0) + 5(4) = 18$$
$$2x + 20 = 18$$
$$2x = -2$$
$$x = -1$$

The solution is $(-1, 0, 4)$.

27. Let $c$ = the speed of the cyclist in still air and $w$ = the speed of the wind.

|  | Rate (mph) | Time (hours) | Distance (miles) |
|---|---|---|---|
| Against | $c - w$ | 2 | $2(c-w)$ |
| With | $c + w$ | $1\frac{1}{3}$ | $\frac{4}{3}(c+w)$ |

Because the cyclist traveled 32 miles against the wind and 32 miles with the wind, we get the following system of equations.

$$2(c - w) = 32 \quad \rightarrow \quad c - w = 16 \quad (1)$$
$$\tfrac{4}{3}(c + w) = 32 \quad \rightarrow \quad c + w = 24 \quad (2)$$

Multiply equation (1) by $-1$ and add the two equations to eliminate $c$.
$$-c + w = -16$$
$$\underline{\phantom{-}c + w = \phantom{-}24}$$
$$2w = 8$$
$$w = 4$$

The speed of the wind is 4 mph.

29.  $x + y \leq 9 \quad \rightarrow \quad y \leq -x + 9$
$2x - y \leq 0 \quad \rightarrow \quad y \geq 2x$
$y \geq 2 \quad \rightarrow \quad y \geq 2$

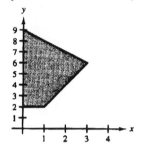

# Chapter 6

# Exponents and Polynomials

## Section 6.1 Properties of Exponents

1. Any two monomials can be multiplied, but monomials must be like terms in order to add them.

3. (a) $-(-b)^3 = -(-b^3) = b^3$
   (b) $-b^3$
   (c) $(-b)^3 = -b^3$
   (d) $b^3$
   (e) $-(b)^3 = -b^3$
   
   Parts (a) and (d) are equivalent, and parts (b), (c), and (e) are equivalent.

5. $2^4 \cdot 2^7 = 2^{4+7} = 2^{11}$

7. $-5x^4 \cdot x^2 = -5 \cdot x^{4+2} = -5x^6$

9. $6t \cdot 3t^7 \cdot t^2 = (6 \cdot 3)(t^1 \cdot t^7 \cdot t^2)$
   $= 18 t^{1+7+2}$
   $= 18 t^{10}$

11. $(6+t)^5 (6+t)^4 = (6+t)^{5+4} = (6+t)^9$

13. $y^5(3y^2) = 3 \cdot y^{5+2} = 3y^7$

15. $(-2x^7)(3x) = (-2 \cdot 3)(x^7 \cdot x^1)$
    $= -6x^{7+1} = -6x^8$

17. $(2xy^2)(-3x^3 y) = [2 \cdot (-3)](x^1 \cdot x^3)(y^2 \cdot y^1)$
    $= -6 \cdot x^{1+3} \cdot y^{2+1}$
    $= -6x^4 y^3$

19. $8y^5,\ 8y^4$
    (a) It is not possible to add the terms because they are not like terms.
    (b) $8y^5 \cdot 8y^4 = (8 \cdot 8)(y^5 \cdot y^4)$
    $= 64 y^{5+4} = 64 y^9$

21. $n,\ 3n,\ -6n$
    (a) $n + 3n + (-6n) = -2n$
    (b) $(n)(3n)(-6n) = [3 \cdot (-6)](n^1 \cdot n^1 \cdot n^1)$
    $= -18 \cdot n^{1+1+1}$
    $= -18 n^3$

23. $y^3,\ 5y^3,\ -4y^3$
    (a) $y^3 + 5y^3 + (-4y^3) = 2y^3$
    (b) $(y^3)(5y^3)(-4y^3) = -(5 \cdot 4)(y^3 \cdot y^3 \cdot y^3)$
    $= -20 \cdot y^{3+3+3}$
    $= -20 y^9$

25. (a) $x^2(3x^3 + 5x^3) = x^2(8x^3)$
    $= 8 \cdot x^{2+3} = 8x^5$
    (b) $x^2(3x^3 + 5x^3) = x^2(3x^3) + x^2(5x^3)$
    $= 3 \cdot x^{2+3} + 5 \cdot x^{2+3}$
    $= 3x^5 + 5x^5$
    $= 8x^5$

27. When we multiply, we add exponents: $x^3 \cdot x^3 = x^6$. When we combine terms, we do nothing with the exponents: $x^3 + x^3 = 2x^3$. When we raise a power to a power, we multiply the exponents: $(x^3)^3 = x^9$.

29. $(x^3)^4 = x^{3 \cdot 4} = x^{12}$

31. $[(x-3)^3]^4 = (x-3)^{3 \cdot 4} = (x-3)^{12}$

33. $(xy)^5 = x^5 y^5$

35. $(-x^2 y)^3 = (-1)^3 (x^2)^3 (y)^3$
$= (-1)(x^{2 \cdot 3})(y^3)$
$= -x^6 y^3$

37. $(-2x^2 y^4)^3 = (-2)^3 (x^2)^3 (y^4)^3$
$= -8 x^6 y^{12}$

39. $\left(\dfrac{a}{b}\right)^6 = \dfrac{a^6}{b^6}$

41. $\left(\dfrac{4x^6}{y^5}\right)^2 = \dfrac{(4x^6)^2}{(y^5)^2} = \dfrac{4^2 (x^6)^2}{(y^5)^2} = \dfrac{16 x^{12}}{y^{10}}$

43. $\left(\dfrac{7x^3}{5y^5}\right)^2 = \dfrac{(7x^3)^2}{(5y^5)^2} = \dfrac{7^2 (x^3)^2}{5^2 (y^5)^2} = \dfrac{49 x^6}{25 y^{10}}$

45. $\dfrac{-7^{11}}{7^5} = (-1) \cdot 7^{11-5} = -7^6$

47. $\dfrac{x^{12}}{x} = x^{12-1} = x^{11}$

49. $\dfrac{(x+1)^5}{(x+1)^3} = (x+1)^{5-3} = (x+1)^2$

51. $\dfrac{-5x^4}{10x^3} = -\dfrac{5}{5} \cdot \dfrac{1}{2} \cdot x^{4-3} = -\dfrac{1}{2} x^1 = -\dfrac{x}{2}$

53. $\dfrac{-3z^5}{-4z^2} = \dfrac{3}{4} \cdot z^{5-2} = \dfrac{3z^3}{4}$

55. $\dfrac{a^4 b^5}{a^2 b^4} = a^{4-2} b^{5-4} = a^2 b$

57. $\dfrac{a^4 b^5}{b^2} = a^4 b^{5-2} = a^4 b^3$

59. $\dfrac{6 x^{12} y^8}{12 x^6 y^4} = \dfrac{6}{12} \cdot x^{12-6} y^{8-4} = \dfrac{x^6 y^4}{2}$

61. $(7-2)^2 = 5^2 = 25$

63. $(8 \cdot 5)^2 = 40^2 = 1600$

65. $\dfrac{x^{12}}{x^3} = x^{12-3} = x^9 \neq x^4$
The statement is false.

67. $(xy)^2 = x^2 y^2$
The statement is true.

69. $\dfrac{x^{20}}{x^{19}} = x^{20-19} = x^1 > 0$

Because $x$ is defined as being positive, the statement is true.

71. $(-5x^3)(4x^5) = (-5 \cdot 4)(x^3 \cdot x^5)$
$= -20 x^{3+5}$
$= -20 x^8$

73. $(3s^3 t^8)^3 = 3^3 (s^3)^3 (t^8)^3 = 27 s^9 t^{24}$

75. $(-t)^3 (t^2)^4 = -t^3 \cdot t^8 = -1 \cdot t^{3+8} = -t^{11}$

77. $(-2a^4)(a^2)^3 = (-2a^4)(a^6)$
$= -2 \cdot a^{4+6}$
$= -2 a^{10}$

79. $(xy^2)^2 (xy)^2 = (x^2 y^4)(x^2 y^2)$
$= x^{2+2} y^{4+2}$
$= x^4 y^6$

81. $\dfrac{-3 x^{12}}{12 x^4} = -\dfrac{1}{4} x^{12-4} = -\dfrac{x^8}{4}$

83. $\dfrac{(-2x^3)^4}{8x^6} = \dfrac{(-2)^4 (x^3)^4}{8x^6}$
$= \dfrac{16 x^{12}}{8 x^6}$
$= 2 x^{12-6}$
$= 2 x^6$

Section 6.1   Properties of Exponents

**85.** $\dfrac{(2x^3)^2}{10x^2} = \dfrac{2^2(x^3)^2}{10x^2} = \dfrac{4x^6}{10x^2} = \dfrac{2}{5}x^{6-2} = \dfrac{2x^4}{5}$

**87.** $\dfrac{(a^5b^7)^2}{(a^2b^4)^3} = \dfrac{(a^5)^2(b^7)^2}{(a^2)^3(b^4)^3}$
$= \dfrac{a^{10}b^{14}}{a^6b^{12}}$
$= a^{10-6}b^{14-12}$
$= a^4b^2$

**89.** We can write $x^6 = (x^2)^3 = (5)^3 = 125$. Thus $x^6 = 125$.

**91.** $(x^n)^3 = x^{12}$
$x^{3n} = x^{12}$
By setting the exponents equal, we can solve for $n$:
$3n = 12$
$n = 4$

**93.** $\dfrac{x^n}{x^4} = x^2$
$x^{n-4} = x^2$
By setting the exponents equal, we can solve for $n$:
$n - 4 = 2$
$n = 6$

**95.** Let $y_1 = 0.008t^3$ and $y_2 = 0.35t^2$.

Per capita federal taxes rose much more rapidly than per capita taxes for state and local governments.

**97.** Let $y_1 = 0.023x$.

The rising graph shows that the ratio of federal to state/local taxes increased.

**99.** Because $9 = 3^2$, we have
$9^4 \cdot 3^5 = (3^2)^4 \cdot 3^5 = 3^8 \cdot 3^5 = 3^{8+5} = 3^{13}$

**101.** Because $8 = 2^3$, we have
$\dfrac{8^5}{2^{10}} = \dfrac{(2^3)^5}{2^{10}} = \dfrac{2^{15}}{2^{10}} = 2^{15-10} = 2^5$

**103.** $\dfrac{x^5}{x^4} + \dfrac{x^4}{x^3} + \dfrac{x^3}{x^2} + \dfrac{x^2}{x} + x$
$= x^{5-4} + x^{4-3} + x^{3-2} + x^{2-1} + x$
$= x^1 + x^1 + x^1 + x^1 + x^1$
$= 5x$

**105.** $(2x)^3 - x^3 \cdot x^2 - 8x^3 + x^5$
$= 2^3x^3 - x^{3+2} - 8x^3 + x^5$
$= 8x^3 - x^5 - 8x^3 + x^5$
$= 0$

**107.** $y^{3n} \cdot y^2 = y^{3n+2}$

**109.** $(a^{3n})^{4n} = a^{3n \cdot 4n} = a^{12n^2}$

## Section 6.2 Zero and Negative Exponents

1. If $0^0$ were defined, then according to the Quotient Rule for Exponents,
$\dfrac{0^5}{0^5} = 0^{5-5} = 0^0$. But $0^5 = 0$ and division by 0 is undefined.

3. $4^0 = 1$

5. $3t^0 = 3 \cdot t^0 = 3 \cdot 1 = 3$

7. $-6^0 = -1 \cdot 6^0 = -1 \cdot 1 = -1$

9. $5^{-2} = \dfrac{1}{5^2} = \dfrac{1}{25}$

11. $(-2)^{-3} = \dfrac{1}{(-2)^3} = \dfrac{1}{-8} = -\dfrac{1}{8}$

13. $\dfrac{1}{15^{-1}} = 15^1 = 15$

15. $\left(\dfrac{1}{4}\right)^{-1} = \left(\dfrac{4}{1}\right)^1 = 4$

17. $\dfrac{2^{-2}}{2} = \dfrac{2^{-2}}{1} \cdot \dfrac{1}{2} = \dfrac{1}{2^2} \cdot \dfrac{1}{2} = \dfrac{1}{2^3} = \dfrac{1}{8}$

19. $3^{-1} + 2^{-2} = \dfrac{1}{3} + \dfrac{1}{2^2}$
$= \dfrac{1}{3} + \dfrac{1}{4}$
$= \dfrac{4}{12} + \dfrac{3}{12}$
$= \dfrac{7}{12}$

21. If $b$ were 0, $b^{-n}$ would be $0^{-n} = \dfrac{1}{0^n} = \dfrac{1}{0}$, which is undefined.

23. $x^{-7} = \dfrac{1}{x^7}$

25. $4x^{-4} = 4 \cdot \dfrac{1}{x^4} = \dfrac{4}{x^4}$

27. $2^{-5}x^3 = \dfrac{1}{2^5} \cdot x^3 = \dfrac{1}{32} \cdot x^3 = \dfrac{x^3}{32}$

29. $-x^{-3}y^{-4} = (-1) \cdot \dfrac{1}{x^3} \cdot \dfrac{1}{y^4} = -\dfrac{1}{x^3 y^4}$

31. $\dfrac{1}{-5k^{-6}} = -\dfrac{1}{5} \cdot \dfrac{1}{k^{-6}} = -\dfrac{1}{5} \cdot k^6 = -\dfrac{k^6}{5}$

33. $(2+y)^{-3} = \dfrac{1}{(2+y)^3}$

35. $\dfrac{-3a^{-1}}{2b^{-1}} = -\dfrac{3}{2} \cdot \dfrac{a^{-1}}{1} \cdot \dfrac{1}{b^{-1}}$
$= -\dfrac{3}{2} \cdot \dfrac{1}{a^1} \cdot b^1$
$= -\dfrac{3}{2} \cdot \dfrac{b}{a}$
$= -\dfrac{3b}{2a}$

37. Because $-3^0 = (-1)(3^0) = -1(1) = -1$ and $(-3)^0 = 1$, $-3^0 \underline{\;<\;} (-3)^0$.

39. Because $-5^{-2} = (-1) \cdot \dfrac{1}{5^2} = -\dfrac{1}{25}$ and $(-5)^{-2} = \dfrac{1}{(-5)^2} = \dfrac{1}{25}$, $-5^{-2} \underline{\;<\;} (-5)^{-2}$.

41. Because $(-10)^0 = 1$, $-|8|^0 = -1$, $-1 - 4^0 = -1 - 1 = -2$, and $4^0 + x^0 = 1 + 1 = 2$, the given expressions in order from least to greatest are: $-1 - 4^0$, $-|8|^0$, $0$, $(-10)^0$, and $4^0 + x^0$.

43. $x^5 \cdot x^{-5} = x^{5+(-5)} = x^0 = 1$

Section 6.2  Zero and Negative Exponents

**45.** 
$$(-4x^{-4})(3x) = (-4 \cdot 3)(x^{-4} \cdot x^1)$$
$$= -12x^{-4+1}$$
$$= -12x^{-3}$$
$$= -\frac{12}{x^3}$$

**47.**
$$7x^{-4}y^5(2x^{-3}y^{-4})$$
$$= (7 \cdot 2)(x^{-4} \cdot x^{-3})(y^5 \cdot y^{-4})$$
$$= 14x^{-4-3}y^{5+(-4)}$$
$$= 14x^{-7}y^1$$
$$= \frac{14y}{x^7}$$

**49.** $\dfrac{z^{-2}}{z^5} = z^{-2-5} = z^{-7} = \dfrac{1}{z^7}$

**51.** $\dfrac{y^{-8}}{y^{-3}} = y^{-8-(-3)} = y^{-5} = \dfrac{1}{y^5}$

**53.** $\dfrac{3z^{-8}}{2z^{-5}} = \dfrac{3}{2} \cdot z^{-8-(-5)} = \dfrac{3}{2} \cdot z^{-3} = \dfrac{3}{2z^3}$

**55.** $\dfrac{x^{-3}y^8}{x^2 y^{-2}} = x^{-3-2}y^{8-(-2)} = x^{-5}y^{10} = \dfrac{y^{10}}{x^5}$

**57.** 
$$\dfrac{4^{-1}x^{-3}}{-4^2 x^{-2}} = (-1) \cdot 4^{-1-2} \cdot x^{-3-(-2)}$$
$$= (-1) \cdot 4^{-3} x^{-1}$$
$$= -\dfrac{1}{4^3 x}$$
$$= -\dfrac{1}{64x}$$

**59.**
$$\dfrac{3^{-2}x^{-5}y^{-7}}{4x^5 y^{-9}} = \dfrac{1}{3^2} \cdot \dfrac{1}{4} \cdot x^{-5-5} \cdot y^{-7-(-9)}$$
$$= \dfrac{1}{9} \cdot \dfrac{1}{4} \cdot x^{-10} \cdot y^2$$
$$= \dfrac{1}{36} \cdot \dfrac{1}{x^{10}} \cdot y^2$$
$$= \dfrac{y^2}{36x^{10}}$$

**61.**
$$\dfrac{x^{21}y^{-22}}{x^{-18}y^{26}} = x^{21-(-18)}y^{-22-26}$$
$$= x^{39}y^{-48}$$
$$= \dfrac{x^{39}}{y^{48}}$$

**63.** Because $\dfrac{x^{-8}}{x^{-3}} = x^{-8-(-3)} = x^{-5}$, the unknown expression is $x^{-8}$.

**65.** Because $x^4 \cdot x^{-4} = x^{4-4} = x^0 = 1$, the unknown expression is $x^4$.

**67.** $(x^{-6})^5 = x^{-30} = \dfrac{1}{x^{30}}$

**69.** $(2a^{-5})^{-4} = 2^{-4}a^{20} = \dfrac{a^{20}}{2^4} = \dfrac{a^{20}}{16}$

**71.**
$$(-2x^{-2}y^4)^3 = (-2)^3(x^{-2})^3(y^4)^3$$
$$= -8x^{-6}y^{12}$$
$$= -\dfrac{8y^{12}}{x^6}$$

**73.** $\left(-\dfrac{2}{k}\right)^{-5} = \left(-\dfrac{k}{2}\right)^5 = (-1)^5 \cdot \dfrac{k^5}{2^5} = -\dfrac{k^5}{32}$

**75.**
$$\left(\dfrac{-4x^{-3}}{5^{-1}y^4}\right)^3 = \dfrac{(-4)^3(x^{-3})^3}{(5^{-1})^3(y^4)^3}$$
$$= \dfrac{-64x^{-9}}{5^{-3}y^{12}}$$
$$= -\dfrac{(64)(125)}{x^9 y^{12}}$$
$$= -\dfrac{8000}{x^9 y^{12}}$$

**77.** $(4-3)^{-9} = (1)^{-9} = \dfrac{1}{1^9} = \dfrac{1}{1} = 1$

**79.** $(2^5 + 1)^0 = (32+1)^0 = 33^0 = 1$

**81.** $2(xy)^{-2} = 2x^{-2}y^{-2} = \dfrac{2}{x^2 y^2}$

**83.** 
$$(3x^{-2})^0(2x^{-3})^{-4} = 1 \cdot (2)^{-4}(x^{-3})^{-4}$$
$$= 1 \cdot 2^{-4} \cdot x^{12}$$
$$= \frac{x^{12}}{2^4}$$
$$= \frac{x^{12}}{16}$$

**85.** 
$$-4x^{-2}(x^3)^{-1} = -4x^{-2}x^{-3}$$
$$= -4x^{-5}$$
$$= -\frac{4}{x^5}$$

**87.** 
$$-6x^3(2x^{-3}y^{-2}) = -12x^{3-3}y^{-2}$$
$$= -12x^0 y^{-2}$$
$$= -\frac{12}{y^2}$$

**89.** 
$$(a^{-5}b^{-6})^5 (a^{-2}b^5)^{-4}$$
$$= (a^{-5})^5 (b^{-6})^5 (a^{-2})^{-4}(b^5)^{-4}$$
$$= a^{-25}b^{-30}a^8 b^{-20}$$
$$= a^{-25+8}b^{-30-20}$$
$$= a^{-17}b^{-50}$$
$$= \frac{1}{a^{17}b^{50}}$$

**91.** 
$$\frac{-4x^2 y^{-2}}{8y^5} = -\frac{1}{2} \cdot x^2 \cdot y^{-2-5}$$
$$= -\frac{1}{2} \cdot x^2 \cdot y^{-7}$$
$$= -\frac{x^2}{2y^7}$$

**93.** 
$$\left(\frac{-2x^{-2}}{y^5}\right)^{-3} = \left(\frac{y^5}{-2x^{-2}}\right)^3$$
$$= \frac{(y^5)^3}{(-2x^{-2})^3}$$
$$= \frac{y^{15}}{(-2)^3(x^{-2})^3}$$
$$= \frac{y^{15}}{-8x^{-6}}$$
$$= -\frac{x^6 y^{15}}{8}$$

**95.** 
$$\frac{(x^{-2})^0 x^{-3}}{x^5} = \frac{1 \cdot x^{-3}}{x^5} = x^{-3-5} = x^{-8} = \frac{1}{x^8}$$

**97.** 
$$\frac{(a^2 b^{-3})^{-4}}{(a^{-3}b^5)^2} = \frac{(a^2)^{-4}(b^{-3})^{-4}}{(a^{-3})^2(b^5)^2}$$
$$= \frac{a^{-8}b^{12}}{a^{-6}b^{10}}$$
$$= a^{-8-(-6)}b^{12-10}$$
$$= a^{-2}b^2$$
$$= \frac{b^2}{a^2}$$

**99.** Because $\left(\frac{4}{3}\right)^{-2} = \left(\frac{3}{4}\right)^2 = \frac{9}{16}$, the unknown number is $\frac{4}{3}$.

**101.** 
$$\frac{x^{-m}}{y^{-n}} - \frac{y^n}{x^m} = \left(\frac{x^m}{y^n}\right)^{-1} - \frac{y^n}{x^m}$$
$$= \left(\frac{y^n}{x^m}\right) - \frac{y^n}{x^m}$$
$$= 0$$

**103.** 
$$\frac{x^{-4}}{x^{-5}} - \frac{x^{-3}}{x^{-4}} + \frac{x^{-2}}{x^{-3}} - \frac{x^{-1}}{x^{-2}} + \frac{x^0}{x^{-1}} - x$$
$$= x^{-4-(-5)} - x^{-3-(-4)} + x^{-2-(-3)}$$
$$\quad - x^{-1-(-2)} + x^{0-(-1)} - x$$
$$= x^1 - x^1 + x^1 - x^1 + x^1 - x$$
$$= 0$$

**105.** 
$$\frac{a^3 b^n}{a^5 b^2} \cdot \frac{a^4 b^3}{a^m b^{-5}} = \frac{a^7 b^{n+3}}{a^{m+5}b^{-3}}$$
$$= a^{7-(m+5)}b^{n+3-(-3)}$$
$$= a^{2-m}b^{n+6}$$

For this expression to equal 1, each exponent must be 0.

$$2 - m = 0 \qquad n + 6 = 0$$
$$2 = m \qquad n = -6$$

For the statement to be true, $m = 2$ and $n = -6$.

## Section 6.3  Scientific Notation

1. The number $n$ is such that $1 \leq n < 10$, and $p$ is an integer.

3. For 125,000, $n = 1.25$. The decimal point in the original number was 5 places to the right, so $p = 5$.
$125,000 = 1.25 \cdot 10^5$

5. For 2570.4, $n = 2.5704$. The decimal point in the original number was 3 places to the right, so $p = 3$.
$2570.4 = 2.5704 \cdot 10^3$

7. For $-537,600,000$, $n = -5.376$. The decimal point in the original number was 8 places to the right, so $p = 8$.
$-537,600,000 = -5.376 \cdot 10^8$

9. For 456,700,000,000, $n = 4.567$. The decimal point in the original number was 11 places to the right, so $p = 11$.
$456,700,000,000 = 4.567 \cdot 10^{11}$

11. For $-0.4$, $n = -4$. The decimal point in the original number was 1 place to the left, so $p = -1$.
$-0.4 = -4 \cdot 10^{-1}$

13. For 0.00000645, $n = 6.45$. The decimal point in the original number was 6 places to the left, so $p = -6$.
$0.00000645 = 6.45 \cdot 10^{-6}$

15. For 0.0000003749, $n = 3.749$. The decimal point in the original number was 7 places to the left, so $p = -7$.
$0.0000003749 = 3.749 \cdot 10^{-7}$

17. Move the decimal point 6 places to the right.  $1.34 \cdot 10^6 = 1,340,000$

19. Move the decimal point 7 places to the left.  $-4.214 \cdot 10^{-7} = -0.0000004214$

21. Move the decimal point 6 places to the left.  $10^{-6} = 1 \cdot 10^{-6} = 0.000001$

23. Move the decimal point 4 places to the right.  $3 \cdot 10^4 = 30,000$

25. Move the decimal point 9 places to the right.  $-5.72 \cdot 10^9 = -5,720,000,000$

27. Move the decimal point 4 places to the left.  $8.9 \cdot 10^{-4} = 0.00089$

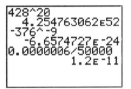

**Figure for Exercises 29, 31, and 33**

29. $428^{20} \approx 4.3 \cdot 10^{52}$
(see first entry in figure)

31. $-376^{-9} \approx -6.7 \cdot 10^{-24}$
(see second entry in figure)

33. $0.0000006 \div 50,000 \approx 1.2 \cdot 10^{-11}$
(see third entry in figure)

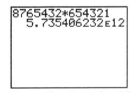

**Figure for Exercise 35**

35. $8,765,432 \cdot 654,321 \approx 5.7 \cdot 10^{12}$
(see figure)

37. Because the number on the left is 10 times as large as the number on the right, $2.84 \cdot 10^9 \; > \; 2.84 \cdot 10^8$.

39. Because the number on the left is 10 times as large as the number on the right, $6.58 \cdot 10^{-3} \; > \; 6.58 \cdot 10^{-4}$.

**41.** $7 \cdot 10^3 + 1 \cdot 10^2 + 5 \cdot 10^1 + 3$
$= 7000 + 100 + 50 + 3$
$= 7153$

**43.** $2 + 3 \cdot 10^{-1} + 4 \cdot 10^{-2} = 2 + 0.3 + 0.04$
$= 2.34$

**45.** $(2 \cdot 10^2)(3 \cdot 10^1) = 6000$

**47.** $\dfrac{4.2 \cdot 10^8}{2.1 \cdot 10^6} = 200$

**49.** $(2 \cdot 10^{-1})^{-1} = 5$

**51.** $(3.72 \cdot 10^7)(9.8 \cdot 10^3) \approx 3.65 \cdot 10^{11}$

**53.** $\dfrac{4.23 \cdot 10^8}{5.2 \cdot 10^{-5}} \approx 8.13 \cdot 10^{12}$

**55.** $(5.7 \cdot 10^3)^4 \approx 1.06 \cdot 10^{15}$

**57.** $(8.31 \cdot 10^5)^{-1} \approx 1.20 \cdot 10^{-6}$

**59.** $\dfrac{(4.9 \cdot 10^5)(3 \cdot 10^{-7})}{3.79 \cdot 10^8} \approx 3.88 \cdot 10^{-10}$

```
3456E6
          3.456E9
0.0001234E12
          1.234E8
0.0000975E-7
          9.75E-12
```

**Figure for Exercises 61, 63, and 65**

**61.** $3456 \cdot 10^6 = 3.456 \cdot 10^9$
(see first entry in figure)

**63.** $0.0001234 \cdot 10^{12} = 1.234 \cdot 10^8$
(see second entry in figure)

**65.** $0.0000975 \cdot 10^{-7} = 9.75 \cdot 10^{-12}$
(see third entry in figure)

**67.**
```
85962
         8.5962E4
```

**69.**
```
2.56/1250
         2.048E-3
```

**71.**
```
2^-36
         1.45519152E-11
```

**73.** The area of Georgia is $3.77 \cdot 10^7 = 37{,}700{,}000$ acres.

**75.** The radius of the nucleus of a heavy atom is $0.000000000007 = 7 \cdot 10^{-12}$ millimeter.

**77.** A computer can do $\dfrac{30}{6 \cdot 10^{-6}} = 5 \cdot 10^6 = 5{,}000{,}000$ addition problems in 30 seconds.

**79.** Cable subscribers pay a total of $50{,}455{,}000 \times 17.58 \times 12 \approx 1.06 \cdot 10^{10}$ dollars in one year.

**81.** A typical cell is $\dfrac{1}{4000} = 2.5 \cdot 10^{-4}$ inch wide.

**83.** AIDS: $1 \cdot 10^6$
Measles: $1 \cdot 10^6$
Malaria: $(1 \cdot 10^6)(1 + 1.1) = 2.1 \cdot 10^6$
Tuberculosis: $(1 \cdot 10^6)(1 + 2.1) = 3.1 \cdot 10^6$
Total deaths:
$1 \cdot 10^6 + 1 \cdot 10^6 + 2.1 \cdot 10^6 + 3.1 \cdot 10^6 =$
$7.2 \cdot 10^6$ deaths

**85.** $\dfrac{1 \text{ ounce}}{78 \text{ eggs}} \cdot \dfrac{1 \text{ pound}}{16 \text{ ounces}} \approx 8.0 \cdot 10^{-4}$ pound

**87.** Total budget: $\$1.3653 \cdot 10^{10}$
Shuttle budget: $\$2.986 \cdot 10^9$

**89.** The total budget in 2000 would be:
$(\$1.3653 \cdot 10^{10})(1.02) = \$1.392606 \cdot 10^{10}$

Section 6.4   Polynomials

The shuttle budget in 2000 would be:
($2.986 \cdot 10^9)(0.985) = 2.94121 \cdot 10^9$
The amount available for NASA programs other than the space shuttle is
$(1.392606 \cdot 10^{10}) - (2.94121 \cdot 10^9) \approx$
$11 billion

**91.** $\dfrac{(4 \cdot 10^7)(8 \cdot 10^{-3})}{(16 \cdot 10^{-2})(2 \cdot 10^n)} = \dfrac{32 \cdot 10^4}{32 \cdot 10^{n-2}} = 1 \cdot 10^{6-n}$

For this expression to equal 1, the exponent must equal 0.
$6 - n = 0$
$6 = n$
The given statement is true if $n = 6$.

**93.** $\dfrac{(15 \cdot 10^{-8})(4 \cdot 10^n)}{(24 \cdot 10^{12})(5 \cdot 10^{-7})} = \dfrac{60 \cdot 10^{n-8}}{120 \cdot 10^5}$
$= \dfrac{1 \cdot 10^{n-8}}{2 \cdot 10^5}$
$= \dfrac{1}{2} \cdot 10^{n-13}$

For this expression to equal $\dfrac{1}{2}$, the exponent must equal 0.
$n - 13 = 0$
$n = 13$
The given statement is true if $n = 13$.

**95.** `1234`
`        1.234e3`

**97.** `3^-28`
`        43.7124217e-15`

**99.** `5^30`
`        931.3225746e18`

## Section 6.4  Polynomials

**1.** (a) The variable is in the denominator.
(b) The exponent is negative.

**3.** Yes, because $5x^2 + 4x - 6$ is the finite sum of monomials, it is a polynomial.

**5.** Yes, because $\sqrt{7} - 4x$ is the finite sum of monomials, it is a polynomial.

**7.** Term: 5
Coefficient: 5
Degree: 0

**9.** Term: $-4x^3$
Coefficient: $-4$
Degree: 3

**11.** Term: $xy$
Coefficient: 1
Degree: 2

**13.** The degree of a term is the value of the exponent on the variable (or the sum of the exponents if there is more than one variable). The degree of a polynomial is the degree of the term with the highest degree.

**15.** Polynomial: $2x + 3y$

Because the highest degree of a term is 1, the degree of the polynomial is 1.

**17.** Polynomial: $2x^4 + xy^2 - x^2y^3$

Because the highest degree of a term is 5, the degree of the polynomial is 5.

**19.** $7x + 3x^5 - 9 = 3x^5 + 7x - 9$

Degree: 5

Because the polynomial has three terms, it is a trinomial.

**21.** $5x^3 + 7x - 2x^2 - 3 = 5x^3 - 2x^2 + 7x - 3$

Degree: 3

Because the polynomial has four terms, it is none of these.

**23.** $x^2y^3 + xy^2$

Degree: 5

Because the polynomial has two terms, it is a binomial.

**25.** Degree: 2

$6 - 3y^2 + 6y = -3y^2 + 6y + 6$
$\phantom{6 - 3y^2 + 6y} = 6 + 6y - 3y^2$

**27.** Degree: 9

$x^3 - 5x^6 + 4x^9 - 7x^8$
$= 4x^9 - 7x^8 - 5x^6 + x^3$
$= x^3 - 5x^6 - 7x^8 + 4x^9$

**29.** Degree: 12

$4n^8 - 3n^4 + n^{12} - 2n^6$
$= n^{12} + 4n^8 - 2n^6 - 3n^4$
$= -3n^4 - 2n^6 + 4n^8 + n^{12}$

**31.** $x = -1, \; y = 2$

$2x^2 + 3xy = 2(-1)^2 + 3(-1)(2)$
$\phantom{2x^2 + 3xy} = 2(1) + 3(-2)$
$\phantom{2x^2 + 3xy} = 2 - 6$
$\phantom{2x^2 + 3xy} = -4$

**33.** $a = 1, \; b = -3$

$ab^2 - 2ab = (1)(-3)^2 - 2(1)(-3)$
$\phantom{ab^2 - 2ab} = (1)(9) - 2(-3)$
$\phantom{ab^2 - 2ab} = 9 + 6$
$\phantom{ab^2 - 2ab} = 15$

**35.** $x = -2, \; y = 3$

$x^2y + 3xy - 2y^2$
$= (-2)^2(3) + 3(-2)(3) - 2(3)^2$
$= (4)(3) + 3(-6) - 2(9)$
$= 12 - 18 - 18$
$= -24$

**37.** $(-a^2 + 2a - 1) + (a^2 + 2a + 1)$
$= -a^2 + 2a - 1 + a^2 + 2a + 1$
$= 4a$

**39.** $(2xy^2 - y^3) + (y^3 + xy^2)$
$= 2xy^2 - y^3 + y^3 + xy^2$
$= 3xy^2$

**41.** $(-5x - 3) - (2x - 9) = -5x - 3 - 2x + 9$
$\phantom{(-5x - 3) - (2x - 9)} = -7x + 6$

**43.** $(2x^2 - 4x + 7) - (-2x^2 - 3x - 7)$
$= 2x^2 - 4x + 7 + 2x^2 + 3x + 7$
$= 4x^2 - x + 14$

**45.** $(-3t + 7) - (2t^2 + 6) = -3t + 7 - 2t^2 - 6$
$\phantom{(-3t + 7) - (2t^2 + 6)} = -2t^2 - 3t + 1$

**47.** $(4mn + n^2) - (m^2 - 3mn)$
$= 4mn + n^2 - m^2 + 3mn$
$= n^2 - m^2 + 7mn$

**49.** $(x^2y + 2xy + 3y^2) + (-x^2y - 4xy + 2y^2)$
$= x^2y + 2xy + 3y^2 - x^2y - 4xy + 2y^2$
$= -2xy + 5y^2$

**51.** $(9a^2b + 2ab - 5a^2) - (2ab - 3a^2 + 4a^2b)$
$= 9a^2b + 2ab - 5a^2 - 2ab + 3a^2 - 4a^2b$
$= -2a^2 + 5a^2b$

**53.** Subtract:
$$\begin{array}{r} 4x^3 - 10x \\ -\ 6x^3 + 2x \\ \hline -2x^3 - 8x \end{array}$$

**55.** Add:
$$\begin{array}{r} x^2 + 3x - 7 \\ -5x^2 + 2x - 1 \\ \hline -4x^2 + 5x - 8 \end{array}$$

**57.** Subtract:
$$\begin{array}{r} 4x^2 - x - 6 \\ -3x^2 + x - 6 \\ \hline x^2 \phantom{+ x} - 12 \end{array}$$

**59.** Add:
$$\begin{array}{r} x^2 + 3x - 2 \\ -2x^2 - x \\ 3x^2 \phantom{+ x} + 4 \\ \hline 2x^2 + 2x + 2 \end{array}$$

**61.** (ii) $(3x+4)-(x-2)$ is correct because the minuend is $3x+4$ and the subtrahend is $x-2$.

**63.** $3x^2 - 6x + 4x + 7 = 3x^2 - 2x + 7$

**65.** $-5x - 2x^2 - (7x^2 - 2x + 9)$
$= -5x - 2x^2 - 7x^2 + 2x - 9$
$= -9x^2 - 3x - 9$

**67.** Because $x-7+(-x+7) = x-7-x+7 = 0$, $-x+7$ must be added to $x-7$ to obtain 0.

**69.** Because $x-3-(x-3) = x-3-x+3 = 0$, $x-3$ must be subtracted from $x-3$ to obtain 0.

**71.** $(2x-3) - (3x^2 + 4x - 5) - (3x - 4x^2)$
$= 2x - 3 - 3x^2 - 4x + 5 - 3x + 4x^2$
$= x^2 - 5x + 2$

**73.** $-[3x - (2x-1)] = -3x + (2x-1)$
$= -3x + 2x - 1$
$= -x - 1$

**75.** $(x-3) + [2x - (3x-1)]$
$= x - 3 + (2x - 3x + 1)$
$= x - 3 + 2x - 3x + 1$
$= -2$

**77.** $[3 - (y-1)] - (2y+1) = 3 - y + 1 - 2y - 1$
$= -3y + 3$

**79.** $(3x^2 - 2xy + y^2) - (4xy + 3x^2 - y^2) + (4xy^2 - 2x^2)$
$= 3x^2 - 2xy + y^2 - 4xy - 3x^2 + y^2 + 4xy^2 - 2x^2$
$= -2x^2 - 6xy + 2y^2 + 4xy^2$

**81.** $P(x) = 3 - x$
$P(-2) = 3 - (-2) = 5$
$P(3) = 3 - 3 = 0$

**83.** $Q(x) = -x^2 + 3x - 1$
$Q(0) = -(0)^2 + 3(0) - 1 = -1$
$Q(-2) = -(-2)^2 + 3(-2) - 1$
$= -4 - 6 - 1$
$= -11$

**85.** $R(x) = x^3 - 4x^5 + 5x^2 + 2$
$R(2.1) = (2.1)^3 - 4(2.1)^5 + 5(2.1)^2 + 2$
$\approx -130.05$
$R(-1.7) = (-1.7)^3 - 4(-1.7)^5 + 5(-1.7)^2 + 2$
$\approx 68.33$

**87.** $P(x) = 4x + 3$
$Q(x) = x^2 - 5$
$P(x) + Q(x) = 4x + 3 + x^2 - 5$
$= x^2 + 4x - 2$

**89.** $P(x) = 4x + 3$
$Q(x) = x^2 - 5$
$R(x) = 2x^2 - 7x - 8$
$R(x) - [P(x) - Q(x)]$
$= 2x^2 - 7x - 8 - [4x + 3 - (x^2 - 5)]$
$= 2x^2 - 7x - 8 - (4x + 3 - x^2 + 5)$
$= 2x^2 - 7x - 8 - (-x^2 + 4x + 8)$
$= 2x^2 - 7x - 8 + x^2 - 4x - 8$
$= 3x^2 - 11x - 16$

**91.** $A(t) = -2t^2 + 280t - 8000$

(a) $A(80) = -2(80)^2 + 280(80) - 8000$
$= 1600$

(b) $A(50) = -2(50)^2 + 280(50) - 8000$
$= 1000$

(c) $A(98) = -2(98)^2 + 280(98) - 8000$
$= 232$

(d) $A(40) = -2(40)^2 + 280(40) - 8000$
$= 0$

(e) $A(100) = -2(100)^2 + 280(100) - 8000$
$= 0$

(f) The domain of $A(t)$ is $40 \leq t \leq 100$.

**93.** $C(x) = 20x + 12$

$M(x) = x^2 - 3x + 10$

$R(x) = 2x^2 + 8x$

(a) $C(100) = 20(100) + 12 = \$2012$

(b) $M(100) = (100)^2 - 3(100) + 10 = \$9710$

(c) $T(x) = C(x) + M(x)$
$= 20x + 12 + x^2 - 3x + 10$
$= x^2 + 17x + 22$

(d) $T(100) = (100)^2 + 17(100) + 22 = \$11{,}722$

(e) From part (a), we have $C(100) = \$2012$ and from part (b), we have $M(100) = \$9710$. Thus, $C(100) + M(100) = 2012 + 9710 = \$11{,}722$. From part (d), we have $T(100) = \$11{,}722$. Thus, $C(100) + M(100) = T(100)$.

(f) $R(100) = 2(100)^2 + 8(100) = \$20{,}800$

(g) $P(x) = R(x) - T(x)$
$= 2x^2 + 8x - (x^2 + 17x + 22)$
$= 2x^2 + 8x - x^2 - 17x - 22$
$= x^2 - 9x - 22$

(h) $P(100) = (100)^2 - 9(100) - 22 = \$9078$

(i) From part (f), we have $R(100) = \$20{,}800$ and from part (d), we have $T(100) = \$11{,}722$. Thus, $R(100) - T(100) = 20{,}800 - 11{,}722 = \$9078$. From part (h), we have $P(100) = \$9078$. Thus, $P(100) = R(100) - T(100)$.

**95.** $(ax^2 + bx - 3) + (x^2 + 2x + c)$
$= ax^2 + bx - 3 + x^2 + 2x + c$
$= (a + 1)x^2 + (b + 2)x + (c - 3)$

For this polynomial to be equal to $3x + 1$, the leading coefficient must be 0, the $x$-term coefficient must be 3, and the constant must be 1.

$a + 1 = 0 \qquad b + 2 = 3 \qquad c - 3 = 1$
$a = -1 \qquad\quad b = 1 \qquad\quad c = 4$

**97.** $(x^2 + bx - c) + ax^2 = (a + 1)x^2 + bx - c$

For this polynomial to be equal to $x^2 + 6x + 9$, the leading coefficient must be 1, the $x$-term coefficient must be 6, and the constant must be 9.

$a + 1 = 1 \qquad b = 6 \qquad -c = 9$
$a = 0 \qquad\qquad\qquad\quad c = -9$

**99.** When we simplify the left side, we obtain $(a + b)x^2 - 2 = 4x^2 - 2$, which means that $a + b = 4$. There are infinitely many values of $a$ and $b$ such that $a + b = 4$.

**101.** $P(x) = x^3 - 5x^2 + 6x + 10$

$P(0) = (0)^3 - 5(0)^2 + 6(0) + 10 = 10$

With the given information, we can form the following system of equations.

$\begin{aligned} a + b + c &= 10 \\ a + b - c &= 0 \\ 2a - b \phantom{{}+c} &= 1 \end{aligned}$

Let $A =$ the augmented matrix.

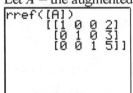

The constants that satisfy the given conditions are $a = 2$, $b = 3$, and $c = 5$.

## Section 6.5 Multiplication and Special Products

1. Statement (ii) is false. For example, $(x+2)(x-2) = x^2 - 4$, which is a binomial.

3. $3x^2(2x-3) = 3x^2(2x) - 3x^2(3)$
   $= 6x^3 - 9x^2$

5. $-4x^3(2x^2 + 5) = -4x^3(2x^2) + (-4x^3)(5)$
   $= -8x^5 - 20x^3$

7. $-xy^2(x^2 - xy - 2)$
   $= -xy^2(x^2) - (-xy^2)(xy) - (-xy^2)(2)$
   $= -x^3y^2 + x^2y^3 + 2xy^2$

9. $(x-6)(x+4) = x(x+4) - 6(x+4)$
   $= x^2 + 4x - 6x - 24$
   $= x^2 - 2x - 24$

11. $(y-7)(y-2) = y(y-2) - 7(y-2)$
    $= y^2 - 2y - 7y + 14$
    $= y^2 - 9y + 14$

13. $(y+5)(3-y) = y(3-y) + 5(3-y)$
    $= 3y - y^2 + 15 - 5y$
    $= -y^2 - 2y + 15$

15. Using the FOIL pattern:
    $(5x-6)(4x+3) = 20x^2 + 15x - 24x - 18$
    $= 20x^2 - 9x - 18$

17. Using the FOIL pattern:
    $(9x+2)(x+3) = 9x^2 + 27x + 2x + 6$
    $= 9x^2 + 29x + 6$

19. Using the FOIL pattern:
    $(4-5y)(2y+1) = 8y + 4 - 10y^2 - 5y$
    $= -10y^2 + 3y + 4$

21. Using the FOIL pattern:
    $(x-9y)(x-4y) = x^2 - 4xy - 9xy + 36y^2$
    $= x^2 - 13xy + 36y^2$

23. Using the FOIL pattern:
    $(a+5b)(a-2b) = a^2 - 2ab + 5ab - 10b^2$
    $= a^2 + 3ab - 10b^2$

25. Using the FOIL pattern:
    $(6x-5y)(2x-3y)$
    $= 12x^2 - 18xy - 10xy + 15y^2$
    $= 12x^2 - 28xy + 15y^2$

27. Using the FOIL pattern:
    $(2x^2 - 3)(3x^2 - 2)$
    $= 6x^4 - 4x^2 - 9x^2 + 6$
    $= 6x^4 - 13x^2 + 6$

29. $(x+2)(x^2 - 4x + 5)$
    $= x(x^2 - 4x + 5) + 2(x^2 - 4x + 5)$
    $= x^3 - 4x^2 + 5x + 2x^2 - 8x + 10$
    $= x^3 - 2x^2 - 3x + 10$

31. $(2x-3)(3x^2 + x - 1)$
    $= 2x(3x^2 + x - 1) - 3(3x^2 + x - 1)$
    $= 6x^3 + 2x^2 - 2x - 9x^2 - 3x + 3$
    $= 6x^3 - 7x^2 - 5x + 3$

33. $(x+y)(x^2 - xy + y^2)$
    $= x(x^2 - xy + y^2) + y(x^2 - xy + y^2)$
    $= x^3 - x^2y + xy^2 + x^2y - xy^2 + y^3$
    $= x^3 + y^3$

35. $(2x-1)(2x^3 - x^2 + 3x - 4)$
    $= 2x(2x^3 - x^2 + 3x - 4) - 1(2x^3 - x^2 + 3x - 4)$
    $= 4x^4 - 2x^3 + 6x^2 - 8x - 2x^3 + x^2 - 3x + 4$
    $= 4x^4 - 4x^3 + 7x^2 - 11x + 4$

**37.** $(x^2 - 2x + 1)(x^2 + 2x + 1)$
$= x^2(x^2 + 2x + 1) - 2x(x^2 + 2x + 1)$
$\quad + 1(x^2 + 2x + 1)$
$= x^4 + 2x^3 + x^2 - 2x^3 - 4x^2 - 2x$
$\quad + x^2 + 2x + 1$
$= x^4 - 2x^2 + 1$

**39.**
$$\begin{array}{r} x^2 + 2x + 4 \\ \times \quad\quad x - 2 \\ \hline -2x^2 - 4x - 8 \\ x^3 + 2x^2 + 4x \quad\quad\quad \\ \hline x^3 \quad\quad\quad\quad\quad - 8 \end{array}$$

**41.**
$$\begin{array}{r} 2x^2 - 4x + 5 \\ \times \quad x^2 - 2x + 3 \\ \hline 6x^2 - 12x + 15 \\ -4x^3 + 8x^2 - 10x \quad\quad \\ 2x^4 - 4x^3 + 5x^2 \quad\quad\quad\quad \\ \hline 2x^4 - 8x^3 + 19x^2 - 22x + 15 \end{array}$$

**43.** Yes, but recognizing the special pattern saves the step of determining the inner and outer products.

**45.** Using the special Square of a Binomial pattern:
$(x + 3)^2 = (x)^2 + 2(x)(3) + (3)^2$
$\quad\quad\quad = x^2 + 6x + 9$

**47.** Using the special Square of a Binomial pattern:
$(2 - y)^2 = 2^2 - 2(2)(y) + y^2$
$\quad\quad\quad = 4 - 4y + y^2$

**49.** Using the special Square of a Binomial pattern:
$(7x - 3)^2 = 49x^2 - 42x + 9$

**51.** Using the special Square of a Binomial pattern:
$(3a - b)^2 = 9a^2 - 6ab + b^2$

**53.** Using the special Square of a Binomial pattern:
$(2x + 5y)^2 = 4x^2 + 20xy + 25y^2$

**55.** Using the Product of the Sum and Difference of the Same Two terms pattern:
$(x - 4)(x + 4) = x^2 - 4^2 = x^2 - 16$

**57.** Using the Product of the Sum and Difference of the Same Two terms pattern:
$(7 - y)(7 + y) = 7^2 - y^2 = 49 - y^2$

**59.** Using the Product of the Sum and Difference of the Same Two terms pattern:
$(2x - 3)(2x + 3) = (2x)^2 - 3^2 = 4x^2 - 9$

**61.** Using the Product of the Sum and Difference of the Same Two terms pattern:
$(4x + 5y)(4x - 5y) = 16x^2 - 25y^2$

**63.** Using the Product of the Sum and Difference of the Same Two terms pattern:
$(9n - 4m)(9n + 4m) = 81n^2 - 16m^2$

**65.** The Associative Property of Multiplication guarantees that either grouping is correct. Method (i) is easier because it can be done in one step.

**67.** Using the Product of the Sum and Difference of the Same Two terms pattern:
$(a^3 - b^3)(a^3 + b^3) = a^6 - b^6$

**69.** Using the FOIL pattern:
$(4 - pq^2)(4 + p^2q)$
$\quad = 16 + 4p^2q - 4pq^2 - p^3q^3$

**71.** Using the special Square of a Binomial pattern:
$(x^2y^2 + 2)^2 = (x^2y^2)^2 + 2(2)x^2y^2 + 2^2$
$\quad\quad\quad\quad = x^4y^4 + 4x^2y^2 + 4$

**73.** Begin by multiplying the first binomial by the monomial and then use the FOIL pattern:
$$2x^3(2x-5)(3x+1)$$
$$=(4x^4-10x^3)(3x+1)$$
$$=12x^5+4x^4-30x^4-10x^3$$
$$=12x^5-26x^4-10x^3$$

**75.** $5x-6(4x+3)=5x-24x-18$
$$=-19x-18$$

**77.** $(b-3)(2b+5)(b-2)$
$$=(b-3)(2b^2-4b+5b-10)$$
$$=(b-3)(2b^2+b-10)$$
$$=b(2b^2+b-10)-3(2b^2+b-10)$$
$$=2b^3+b^2-10b-6b^2-3b+30$$
$$=2b^3-5b^2-13b+30$$

**79.** $(2x+1)(4x^2+1)(2x-1)$
$$=(2x+1)(2x-1)(4x^2+1)$$
$$=(4x^2-1)(4x^2+1)$$
$$=(4x^2)^2-1^2$$
$$=16x^4-1$$

**81.** $(2x+5)(2x-5)-(2x+5)^2$
$$=(4x^2-25)-(4x^2+20x+25)$$
$$=4x^2-25-4x^2-20x-25$$
$$=-20x-50$$

**83.** $(x+6)(x-4)-(2x-3)(x+2)$
$$=(x^2-4x+6x-24)-(2x^2+4x-3x-6)$$
$$=(x^2+2x-24)-(2x^2+x-6)$$
$$=x^2+2x-24-2x^2-x+6$$
$$=-x^2+x-18$$

**85.** $(x+3)^2-(x-3)^2$
$$=(x^2+6x+9)-(x^2-6x+9)$$
$$=x^2+6x+9-x^2+6x-9$$
$$=12x$$

**87.** $2x^3(4x-3)^2$
$$=2x^3(16x^2-24x+9)$$
$$=32x^5-48x^4+18x^3$$

**89.** $(3y+2)(3y-2)-(5-2y)^2$
$$=(9y^2-4)-(25-20y+4y^2)$$
$$=9y^2-4-25+20y-4y^2$$
$$=5y^2+20y-29$$

**91.** $(ax+3)(2x+b)=2ax^2+abx+6x+3b$
$$=2ax^2+(ab+6)x+3b$$
For this expression to be equal to $8x^2-2x-6$, equate the coefficients of corresponding terms and solve for $a$ and $b$.
$$2a=8 \qquad 3b=-6$$
$$a=4 \qquad b=-2$$
Verify that these values of $a$ and $b$ yield the desired coefficient of $x$.
$$ab+6=(4)(-2)+6=-2$$

**93.** $(ax-2)(3x+b)=3ax^2+abx-6x-2b$
$$=3ax^2+(ab-6)x-2b$$
For this expression to be equal to $9x^2-4$, equate the coefficients of corresponding terms and solve for $a$ and $b$.
$$3a=9 \qquad -2b=-4$$
$$a=3 \qquad b=2$$
Verify that these values of $a$ and $b$ yield the desired coefficient of $x$.
$$ab-6=(3)(2)-6=0$$

**95.** Distance traveled by the car:
(rate)(time) = $rt$
Distance traveled by the van:
(rate)(time) = $(r+5)(t-1)$
Total distance between the car and the van at the time that the car reached its destination is:
$$rt+(r+5)(t-1)=rt+(rt-r+5t-5)$$
$$=rt+rt-r+5t-5$$
$$=2rt-r+5t-5$$

**97.** Let $w$ = the width of the garden and $300-2w$ = the length of the garden. The area of the garden is
(width)(length) = $w(300-2w)$
$$=300w-2w^2$$

**99.** Because the rate of change is not approximately constant, a linear model does not appear to be appropriate.

**101.**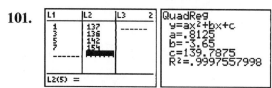

The model is
$B(x) = 0.8125x^2 - 3.65x + 139.7875$ or
$B(x) = 0.8x^2 - 3.7x + 139.8$

**103.** $(x^n + 2)(x^n + 4)$
$= x^n \cdot x^n + 4x^n + 2x^n + 8$
$= x^{2n} + 6x^n + 8$

**105.** $(x^n + 4)(x^n - 4) = (x^n)^2 - 4^2$
$= x^{2n} - 16$

**107.** The length of a side of the large rectangle is $a + b$. The area of the large rectangle is $(a+b)(a+b) = (a+b)^2$. The areas of the smaller rectangles (from upper left to lower right) are $a^2$, $ab$, $ab$, and $b^2$.
Thus,
$(a+b)^2 = a^2 + ab + ab + b^2$
$= a^2 + 2ab + b^2$

**109.** Write the expression as $[(x+3y)-4][(x+3y)+4]$ and use the special product form $(A-B)(A+B)$. The result is
$[(x+3y)-4][(x+3y)+4]$
$= (x+3y)^2 - 4^2$
$= x^2 + 6xy + 9y^2 - 16$

**111.** $(x^2 + 5x + 3)(x^2 + 5x - 3)$ can be rewritten as
$[(x^2 + 5x) + 3][(x^2 + 5x) - 3]$
$= (x^2 + 5x)^2 - 3^2$
$= x^4 + 10x^3 + 25x^2 - 9$

**113.** $[(x-3) + 5y][(x-3) - 5y]$
$= (x-3)^2 - (5y)^2$
$= x^2 - 6x + 9 - 25y^2$

## Chapter 6 Project

**1.** 6,370,000 kilometers = $6.37 \cdot 10^6$ kilometers

**3.** $170(1.0 \cdot 10^{-2}) = 1.7$ pounds

**5.** $\dfrac{d}{r} = t = \dfrac{24,900}{17,640} \approx 1.4116$ hours or $1.4116 \times 60 = 84.7$ minutes

## Chapter 6 Review Exercises

1. In the expression $x^2 \cdot x^3$, there is a total of 5 factors of $x$. This total can be found by adding the exponents according to the Product Rule for Exponents.

3. $-3 \cdot 3^4 \cdot 3^7 = (-1) \cdot 3^1 \cdot 3^4 \cdot 3^7$
   $= (-1)3^{1+4+7}$
   $= -3^{12}$

5. $\dfrac{4^{11}}{4^7} = 4^{11-7} = 4^4$

7. $\left(-\dfrac{2}{5}\right)^3 = \dfrac{(-2)^3}{5^3} = \dfrac{-8}{125}$

9. $\left(\dfrac{2h^3}{k^2}\right)^3 = \dfrac{(2h^3)^3}{(k^2)^3} = \dfrac{8h^9}{k^6}$

11. $(-7x^5 y)(5x^7 y^8) = -35x^{5+7} y^{1+8}$
    $= -35x^{12} y^9$

13. Any nonzero number divided by itself is 1.

15. $1776^0 = 1$

17. $(-3)^{-3} = \dfrac{1}{(-3)^3} = \dfrac{1}{-27} = -\dfrac{1}{27}$

19. $(5+3)^{-1} = \dfrac{1}{(5+3)} = \dfrac{1}{8}$

21. $\left(-\dfrac{h}{k}\right)^{-1} = -\dfrac{k}{h}$

23. $r^0 \cdot r^{-3} \cdot r \cdot r^{-7} = r^0 \cdot r^{-3} \cdot r^1 \cdot r^{-7}$
    $= r^{0-3+1-7}$
    $= r^{-9}$
    $= \dfrac{1}{r^9}$

25. $\dfrac{y^{-7}}{y^{-5}} = y^{-7-(-5)} = y^{-2} = \dfrac{1}{y^2}$

27. $\left(\dfrac{c^2}{d}\right)^{-7} = \left(\dfrac{d}{c^2}\right)^7 = \dfrac{d^7}{(c^2)^7} = \dfrac{d^7}{c^{14}}$

29. $\left(\dfrac{2h^{-5}}{k^3}\right)^{-4} = \left(\dfrac{k^3}{2h^{-5}}\right)^4$
    $= \dfrac{(k^3)^4}{(2h^{-5})^4}$
    $= \dfrac{k^{12}}{16h^{-20}}$
    $= \dfrac{k^{12} h^{20}}{16}$

31. The number $p$ is the number of places to move the decimal point in 2.3 to obtain 23,000.

33. $24{,}500{,}000{,}000 = 2.45 \cdot 10^{10}$

35. $6.83 \cdot 10^{-7} = 0.000000683$

37. $(-439)^{33} \approx -1.6 \cdot 10^{87}$

39. (a) The land area of Missouri is $68{,}945 \approx 6.9 \cdot 10^4$ square miles. The population of Missouri in 1995 was $5{,}137{,}804 \approx 5.1 \cdot 10^6$.
    (b) The average population is $\dfrac{5.1 \cdot 10^6}{6.9 \cdot 10^4} \approx 74$ people per square mile.

41. $3x - 2x^2$
    (a) Because there are two terms, this is a binomial.
    (b) The degree of the binomial is 2.
    (c) Descending order: $-2x^2 + 3x$
    (d) Coefficient of $x^2$ term: $-2$

**43.** $x^2 + 9x^5 - 3$
   (a) Because there are three terms, this is a trinomial.
   (b) The degree of the trinomial is 5.
   (c) Descending order: $9x^5 + x^2 - 3$
   (d) Coefficient of $x^2$ term: 1

**45.** $4x^2 + y^2 + x^2y^4$
   Because the highest degree of a term is 6, the degree of the polynomial is 6.

**47.** $P(x) = x^5 - x^3 + 2x^2$
   $P(1) = (1)^5 - (1)^3 + 2(1)^2$
   $= 1 - 1 + 2$
   $= 2$

**49.** $x = 1, y = 2$
   $x^5y^3 - 4x^3y^4 + x^2$
   $(1)^5(2)^3 - 4(1)^3(2)^4 + (1)^2$
   $= 8 - 64 + 1$
   $= -55$

**51.** $(5x^2 - 3x + 4) + (-5x - 3x^2 + 9)$
   $= 5x^2 - 3x + 4 - 5x - 3x^2 + 9$
   $= 2x^2 - 8x + 13$

**53.** $(3x^2 - 6x) + (4x + 7) = 3x^2 - 6x + 4x + 7$
   $= 3x^2 - 2x + 7$

**55.** $4x^5(-2x^3) = -8x^8$

**57.** $2a^2b(5 - b^2 + 3b)$
   $= 2a^2b(5) - 2a^2b(b^2) + 2a^2b(3b)$
   $= 10a^2b - 2a^2b^3 + 6a^2b^2$

**59.** $(2x + 3)(8 - 3x) = 16x - 6x^2 + 24 - 9x$
   $= -6x^2 + 7x + 24$

**61.** $(x + 3)(2x - 5) - (3x + 1)(x + 4)$
   $= (2x^2 - 5x + 6x - 15) - (3x^2 + 12x + x + 4)$
   $= (2x^2 + x - 15) - (3x^2 + 13x + 4)$
   $= 2x^2 + x - 15 - 3x^2 - 13x - 4$
   $= -x^2 - 12x - 19$

**63.** Area $= (4x - 5)(2x - 3)$
   $= 8x^2 - 12x - 10x + 15$
   $= 8x^2 - 22x + 15$

**65.** A difference of two squares consists of two perfect square terms separated by a minus sign.

**67.** $(3x - 2)(3x + 2) = (3x)^2 - 2^2 = 9x^2 - 4$

**69.** $(3y - 2)^2 = (3y)^2 - 2(3y)(2) + 2^2$
   $= 9y^2 - 12y + 4$

**71.** $(2x - 1)(x - 3) - (x - 3)^2$
   $= (2x^2 - 6x - x + 3) - (x^2 - 6x + 9)$
   $= (2x^2 - 7x + 3) - (x^2 - 6x + 9)$
   $= 2x^2 - 7x + 3 - x^2 + 6x - 9$
   $= x^2 - x - 6$

**73.** Let $x =$ the second consecutive integer, then $x - 1 =$ the first consecutive integer and $x + 1 =$ the third consecutive integer.
   $(x - 1)(x)(x + 1) = x(x - 1)(x + 1)$
   $= x(x^2 - 1)$
   $= x^3 - x$

## Chapter 6 Looking Back

1. (a) $5(2-3y) = 10 - 15y$
   (b) $3x + 3y = 3(x+y)$

3. $(y+5)(y-5) = y^2 - 25$

5. $(a+1)^2 = a^2 + 2a + 1$

7. $(3x+1)(x-5) = 3x^2 - 15x + x - 5$
   $= 3x^2 - 14x - 5$

9. (a) Because $4 + 5 = 9$ and $4(5) = 20$, the numbers are 4 and 5.
   (b) Because $-6 + 1 = -5$ and $-6(1) = -6$, the numbers are $-6$ and 1.
   (c) Because $-3 + (-4) = -7$ and $-3(-4) = 12$, the numbers are $-3$ and $-4$.
   (d) Because $-2 + 8 = 6$ and $-2(8) = -16$, the numbers are $-2$ and 8.

11. $|x - 7| = 8$
    $x - 7 = 8$   or   $x - 7 = -8$
    $x = 15$          $x = -1$

## Chapter 6 Test

1. $(x^3 y^2)(xy^4) = x^{3+1} y^{2+4} = x^4 y^6$

3. $(2a^{-5} b^4)^2 = 2^2 (a^{-5})^2 (b^4)^2$
   $= 4a^{-10} b^8$
   $= \dfrac{4b^8}{a^{10}}$

5. $\dfrac{4a^{-3}}{5b^{-2}} = \dfrac{4}{5}\left(\dfrac{a^3}{b^2}\right)^{-1} = \dfrac{4}{5}\left(\dfrac{b^2}{a^3}\right) = \dfrac{4b^2}{5a^3}$

7. $\dfrac{(-2)^{-3}}{(-2)^2} = (-2)^{-3-2} = (-2)^{-5} = \dfrac{1}{(-2)^5} = -\dfrac{1}{32}$

9. $(-2)^{-3} = \dfrac{1}{(-2)^3} = -\dfrac{1}{8}$

11. $4.56 \cdot 10^{-5} = 0.0000456$

13. $0.0008 = 8 \cdot 10^{-4}$
    $260,000,000 = 2.6 \cdot 10^8$
    $(8 \cdot 10^{-4})(2.6 \cdot 10^8) = 2.08 \cdot 10^5$
    This represents the president's salary.

15. $p(x) = 3x^4 - 4x^3 - 5x^2 + 6x - 7$
    $q(x) = 5x^3 - 8x + 2$
    (a) $(p+q)(x) = 3x^4 - 4x^3 - 5x^2 + 6x$
    $\phantom{(p+q)(x) = } -7 + 5x^3 - 8x + 2$
    $= 3x^4 + x^3 - 5x^2 - 2x - 5$
    $(p+q)(1) = 3(1)^4 + (1)^3 - 5(1)^2$
    $\phantom{(p+q)(1) = } -2(1) - 5$
    $= 3 + 1 - 5 - 2 - 5$
    $= -8$
    or
    $p(1) = 3(1)^4 - 4(1)^3 - 5(1)^2 + 6(1) - 7$
    $= 3 - 4 - 5 + 6 - 7 = -7$
    $q(1) = 5(1)^3 - 8(1) + 2$
    $= 5 - 8 + 2 = -1$
    $p(1) + q(1) = -7 + (-1) = -8$
    (b) $(p-q)(x) = 3x^4 - 4x^3 - 5x^2 + 6x$
    $\phantom{(p-q)(x) = } -7 - 5x^3 + 8x - 2$
    $= 3x^4 - 9x^3 - 5x^2 + 14x$
    $\phantom{= } -9$
    $(p-q)(-1) = 3(-1)^4 - 9(-1)^3 - 5(-1)^2$
    $\phantom{(p-q)(-1) = } + 14(-1) - 9$
    $= 3 + 9 - 5 - 14 - 9$
    $= -16$

or
$$p(-1) = 3(-1)^4 - 4(-1)^3 - 5(-1)^2 \\ + 6(-1) - 7 \\ = 3 + 4 - 5 - 6 - 7 = -11$$
$$q(-1) = 5(-1)^3 - 8(-1) + 2 \\ = -5 + 8 + 2 = 5$$
$$p(-1) - q(-1) = -11 - 5 = -16$$

**17.** $(2t + s)(3t + 4s) = 6t^2 + 8st + 3st + 4s^2$
$\phantom{(2t + s)(3t + 4s)} = 6t^2 + 11st + 4s^2$

**19.** $(x^2 + 3y)^2 = (x^2)^2 + 2(x^2)(3y) + (3y)^2$
$\phantom{(x^2 + 3y)^2} = x^4 + 6x^2y + 9y^2$

**21.** $(-3a^2b^5)(2a^3b) = -6a^{2+3}b^{5+1}$
$\phantom{(-3a^2b^5)(2a^3b)} = -6a^5b^6$

# Chapter 7

# Factoring

## Section 7.1  Common Factors and Grouping

**1.** Choose the smallest exponent for each variable.

**3.** $24 = 2 \cdot 2 \cdot 2 \cdot 3$
$60x = 2 \cdot 2 \cdot 3 \cdot 5 \cdot x$
Because the only factors common to both expressions are $2^2$ and 3, the GCF is 12.

**5.** $12x = 2 \cdot 2 \cdot 3 \cdot x$
$16y = 2 \cdot 2 \cdot 2 \cdot 2 \cdot y$
$24z = 2 \cdot 2 \cdot 2 \cdot 3 \cdot z$
Because the only factor common to the expressions is $2^2$, the GCF is 4.

**7.** $3x^5 = 3 \cdot x^5$
$4x^7 = 2 \cdot 2 \cdot x^7$
$5x^3 = 5 \cdot x^3$
The GCF of the numerical coefficients is 1 and the smallest exponent on $x$ is 3. Thus the GCF is $x^3$.

**9.** $m^5 n^3 = m^5 \cdot n^3$, $mn^6 = m \cdot n^6$
The smallest exponents on $m$ and $n$ are 1 and 3, respectively. The GCF is $mn^3$.

**11.** $-6xy + 3y = -3y(2x) - 3y(-1)$
$= -3y(2x - 1)$

**13.** $5x^2 + 10x + 15 = 5(x^2) + 5(2x) + 5(3)$
$= \underline{5}(x^2 + 2x + 3)$

**15.** Begin by factoring out the GCF of 3.
$3x + 12 = 3(x) + 3(4) = 3(x + 4)$

**17.** Begin by factoring out the GCF of 5.
$5x + 10y - 30 = 5(x) + 5(2y) - 5(6)$
$= 5(x + 2y - 6)$

**19.** Begin by factoring out the GCF of $4x$.
$12x^2 - 4x = 4x(3x) - 4x(1) = 4x(3x - 1)$

**21.** $7x^2 - 10y$ is prime. The GCF is 1.

**23.** Begin by factoring out the GCF of $x^2$.
$x^3 - x^2 = x^2(x) - x^2(1) = x^2(x - 1)$

**25.** Begin by factoring out the GCF of $3x$.
$3x^3 - 9x^2 + 12x$
$= 3x(x^2) - 3x(3x) + 3x(4)$
$= 3x(x^2 - 3x + 4)$

**27.** Begin by factoring out the GCF of $xy^2$.
$x^5 y^4 - x^4 y^3 + xy^2$
$= xy^2(x^4 y^2) - xy^2(x^3 y) + xy^2(1)$
$= xy^2(x^4 y^2 - x^3 y + 1)$

**29.** Begin by factoring out the GCF of $(a + by)$.
$3x(a + by) - 2y(a + by)$
$= (a + by)(3x - 2y)$

**31.** Begin by factoring out the GCF of $(2x + 3)$.
$4x^2(2x + 3) - 7y^2(2x + 3)$
$= (2x + 3)(4x^2 - 7y^2)$

**33.** (a) $4(x-3) = 4x - 12$
(b) $-4(-x+3) = 4x - 12$
(c) $4(-3+x) = -12 + 4x = 4x - 12$
(d) $-4(3-x) = -12 + 4x = 4x - 12$
All of these are correct because the product in each case is $4x - 12$.

**35.** The GCF is 4.
$4x - 12 = 4(x-3)$
$4x - 12 = -4(-x+3) = -4(3-x)$

**37.** The GCF is $y$.
$6y + 3xy - x^2y = y(6 + 3x - x^2)$
$6y + 3xy - x^2y = -y(-6 - 3x + x^2)$

**39.** $y(y-3) + 2(3-y) = y(y-3) - 2(y-3)$
$= (y-3)(y-2)$

**41.** $5(3-z) - z(z-3) = 5(3-z) + z(3-z)$
$= (3-z)(5+z)$

**43.** $(x+3)(x-4) - (x+2)(4-x)$
$= (x+3)(x-4) + (x+2)(x-4)$
$= (x-4)[(x+3) + (x+2)]$
$= (x-4)(2x+5)$

**45.** The answer is obtained from multiplying the factors.
$(2x-5)(x+3) = 2x^2 + 6x - 5x - 15$

**47.** $ab + bx + ay + xy = (ab + bx) + (ay + xy)$
$= b(a+x) + y(a+x)$
$= (a+x)(b+y)$

**49.** $cd + 3d - 4c - 12 = (cd + 3d) - (4c + 12)$
$= d(c+3) - 4(c+3)$
$= (c+3)(d-4)$

**51.** $ab - ay - bx - xy$ is prime.

**53.** $ax^2 + 3x^2 + ay + 3y$
$= (ax^2 + 3x^2) + (ay + 3y)$
$= x^2(a+3) + y(a+3)$
$= (a+3)(x^2+y)$

**55.** $x^3 + 2x^2 + 3x + 6$
$= (x^3 + 2x^2) + (3x+6)$
$= x^2(x+2) + 3(x+2)$
$= (x+2)(x^2+3)$

**57.** $3y^3 + 9y^2 + 12y + 36$
$= (3y^3 + 9y^2) + (12y + 36)$
$= 3y^2(y+3) + 12(y+3)$
$= (y+3)(3y^2 + 12)$
$= 3(y+3)(y^2+4)$

**59.** $a^2b + a^2y + b + y = (a^2b + a^2y) + (b+y)$
$= a^2(b+y) + 1(b+y)$
$= (b+y)(a^2+1)$

**61.** $x^3 - 15 + 3x - 5x^2$
$= x^3 - 5x^2 + 3x - 15$
$= (x^3 - 5x^2) + (3x - 15)$
$= x^2(x-5) + 3(x-5)$
$= (x-5)(x^2+3)$

**63.** $6x^3 - 12 + 8x - 9x^2$
$= 6x^3 - 9x^2 + 8x - 12$
$= (6x^3 - 9x^2) + (8x - 12)$
$= 3x^2(2x-3) + 4(2x-3)$
$= (2x-3)(3x^2+4)$

**65.** $18x^3 + 6x^2 - 45x - 15$
$= 3(6x^3 + 2x^2 - 15x - 5)$
$= 3[(6x^3 + 2x^2) - (15x + 5)]$
$= 3[2x^2(3x+1) - 5(3x+1)]$
$= 3(3x+1)(2x^2 - 5)$

**67.** $3xy^3 - 6xy^2 + 7xy - 14x$
$= x(3y^3 - 6y^2 + 7y - 14)$
$= x[(3y^3 - 6y^2) + (7y - 14)]$
$= x[3y^2(y-2) + 7(y-2)]$
$= x(y-2)(3y^2 + 7)$

**69.** $\frac{2}{3}y - \frac{5}{6} = \frac{4}{6}y - \frac{5}{6} = \frac{1}{6}(4y-5)$

Section 7.2   Special Factoring

71. $\frac{1}{2}x^2 + \frac{1}{3}x + 1 = \frac{3}{6}x^2 + \frac{2}{6}x + \frac{6}{6}$
$= \frac{1}{6}(3x^2 + 2x + 6)$

73. $\frac{1}{3}x + 5 = \frac{1}{3}x + \frac{15}{3} = \frac{1}{3}(x + 15)$

75. $\frac{1}{6}x^2 + \frac{1}{4}x + 3 = \frac{2}{12}x^2 + \frac{3}{12}x + \frac{36}{12}$
$= \frac{1}{12}(2x^2 + 3x + 36)$

77. The GCF is $d^{15}$.
$d^{20} - d^{15} = d^{15}(d^5 - 1)$

79. The GCF is $11x^{48}$.
$88x^{48} + 11x^{51} = 11x^{48}(8 + x^3)$

81. The GCF is $6ab^2$.
$24a^5b^3 - 6ab^4 + 12a^3b^2$
$= 6ab^2(4a^4b - b^2 + 2a^2)$

83. The GCF is $4a^2b^2(x-4)$.
$4a^2b^3(x-4) + 12a^3b^2(x-4)$
$= 4a^2b^2(x-4)(b+3a)$

85. $A = P + Pr = P(1+r)$

87. $A = 2\pi rh + 2\pi r^2$
$= 2\pi r(h+r)$

89. Use the Distributive Property to multiply $x^{-6}(3x^2 + 2)$.

91. $2x^{-3} - 5x^{-4} = x^{-4}(2x - 5)$

93. $2x^3 - 10x^{-4} = 2x^{-4}(x^7 - 5)$

95. $5x^{-4}y^7 - 15x^3y^{-2} = 5x^{-4}y^{-2}(y^9 - 3x^7)$

97. $a^2x + 3a^2 + bx + 3b - 5x - 15$
$= (a^2x + 3a^2) + (bx + 3b) - (5x + 15)$
$= a^2(x+3) + b(x+3) - 5(x+3)$
$= (x+3)(a^2 + b - 5)$

## Section 7.2  Special Factoring

1. A difference of two squares consists of two perfect square terms separated by a minus sign.

3. $x^2 - 9 = x^2 - 3^2 = (x+3)(x-3)$

5. $1 - 4x^2 = 1^2 - (2x)^2 = (1+2x)(1-2x)$

7. $16x^2 - 25y^2 = (4x)^2 - (5y)^2$
$= (4x+5y)(4x-5y)$

9. $\frac{1}{4}x^2 - 1 = \left(\frac{1}{2}x\right)^2 - 1^2$
$= \left(\frac{1}{2}x + 1\right)\left(\frac{1}{2}x - 1\right)$

11. $a^6 - b^{16} = (a^3)^2 - (b^8)^2$
$= (a^3 + b^8)(a^3 - b^8)$

13. $(a+b)^2 - 4 = (a+b)^2 - 2^2$
$= (a+b+2)(a+b-2)$

15. $16 - (x-y)^2 = 4^2 - (x-y)^2$
$= (4+x-y)(4-x+y)$

17. A perfect square trinomial is a trinomial with two terms that are perfect squares and a third term that is twice the product of the square roots of the other two terms.

19. $a^2 - 4a + 4 - a^2 - 2(2)a + 2^2$
$= (a-2)^2$

21. $4x^2 + 12x + 9 = (2x)^2 + 2(3)(2x) + 3^2$
$= (2x+3)^2$

23. $x^2 - 14xy + 49y^2 = x^2 - 2(x)(7y) + (7y)^2$
$= (x-7y)^2$

25. $x^2 + x + \dfrac{1}{4} = x^2 + 2\left(\dfrac{1}{2}\right)(x) + \left(\dfrac{1}{2}\right)^2$
$= \left(x + \dfrac{1}{2}\right)^2$

27. $x^4 - 10x^2 + 25 = (x^2)^2 - 2(5)x^2 + 5^2$
$= (x^2 - 5)^2$

29. $x^3 + 1 = x^3 + 1^3$
$= (x+1)(x^2 - x + 1)$

31. $27y^3 - 8 = (3y)^3 - 2^3$
$= (3y-2)(9y^2 + 6y + 4)$

33. $125 - 8x^3 = 5^3 - (2x)^3$
$= (5 - 2x)(25 + 10x + 4x^2)$

35. $27x^3 + y^3 = (3x)^3 + y^3$
$= (3x + y)(9x^2 - 3xy + y^2)$

37. $x^6 + 1 = (x^2)^3 + 1^3$
$= (x^2 + 1)(x^4 - x^2 + 1)$

39. $x^2 + 2x + 1 = x^2 + 2(1)x + 1$
$= (x+1)^2$

41. $0.25x^2 - 0.49 = (0.5x)^2 - (0.7)^2$
$= (0.5x + 0.7)(0.5x - 0.7)$

43. $x^3 - 1 = x^3 - 1^3$
$= (x-1)(x^2 + x + 1)$

45. $c^2 + 100 - 20c = c^2 - 20c + 100$
$= c^2 - 2(10)c + 10^2$
$= (c - 10)^2$

47. $2x^3 + 54 = 2(x^3 + 27)$
$= 2(x^3 + 3^3)$
$= 2(x+3)(x^2 - 3x + 9)$

49. $81 + y^2$ is prime.

51. $36 - 84y + 49y^2 = 6^2 - 2(6)(7y) + (7y)^2$
$= (6 - 7y)^2$

53. $4x^2 - 64 = 4(x^2 - 16)$
$= 4(x+4)(x-4)$

55. $25x^2 - 40xy + 16y^2$
$= (5x)^2 - 2(5x)(4y) + (4y)^2$
$= (5x - 4y)^2$

57. $x^4 - 16 = (x^2)^2 - 4^2$
$= (x^2 + 4)(x^2 - 4)$
$= (x^2 + 4)(x+2)(x-2)$

59. $8 + 125x^6 = 2^3 + (5x^2)^3$
$= (2 + 5x^2)(4 - 10x^2 + 25x^4)$

61. $y^8 - 6y^4 + 9 = (y^4)^2 - 2(3)(y^4) + 3^2$
$= (y^4 - 3)^2$

63. $216x^3 - y^3 = (6x)^3 - y^3$
$= (6x - y)(36x^2 + 6xy + y^2)$

65. $x^6 - 16x^3 + 64 = (x^3)^2 - 2(8)x^3 + 8^2$
$= (x^3 - 8)^2$
$= (x-2)^2(x^2 + 2x + 4)^2$

67. $x^{12} - 4y^{20} = (x^6)^2 - (2y^{10})^2$
$= (x^6 + 2y^{10})(x^6 - 2y^{10})$

69. $x^6 + 4 - 4x^3 = x^6 - 4x^3 + 4$
$= (x^3)^2 - 2(2)x^3 + 2^2$
$= (x^3 - 2)^2$

**71.** $98 - 2y^6 = 2(49 - y^6)$
$= 2[7^2 - (y^3)^2]$
$= 2(7 + y^3)(7 - y^3)$

**73.** $125x^5y^3 - x^2$
$= x^2(125x^3y^3 - 1)$
$= x^2[(5xy)^3 - 1]$
$= x^2(5xy - 1)(25x^2y^2 + 5xy + 1)$

**75.** $49x^7 - 70x^5 + 25x^3$
$= x^3(49x^4 - 70x^2 + 25)$
$= x^3[(7x^2)^2 - 2(5)(7x^2) + 5^2]$
$= x^3(7x^2 - 5)^2$

**77.** $x^4 + 36y^4$ is prime.

**79.** $64 + 27x^3y^9$
$= 4^3 + (3xy^3)^3$
$= (4 + 3xy^3)(16 - 12xy^3 + 9x^2y^6)$

**81.** $x^4 - 8x^2 + 16$
$= (x^2)^2 - 2(4)(x^2) + 4^2$
$= (x^2 - 4)^2$
$= (x^2 - 4)(x^2 - 4)$
$= (x + 2)(x - 2)(x + 2)(x - 2)$
$= (x + 2)^2(x - 2)^2$

**83.** $16x^4 - 25y^2 = (4x^2)^2 - (5y)^2$
$= (4x^2 + 5y)(4x^2 - 5y)$

**85.** $64x^4 + 80x^2y + 25y^2$
$= (8x^2)^2 + 2(8x^2)(5y) + (5y)^2$
$= (8x^2 + 5y)^2$

**87.** $16a^4 - 81d^4$
$= (4a^2)^2 - (9d^2)^2$
$= (4a^2 + 9d^2)(4a^2 - 9d^2)$
$= (4a^2 + 9d^2)(2a + 3d)(2a - 3d)$

**89.** $2x^4 + 16x = 2x(x^3 + 8)$
$= 2x(x + 2)(x^2 - 2x + 4)$

**91.** $125x^6 - 27y^{12}$
$= (5x^2)^3 - (3y^4)^3$
$= (5x^2 - 3y^4)(25x^4 + 15x^2y^4 + 9y^8)$

**93.** $4a(x^2 - y^2) + 8b(x^2 - y^2)$
$= (x^2 - y^2)(4a + 8b)$
$= (x + y)(x - y)4(a + 2b)$
$= 4(x + y)(x - y)(a + 2b)$

**95.** $x^2(a + b) - y^2(a + b)$
$= (a + b)(x^2 - y^2)$
$= (a + b)(x + y)(x - y)$

**97.** $(2x + 1)^2 - 10(2x + 1) + 25$
$= u^2 - 10u + 25$
$= u^2 - 2(5)u + 5^2$
$= (u - 5)^2$
$= [(2x + 1) - 5]^2$
$= (2x - 4)^2$
$= [2(x - 2)]^2$
$= 4(x - 2)^2$

**99.** $(x^2 - 2)^2 - 4(x^2 - 2) + 4$
$= u^2 - 4u + 4$
$= u^2 - 2(2)u + 2^2$
$= (u - 2)^2$
$= [(x^2 - 2) - 2]^2$
$= (x^2 - 4)^2$
$= (x^2 - 4)(x^2 - 4)$
$= (x + 2)(x - 2)(x + 2)(x - 2)$
$= (x + 2)^2(x - 2)^2$

**101.** Because $x^6 - y^6 = (x^3)^2 - (y^3)^2$, the expression is a difference of two squares. Because $x^6 - y^6 = (x^2)^3 - (y^2)^3$, the expression is a difference of two cubes.

**103.** $9x^2 + bx + 1 = (3x)^2 + bx + (\pm 1)^2$
For this trinomial to be a perfect square, the middle term should be $2(\pm 1)(3x) = \pm 6x$. Thus, $b = 6$ or $-6$.

**105.** $ax^2 - 10x + 25 = ax^2 - 2(5)x + 5^2$
For this trinomial to be a perfect square, $a$ should be 1.

**107.** $9x^2 - 24x + c = (3x)^2 - 2(4)(3x) + c$
For this trinomial to be a perfect square, $c$ should be $4^2 = 16$.

**109.** The area of the pool is $\pi r^2$ and the area of the region (pool plus deck) is $\pi R^2$. Therefore, the area of just the deck is
$$A = \pi R^2 - \pi r^2$$
$$= \pi(R^2 - r^2)$$
$$= \pi(R+r)(R-r)$$

**111.** The kinetic energy of the first object is $\frac{1}{2}mv_1^2$. The kinetic energy of the second object is $\frac{1}{2}mv_2^2$. The difference in their kinetic energies is:
$$\frac{1}{2}mv_1^2 - \frac{1}{2}mv_2^2 = \frac{1}{2}m(v_1^2 - v_2^2)$$
$$= \frac{1}{2}m(v_1 + v_2)(v_1 - v_2)$$

**113.** $V(x) = 560x^2 + 44{,}400$
$= 80(7x^2 + 555)$

**115.** $F(x) = \dfrac{V(x)}{1000} = \dfrac{80(7x^2 + 555)}{1000}$
$= 0.08(7x^2 + 555)$

**117.** $x^6 - 64$
$= (x^3)^2 - 8^2$
$= (x^3 + 8)(x^3 - 8)$
$= (x^3 + 2^3)(x^3 - 2^3)$
$= (x+2)(x^2 - 2x + 4)(x-2)(x^2 + 2x + 4)$
$= (x+2)(x-2)(x^2 - 2x + 4)(x^2 + 2x + 4)$

**119.** $a^6 - b^6$
$= (a^3)^2 - (b^3)^2$
$= (a^3 + b^3)(a^3 - b^3)$
$= (a+b)(a^2 - ab + b^2)(a-b)(a^2 + ab + b^2)$
$= (a+b)(a-b)(a^2 - ab + b^2)(a^2 + ab + b^2)$

**121.** $4x^5 - x^3 - 32x^2 + 8$
$= 4x^5 - 32x^2 - x^3 + 8$
$= (4x^5 - 32x^2) - (x^3 - 8)$
$= 4x^2(x^3 - 8) - (x^3 - 8)$
$= (x^3 - 8)(4x^2 - 1)$
$= (x^3 - 2^3)(4x^2 - 1)$
$= (x-2)(x^2 + 2x + 4)(2x + 1)(2x - 1)$

**123.** $x^2y^3 - x^2z^3 - y^3 + z^3$
$= (x^2y^3 - x^2z^3) - (y^3 - z^3)$
$= x^2(y^3 - z^3) - (y^3 - z^3)$
$= (y^3 - z^3)(x^2 - 1)$
$= (y-z)(y^2 + yz + z^2)(x+1)(x-1)$

## Section 7.3 Factoring Trinomials of the Form $ax^2 + bx + c$

**1.** The Commutative Property of Multiplication allows us to write the factors in either order.

**3.** Because $1 \cdot 2 = 2$ and $1 + 2 = 3$, we will use 1 and 2.
$x^2 + 3x + 2 = (x+1)(x+2)$

**5.** Because $(-3) \cdot (-6) = 18$ and $-3 + (-6) = -9$, we will use $-3$ and $-6$.
$x^2 + 18 - 9x = x^2 - 9x + 18$
$= (x-3)(x-6)$

**7.** Because $(2) \cdot (7) = 14$ and $2 + 7 = 9$, we will use 2 and 7.
$14 + 9x + x^2 = x^2 + 9x + 14$
$= (x+2)(x+7)$

Section 7.3  Factoring Trinomials of the Form $ax^2 + bx + c$

9. Because there are no integers that, when multiplied, yield 2 and, when added, yield 2, $x^2 + 2x + 2$ is prime.

11. Because $2 \cdot (-8) = -16$ and $2 + (-8) = -6$, we will use 2 and $-8$.
$x^2 - 6xy - 16y^2 = (x + 2y)(x - 8y)$

13. Because $b$ and $c$ are both positive, we need two positive factors. Use $2x^2 = 2x \cdot 1x$ and $2 = 2 \cdot 1$, because $(2)(2) + (1)(1) = 5$, to obtain
$2x^2 + 5x + 2 = (2x + 1)(x + 2)$

15. Use $12 = 6 \cdot 2$ and $-x^2 = -1x \cdot 1x$, because $(6)(1) + (2)(-1) = 4$, to obtain
$12 + 4x - x^2 = (6 - x)(2 + x)$

17. Use $3x^2 = 3x \cdot 1x$ and $-3 = 1 \cdot (-3)$, because $(3)(-3) + (1)(1) = -8$, to obtain
$3x^2 - 8x - 3 = (3x + 1)(x - 3)$

19. Use $3x^2 = 4x \cdot 1x$ and $-15 = 3 \cdot (-5)$, because $(4)(-5) + (1)(3) = -17$, to obtain
$4x^2 - 17x - 15 = (4x + 3)(x - 5)$

21. $12x^2 + 19x + 21$ is prime.

23. Because $-1 \cdot 2 = -2$ and $2 - 1 = 1$, we will use $-1$ and 2.
$c^2 + c - 2 = (c - 1)(c + 2)$

25. Use $6x^2 = 6x \cdot 1x$ and $-15 = 3 \cdot (-5)$, because $(6)(3) + (1)(-5) = 13$, to obtain
$6x^2 + 13x - 15 = (6x - 5)(x + 3)$

27. Because $-3 \cdot 8 = -24$ and $8 - 3 = 5$, we will use $-3$ and 8.
$x^2 + 5x - 24 = (x + 8)(x - 3)$

29. Use $24 = 6 \cdot 4$ and $-x^2 = -1x \cdot 1x$, because $(6)(1) + (4)(-1) = 2$, to obtain
$24 + 2x - x^2 = (6 - x)(4 + x)$

31. Because $-5 \cdot 3 = -15$ and $-5 + 3 = -2$, we will use $-5$ and 3.
$b^2 - 2b - 15 = (b - 5)(b + 3)$

33. Use $3x^2 = 3x \cdot 1x$ and $2 = 2 \cdot 1$, because $(3)(1) + (2)(1) = 5$, to obtain
$3x^2 + 5x + 2 = (3x + 2)(x + 1)$

35. Use $6x^2 = 3x \cdot 2x$ and $15 = (-5) \cdot (-3)$, because $(3)(-3) + (-5)(2) = -19$, to obtain
$15 + 6x^2 - 19x = 6x^2 - 19x + 15$
$= (3x - 5)(2x - 3)$

37. $x^2 + 25 + 10x = x^2 + 10x + 25$
$= x^2 + 2(5)x + 5^2$
$= (x + 5)^2$

39. Use $2 = 2 \cdot 1$ and $-15x^2 = -5x \cdot 3x$, because $(2)(3) + (1)(-5) = 1$, to obtain
$2 + x - 15x^2 = (2 - 5x)(1 + 3x)$

41. Use $18 = 2 \cdot 9$ and $-10x^2 = -10x \cdot 1x$, because $(9)(1) + (2)(-10) = -11$, to obtain
$18 - 11x - 10x^2 = (9 - 10x)(2 + x)$

43. Because $-4 \cdot (-6) = 24$ and $-4 + (-6) = -10$, we will use $-4$ and $-6$.
$24 + a^2 - 10a = a^2 - 10a + 24$
$= (a - 4)(a - 6)$

45. Use $9x^2 = 3x \cdot 3x$ and $-10 = 2 \cdot (-5)$, because $(3)(-5) + (2)(3) = -9$, to obtain
$9x^2 - 9x - 10 = (3x + 2)(3x - 5)$

47. Because $-7 \cdot (-9) = -63$ and $-7 + (-9) = -16$, we will use $-7$ and $-9$.
$x^2 - 16x + 63 = (x - 7)(x - 9)$

**49.** $9x^2 - 9x - 18 = 9(x^2 - x - 2)$
Use $x^2 = 1x \cdot 1x$ and $-2 = 1 \cdot (-2)$,
because $(1)(1) + (1)(-2) = -1$, to obtain
$9(x^2 - x - 2) = 9(x - 2)(x + 1)$

**51.** Use $12x^2 = 2x \cdot 6x$ and $-3 = 3 \cdot (-1)$,
because $(2)(-1) + (3)(6) = 16$, to obtain
$12x^2 + 16x - 3 = (2x + 3)(6x - 1)$

**53.** Use $8x^2 = 2x \cdot 4x$ and
$-21y^2 = 3y \cdot (-7y)$, because $(4)(3) + (-7)(2) = -2$, to obtain
$8x^2 - 2xy - 21y^2 = (4x - 7y)(2x + 3y)$

**55.** $2x^2 - 12 + 2x = 2x^2 + 2x - 12$
$\qquad = 2(x^2 + x - 6)$
Because $3 \cdot (-2) = -6$ and $3 + (-2) = 1$,
we will use 3 and $-2$.
$2(x^2 + x - 6) = 2(x + 3)(x - 2)$

**57.** $24x + 36 + 4x^2 = 4x^2 + 24x + 36$
$\qquad = 4(x^2 + 6x + 9)$
$\qquad = 4[x^2 + 2(3)x + 3^2]$
$\qquad = 4(x + 3)^2$

**59.** There are fewer combinations to try because 17 and 23 have fewer factors than 24 and $-18$.

**61.** $8x^3y - 24x^2y^2 - 80xy^3$
$\qquad = 8xy(x^2 - 3xy - 10y^2)$
$\qquad = 8xy(x - 5y)(x + 2y)$

**63.** $45x + 12x^2 - 12x^3$
$\qquad = -12x^3 + 12x^2 + 45x$
$\qquad = -3x(4x^2 - 4x - 15)$
$\qquad = -3x(2x - 5)(2x + 3)$

**65.** $28x^2y - 30xy - 18y$
$\qquad = 2y(14x^2 - 15x - 9)$
$\qquad = 2y(2x - 3)(7x + 3)$

**67.** $a^2b^2 - 3ab - 10 = (ab - 5)(ab + 2)$

**69.** $12x^2y + 36xy + x^3y$
$\qquad = x^3y + 12x^2y + 36xy$
$\qquad = xy(x^2 + 12x + 36)$
$\qquad = xy(x + 6)^2$

**71.** $x^2y^2 - 2xy - 48 = (xy - 8)(xy + 6)$

**73.** $18x^2 - 12xy + 2y^2 = 2(9x^2 - 6xy + y^2)$
$\qquad\qquad\qquad\qquad = 2(3x - y)^2$

**75.** $6m^2 + 5mn - n^2 = (6m - n)(m + n)$

**77.** $16x^3y^2 - 12x^2y^3 + 2xy^4$
$\qquad = 2xy^2(8x^2 - 6xy + y^2)$
$\qquad = 2xy^2(4x - y)(2x - y)$

**79.** $x^3(x - 1) + 11x^2(x - 1) - 42x(x - 1)$
$\qquad = x(x - 1)(x^2 + 11x - 42)$
$\qquad = x(x - 1)(x + 14)(x - 3)$

**81.** $x^4(x^2 - 25) - 10x^2(x^2 - 25) + 9(x^2 - 25)$
$\qquad = (x^2 - 25)(x^4 - 10x^2 + 9)$
$\qquad = (x + 5)(x - 5)(x^4 - 10x^2 + 9)$
$\qquad = (x + 5)(x - 5)(x^2 - 9)(x^2 - 1)$
$\qquad = (x + 5)(x - 5)(x + 3)(x - 3)(x + 1)(x - 1)$

**83.** $2x^2y^2(xy + 1) + 3xy(xy + 1) - 9(xy + 1)$
$\qquad = (xy + 1)(2x^2y^2 + 3xy - 9)$
$\qquad = (xy + 1)(2xy - 3)(xy + 3)$

**85.** No, the factor $x^2 - 4$ can be factored as $(x + 2)(x - 2)$.

**87.** $x^4 + 28x^2 - 60 = (x^2 + 30)(x^2 - 2)$

**89.** $x^6 + 17x^3 + 60 = (x^3 + 12)(x^3 + 5)$

**91.** $x^4 - 13x^2 + 36 = (x^2 - 9)(x^2 - 4)$
$\qquad\qquad\qquad\quad = (x + 3)(x - 3)(x + 2)(x - 2)$

**93.** $5x^8 + 3x^4 - 14 = (5x^4 - 7)(x^4 + 2)$

**95.** $4x^4 - 37x^2 + 9$
$\qquad = (4x^2 - 1)(x^2 - 9)$
$\qquad = (2x + 1)(2x - 1)(x + 3)(x - 3)$

Section 7.3 Factoring Trinomials of the Form $ax^2 + bx + c$

**97.** $9x^4 - 13x^2 + 4$
$= (9x^2 - 4)(x^2 - 1)$
$= (3x + 2)(3x - 2)(x + 1)(x - 1)$

**99.** $(x-1)^2 + 11(x-1) + 28$
$= u^2 + 11u + 28$
$= (u + 7)(u + 4)$
$= (x - 1 + 7)(x - 1 + 4)$
$= (x + 6)(x + 3)$

**101.** $(2x+1)^2 - 15(2x+1) + 56$
$= u^2 - 15u + 56$
$= (u - 7)(u - 8)$
$= (2x + 1 - 7)(2x + 1 - 8)$
$= (2x - 6)(2x - 7)$
$= 2(x - 3)(2x - 7)$

**103.** $2x^2(x+2)^2 + x(x+2) - 6$
$= 2x^2 u^2 + xu - 6$
$= (2xu - 3)(xu + 2)$
$= [2x(x+2) - 3][x(x+2) + 2]$
$= (2x^2 + 4x - 3)(x^2 + 2x + 2)$

**105.** $1.01x^2 + 4.99x - 24.05 \approx x^2 + 5x - 24$
$= (x + 8)(x - 3)$

**107.** $f(t) = 78t + 3900$
$= 52(1.5t + 75)$

**109.** The coefficient of $t$ indicates that food costs were increasing at an annual rate of $78 during this period. In the factored form, the coefficient of $t$ indicates that food costs were increasing at a weekly rate of $1.50.

**111.** $x^2 + kx + 10$
The factors of 10 are 1 and 10, −1 and − 10, 2 and 5, and − 2 and − 5.
Therefore, $x^2 + kx + 10$ is factorable if $k$ is the sum of any factor pair of 10: 11, − 11, 7, − 7

**113.** $3x^2 + kx + 5$
The factors of 3 are 1 and 3 or − 1 and − 3. The factors of 5 are 1 and 5 or − 1 and − 5. Therefore, $3x^2 + kx + 5$ is factorable if $k$ is any of the following:
$(1)(5) + (3)(1) = 8$
$(1)(-5) + (3)(-1) = -8$
$(3)(5) + (1)(1) = 16$
$(3)(-5) + (1)(-1) = -16$

**115.** $x^2 + 3x + k$
For this to be factorable, the factors of $k$ must sum to 3 and be positive, so $k = 2$.

**117.** $x^{2n} + 11x^n + 24 = (x^n + 8)(x^n + 3)$

**119.** $y^{4n} - 20y^{2n} + 99$
$= (y^{2n} - 9)(y^{2n} - 11)$
$= (y^n + 3)(y^n - 3)(y^{2n} - 11)$

**121.** $5x^{2n} + 2x^n - 3 = (5x^n - 3)(x^n + 1)$

**123.** $3x^{4n} + 5x^{3n} + 2x^{2n}$
$= x^{2n}(3x^{2n} + 5x^n + 2)$
$= x^{2n}(3x^n + 2)(x^n + 1)$

## Section 7.3 Supplementary Exercises

**1.** $5x^2 - 7x - 6 = (5x + 3)(x - 2)$

**3.** $25 - 81y^2 = (5 + 9y)(5 - 9y)$

**5.** $2x^3 + x^2 - 6x - 3$
$= (2x^3 + x^2) - (6x + 3)$
$= x^2(2x + 1) - 3(2x + 1)$
$= (2x + 1)(x^2 - 3)$

7. $3x(x-2) + 5y(x-2)$
   $= (x-2)(3x+5y)$

9. $125x^3 + 1 = (5x)^3 + 1^3$
   $= (5x+1)(25x^2 - 5x + 1)$

11. $6x + x^2 + 8 = x^2 + 6x + 8$
    $= (x+4)(x+2)$

13. $x^2 + 9$ is prime.

15. $4x^2 + 20x + 25 = (2x)^2 + 2(5)(2x) + 5^2$
    $= (2x+5)^2$

17. $ax - 3a - bx + 3b$
    $= (ax - 3a) - (bx - 3b)$
    $= a(x-3) - b(x-3)$
    $= (x-3)(a-b)$

19. $25y^3 - 25 = 25(y^3 - 1)$
    $= 25(y-1)(y^2 + y + 1)$

21. $y^4 - 16x^4 = (y^2 + 4x^2)(y^2 - 4x^2)$
    $= (y^2 + 4x^2)(y + 2x)(y - 2x)$

23. $a^2 b^7 + 3a^3 b^5 - a^2 b^4$
    $= a^2 b^4 (b^3 + 3ab - 1)$

25. $x + 21x^2 - 2 = 21x^2 + x - 2$
    $= (7x - 2)(3x + 1)$

27. $16 - 8a + a^2 = 4^2 - 2(4)a + a^2$
    $= (4-a)^2$

29. $50 - 18x^4 = 2(25 - 9x^4)$
    $= 2(5 + 3x^2)(5 - 3x^2)$

31. $2y + 10 - 3xy - 15x$
    $= (2y + 10) - (3xy + 15x)$
    $= 2(y+5) - 3x(y+5)$
    $= (y+5)(2-3x)$

33. $24x^4 - 24x^2 + 20x^3$
    $= 24x^4 + 20x^3 - 24x^2$
    $= 4x^2(6x^2 + 5x - 6)$
    $= 4x^2(3x - 2)(2x + 3)$

35. $c^{20} - d^8 = (c^{10} + d^4)(c^{10} - d^4)$
    $= (c^{10} + d^4)(c^5 + d^2)(c^5 - d^2)$

37. $x^2 + xy + y^2$ is prime.

39. $a^2 x^3 - a^2 y^3 - 4x^3 + 4y^3$
    $= (a^2 x^3 - a^2 y^3) - (4x^3 - 4y^3)$
    $= a^2(x^3 - y^3) - 4(x^3 - y^3)$
    $= (x^3 - y^3)(a^2 - 4)$
    $= (x-y)(x^2 + xy + y^2)(a+2)(a-2)$

41. $a^5 b^4 - a^7 b^8 + a^6 b^3$
    $= a^5 b^3(b - a^2 b^5 + a)$

43. $m^2 - 10m + 16 = (m-2)(m-8)$

45. $4x^4 - 64y^4$
    $= 4(x^4 - 16y^4)$
    $= 4(x^2 + 4y^2)(x^2 - 4y^2)$
    $= 4(x^2 + 4y^2)(x + 2y)(x - 2y)$

47. $(a+b)^3 - 8 = u^3 - 8$
    $= u^3 - 2^3$
    $= (u-2)(u^2 + 2u + 4)$
    $= (a+b-2)[(a+b)^2 + 2(a+b) + 4]$
    $= (a+b-2)(a^2 + 2ab + b^2 + 2a + 2b + 4)$

49. $20x^2 - 60x + 45$
    $= 5(4x^2 - 12x + 9)$
    $= 5[(2x)^2 - 2(3)(2x) + 3^2]$
    $= 5(2x-3)^2$

51. $30a^6 - 104a^5 + 90a^4$
    $= 2a^4(15a^2 - 52a + 45)$
    $= 2a^4(3a-5)(5a-9)$

53. $c^5 - 4c^3 - c^2 + 4$
    $= (c^5 - 4c^3) - (c^2 - 4)$
    $= c^3(c^2 - 4) - (c^2 - 4)$
    $= (c^2 - 4)(c^3 - 1)$
    $= (c+2)(c-2)(c-1)(c^2 + c + 1)$

**55.** $x^4 + 2x^2y^2 - 99y^4$
$= (x^2 + 11y^2)(x^2 - 9y^2)$
$= (x^2 + 11y^2)(x + 3y)(x - 3y)$

## Section 7.4 Solving Equations by Factoring

1. Produce the graph and estimate the x-intercepts.

3. (a) $x^2 - 4x - 12 = 0$
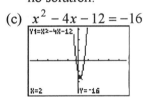
The solutions are $-2$ and $6$.

   (b) $x^2 - 4x - 12 = -18$
   The lowest point on the graph is $-16$ so $-18$ cannot be reached. There is no solution.

   (c) $x^2 - 4x - 12 = -16$
   
   The solution is $2$.

   (d) $x^2 - 4x - 12 = 20$
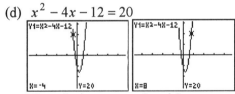
The solutions are $-4$ and $8$.

5. $(x - 4)(x + 3) = 0$
$x - 4 = 0$ or $x + 3 = 0$
$x = 4$ or $x = -3$
The solutions are $4$ and $-3$.

7. $5x(6 - x) = 0$
$5x = 0$ or $6 - x = 0$
$x = 0$ or $x = 6$
The solutions are $0$ and $6$.

9. $(x - 5)^2 = 0$
$(x - 5)(x - 5) = 0$
$x - 5 = 0$
$x = 5$
The solution is $5$.

11. $(x + 5)(2x - 6)(3x + 3) = 0$
$x + 5 = 0$ or $2x - 6 = 0$ or $3x + 3 = 0$
$x = -5$ or $x = 3$ or $x = -1$
The solutions are $-5$, $-1$, and $3$.

13. Assuming that the equation has only one variable, the largest exponent on the variable is 1 for a first-degree equation and 2 for a second-degree equation.

15. $x^2 + 7x = 0$
$x(x + 7) = 0$
$x = 0$ or $x + 7 = 0$
$x = -7$
The solutions are $0$ and $-7$.

17. $2g^2 - 18 = 0$
$2(g^2 - 9) = 0$
$2(g + 3)(g - 3) = 0$
$g + 3 = 0$ or $g - 3 = 0$
$g = -3$ or $g = 3$
The solutions are $3$ and $-3$.

19. $x^2 - 5x + 6 = 0$
$(x - 3)(x - 2) = 0$
$x - 3 = 0$ or $x - 2 = 0$
$x = 3$ or $x = 2$
The solutions are $3$ and $2$.

21. $\frac{2}{5}x^2 - \frac{3}{5}x - 1 = 0$
$2x^2 - 3x - 5 = 0$
$(2x - 5)(x + 1) = 0$
$2x - 5 = 0$ or $x + 1 = 0$
$x = 2.5$ or $x = -1$
The solutions are $2.5$ and $-1$.

**23.**
$$a^2 = 5a$$
$$a^2 - 5a = 0$$
$$a(a-5) = 0$$
$a = 0$   or   $a - 5 = 0$
$\phantom{a = 0 \text{ or } a} a = 5$
The solutions are 0 and 5.

**25.**
$$(x-2)^2 = 25$$
$$(x-2)^2 - 25 = 0$$
$$[(x-2)+5][(x-2)-5] = 0$$
$x - 2 + 5 = 0$   or   $x - 2 - 5 = 0$
$x + 3 = 0$   or   $x - 7 = 0$
$x = -3$   or   $x = 7$
The solutions are $-3$ and 7.

**27.**
$$x^2 + 5x = 6$$
$$x^2 + 5x - 6 = 0$$
$$(x+6)(x-1) = 0$$
$x + 6 = 0$   or   $x - 1 = 0$
$x = -6$   or   $x = 1$
The solutions are 1 and $-6$.

**29.**
$$x^2 + 35 = 12x$$
$$x^2 - 12x + 35 = 0$$
$$(x-5)(x-7) = 0$$
$x - 5 = 0$   or   $x - 7 = 0$
$x = 5$   or   $x = 7$
The solutions are 5 and 7.

**31.**
$$12 = 11d - 2d^2$$
$$2d^2 - 11d + 12 = 0$$
$$(2d-3)(d-4) = 0$$
$2d - 3 = 0$   or   $d - 4 = 0$
$d = 1.5$   or   $d = 4$
The solutions are 1.5 and 4.

**33.**
$$-8 = -9y^2 - 14y$$
$$9y^2 + 14y - 8 = 0$$
$$(9y-4)(y+2) = 0$$
$9y - 4 = 0$   or   $y + 2 = 0$
$y = \dfrac{4}{9}$   or   $y = -2$
The solutions are 4/9 and $-2$.

**35.**
$$x(x+1) = 30$$
$$x^2 + x = 30$$
$$x^2 + x - 30 = 0$$
$$(x+6)(x-5) = 0$$
$x + 6 = 0$   or   $x - 5 = 0$
$x = -6$   or   $x = 5$
The solutions are $-6$ and 5.

**37.**
$$a(a-32) + 60 = 0$$
$$a^2 - 32a + 60 = 0$$
$$(a-30)(a-2) = 0$$
$a - 30 = 0$   or   $a - 2 = 0$
$a = 30$   or   $a = 2$
The solutions are 30 and 2.

**39.**
$$3x(2+x) = 24$$
$$6x + 3x^2 = 24$$
$$3x^2 + 6x - 24 = 0$$
$$3(x^2 + 2x - 8) = 0$$
$$3(x-2)(x+4) = 0$$
$x - 2 = 0$   or   $x + 4 = 0$
$x = 2$   or   $x = -4$
The solutions are 2 and $-4$.

**41.**
$$\frac{2}{5}x(x-2) = x - 2$$
$$2x(x-2) = 5x - 10$$
$$2x^2 - 4x = 5x - 10$$
$$2x^2 - 9x + 10 = 0$$
$$(2x-5)(x-2) = 0$$
$2x - 5 = 0$   or   $x - 2 = 0$
$x = 2.5$   or   $x = 2$
The solutions are 2.5 and 2.

**43.**
$$(x-3)(x+2) = 6$$
$$x^2 - x - 6 = 6$$
$$x^2 - x - 12 = 0$$
$$(x-4)(x+3) = 0$$
$x - 4 = 0$   or   $x + 3 = 0$
$x = 4$   or   $x = -3$
The solutions are $-3$ and 4.

**45.** 
$(3x-2)(x+2) = (x-2)(x+1) + 10$
$3x^2 + 4x - 4 = x^2 - x - 2 + 10$
$3x^2 + 4x - 4 = x^2 - x + 8$
$2x^2 + 5x - 12 = 0$
$(2x - 3)(x + 4) = 0$
$2x - 3 = 0$ or $x + 4 = 0$
$x = 1.5$ or $x = -4$
The solutions are $-4$ and $1.5$.

**47.** 
$(4x - 3)^2 + (4x - 3) = 6$
$u^2 + u = 6$
$u^2 + u - 6 = 0$
$(u + 3)(u - 2) = 0$
$[(4x - 3) + 3][(4x - 3) - 2] = 0$
$4x - 3 + 3 = 0$ or $4x - 3 - 2 = 0$
$4x = 0$ or $4x = 5$
$x = 0$ or $x = 1.25$
The solutions are $0$ and $1.25$.

**49.** 
$(x - 5)(x^2 + 3x - 18) = 0$
$(x - 5)(x + 6)(x - 3) = 0$
$x - 5 = 0$ or $x + 6 = 0$ or $x - 3 = 0$
$x = 5$ or $x = -6$ or $x = 3$
The solutions are $-6$, $3$, and $5$.

**51.** 
$3x^3 - 27x = 0$
$3x(x^2 - 9) = 0$
$3x(x + 3)(x - 3) = 0$
$3x = 0$ or $x + 3 = 0$ or $x - 3 = 0$
$x = 0$ or $x = -3$ or $x = 3$
The solutions are $-3$, $0$, and $3$.

**53.** 
$6x^3 + 2x^2 = 4x$
$6x^3 + 2x^2 - 4x = 0$
$2x(3x^2 + x - 2) = 0$
$2x(3x - 2)(x + 1) = 0$
$2x = 0$ or $3x - 2 = 0$ or $x + 1 = 0$
$x = 0$ or $x = \dfrac{2}{3}$ or $x = -1$
The solutions are $-1$, $0$, and $2/3$.

**55.** 
$x^3 - 5x^2 - x + 5 = 0$
$(x^3 - 5x^2) - (x - 5) = 0$
$x^2(x - 5) - (x - 5) = 0$
$(x - 5)(x^2 - 1) = 0$
$(x - 5)(x + 1)(x - 1) = 0$
$x - 5 = 0$ or $x + 1 = 0$ or $x - 1 = 0$
$x = 5$ or $x = -1$ or $x = 1$
The solutions are $-1$, $1$, and $5$.

**57.** 
$2x^3 - 3x^2 - 8x + 12 = 0$
$(2x^3 - 3x^2) - (8x - 12) = 0$
$x^2(2x - 3) - 4(2x - 3) = 0$
$(2x - 3)(x^2 - 4) = 0$
$(2x - 3)(x + 2)(x - 2) = 0$
$2x - 3 = 0$ or $x + 2 = 0$ or $x - 2 = 0$
$x = 1.5$ or $x = -2$ or $x = 2$
The solutions are $-2$, $1.5$, and $2$.

**59.** 
$x^4 + 36 = 13x^2$
$x^4 - 13x^2 + 36 = 0$
$(x^2 - 9)(x^2 - 4) = 0$
$(x + 3)(x - 3)(x + 2)(x - 2) = 0$
$x + 3 = 0$ $x - 3 = 0$ $x + 2 = 0$, or
$x = -3$, $x = 3$, $x = -2$,
$x - 2 = 0$
$x = 2$
The solutions are $-3$, $-2$, $2$, and $3$.

**61.** 
$6(y - 2)^3 - (y - 2)^2 - (y - 2) = 0$
$6u^3 - u^2 - u = 0$
$u(6u^2 - u - 1) = 0$
$u(3u + 1)(2u - 1) = 0$
$(y - 2)[3(y - 2) + 1][2(y - 2) - 1] = 0$
$3(y - 2) + 1 = 0$     $2(y - 2) - 1 = 0$
$3y - 6 + 1 = 0$        $2y - 4 - 1 = 0$
$3y = 5$ or             $2y = 5$
$y = \dfrac{5}{3}$                $y = \dfrac{5}{2}$
or $y - 2 = 0$
$y = 2$
The solutions are $2$, $5/2$, and $5/3$.

**63.**
$$3x^2(3-x) + 3x(3-x) + 18(x-3) = 0$$
$$3x^2(3-x) + 3x(3-x) - 18(3-x) = 0$$
$$3(3-x)(x^2 + x - 6) = 0$$
$$3(3-x)(x+3)(x-2) = 0$$
$$-3(x-3)(x+3)(x-2) = 0$$
$$x - 3 = 0 \quad \text{or} \quad x + 3 = 0 \quad \text{or} \quad x - 2 = 0$$
$$x = 3 \quad \text{or} \quad x = -3 \quad \text{or} \quad x = 2$$
The solutions are $-3, 2,$ and $3$.

**65.**
$$2x^2 + bx + 2 = 0, \quad x = -2$$
$$2(-2)^2 + b(-2) + 2 = 0$$
$$-2b + 10 = 0$$
$$-2b = -10$$
$$b = 5$$
The equation is:
$$2x^2 + 5x + 2 = 0$$
$$(2x+1)(x+2) = 0$$
Another solution of the equation:
$$2x + 1 = 0$$
$$2x = -1$$
$$x = -0.5$$

**67.**
$$2x^2 + x + (3 - 4c) = 0, \quad x = 3$$
$$2(3)^2 + 3 + (3 - 4c) = 0$$
$$3 - 4c = -21$$
$$-4c = -24$$
$$c = 6$$
The equation is:
$$2x^2 + x + (3 - 4 \cdot 6) = 0$$
$$2x^2 + x - 21 = 0$$
$$(2x+7)(x-3) = 0$$
Another solution of the equation:
$$2x + 7 = 0$$
$$2x = -7$$
$$x = -3.5$$

**69.** If 3 and $-4$ are solutions, then $(x-3)$ and $(x+4)$ are factors.
$$(x-3)(x+4) = 0$$
$$x^2 + 4x - 3x - 12 = 0$$
$$x^2 + x - 12 = 0$$

**71.** If 4 is a double root, then $(x-4)$ and $(x-4)$ are factors.
$$(x-4)(x-4) = 0$$
$$x^2 - 4x - 4x + 16 = 0$$
$$x^2 - 8x + 16 = 0$$

**73.** If 0 and 1/2 are solutions, then $x$ and $(2x-1)$ are factors.
$$x(2x-1) = 0$$
$$2x^2 - x = 0$$

**75.**
$$x^2 - cx - 2c^2 = 0$$
$$(x - 2c)(x + c) = 0$$
$$x - 2c = 0 \quad \text{or} \quad x + c = 0$$
$$x = 2c \quad \text{or} \quad x = -c$$

**77.**
$$3y^2 - 4xy = 4x^2$$
$$-4x^2 - 4xy + 3y^2 = 0$$
$$4x^2 + 4xy - 3y^2 = 0$$
$$(2x - y)(2x + 3y) = 0$$
$$2x - y = 0 \quad \text{or} \quad 2x + 3y = 0$$
$$x = \frac{y}{2} \quad \text{or} \quad x = -\frac{3y}{2}$$

**79.**
$$x^2 + 2ax = 3a^2$$
$$x^2 + 2ax - 3a^2 = 0$$
$$(x - a)(x + 3a) = 0$$
$$x - a = 0 \quad \text{or} \quad x + 3a = 0$$
$$x = a \quad \text{or} \quad x = -3a$$

**81.** $|x^2 - 4x| = 3$

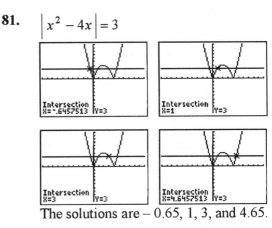

The solutions are $-0.65, 1, 3,$ and $4.65$.

**83.** $|2x^2 - 3x| = 2$

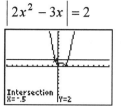 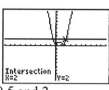

The solutions are $-0.5$ and $2$.

**85.** $|6x^2 - 5x - 2| = 2$

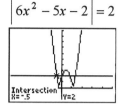 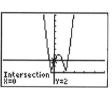

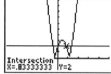 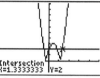

The solutions are $-0.5, 0, 0.83,$ and $1.33$.

**87.** $|x^2 + 5x| = 6$ can be rewritten as

$x^2 + 5x = 6$  or  $x^2 + 5x = -6$

$x^2 + 5x = 6$
$x^2 + 5x - 6 = 0$
$(x + 6)(x - 1) = 0$
$x + 6 = 0$  or  $x - 1 = 0$
$x = -6$  or  $x = 1$

$x^2 + 5x = -6$
$x^2 + 5x + 6 = 0$
$(x + 3)(x + 2) = 0$
$x + 3 = 0$  or  $x + 2 = 0$
$x = -3$  or  $x = -2$

The solutions are $-6, -3, -2,$ and $1$.

**89.** $|x^2 + 10x + 15| = 6$ can be rewritten as

$x^2 + 10x + 15 = 6$ or
$x^2 + 10x + 15 = -6$

$x^2 + 10x + 15 = 6$
$x^2 + 10x + 9 = 0$
$(x + 9)(x + 1) = 0$

$x + 9 = 0$  or  $x + 1 = 0$
$x = -9$  or  $x = -1$

$x^2 + 10x + 15 = -6$
$x^2 + 10x + 21 = 0$
$(x + 7)(x + 3) = 0$
$x + 7 = 0$  or  $x + 3 = 0$
$x = -7$  or  $x = -3$

The solutions are $-9, -7, -3,$ and $-1$.

**91.** $|x^2 - 3x - 1| = 3$ can be rewritten as

$x^2 - 3x - 1 = 3$  or  $x^2 - 3x - 1 = -3$

$x^2 - 3x - 1 = 3$
$x^2 - 3x - 4 = 0$
$(x - 4)(x + 1) = 0$
$x - 4 = 0$  or  $x + 1 = 0$
$x = 4$  or  $x = -1$

$x^2 - 3x - 1 = -3$
$x^2 - 3x + 2 = 0$
$(x - 2)(x - 1) = 0$
$x - 2 = 0$  or  $x - 1 = 0$
$x = 2$  or  $x = 1$

The solutions are $-1, 1, 2,$ and $4$.

**93.** Let $x$ represent the number.
$x + x^2 = 72$
$x^2 + x - 72 = 0$
$(x + 9)(x - 8) = 0$
$x + 9 = 0$  or  $x - 8 = 0$
$x = -9$  or  $x = 8$

The number is $8$ or $-9$.

**95.** Let $x =$ the first number, $x + 1 =$ the second consecutive number, and $x + 2 =$ the third consecutive number.

$x^2 + (x+1)^2 + (x+2)^2 = 50$
$x^2 + x^2 + 2x + 1 + x^2 + 4x + 4 = 50$
$3x^2 + 6x + 5 = 50$
$3x^2 + 6x - 45 = 0$
$3(x^2 + 2x - 15) = 0$
$3(x + 5)(x - 3) = 0$
$x + 5 = 0$  or  $x - 3 = 0$
$x = -5$  or  $x = 3$

They are $-5, -4,$ and $-3$ or $3, 4,$ and $5$.

**97.** Let $x$ = the number.
$$x^3 + 6x = 5x^2$$
$$x^3 - 5x^2 + 6x = 0$$
$$x(x^2 - 5x + 6) = 0$$
$$x(x-3)(x-2) = 0$$
$x = 0$ or $x - 3 = 0$ or $x - 2 = 0$
$\quad\quad\quad\quad x = 3$ or $x = 2$
The number is 0, 2, or 3.

**99.** Let $x$ = the first number and $3 - x$ = the second number.
$$x(3 - x) = -10$$
$$3x - x^2 = -10$$
$$10 + 3x - x^2 = 0$$
$$(5 - x)(2 + x) = 0$$
$5 - x = 0$ or $2 + x = 0$
$x = 5$ or $x = -2$
The numbers are 5 and $-2$.

**101.** Let $x$ = the width and $x + 5$ = the length.
$$x(x + 5) = 594$$
$$x^2 + 5x = 594$$
$$x^2 + 5x - 594 = 0$$
$$(x - 22)(x + 27) = 0$$
$x - 22 = 0$ or $x + 27 = 0$
$x = 22$ or $x = -27$
Discard the negative solution. The width is 22 meters and the length is $22 + 5 = 27$ meters.

**103.** Let $x$ = the length of one leg and $x + 7$ = the length of the other leg.
$$0.5x(x + 7) = 60$$
$$x(x + 7) = 120$$
$$x^2 + 7x - 120 = 0$$
$$(x + 15)(x - 8) = 0$$
$x + 15 = 0$ or $x - 8 = 0$
$x = -15$ or $x = 8$
Discard the negative solution. The length of one leg is 8, the length of the other leg is $8 + 7 = 15$. The length of the third side, the hypotenuse, can be found using the Pythagorean Theorem:
$$\sqrt{8^2 + 15^2} = \sqrt{289} = 17$$

**105.**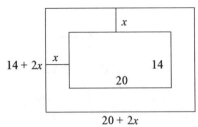

Using the figure and the fact that the area of the path is 152 square feet:
$$(14 + 2x)(20 + 2x) - (20)(14) = 152$$
$$280 + 68x + 4x^2 - 280 = 152$$
$$4x^2 + 68x - 152 = 0$$
$$4(x^2 + 17x - 38) = 0$$
$$4(x - 2)(x + 19) = 0$$
$x - 2 = 0$ or $x + 19 = 0$
$x = 2$ or $x = -19$
Discard the negative solution. The width of the path is 2 feet.

**107.** If the original dimensions are $x + 5$ and $x$, the dimensions of the bottom of the tray are $x + 5 - 4 = x + 1$ and $x - 4$.
$$(x + 1)(x - 4) = 176$$
$$x^2 - 3x - 4 = 176$$
$$x^2 - 3x - 180 = 0$$
$$(x + 12)(x - 15) = 0$$
$x + 12 = 0$ or $x - 15 = 0$
$x = -12$ or $x = 15$
Discard the negative solution. The original width is 15 inches and the original length is $15 + 5 = 20$ inches.

**109.** $C(x) = -0.05x^2 + 4.4x - 32$

(a) $C(x) = -0.05(x^2 - 88x + 640)$

(b) $C(x) = -0.05(x - 80)(x - 8)$

(c) Using the Zero Factor Property with $x - 80$ gives $x = 80$, and using the Zero Factor Property with $x - 8$ gives $x = 8$.

(d) No person of age 8 or age 80 becomes upset about not having coffee at a regular time.

Section 7.5 Division

**111.**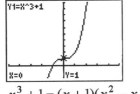

$x^3 + 1 = (x+1)(x^2 - x + 1) = 0$

From the Zero Factor Property, each factor can be set equal to 0. Doing so with the first factor, we find that the value of $c$ for which $p(c) = 0$ is $-1$.
The quadratic factor cannot be factored.

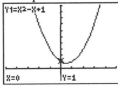

The graph of the quadratic factor has no $x$-intercepts.

**113.**
$(x+y+1)(1-x) + (x+y-1)(x-1) = 0$
$-1(x+y+1)(x-1) + (x+y-1)(x-1) = 0$
$(x-1)[-1(x+y+1) + (x+y-1)] = 0$
$(x-1)(-x-y-1+x+y-1) = 0$
$(x-1)(-2) = 0$
$x - 1 = 0$
$x = 1$

The solution is $x = 1$ for any value of $y$.

**115.**
$\dfrac{2}{x^2} - \dfrac{1}{x} - 1 = 0$

$2\left(\dfrac{1}{x}\right)^2 - \dfrac{1}{x} - 1 = 0$

$2u^2 - u - 1 = 0$
$(2u+1)(u-1) = 0$

$2u + 1 = 0$ or $u - 1 = 0$
$u = -\dfrac{1}{2}$ or $u = 1$
$\dfrac{1}{x} = -\dfrac{1}{2}$  $\dfrac{1}{x} = 1$
$x = -2$  $x = 1$

## Section 7.5 Division

**1.** The dividend is $x + 7$, the divisor is $x$, the quotient is 1, and the remainder is 7.

**3.** $\dfrac{10x^4 + 20x^3}{5x^2} = \dfrac{10x^4}{5x^2} + \dfrac{20x^3}{5x^2}$
$= 2x^2 + 4x$

**5.** $\dfrac{4x^5 + 6x^3 + x^2}{2x^2} = \dfrac{4x^5}{2x^2} + \dfrac{6x^3}{2x^2} + \dfrac{x^2}{2x^2}$
$= 2x^3 + 3x + \dfrac{1}{2}$

**7.** $\dfrac{16a^4 + 12a^3 - 6a^2 + 8a}{2a}$
$= \dfrac{16a^4}{2a} + \dfrac{12a^3}{2a} - \dfrac{6a^2}{2a} + \dfrac{8a}{2a}$
$= 8a^3 + 6a^2 - 3a + 4$

**9.** $\dfrac{20x^5 - 12x^4 + 6x^2 - 8x}{4x^2}$
$= \dfrac{20x^5}{4x^2} - \dfrac{12x^4}{4x^2} + \dfrac{6x^2}{4x^2} - \dfrac{8x}{4x^2}$
$= 5x^3 - 3x^2 + \dfrac{3}{2} - \dfrac{2}{x}$

**11.** 
$$\frac{15c^5d^7 + 6c^4d^5 - 9cd^3}{3c^2d^2}$$
$$= \frac{15c^5d^7}{3c^2d^2} + \frac{6c^4d^5}{3c^2d^2} - \frac{9cd^3}{3c^2d^2}$$
$$= 5c^3d^5 + 2c^2d^3 - \frac{3d}{c}$$

**13.**
$$\frac{6x^3(3x+1) + 4x^2(2x-3)}{2x^2}$$
$$= \frac{6x^3(3x+1)}{2x^2} + \frac{4x^2(2x-3)}{2x^2}$$
$$= 3x(3x+1) + 2(2x-3)$$
$$= 9x^2 + 3x + 4x - 6$$
$$= 9x^2 + 7x - 6$$

**15.**
$$\frac{(x+4)^2 - (x-4)^2}{2x}$$
$$= \frac{(x^2+8x+16) - (x^2-8x+16)}{2x}$$
$$= \frac{x^2 + 8x + 16 - x^2 + 8x - 16}{2x}$$
$$= \frac{16x}{2x} = 8$$

**17.** We stop dividing when the remainder is 0 or when the degree of the remainder is less than the degree of the divisor.

**19.**
$$\require{enclose}\begin{array}{r}2\phantom{x+1}\\x-3\enclose{longdiv}{2x+1}\\\underline{2x-6}\phantom{x}\\7\phantom{x}\end{array}$$
$Q(x) = 2$ and $R(x) = 7$

**21.**
$$\begin{array}{r}x+2\phantom{x+3}\\x+2\enclose{longdiv}{x^2+4x+3}\\\underline{x^2+2x}\phantom{xxxx}\\2x+3\phantom{x}\\\underline{2x+4}\phantom{x}\\-1\phantom{x}\end{array}$$
$Q(x) = x + 2$ and $R(x) = -1$

**23.**
$$\begin{array}{r}3x+3\phantom{xx}\\x-1\enclose{longdiv}{3x^2+0x+2}\\\underline{3x^2-3x}\phantom{xxxx}\\3x+2\phantom{x}\\\underline{3x-3}\phantom{x}\\5\phantom{x}\end{array}$$
$Q(x) = 3x + 3$ and $R(x) = 5$

**25.**
$$\begin{array}{r}a^2-1a+2\phantom{xx}\\4a+3\enclose{longdiv}{4a^3-1a^2+5a+4}\\\underline{4a^3+3a^2}\phantom{xxxxxxx}\\-4a^2+5a\phantom{xx}\\\underline{-4a^2-3a}\phantom{xx}\\8a+4\phantom{x}\\\underline{8a+6}\phantom{x}\\-2\phantom{x}\end{array}$$
$Q(a) = a^2 - a + 2$ and $R(a) = -2$

**27.**
$$\begin{array}{r}3x^2+4x+1\phantom{xx}\\4x-1\enclose{longdiv}{12x^3+13x^2+0x-1}\\\underline{12x^3-3x^2}\phantom{xxxxxxx}\\16x^2+0x\phantom{xx}\\\underline{16x^2-4x}\phantom{xx}\\4x-1\phantom{x}\\\underline{4x-1}\phantom{x}\\0\phantom{x}\end{array}$$
$Q(x) = 3x^2 + 4x + 1$ and $R(x) = 0$

**29.**
$$\begin{array}{r}1\phantom{xxxx}\\x^2+0x-1\enclose{longdiv}{x^2+0x+3}\\\underline{x^2+0x-1}\\4\end{array}$$
$Q(x) = 1$ and $R(x) = 4$

**31.**
$$\begin{array}{r}1\phantom{xxxx}\\x^2+2x+0\enclose{longdiv}{x^2+3x-4}\\\underline{x^2+2x+0}\\x-4\end{array}$$
$Q(x) = 1$ and $R(x) = x - 4$

**33.**
$$a^2 - 2a + 3 \overline{\smash{\big)}\, a^4 + 0a^3 - 1a^2 + 7a - 3}$$
$$\underline{a^4 - 2a^3 + 3a^2}$$
$$2a^3 - 4a^2 + 7a$$
$$\underline{2a^3 - 4a^2 + 6a}$$
$$a - 3$$

quotient on top: $a^2 + 2a$

$Q(a) = a^2 + 2a$ and $R(x) = a - 3$

**35.**
$$x - 2y \overline{\smash{\big)}\, x^2 + 4xy + 3y^2}$$
$$\underline{x^2 - 2xy}$$
$$6xy + 3y^2$$
$$\underline{6xy - 12y^2}$$
$$15y^2$$

quotient: $x + 6y$

$Q: x + 6y$ and $R: 15y^2$

**37.**
$$x + y \overline{\smash{\big)}\, 4x^3 + 0x^2y - 3xy^2 - 2y^3}$$
$$\underline{4x^3 + 4x^2y}$$
$$-4x^2y - 3xy^2$$
$$\underline{-4x^2y - 4xy^2}$$
$$xy^2 - 2y^3$$
$$\underline{xy^2 + 1y^3}$$
$$-3y^3$$

quotient: $4x^2 - 4xy + y^2$

$Q: 4x^2 - 4xy + y^2$ and $R: -3y^3$

**39.** Write the divisor as $x - (-3)$. Then the divider is $-3$.

**41.** $\dfrac{3x^2 - 10x - 8}{x - 4}$

$$\begin{array}{c|rrr} 4 & 3 & -10 & -8 \\ & & 12 & 8 \\ \hline & 3 & 2 & 0 \end{array}$$

$Q(x) = 3x + 2$ and $R(x) = 0$

**43.** $(6x^2 + x + 1) \div \left(x - \dfrac{1}{2}\right)$

$$\begin{array}{c|rrr} 1/2 & 6 & 1 & 1 \\ & & 3 & 2 \\ \hline & 6 & 4 & 3 \end{array}$$

$Q(x) = 6x + 4$ and $R(x) = 3$

**45.** $(2x^4 - 3x^3 + x^2 - 3x + 4) \div (x - 2)$

$$\begin{array}{c|rrrrr} 2 & 2 & -3 & 1 & -3 & 4 \\ & & 4 & 2 & 6 & 6 \\ \hline & 2 & 1 & 3 & 3 & 10 \end{array}$$

$Q(x) = 2x^3 + x^2 + 3x + 3$ and $R(x) = 10$

**47.** $(x^2 + x^4 - 14) \div (x + 2)$
$= (x^4 + 0x^3 + x^2 + 0x - 14) \div [x - (-2)]$

$$\begin{array}{c|rrrrr} -2 & 1 & 0 & 1 & 0 & -14 \\ & & -2 & 4 & -10 & 20 \\ \hline & 1 & -2 & 5 & -10 & 6 \end{array}$$

$Q(x) = x^3 - 2x^2 + 5x - 10$ and $R(x) = 6$

**49.** $\dfrac{2x^2 - 6 + x^3 - 3x}{x - 2} = \dfrac{x^3 + 2x^2 - 3x - 6}{x - 2}$

$$\begin{array}{c|rrrr} 2 & 1 & 2 & -3 & -6 \\ & & 2 & 8 & 10 \\ \hline & 1 & 4 & 5 & 4 \end{array}$$

$Q(x) = x^2 + 4x + 5$ and $R(x) = 4$

**51.** No, the divisor cannot be second degree. It must be in the form $x - c$.

**53.** By inspecting the following,

$$\begin{array}{c|rrrrr} 2 & a & b & c & d & e \\ & & 2 & 0 & 0 & 2 \\ \hline & 1 & 0 & 0 & 1 & 0 \end{array}$$

we see that $a = 1$ (because $a$ is brought down as 1), $b = -2$ (because $b + 2 = 0$), $c = 0$ (because $c + 0 = 0$), $d = 1$ (because $d + 0 = 1$), and $e = -2$ (because $e + 2 = 0$).

**55.**
$$\begin{array}{c|rrrr} ? & 1 & 3 & -2 & -4 \\ & & -1 & -2 & 4 \\ \hline & 1 & 2 & -4 & 0 \end{array}$$

The values in the first row give the coefficients of the dividend. The dividend is $x^3 + 3x^2 - 2x - 4$. The divider is $-1$. So, the divisor is $x - (-1) = x + 1$.

**57.** $\dfrac{x^2 - 2x - 9}{x - 6}$

$$\begin{array}{r|rrr} 6 & 1 & -2 & -9 \\ & & 6 & 24 \\ \hline & 1 & 4 & 15 \end{array}$$

$\dfrac{x^2 - 2x - 9}{x - 6} = x + 4 + \dfrac{15}{x - 6}$

**59.** $(3x^2 + 5x + 2) \div (x - 3)$

$$\begin{array}{r|rrr} 3 & 3 & 5 & 2 \\ & & 9 & 42 \\ \hline & 3 & 14 & 44 \end{array}$$

$\dfrac{3x^2 + 5x + 2}{x - 3} = 3x + 14 + \dfrac{44}{x - 3}$

**61.** $\dfrac{x^2 + 9x + 20}{x + 4}$

$$\begin{array}{r|rrr} -4 & 1 & 9 & 20 \\ & & -4 & -20 \\ \hline & 1 & 5 & 0 \end{array}$$

$\dfrac{x^2 + 9x + 20}{x + 4} = x + 5$

**63.**
$$\begin{array}{r}
2a^2 + 3a + 2 \\
3a - 2 \overline{\smash{)}6a^3 + 5a^2 + 0a + 3} \\
\underline{6a^3 - 4a^2} \\
9a^2 + 0a \\
\underline{9a^2 - 6a} \\
6a + 3 \\
\underline{6a - 4} \\
7
\end{array}$$

$\dfrac{6a^3 + 5a^2 + 3}{3a - 2} = 2a^2 + 3a + 2 + \dfrac{7}{3a - 2}$

**65.**
$$\begin{array}{r}
2x^2 + x - 5 \\
x + 2 \overline{\smash{)}2x^3 + 5x^2 - 3x + 2} \\
\underline{2x^3 + 4x^2} \\
x^2 - 3x \\
\underline{x^2 + 2x} \\
-5x + 2 \\
\underline{-5x - 10} \\
12
\end{array}$$

$\dfrac{2x^3 + 5x^2 - 3x + 2}{x + 2} = 2x^2 + x - 5 + \dfrac{12}{x + 2}$

**67.** $\dfrac{x + 5 - 2x^3 + x^5}{x + 2}$

$= \dfrac{x^5 + 0x^4 - 2x^3 + 0x^2 + x + 5}{x - (-2)}$

$$\begin{array}{r|rrrrrr} -2 & 1 & 0 & -2 & 0 & 1 & 5 \\ & & -2 & 4 & -4 & 8 & -18 \\ \hline & 1 & -2 & 2 & -4 & 9 & -13 \end{array}$$

$\dfrac{x + 5 - 2x^3 + x^5}{x + 2}$
$= x^4 - 2x^3 + 2x^2 - 4x + 9 - \dfrac{13}{x + 2}$

**69.** $\dfrac{x^3 - 13x + 12}{x - 1}$

$$\begin{array}{r|rrrr} 1 & 1 & 0 & -13 & 12 \\ & & 1 & 1 & -12 \\ \hline & 1 & 1 & -12 & 0 \end{array}$$

$x^3 - 13x + 12 = (x - 1)(x^2 + x - 12)$
$\phantom{x^3 - 13x + 12} = (x - 1)(x + 4)(x - 3)$

**71.** $\dfrac{x^3 - 3x^2 - 10x + 24}{x + 3}$

$$\begin{array}{r|rrrr} -3 & 1 & -3 & -10 & 24 \\ & & -3 & 18 & -24 \\ \hline & 1 & -6 & 8 & 0 \end{array}$$

$x^3 - 3x^2 - 10x + 24 = (x + 3)(x^2 - 6x + 8)$
$\phantom{x^3 - 3x^2 - 10x + 24} = (x + 3)(x - 4)(x - 2)$

**73.**
$$\begin{array}{r}
x^2 - 2x + 1 \\
2x + 1 \overline{\smash{)}2x^3 - 3x^2 + 0x + 1} \\
\underline{2x^3 + 1x^2} \\
-4x^2 + 0x \\
\underline{-4x^2 - 2x} \\
2x + 1 \\
\underline{2x + 1} \\
0
\end{array}$$

$2x^3 - 3x^2 + 1 = (2x + 1)(x^2 - 2x + 1)$
$\phantom{2x^3 - 3x^2 + 1} = (2x + 1)(x - 1)^2$

**75.** $(2x + 1)(x - 5) = 2x^2 - 9x - 5$

$$\begin{array}{r|rrr} -2 & 2 & -9 & -5 \\ & & -4 & 26 \\ \hline & 2 & -13 & 21 \end{array}$$

$\dfrac{(2x + 1)(x - 5)}{x + 2} = 2x - 13 + \dfrac{21}{x + 2}$

Section 7.5   Division

**77.** $(3x+1)(3x+1) = 9x^2 + 6x + 1$

$$\begin{array}{r|rrr} 1 & 9 & 6 & 1 \\ & & 9 & 15 \\ \hline & 9 & 15 & 16 \end{array}$$

$$\frac{(3x+1)^2}{x-1} = 9x + 15 + \frac{16}{x-1}$$

**79.** Use the relation: (Quotient)(Divisor) + Remainder = Dividend.

$(x+3)(2x-1) + 2 = 2x^2 - x + 6x - 3 + 2$
$= 2x^2 + 5x - 1$

**81.** Use the relation:

$$\text{Divisor} = \frac{\text{Dividend} - \text{Remainder}}{\text{Quotient}}.$$

$$\text{Divisor} = \frac{x^2 - x - 7 - (-1)}{x - 3}$$

$$= \frac{x^2 - x - 6}{x - 3}$$

$$\begin{array}{r|rrr} 3 & 1 & -1 & -6 \\ & & 3 & 6 \\ \hline & 1 & 2 & 0 \end{array}$$

The divisor is $x + 2$.

**83.** $\dfrac{x - 12 - 17x^2 + x^5 - 3x^3}{x - 3}$

$= \dfrac{x^5 + 0x^4 - 3x^3 - 17x^2 + x - 12}{x - 3}$

$$\begin{array}{r|rrrrrr} 3 & 1 & 0 & -3 & -17 & 1 & -12 \\ & & 3 & 9 & 18 & 3 & 12 \\ \hline & 1 & 3 & 6 & 1 & 4 & 0 \end{array}$$

$Q(x) = x^4 + 3x^3 + 6x^2 + x + 4$
and $R(x) = 0$

**85.** $\dfrac{x^3 + 2x - 3}{x(x-3)} = \dfrac{x^3 + 0x^2 + 2x - 3}{x^2 - 3x}$

$$\begin{array}{r} x + 3 \\ x^2 - 3x \overline{\smash{)}x^3 + 0x^2 + 2x - 3} \\ \underline{x^3 - 3x^2} \\ 3x^2 + 2x \\ \underline{3x^2 - 9x} \\ 11x - 3 \end{array}$$

$Q(x) = x + 3$, $R(x) = 11x - 3$

**87.** Use the relation: (Quotient)(Divisor) = Dividend.

$3x^2(3x^3 + 4x^2 + x - 4)$
$= 9x^5 + 12x^4 + 3x^3 - 12x^2$

**89.** Divide the distance by the time in seconds to find the average speed:

$\text{Speed} = \dfrac{x^2 + 5x - 24}{x - 3}$

$= \dfrac{(x-3)(x+8)}{x-3}$

$= x + 8$

The average speed of the particle is $(x + 8)$ meters per second.

**91.** (a) If the number of posts is $x + 9$, the number of spans is $x + 9 - 1 = x + 8$.

Divide the length $(x^3 - 65x - 6)$ by $x + 8$:

$$\begin{array}{r|rrrr} -8 & 1 & 0 & -65 & -6 \\ & & -8 & 64 & 8 \\ \hline & 1 & -8 & -1 & 2 \end{array}$$

The distance between the posts is $x^2 - 8x - 1$.

(b) The amount of fencing left over is given by the remainder from the division carried out in part (a). Thus, there will be 2 feet of fencing left over.

**93.** $(x^2 + 4x + k) \div (x - 2)$

$$\begin{array}{r|rrr} 2 & 1 & 4 & k \\ & & 2 & 12 \\ \hline & 1 & 6 & 12 + k \end{array}$$

If the remainder is 4 then
$12 + k = 4$
$k = -8$

**95.** $(x^3 + kx^2 - 2k + 3) \div (x - 1)$

$$\begin{array}{r|rrrr} 1 & 1 & k & 0 & -2k+3 \\ & & 1 & k+1 & k+1 \\ \hline & 1 & k+1 & k+1 & -k+4 \end{array}$$

If the remainder is 0 then
$-k + 4 = 0$
$k = 4$

**97.** $\dfrac{16x^{4n}-64x^{2n}}{4x^{2n}} = \dfrac{16x^{4n}}{4x^{2n}} - \dfrac{64x^{2n}}{4x^{2n}}$
$= 4x^{2n} - 16$

**99.**
$$\begin{array}{r} x^{3n}-x^n+5 \\ x^n+3\overline{\smash{)}x^{4n}+3x^{3n}-x^{2n}+2x^n+1} \\ \underline{x^{4n}+3x^{3n}} \\ 0\ -x^{2n}+2x^n \\ \underline{-x^{2n}-3x^n} \\ 5x^n+1 \\ \underline{5x^n+15} \\ -14 \end{array}$$
$Q(x) = x^{3n}-x^n+5,\ \ R(x) = -14$

**101.** $(x^2+bx+b^2) \div (x-b)$

$$\begin{array}{r|rrr} b & 1 & b & b^2 \\ & & b & 2b^2 \\ \hline & 1 & 2b & 3b^2 \end{array}$$

If the remainder is 12, then
$3b^2 = 12$
$3b^2 - 12 = 0$
$3(b^2-4) = 0$
$3(b+2)(b-2) = 0$
$b+2 = 0$ or $b-2 = 0$
$b = -2$ or $b = 2$
The positive value of $b = 2$ yields a remainder of 12.

## Chapter 7 Project

**1.** For each country, multiply $1534 by the exchange rate.

| Country | One Dollar Exchange |
|---|---|
| England | $1534(0.65 pound) = 997.10 pounds |
| Italy | $1534(1517.00 lire) = 2,327,078 lire |
| Germany | $1534(1.48 marks) = 2270.32 marks |
| Norway | $1534(6.41 crowns) = 9832.94 crowns |
| Spain | $1534(125.00 pesetas) = 191,750 pesetas |

**3.** Solving the equation $c = 6.41d$ (from Exercise 2) for $d$, we obtain $d = \dfrac{c}{6.41}$ or approximately $d = 0.16c$. Solving the equation $p = 0.65d$ (from Exercise 2) for $d$, we obtain $d = \dfrac{p}{0.65}$ or approximately $d = 1.54p$. To obtain an equation of the form $c = k_3 p$, set the two equations $d = 0.16c$ and $d = 1.54p$ equal and solve for $c$.
$0.16c = 1.54p$
$c = \dfrac{1.54}{0.16}p$
$c \approx 9.63p$

This formula can be used to convert pounds to crowns.

**5.** If you have 8667 crowns when you arrive in Norway and then spend 6000 crowns, you will return to the United States with 2667 crowns. If you exchange this for U.S. dollars, you will receive $\dfrac{2667}{6.41} \approx \$416.07$.

## Chapter 7 Review Exercises

1. To obtain a binomial with a positive leading coefficient, choose the common factor $-x^2y^2$.

3. $x^{25} + x^{35} = x^{25}(1 + x^{10})$

5. $x^6y^3 - x^7y^4 - x^3y^5$
   $= x^3y^3(x^3 - x^4y - y^2)$

7. $x^3 + 3x^2 - 2x - 6 = (x^3 + 3x^2) - (2x + 6)$
   $= x^2(x+3) - 2(x+3)$
   $= (x+3)(x^2 - 2)$

9. $14x^7 - 35x^5 + 42x^3$
   $= 7x^3(2x^4 - 5x^2 + 6)$

11. The difference between the square of an odd number and the odd number itself is written as
    $(2n-1)^2 - (2n-1)$
    $= (2n-1)[(2n-1) - 1]$
    $= (2n-1)(2n-2)$
    $= 2(2n-1)(n-1)$

13. You can quickly tell that $9x^2 - 30x + 24$ is not a perfect square trinomial because the constant term is not a perfect square.

15. $49x^2 - 16y^2 = (7x)^2 - (4y)^2$
    $= (7x + 4y)(7x - 4y)$

17. $16 - a^4 = 4^2 - (a^2)^2$
    $= (4 + a^2)(4 - a^2)$
    $= (4 + a^2)(2 + a)(2 - a)$

19. $27b^6 + 8a^3$
    $= (3b^2)^3 + (2a)^3$
    $= (3b^2 + 2a)(9b^4 - 6ab^2 + 4a^2)$

21. $c^2 + 64 + 16c = c^2 + 16c + 64$
    $= c^2 + 2(8)c + 8^2$
    $= (c+8)^2$

23. $a^2 - 16 = (a+4)(a-4)$
    If the base is $(a+4)$ feet and the height is $(a-4)$ feet, the difference between the base and height is
    $(a+4) - (a-4) = a + 4 - a + 4 = 8$ feet

25. $x^2 - 5x - 6 = (x-6)(x+1)$

27. $2x^2 + 12x - 14 = 2(x^2 + 6x - 7)$
    $= 2(x+7)(x-1)$

29. Because the numbers 6 and 24 have several factors each, they lead to a large number of possible factorizations to try.

31. $3x^2 + 14x + 15 = (3x+5)(x+3)$

33. $12x^2 + 8x - 15 = (6x-5)(2x+3)$

35. $2x - 8 + x^2 = x^2 + 2x - 8$
    $= (x+4)(x-2)$

37. $125x^3 - 64$
    $= (5x)^3 - 4^3$
    $= (5x-4)(25x^2 + 20x + 16)$

39. No, but you can factor out the 2 to obtain $[2(x+2y)]^2 = 4(x+2y)^2$

41. $3x^2 + 4x - 15 = (3x-5)(x+3)$

43. $x^9 + 27y^6$
    $= (x^3)^3 + (3y^2)^3$
    $= (x^3 + 3y^2)(x^6 - 3x^3y^2 + 9y^4)$

45. $x^3y^4 - x^4y^3 = x^3y^3(y - x)$

47. $6x^3 + 25x^2 + 21x = x(6x^2 + 25x + 21)$
    $= x(6x+7)(x+3)$

49. In $a + b = 0$, we know that $a$ and $b$ are opposites, but we do not know what the numbers are. In $ab = 0$, either $a$ or $b$ must be 0.

**51.**
$$x^2 = 8x$$
$$x^2 - 8x = 0$$
$$x(x-8) = 0$$
$x = 0$ or $x - 8 = 0$
$\qquad\qquad\quad x = 8$
The solutions are 0 and 8.

**53.**
$$x^2 + x = 42$$
$$x^2 + x - 42 = 0$$
$$(x-6)(x+7) = 0$$
$x - 6 = 0$ or $x + 7 = 0$
$x = 6$ or $x = -7$
The solutions are 6 and $-7$.

**55.**
$$x(x+2) = 24$$
$$x^2 + 2x = 24$$
$$x^2 + 2x - 24 = 0$$
$$(x+6)(x-4) = 0$$
$x + 6 = 0$ or $x - 4 = 0$
$x = -6$ or $x = 4$
The solutions are 4 and $-6$.

**57.**
$$x^3 + 3x^2 - 4x = 12$$
$$x^3 + 3x^2 - 4x - 12 = 0$$
$$(x^3 + 3x^2) - (4x + 12) = 0$$
$$x^2(x+3) - 4(x+3) = 0$$
$$(x+3)(x^2 - 4) = 0$$
$$(x+3)(x+2)(x-2) = 0$$
$x + 3 = 0$ or $x + 2 = 0$ or $x - 2 = 0$
$x = -3$ or $x = -2$ or $x = 2$
The solutions are $-3, -2,$ and 2.

**59.** $|3x^2 + 7x - 2| = 4$
$3x^2 + 7x - 2 = 4$ or $3x^2 + 7x - 2 = -4$

$3x^2 + 7x - 2 = 4$
$3x^2 + 7x - 6 = 0$
$(3x - 2)(x + 3) = 0$
$3x - 2 = 0$ or $x + 3 = 0$
$x = 2/3$ or $x = -3$

or

$3x^2 + 7x - 2 = -4$
$3x^2 + 7x + 2 = 0$
$(3x + 1)(x + 2) = 0$
$3x + 1 = 0$ or $x + 2 = 0$
$x = -1/3$ or $x = -2$
The solutions are $-3, -2, -1/3,$ and $2/3$.

**61.** Let $x =$ the number.
$$2x + x^2 = 63$$
$$x^2 + 2x - 63 = 0$$
$$(x+9)(x-7) = 0$$
$x + 9 = 0$ or $x - 7 = 0$
$x = -9$ or $x = 7$
The number is $-9$ or 7.

**63.** Multiplying the quotient times the divisor and adding the remainder should result in the dividend.

**65.**
$$\frac{9x^5 - 12x^3}{3x^2} = \frac{9x^5}{3x^2} - \frac{12x^3}{3x^2}$$
$$= 3x^3 - 4x$$

**67.**
$$\begin{array}{r}
3x^2 - 4x - 5 \\
3x - 2 \overline{\smash{\big)}\, 9x^3 - 18x^2 - 7x + 5} \\
\underline{9x^3 - 6x^2\phantom{-7x+5}} \\
-12x^2 - 7x\phantom{+5} \\
\underline{-12x^2 + 8x\phantom{+5}} \\
-15x + 5 \\
\underline{-15x + 10} \\
-5
\end{array}$$
$Q(x) = 3x^2 - 4x - 5, \quad R(x) = -5$

**69.**
$$\begin{array}{r}
x^2 + 3y^2 \\
x - 3y \overline{\smash{\big)}\, x^3 - 3x^2y + 3xy^2 - y^3} \\
\underline{x^3 - 3x^2y\phantom{+3xy^2-y^3}} \\
0 + 3xy^2 - y^3 \\
\underline{3xy^2 - 9y^3} \\
8y^3
\end{array}$$
$Q(x) = x^2 + 3y^2, \quad R(x) = 8y^3$

**71.** Using the relation (Quotient)(Divisor) + Remainder = Dividend,
$$P(x) = (2x^2 - x + 1)(x + 1) + 3$$
$$= x(2x^2 - x + 1) + 1(2x^2 - x + 1) + 3$$
$$= 2x^3 - x^2 + x + 2x^2 - x + 1 + 3$$
$$= 2x^3 + x^2 + 4$$

**73.** $(2x^3 - 3x^2 + 5) \div (x-1)$

$$\begin{array}{r|rrrr} 1 & 2 & -3 & 0 & 5 \\ & & 2 & -1 & -1 \\ \hline & 2 & -1 & -1 & 4 \end{array}$$

$Q(x) = 2x^2 - x - 1$, $R(x) = 4$

**75.** $(x^5 - 32) \div (x - 2)$

$$\begin{array}{r|rrrrr} 2 & 1 & 0 & 0 & 0 & 0 & -32 \\ & & 2 & 4 & 8 & 16 & 32 \\ \hline & 1 & 2 & 4 & 8 & 16 & 0 \end{array}$$

$Q(x) = x^4 + 2x^3 + 4x^2 + 8x + 16$, $R(x) = 0$

# Chapter 7 Looking Ahead

**1.**
(a) $\dfrac{-2}{3} + \dfrac{1}{6} = \dfrac{-4}{6} + \dfrac{1}{6} = -\dfrac{3}{6} = -\dfrac{1}{2}$

(b) $\dfrac{5}{8} - \dfrac{3}{10} = \dfrac{25}{40} - \dfrac{12}{40} = \dfrac{13}{40}$

(c) $\dfrac{-3}{14} \cdot \dfrac{4}{5} = \dfrac{-12}{70} = -\dfrac{6}{35}$

(d) $\dfrac{9}{5} \div 6 = \dfrac{9}{5} \cdot \dfrac{1}{6} = \dfrac{9}{30} = \dfrac{3}{10}$

**3.** $\dfrac{x-3}{x+6}$

(a) 3: $\dfrac{3-3}{3+6} = \dfrac{0}{9} = 0$

(b) $-6$: $\dfrac{-6-3}{-6+6} = \dfrac{-9}{0}$ is undefined

(c) $-5$: $\dfrac{-5-3}{-5+6} = \dfrac{-8}{1} = -8$

**5.** $2x^2 + 6x = 2x(x+3)$

**7.** $6 - x - 2x^2 = (3 - 2x)(2 + x)$

**9.**
$x^2 + 2x - 8 = 0$
$(x+4)(x-2) = 0$
$x + 4 = 0 \quad$ or $\quad x - 2 = 0$
$x = -4 \quad$ or $\quad x = 2$

**11.**
$\dfrac{x}{2} + \dfrac{5}{6}(x-1) = \dfrac{3x-1}{3}$

$6 \cdot \dfrac{x}{2} + 6 \cdot \dfrac{5}{6}(x-1) = 6 \cdot \dfrac{3x-1}{3}$

$3x + 5(x-1) = 2(3x-1)$

$3x + 5x - 5 = 6x - 2$

$8x - 5 - 6x = -2$

$2x - 5 = -2$

$2x = 3$

$x = 3/2$

## Chapter 7 Test

**1.** $18r^3s^5 - 12r^5s^4 = 6r^3s^4(3s - 2r^2)$

**3.** $27 - 12x^2 = 3(9 - 4x^2)$
$= 3(3 + 2x)(3 - 2x)$

**5.** $1 - 8x^3$
$= 1^3 - (2x)^3$
$= (1 - 2x)(1 + 2x + 4x^2)$

**7.** $3x^2 - 4x = 0$
$x(3x - 4) = 0$
$x = 0$ or $3x - 4 = 0$
     or $x = 4/3$

**9.** $2x^3 - x^2 - 8x + 4 = 0$
$(2x^3 - x^2) - (8x - 4) = 0$
$x^2(2x - 1) - 4(2x - 1) = 0$
$(2x - 1)(x^2 - 4) = 0$
$(2x - 1)(x + 2)(x - 2) = 0$
$2x - 1 = 0$ or $x + 2 = 0$
$x = 1/2$ or $x = -2$
   or $x - 2 = 0$
       $x = 2$

**11.** Using the relation: (Quotient)(Divisor) + Remainder = Dividend, we have
$P(x)$
$= Q(x)D(x) + R(x)$
$= (2x^2 - x + 3)(x + 3) - 10$
$= x(2x^2 - x + 3) + 3(2x^2 - x + 3) - 10$
$= 2x^3 - x^2 + 3x + 6x^2 - 3x + 9 - 10$
$= 2x^3 + 5x^2 - 1$
$2x^3 + 5x^2 - 1 = (2x^2 - x + 3)(x + 3) - 10$

**13.** $\dfrac{15x^6y^4 - 21x^5y^8}{-3xy^2} = \dfrac{15x^6y^4}{-3xy^2} - \dfrac{21x^5y^8}{-3xy^2}$
$= -5x^5y^2 + 7x^4y^6$
$Q(x) = -5x^5y^2 + 7x^4y^6$, $R(x) = 0$

**15.** $(x^5 - 1) \div (x + 2)$

$\begin{array}{r|rrrrrr} -2 & 1 & 0 & 0 & 0 & 0 & -1 \\ & & -2 & 4 & -8 & 16 & -32 \\ \hline & 1 & -2 & 4 & -8 & 16 & -33 \end{array}$

$Q(x) = x^4 - 2x^3 + 4x^2 - 8x + 16$,
$R(x) = -33$

**17.** Let $x$ = the first integer, $x + 2$ = the second consecutive even integer, and $x + 4$ = the third consecutive even integer. The sum of the squares of three consecutive even integers is:
$x^2 + (x + 2)^2 + (x + 4)^2$
$= x^2 + x^2 + 4x + 4 + x^2 + 8x + 16$
$= 3x^2 + 12x + 20$

When $3x^2 + 12x + 20$ is divided by $x + 2$, we obtain a remainder of 8.

$\begin{array}{r|rrr} -2 & 3 & 12 & 20 \\ & & -6 & -12 \\ \hline & 3 & 6 & 8 \end{array}$

**19.**
$\dfrac{1}{2}h(b_1 + b_2) = A$
$\dfrac{1}{2}(x)(2x + 2x + 20) = 5500$
$\dfrac{1}{2}(x)(4x + 20) = 5500$
$(x)(4x + 20) = 11{,}000$
$4x^2 + 20x - 11{,}000 = 0$
$4(x^2 + 5x - 2750) = 0$
$4(x + 55)(x - 50) = 0$
$x + 55 = 0$ or $x - 50 = 0$
$x = -55$ or $x = 50$
The lengths are $2(50) = 100$ feet and $2(50) + 20 = 100 + 20 = 120$ feet.

# Chapter 8

# Rational Expressions

## Section 8.1 Introduction to Rational Expressions

1. Any number that makes the denominator zero is excluded from the domain.

3. $y = \dfrac{10}{x-5}$

   The domain is the set of all real numbers except $x = 5$. On the graph, the vertical line indicates that $x = 5$ is to be excluded.

   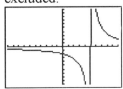

5. $\dfrac{x}{x-3}$

   For $x = -3$: $\dfrac{x}{x-3} = \dfrac{-3}{-3-3} = \dfrac{-3}{-6} = \dfrac{1}{2}$

   For $x = 0$: $\dfrac{x}{x-3} = \dfrac{0}{0-3} = \dfrac{0}{-3} = 0$

   For $x = 3$: $\dfrac{x}{x-3} = \dfrac{3}{3-3} = \dfrac{3}{0}$ is undefined.

7. $\dfrac{t-3}{4-t}$

   For $t = -4$: $\dfrac{-4-3}{4-(-4)} = -\dfrac{7}{8} = -0.875$

   For $t = 3$: $\dfrac{3-3}{4-3} = 0$

   For $t = 4$: $\dfrac{4-3}{4-4} = \dfrac{1}{0}$ is undefined.

9. $y = \dfrac{2x+3}{x-4}$

   To find the restricted values, set the denominator equal to 0 and solve for $x$.
   $$x - 4 = 0$$
   $$x = 4$$
   The domain is $\{x \mid x \neq 4\}$.

11. $y = \dfrac{x-4}{(x+2)^2}$

    To find the restricted values, set the denominator equal to 0 and solve for $x$.
    $$(x+2)^2 = 0$$
    $$x + 2 = 0$$
    $$x = -2$$
    The domain is $\{x \mid x \neq -2\}$.

13. $y = \dfrac{2x-6}{x^2-9}$

    To find the restricted values, set the denominator equal to 0 and solve for $x$.
    $$x^2 - 9 = 0$$
    $$(x+3)(x-3) = 0$$
    $$x + 3 = 0 \quad \text{or} \quad x - 3 = 0$$
    $$x = -3 \quad \text{or} \quad x = 3$$
    The domain is $\{x \mid x \neq -3, 3\}$.

15. $y = \dfrac{3}{x^2+2x-8}$

    To find the restricted values, set the denominator equal to 0 and solve for $x$.
    $$x^2 + 2x - 8 = 0$$
    $$(x+4)(x-2) = 0$$
    $$x + 4 = 0 \quad \text{or} \quad x - 2 = 0$$
    $$x = -4 \quad \text{or} \quad x = 2$$
    The domain is $\{x \mid x \neq -4, 2\}$.

17. $y = \dfrac{2x-1}{x^2+4}$

The denominator is never 0, so the domain is **R**.

19. $y = \dfrac{5}{x \;\underline{\phantom{+}}\; 2}$

The value $-2$ will not be in the domain if the denominator is $x+2$. Fill the blank with a $+$ sign.

21. $y = \dfrac{x}{x^2 \;\underline{\phantom{-25}}\;}$

The values $-5$ and $5$ will not be in the domain if the denominator is $x^2 - 25$ because this factors as $(x+5)(x-5)$. Fill the blank with $-25$.

23. The expression in (i) has a common factor $(x+3)$ that can be divided out. In (ii), there is no common factor in the numerator and the denominator.

25. $\dfrac{6x^3 - 3x^2}{9x^2} = \dfrac{3x^2(2x-1)}{9x^2} = \dfrac{2x-1}{3}$

27. $\dfrac{2(3x+1)}{2+6x} = \dfrac{2(3x+1)}{2(1+3x)} = \dfrac{2(3x+1)}{2(3x+1)} = 1$

29. $\dfrac{x^2 - 81}{x+9} = \dfrac{(x+9)(x-9)}{x+9} = x-9$

31. $\dfrac{a^2 + 5a}{4a-20} = \dfrac{a(a+5)}{4(a-5)}$

Cannot be simplified.

33. $\dfrac{2-x}{2+9x-5x^2} = \dfrac{2-x}{(2-x)(1+5x)} = \dfrac{1}{1+5x}$

35. $\dfrac{x^2-16}{x^2-x-12} = \dfrac{(x+4)(x-4)}{(x-4)(x+3)} = \dfrac{x+4}{x+3}$

37. $\dfrac{a^3-a}{6a-6} = \dfrac{a(a^2-1)}{6(a-1)}$
$= \dfrac{a(a+1)(a-1)}{6(a-1)}$
$= \dfrac{a(a+1)}{6}$

39. $\dfrac{x-3}{2x^2-5x-3} = \dfrac{x-3}{(2x+1)(x-3)} = \dfrac{1}{2x+1}$

41. $\dfrac{x^2-14x+45}{x^2-17x+72} = \dfrac{(x-9)(x-5)}{(x-9)(x-8)} = \dfrac{x-5}{x-8}$

43. $\dfrac{2x^2+6x-1}{2x(x+3)-1} = \dfrac{2x^2+6x-1}{2x^2+6x-1} = 1$

45. $\dfrac{ab-3a}{b^2-9} = \dfrac{a(b-3)}{(b+3)(b-3)} = \dfrac{a}{b+3}$

47. $\dfrac{x+3y}{x^2+2xy-3y^2} = \dfrac{x+3y}{(x+3y)(x-y)} = \dfrac{1}{x-y}$

49. $\dfrac{c^2-5cd+6d^2}{2c^2-5cd-3d^2} = \dfrac{(c-3d)(c-2d)}{(2c+d)(c-3d)}$
$= \dfrac{c-2d}{2c+d}$

51. $\dfrac{2}{a-b} = \dfrac{?}{b-a} = \dfrac{?}{-1(a-b)}$

To get the second fraction's denominator, the first fraction's denominator was multiplied by $-1$. The second fraction's numerator should be $-1(2) = -2$.

53. $\dfrac{x}{x-3} = -\dfrac{x}{?} = \dfrac{-x}{?}$

To get the second fraction, the first fraction's numerator and denominator were multiplied by $-1$. So the second fraction's denominator should be $-1(x-3) = -x+3 = 3-x$.

Section 8.1  Introduction to Rational Expressions

**55.** (a) $\dfrac{x+3}{x-3}$ has neither the value of 1 nor $-1$.

(b) $\dfrac{3+x}{x+3} = 1$

(c) $\dfrac{x-3}{3-x} = \dfrac{x-3}{-(x-3)} = -1$

**57.** (a) $\dfrac{-2x+5}{-(5-2x)} = \dfrac{5-2x}{-(5-2x)} = -1$

(b) $\dfrac{2x-5}{-5+2x} = \dfrac{2x-5}{2x-5} = 1$

(c) $\dfrac{2x-5}{2x+5}$ has neither the value of 1 nor $-1$.

**59.** $\dfrac{5x-2}{2-5x} = \dfrac{5x-2}{-(5x-2)} = -1$

**61.** $\dfrac{5x-3}{-15x+9} = \dfrac{5x-3}{-3(5x-3)} = -\dfrac{1}{3}$

**63.** $\dfrac{a^2-49}{7-a} = \dfrac{(a+7)(a-7)}{-(a-7)} = -(a+7)$

**65.** $\dfrac{z-5w}{25w^2-z^2} = \dfrac{-(5w-z)}{(5w+z)(5w-z)} = \dfrac{-1}{5w+z}$

**67.** $\dfrac{-2x^2-3x+20}{2x^2+x-15} = \dfrac{-(2x^2+3x-20)}{2x^2+x-15}$
$= -\dfrac{(2x-5)(x+4)}{(2x-5)(x+3)}$
$= -\dfrac{x+4}{x+3}$

**69.** $\dfrac{x^3+3x^2-9x-27}{x^2+6x+9}$
$= \dfrac{x^2(x+3)-9(x+3)}{(x+3)(x+3)}$
$= \dfrac{(x^2-9)(x+3)}{(x+3)(x+3)}$
$= \dfrac{(x+3)(x-3)(x+3)}{(x+3)(x+3)}$
$= x-3$

**71.** $\dfrac{x^3+8}{x^2-4} = \dfrac{(x+2)(x^2-2x+4)}{(x+2)(x-2)}$
$= \dfrac{x^2-2x+4}{x-2}$

**73.** $\dfrac{18x^5-39x^4+18x^3}{8x^4-18x^2}$
$= \dfrac{3x^3(6x^2-13x+6)}{2x^2(4x^2-9)}$
$= \dfrac{3x^3(3x-2)(2x-3)}{2x^2(2x+3)(2x-3)}$
$= \dfrac{3x(3x-2)}{2(2x+3)}$

**75.** $\dfrac{3}{2xy^2} = \dfrac{?}{8x^3y^3} = \dfrac{?}{4x^2y(2xy^2)}$

Multiply both the numerator and the denominator of the original expression by $4x^2y$. The unknown numerator is $3(4x^2y) = 12x^2y$.

**77.** $\dfrac{5}{x-5} = \dfrac{?}{5x-25} = \dfrac{?}{5(x-5)}$

Multiply both the numerator and the denominator of the original expression by 5. The unknown numerator is $5(5) = 25$.

**79.** $\dfrac{x+5}{1} = \dfrac{?}{x+5}$

Multiply both the numerator and the denominator of the original expression by $x+5$. The unknown numerator is $(x+5)(x+5) = (x+5)^2$.

**81.** $\dfrac{5}{4x+12} = \dfrac{?}{4x^2-36} = \dfrac{?}{(4x+12)(x-3)}$

Multiply both the numerator and the denominator of the original expression by $x-3$. The unknown numerator is $5(x-3)$.

83. $\dfrac{-3}{x+1} = \dfrac{?}{x^2+5x+4} = \dfrac{?}{(x+1)(x+4)}$

Multiply both the numerator and the denominator of the original expression by $x+4$. The unknown numerator is $-3(x+4)$.

85. $\dfrac{9}{x^2-5x-6} = \dfrac{?}{(x-6)(x-5)(x+1)}$

Multiply both the numerator and the denominator of the original expression by $x-5$. The unknown numerator is $9(x-5)$.

87. $\dfrac{5t}{6} = \dfrac{-10t^2}{?} = \dfrac{-2t(5t)}{?}$

Multiply both the numerator and the denominator of the original expression by $-2t$. The unknown denominator is $6(-2t) = -12t$.

89. $\dfrac{2x-1}{1-x} = \dfrac{1-x-2x^2}{?} = \dfrac{-(2x-1)(x+1)}{?}$

Multiply both the numerator and the denominator of the original expression by $-1(x+1)$. The unknown denominator is $-1(x+1)(1-x) = -1(1+x)(1-x)$.

91. $C(x) = \dfrac{100}{100-x}, \quad 0 \le x < 100$

(a) $C(50) = \dfrac{100}{100-50} = \dfrac{100}{50} = 2$

The cost of cleaning up 50% of the oil spill is $2 million.

(b) No, the function is not defined for $x = 100$.

(c) Using a window setting of Xmin = 0, Xmax = 100, Xscl = 10, Ymin = 0, Ymax = 20, and Yscl = 1, we obtain the graph shown below. At the point at which the cost begins to rise rapidly, 90% of the oil would be cleaned up at a cost of $10 million.

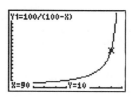

93. Using synthetic division, we have

$28 \;\overline{\big|\; -0.293 \quad 37.077 \quad -808.44}$
$\phantom{28 \;\big|\;}\quad\quad\;\; -8.204 \quad\;\; 808.44$
$\phantom{28 \;\big|\;}\overline{-0.293 \quad 28.873}$

Thus, $\dfrac{N(x)}{x-28} = -0.293x + 28.873$.

95. $N(x) = (x-28)(-0.293x + 28.873)$

97. $\dfrac{2x^{-2} - 3x^{-1}}{3x^2 - 2x} = \dfrac{x^{-2}(2-3x)}{x(3x-2)}$
$= \dfrac{-x^{-2}(3x-2)}{x(3x-2)}$
$= -\dfrac{1}{x^3}$

99. $\dfrac{x^2+3x}{3x^{-4}+x^{-3}} = \dfrac{x(x+3)}{x^{-4}(3+x)} = x^5$

101. $\dfrac{3}{x^n-1}$

If $n$ is even, the solutions of $x^n - 1 = 0$ are 1 and $-1$. If $n$ is odd, the solution of $x^n - 1 = 0$ is 1. Thus, for $n$ even, the domain is $\{x \,|\, x \ne -1, 1\}$, and for $n$ odd, the domain is $\{x \,|\, x \ne 1\}$.

103. $\dfrac{x}{x^{2n}+1}$

The denominator is never 0. The domain is **R**.

**105.** $\dfrac{5251^2 - 5248^2}{5250(5254) - 5251^2}$

Let $x = 5250$. Then we have the following simplification:

$\dfrac{(x+1)^2 - (x-2)^2}{x(x+4) - (x+1)^2}$

$= \dfrac{x^2 + 2x + 1 - (x^2 - 4x + 4)}{x^2 + 4x - (x^2 + 2x + 1)}$

$= \dfrac{x^2 + 2x + 1 - x^2 + 4x - 4}{x^2 + 4x - x^2 - 2x - 1}$

$= \dfrac{6x - 3}{2x - 1}$

$= \dfrac{3(2x - 1)}{2x - 1}$

$= 3$

So the value of the original expression is 3.

## Section 8.2 Multiplication and Division

**1.** An easier method is to factor first and then divide out common factors.

**3.** $\dfrac{5a^2 b}{b^3 c^4} \cdot \dfrac{b^2 c^4}{25a^3} = \dfrac{5a^2 b^3 c^4}{25a^3 b^3 c^4} = \dfrac{1}{5a}$

**5.** $(a+4) \cdot \dfrac{a-4}{3a+12} = \dfrac{a+4}{1} \cdot \dfrac{a-4}{3a+12}$

$= \dfrac{a+4}{1} \cdot \dfrac{a-4}{3(a+4)}$

$= \dfrac{a-4}{3}$

**7.** $\dfrac{5x-7}{12-3x} \cdot \dfrac{3x-12}{7-5x}$

$= \dfrac{5x-7}{-(3x-12)} \cdot \dfrac{3x-12}{-(5x-7)}$

$= \dfrac{1}{-1} \cdot \dfrac{1}{-1}$

$= 1$

**9.** $\dfrac{4c^2}{5c+30} \cdot \dfrac{c^2 - 36}{16c^3} = \dfrac{4c^2}{5(c+6)} \cdot \dfrac{(c+6)(c-6)}{16c^3}$

$= \dfrac{c-6}{20c}$

**11.** $\dfrac{7w^2 - 14w}{w^2 + 3w - 10} \cdot \dfrac{w+5}{21w}$

$= \dfrac{7w(w-2)}{(w+5)(w-2)} \cdot \dfrac{w+5}{21w}$

$= \dfrac{1}{3}$

**13.** $\dfrac{y^2 - 1}{y^2 + 14y + 48} \cdot \dfrac{y^2 + 5y - 6}{(y-1)^2}$

$= \dfrac{(y+1)(y-1)}{(y+6)(y+8)} \cdot \dfrac{(y+6)(y-1)}{(y-1)^2}$

$= \dfrac{y+1}{y+8}$

**15.** $\dfrac{x-3}{3x-21} \cdot \dfrac{7-x}{?} = \dfrac{1}{3}$

$\dfrac{x-3}{3(x-7)} \cdot \dfrac{-(x-7)}{?} = \dfrac{1}{3}$

$\dfrac{x-3}{3} \cdot \dfrac{-1}{?} = \dfrac{1}{3}$

If the ? is replaced by $-(x-3) = 3-x$, the resulting equation is true.

**17.**
$$\frac{3x^2+8x-3}{2x+1} \cdot \frac{?}{3x^2-4x+1} = x+3$$
$$\frac{(3x-1)(x+3)}{2x+1} \cdot \frac{?}{(3x-1)(x-1)} = x+3$$
$$\frac{x+3}{2x+1} \cdot \frac{?}{x-1} = x+3$$

If ? is replaced by $(2x+1)(x-1) = 2x^2-x-1$, the resulting equation is true.

**19.** The opposite of $x+3$ is $-(x+3) = -x-3$. The reciprocal of $x+3$ is $\frac{1}{x+3}$.

**21.**
$$\frac{24x^3y^7}{16x^4y^3} \div \frac{9y^6x}{x^5} = \frac{24x^3y^7}{16x^4y^3} \cdot \frac{x^5}{9y^6x}$$
$$= \frac{24x^8y^7}{144x^5y^9}$$
$$= \frac{x^3}{6y^2}$$

**23.**
$$\frac{2x+6}{5x^2} \div \frac{x+3}{15x} = \frac{2x+6}{5x^2} \cdot \frac{15x}{x+3}$$
$$= \frac{2(x+3)}{5x^2} \cdot \frac{15x}{x+3}$$
$$= \frac{6}{x}$$

**25.**
$$\frac{5x-x^2}{x+2} \div \frac{x^2-6x+5}{x^2-3x+2}$$
$$= \frac{5x-x^2}{x+2} \cdot \frac{x^2-3x+2}{x^2-6x+5}$$
$$= \frac{-x(x-5)}{x+2} \cdot \frac{(x-2)(x-1)}{(x-5)(x-1)}$$
$$= \frac{-x(x-2)}{x+2}$$

**27.**
$$\frac{x^2+3x-4}{8x} \div \frac{x^2+10x+24}{x+6}$$
$$= \frac{x^2+3x-4}{8x} \cdot \frac{x+6}{x^2+10x+24}$$
$$= \frac{(x+4)(x-1)}{8x} \cdot \frac{x+6}{(x+6)(x+4)}$$
$$= \frac{x-1}{8x}$$

**29.**
$$\frac{x^2-3x-10}{x^2-25} \div \frac{x^2-4}{x^2-x-30}$$
$$= \frac{x^2-3x-10}{x^2-25} \cdot \frac{x^2-x-30}{x^2-4}$$
$$= \frac{(x-5)(x+2)}{(x-5)(x+5)} \cdot \frac{(x+5)(x-6)}{(x+2)(x-2)}$$
$$= \frac{x-6}{x-2}$$

**31.**
$$\frac{x^2-16}{2x^2-8x} \div (3x^2+10x-8)$$
$$= \frac{x^2-16}{2x^2-8x} \cdot \frac{1}{3x^2+10x-8}$$
$$= \frac{(x+4)(x-4)}{2x(x-4)} \cdot \frac{1}{(3x-2)(x+4)}$$
$$= \frac{1}{2x(3x-2)}$$

**33.**
$$\frac{?}{5} \div \frac{3x-4}{10} = 4x-2$$
$$\frac{?}{5} \cdot \frac{10}{3x-4} = 2(2x-1)$$
$$? \cdot \frac{2}{3x-4} = 2(2x-1)$$

If the ? is replaced by $(3x-4)(2x-1) = 6x^2-11x+4$, then the resulting equation is true.

**35.**
$$\frac{2x^2+4x}{x^2+2x+4} \div \frac{x^2-4}{?} = 2x$$
$$\frac{2x^2+4x}{x^2+2x+4} \cdot \frac{?}{x^2-4} = 2x$$
$$\frac{2x(x+2)}{x^2+2x+4} \cdot \frac{?}{(x+2)(x-2)} = 2x$$

If the ? is replaced by $(x-2)(x^2+2x+4) = x^3-8$, then the resulting equation is true.

**37.** $\dfrac{32x^5}{3y^4} \cdot \dfrac{5y^3}{-8x^2} = \dfrac{160x^5 y^3}{-24x^2 y^4}$

$= \dfrac{-20x^3}{3y}$

**39.** $-2x \cdot \dfrac{x+3}{4x-x^2} = -2x \cdot \dfrac{x+3}{x(4-x)}$

$= -2x \cdot \dfrac{x+3}{-x(x-4)}$

$= \dfrac{2(x+3)}{x-4}$

**41.** $\dfrac{x^2-4}{x^2-1} \div \dfrac{12x^2-24x}{4x+4}$

$= \dfrac{x^2-4}{x^2-1} \cdot \dfrac{4x+4}{12x^2-24x}$

$= \dfrac{(x+2)(x-2)}{(x+1)(x-1)} \cdot \dfrac{4(x+1)}{12x(x-2)}$

$= \dfrac{x+2}{3x(x-1)}$

**43.** $\dfrac{4x-8}{x^2-5x+6} \div \dfrac{8x+48}{x^2+3x-18}$

$= \dfrac{4x-8}{x^2-5x+6} \cdot \dfrac{x^2+3x-18}{8x+48}$

$= \dfrac{4(x-2)}{(x-3)(x-2)} \cdot \dfrac{(x+6)(x-3)}{8(x+6)}$

$= \dfrac{1}{2}$

**45.** $\dfrac{a^2-81}{a^2-5a-36} \cdot \dfrac{a^2+10a+24}{a^2-3a-54}$

$= \dfrac{(a+9)(a-9)}{(a-9)(a+4)} \cdot \dfrac{(a+6)(a+4)}{(a+6)(a-9)}$

$= \dfrac{a+9}{a-9}$

**47.** $\dfrac{x^2-3x-4}{x^2-10x+24} \div \dfrac{x^2+5x+4}{x^2-5x-6}$

$= \dfrac{x^2-3x-4}{x^2-10x+24} \cdot \dfrac{x^2-5x-6}{x^2+5x+4}$

$= \dfrac{(x-4)(x+1)}{(x-6)(x-4)} \cdot \dfrac{(x-6)(x+1)}{(x+4)(x+1)}$

$= \dfrac{x+1}{x+4}$

**49.** $\dfrac{3x^3+6x^2}{2x^2+x-6} \div (6x^2-15x)$

$= \dfrac{3x^3+6x^2}{2x^2+x-6} \cdot \dfrac{1}{6x^2-15x}$

$= \dfrac{3x^2(x+2)}{(2x-3)(x+2)} \cdot \dfrac{1}{3x(2x-5)}$

$= \dfrac{x}{(2x-3)(2x-5)}$

**51.** $\dfrac{xy+3x-2y-6}{x^2-6x+8} \cdot \dfrac{xy-4y+2x-8}{y^2+5y+6}$

$= \dfrac{x(y+3)-2(y+3)}{(x-4)(x-2)} \cdot \dfrac{y(x-4)+2(x-4)}{(y+2)(y+3)}$

$= \dfrac{(y+3)(x-2)}{(x-4)(x-2)} \cdot \dfrac{(x-4)(y+2)}{(y+2)(y+3)}$

$= 1$

**53.** $\dfrac{x^3-27}{x^2-9} \div \dfrac{x^2+3x+9}{x^2+8x+15}$

$= \dfrac{x^3-27}{x^2-9} \cdot \dfrac{x^2+8x+15}{x^2+3x+9}$

$= \dfrac{(x-3)(x^2+3x+9)}{(x-3)(x+3)} \cdot \dfrac{(x+3)(x+5)}{x^2+3x+9}$

$= x+5$

**55.** $\dfrac{6x-18}{3x-6} \cdot \dfrac{5+5x}{25x+25} \div \dfrac{12-4x}{9x-18}$

$= \dfrac{6x-18}{3x-6} \cdot \dfrac{5+5x}{25x+25} \cdot \dfrac{9x-18}{12-4x}$

$= \dfrac{6(x-3)}{3(x-2)} \cdot \dfrac{5(1+x)}{25(x+1)} \cdot \dfrac{9(x-2)}{-4(x-3)}$

$= -\dfrac{9}{10}$

**57.** $\dfrac{2x+6}{x-5} \div \left[ \dfrac{x}{4x+12} \cdot \dfrac{2(x+3)^2}{x^2-5x} \right]$

$= \dfrac{2x+6}{x-5} \div \left[ \dfrac{x}{4(x+3)} \cdot \dfrac{2(x+3)^2}{x(x-5)} \right]$

$= \dfrac{2x+6}{x-5} \div \dfrac{x+3}{2(x-5)}$

$= \dfrac{2x+6}{x-5} \cdot \dfrac{2(x-5)}{x+3}$

$= \dfrac{2(x+3)}{x-5} \cdot \dfrac{2(x-5)}{x+3}$

$= 4$

**59.** 
$$\frac{x^3-1}{x+1} \cdot \frac{x^2+2x+1}{x^2+x+1} \div \frac{x+1}{3x^2}$$
$$= \frac{x^3-1}{x+1} \cdot \frac{x^2+2x+1}{x^2+x+1} \cdot \frac{3x^2}{x+1}$$
$$= \frac{(x-1)(x^2+x+1)}{x+1} \cdot \frac{(x+1)^2}{x^2+x+1} \cdot \frac{3x^2}{x+1}$$
$$= 3x^2(x-1)$$

**61.**
$$\frac{x^2+x-2}{x^2-1} \div (x+2) \cdot \frac{x^2-9}{x^2+3x}$$
$$= \frac{x^2+x-2}{x^2-1} \cdot \frac{1}{x+2} \cdot \frac{x^2-9}{x^2+3x}$$
$$= \frac{(x+2)(x-1)}{(x+1)(x-1)} \cdot \frac{1}{x+2} \cdot \frac{(x+3)(x-3)}{x(x+3)}$$
$$= \frac{x-3}{x(x+1)}$$

**63.** (a) $\dfrac{B(x)+G(x)}{2} = \dfrac{x+2+\frac{3}{2}x+\frac{5}{3}}{2}$
$$= \frac{\frac{5}{2}x + \frac{11}{3}}{2}$$
$$= \frac{5}{4}x + \frac{11}{6}$$

(b) The graph of the three models is shown below. The graph for the combined data lies between the graphs of B and G.

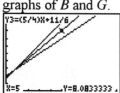

(c) By tracing the graph in part (b), we find that 7 boys, approximately 9 girls, and approximately 8 children combined of age 5 preferred baby-featured bears.

**65.** The average cost per child during the given period is given by $A(t) = \dfrac{C(t)}{N(t)}$.

**67.** 

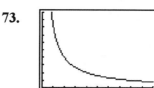

$A(8) - A(4) = 173.49 - 168 = 5.49$

**69.** To find what rational expression multiplied by $\dfrac{x+3}{x-5}$ gives $\dfrac{x^2+x-6}{x^2-2x-15}$, divide $\dfrac{x^2+x-6}{x^2-2x-15}$ by $\dfrac{x+3}{x-5}$.

$$\frac{x^2+x-6}{x^2-2x-15} \div \frac{x+3}{x-5}$$
$$= \frac{x^2+x-6}{x^2-2x-15} \cdot \frac{x-5}{x+3}$$
$$= \frac{(x+3)(x-2)}{(x+3)(x-5)} \cdot \frac{x-5}{x+3}$$
$$= \frac{x-2}{x+3}$$

**71.** To find what rational expression divided by $\dfrac{2x^2+x-15}{2x^2-3x-5}$ gives $\dfrac{2x-3}{x+3}$, multiply $\dfrac{2x^2+x-15}{2x^2-3x-5}$ by $\dfrac{2x-3}{x+3}$.

$$\frac{2x^2+x-15}{2x^2-3x-5} \cdot \frac{2x-3}{x+3}$$
$$= \frac{(2x-5)(x+3)}{(2x-5)(x+1)} \cdot \frac{2x-3}{x+3}$$
$$= \frac{2x-3}{x+1}$$

**73.**

As $x$ approaches 0, the $y$-values become larger. By selecting $x$-values closer and closer to 0, we can make the $y$-value as large as we want.

## Section 8.3 Addition and Subtraction

**1.** Add the numerators and retain the common denominator.

**3.** $\dfrac{5}{a} + \dfrac{3}{a} = \dfrac{5+3}{a} = \dfrac{8}{a}$

**5.** $\dfrac{5x-4}{x} - \dfrac{8x+5}{x} = \dfrac{5x-4-(8x+5)}{x}$
$= \dfrac{5x-4-8x-5}{x}$
$= \dfrac{-3x-9}{x}$

**7.** $\dfrac{4x-45}{x-9} + \dfrac{2x-9}{x-9} = \dfrac{4x-45+2x-9}{x-9}$
$= \dfrac{6x-54}{x-9}$
$= \dfrac{6(x-9)}{x-9} = 6$

**9.** $\dfrac{3c+51d}{c-8d} - \dfrac{9c+3d}{c-8d} = \dfrac{3c+51d-(9c+3d)}{c-8d}$
$= \dfrac{3c+51d-9c-3d}{c-8d}$
$= \dfrac{-6c+48d}{c-8d}$
$= \dfrac{-6(c-8d)}{c-8d} = -6$

**11.** $\dfrac{2x+5}{x^2-2x-8} + \dfrac{x+1}{x^2-2x-8}$
$= \dfrac{2x+5+x+1}{x^2-2x-8} = \dfrac{3x+6}{x^2-2x-8}$
$= \dfrac{3(x+2)}{(x-4)(x+2)} = \dfrac{3}{x-4}$

**13.** $\dfrac{x^3+6x^2-20x}{x(x-2)} - \dfrac{x^2-6x}{x(x-2)}$
$= \dfrac{x^3+6x^2-20x-(x^2-6x)}{x(x-2)}$
$= \dfrac{x^3+6x^2-20x-x^2+6x}{x(x-2)}$
$= \dfrac{x^3+5x^2-14x}{x(x-2)} = \dfrac{x(x^2+5x-14)}{x(x-2)}$
$= \dfrac{x(x-2)(x+7)}{x(x-2)} = x+7$

**15.** $\dfrac{x}{x^2-x-6} - \dfrac{x+1}{x^2-x-6} + \dfrac{x-2}{x^2-x-6}$
$= \dfrac{x-(x+1)+(x-2)}{x^2-x-6}$
$= \dfrac{x-x-1+x-2}{x^2-x-6} = \dfrac{x-3}{x^2-x-6}$
$= \dfrac{x-3}{(x-3)(x+2)} = \dfrac{1}{x+2}$

**17.** $\dfrac{6}{2x+3} - \left(\dfrac{x}{2x+3} - \dfrac{3-x}{2x+3}\right)$
$= \dfrac{6}{2x+3} - \left(\dfrac{x-3+x}{2x+3}\right)$
$= \dfrac{6}{2x+3} - \dfrac{2x-3}{2x+3} = \dfrac{6-(2x-3)}{2x+3}$
$= \dfrac{6-2x+3}{2x+3} = \dfrac{9-2x}{2x+3}$

**19.** $\left(\dfrac{3x+1}{x-2} - \dfrac{2x-2}{x-2}\right) \div (x+3)$
$= \left(\dfrac{3x+1-(2x-2)}{x-2}\right) \div (x+3)$
$= \left(\dfrac{3x+1-2x+2}{x-2}\right) \div (x+3)$
$= \dfrac{x+3}{x-2} \div (x+3) = \dfrac{x+3}{x-2} \cdot \dfrac{1}{x+3} = \dfrac{1}{x-2}$

**21.** $\dfrac{2x-5}{x+4} + \dfrac{?}{x+4} = \dfrac{3x+1}{x+4}$
To find the replacement for the ?, subtract $2x-5$ from $3x+1$.
$3x+1-(2x-5) = 3x+1-2x+5 = x+6$
So, if ? is replaced by $x+6$, then the resulting equation is true.

**23.** $\dfrac{?}{t^2+3} - \dfrac{3t-t^2}{t^2+3} = \dfrac{t^2-3}{t^2+3}$
To find the replacement for the ?, add $t^2-3$ to $3t-t^2$.
$3t-t^2+t^2-3 = 3t-3$
So, if ? is replaced by $3t-3$, then the resulting equation is true.

**25.** $$\frac{?}{(x+2)(x-5)} - \frac{x-x^2}{(x+2)(x-5)} = \frac{3x^2-2x}{(x+2)(x-5)}$$

To find the replacement for the ?, add $x - x^2$ to $3x^2 - 2x$.
$$3x^2 - 2x + x - x^2 = 2x^2 - x$$
So, if ? is replaced by $2x^2 - x$, then the resulting equation is true.

**27.** Multiply the numerator and the denominator of the second fraction by $-1$.

**29.** $\dfrac{a}{3} - \dfrac{5}{-3} = \dfrac{a}{3} + \dfrac{?}{3} = \dfrac{?}{3}$

For the result to be a true equation, replace the first ? with 5 and the second ? with $a + 5$.

**31.** $\dfrac{3x+9}{x-2} - \dfrac{8x}{2-x} = \dfrac{3x+9}{x-2} + \dfrac{?}{x-2} = \dfrac{?}{x-2}$

For the result to be a true equation, replace the first ? with $8x$ and the second ? with $11x + 9$.

**33.** $\dfrac{2x+9}{7x-8} - \dfrac{7x-6}{8-7x} = \dfrac{2x+9}{7x-8} + \dfrac{7x-6}{7x-8}$
$= \dfrac{2x+9+7x-6}{7x-8}$
$= \dfrac{9x+3}{7x-8}$

**35.** $\dfrac{x^2}{x-4} + \dfrac{16}{4-x} = \dfrac{x^2}{x-4} - \dfrac{16}{x-4}$
$= \dfrac{x^2-16}{x-4}$
$= \dfrac{(x+4)(x-4)}{x-4} = x+4$

**37.** $\dfrac{a^2}{a-b} - \dfrac{b^2}{b-a} = \dfrac{a^2}{a-b} + \dfrac{b^2}{a-b}$
$= \dfrac{a^2+b^2}{a-b}$

**39.** $5x^2y$, $10xy^2$, $15x^3y$
$5x^2y = 5 \cdot x \cdot x \cdot y$
$10xy^2 = 2 \cdot 5 \cdot x \cdot y \cdot y$
$15x^3y = 3 \cdot 5 \cdot x \cdot x \cdot x \cdot y$
LCM $= 2 \cdot 3 \cdot 5 \cdot x \cdot x \cdot x \cdot y \cdot y = 30x^3y^2$

**41.** $6x$, $3x - 12$
$6x = 2 \cdot 3 \cdot x$
$3x - 12 = 3 \cdot (x-4)$
LCM $= 2 \cdot 3 \cdot x \cdot (x-4) = 6x(x-4)$

**43.** $t+5$, $-3t-15$, $10t$
$t+5 = t+5$
$-3t-15 = -3(t+5)$
$10t = 2 \cdot 5 \cdot t$
LCM $= 2 \cdot 5 \cdot t \cdot (-3)(t+5) = -30t(t+5)$

**45.** $3x^2 + x - 2$, $4x^2 + 5x + 1$
$3x^2 + x - 2 = (3x-2)(x+1)$
$4x^2 + 5x + 1 = (4x+1)(x+1)$
LCM $= (3x-2)(x+1)(4x+1)$

**47.** $x^2 + 3x + 2$, $x^2 - 4$, $3x + 6$
$x^2 + 3x + 2 = (x+2)(x+1)$
$x^2 - 4 = (x+2)(x-2)$
$3x + 6 = 3(x+2)$
LCM $= 3(x+2)(x+1)(x-2)$

**49.** Multiply the numerator and the denominator by $x$.

**51.** The LCD is $a^3b^2$.
$\dfrac{5}{a^3b} + \dfrac{7}{ab^2} = \dfrac{5 \cdot b}{a^3b \cdot b} + \dfrac{7 \cdot a^2}{ab^2 \cdot a^2}$
$= \dfrac{5b}{a^3b^2} + \dfrac{7a^2}{a^3b^2} = \dfrac{5b+7a^2}{a^3b^2}$

**53.** The LCD is $x$.
$x - \dfrac{3}{x} = \dfrac{x^2}{x} - \dfrac{3}{x} = \dfrac{x^2-3}{x}$

Section 8.3   Addition and Subtraction

**55.** The LCD is $x(x-4)$.

$$\frac{3}{x}+\frac{8}{x-4}=\frac{3(x-4)}{x(x-4)}+\frac{8x}{x(x-4)}$$
$$=\frac{3(x-4)+8x}{x(x-4)}$$
$$=\frac{3x-12+8x}{x(x-4)}=\frac{11x-12}{x(x-4)}$$

**57.** The LCD is $(x-5)(x+7)$.

$$\frac{1}{x-5}-\frac{1}{x+7}$$
$$=\frac{x+7}{(x-5)(x+7)}-\frac{x-5}{(x-5)(x+7)}$$
$$=\frac{x+7-(x-5)}{(x-5)(x+7)}=\frac{x+7-x+5}{(x-5)(x+7)}$$
$$=\frac{12}{(x-5)(x+7)}$$

**59.** The LCD is $t+1$.

$$\frac{2}{t+1}+t+1=\frac{2}{t+1}+\frac{(t+1)(t+1)}{t+1}$$
$$=\frac{2}{t+1}+\frac{t^2+2t+1}{t+1}$$
$$=\frac{2+t^2+2t+1}{t+1}$$
$$=\frac{t^2+2t+3}{t+1}$$

**61.** The LCD is $18(t-5)$.

$$\frac{3}{6(t-5)}-\frac{2-t}{9(t-5)}$$
$$=\frac{3\cdot 3}{3\cdot 6(t-5)}-\frac{2(2-t)}{2\cdot 9(t-5)}$$
$$=\frac{9}{18(t-5)}-\frac{4-2t}{18(t-5)}$$
$$=\frac{9-(4-2t)}{18(t-5)}=\frac{9-4+2t}{18(t-5)}$$
$$=\frac{2t+5}{18(t-5)}$$

**63.** The LCD is $t(t+1)^2$.

$$\frac{5t+2}{t(t+1)^2}-\frac{3}{t(t+1)}$$
$$=\frac{5t+2}{t(t+1)^2}-\frac{3(t+1)}{t(t+1)^2}=\frac{5t+2-3(t+1)}{t(t+1)^2}$$
$$=\frac{5t+2-3t-3}{t(t+1)^2}=\frac{2t-1}{t(t+1)^2}$$

**65.** 
$$\frac{2}{3t^2}+\frac{3}{2t^2+t-1}$$
$$=\frac{2}{3t^2}+\frac{3}{(2t-1)(t+1)}$$
$$=\frac{2(2t-1)(t+1)}{3t^2(2t-1)(t+1)}+\frac{3\cdot 3t^2}{3t^2(2t-1)(t+1)}$$
$$=\frac{2(2t-1)(t+1)+3\cdot 3t^2}{3t^2(2t-1)(t+1)}$$
$$=\frac{2(2t^2+t-1)+9t^2}{3t^2(2t-1)(t+1)}$$
$$=\frac{4t^2+2t-2+9t^2}{3t^2(2t-1)(t+1)}$$
$$=\frac{13t^2+2t-2}{3t^2(2t-1)(t+1)}$$

**67.** 
$$\frac{2x^2-7}{10x^2-3x-4}-\frac{x-3}{5x-4}$$
$$=\frac{2x^2-7}{(5x-4)(2x+1)}-\frac{x-3}{5x-4}$$
$$=\frac{2x^2-7}{(5x-4)(2x+1)}-\frac{(x-3)(2x+1)}{(5x-4)(2x+1)}$$
$$=\frac{2x^2-7}{(5x-4)(2x+1)}-\frac{2x^2-5x-3}{(5x-4)(2x+1)}$$
$$=\frac{2x^2-7-(2x^2-5x-3)}{(5x-4)(2x+1)}$$
$$=\frac{2x^2-7-2x^2+5x+3}{(5x-4)(2x+1)}$$
$$=\frac{5x-4}{(5x-4)(2x+1)}$$
$$=\frac{1}{2x+1}$$

**69.** 
$$\frac{x+9}{4x-36} - \frac{x-9}{x^2-18x+81}$$
$$= \frac{x+9}{4(x-9)} - \frac{x-9}{(x-9)^2}$$
$$= \frac{x+9}{4(x-9)} - \frac{1}{x-9}$$
$$= \frac{x+9}{4(x-9)} - \frac{4}{4(x-9)}$$
$$= \frac{x+9-4}{4(x-9)} = \frac{x+5}{4(x-9)}$$

**71.**
$$\frac{7}{x^2+15x+56} + \frac{1}{x^2-64}$$
$$= \frac{7}{(x+7)(x+8)} + \frac{1}{(x-8)(x+8)}$$
$$= \frac{7(x-8)}{(x+7)(x+8)(x-8)}$$
$$+ \frac{(x+7)}{(x+7)(x+8)(x-8)}$$
$$= \frac{7x-56+x+7}{(x+7)(x+8)(x-8)}$$
$$= \frac{8x-49}{(x+7)(x+8)(x-8)}$$

**73.**
$$\frac{x}{x+6} - \frac{72}{x^2-36}$$
$$= \frac{x}{x+6} - \frac{72}{(x+6)(x-6)}$$
$$= \frac{x(x-6)}{(x+6)(x-6)} - \frac{72}{(x+6)(x-6)}$$
$$= \frac{x(x-6)-72}{(x+6)(x-6)} = \frac{x^2-6x-72}{(x+6)(x-6)}$$
$$= \frac{(x+6)(x-12)}{(x+6)(x-6)} = \frac{x-12}{x-6}$$

**75.**
$$\frac{x}{x^2+4x+3} - \frac{2}{x^2-2x-3}$$
$$= \frac{x}{(x+3)(x+1)} - \frac{2}{(x+1)(x-3)}$$
$$= \frac{x(x-3)}{(x+3)(x+1)(x-3)}$$
$$- \frac{2(x+3)}{(x+3)(x+1)(x-3)}$$
$$= \frac{x(x-3)-2(x+3)}{(x+3)(x+1)(x-3)}$$
$$= \frac{x^2-3x-2x-6}{(x+3)(x+1)(x-3)}$$
$$= \frac{x^2-5x-6}{(x+3)(x+1)(x-3)}$$
$$= \frac{(x-6)(x+1)}{(x+3)(x+1)(x-3)} = \frac{x-6}{(x+3)(x-3)}$$

**77.**
$$\frac{x}{x^2+8x+15} + \frac{15}{x^2+4x-5}$$
$$= \frac{x}{(x+3)(x+5)} + \frac{15}{(x+5)(x-1)}$$
$$= \frac{x(x-1)}{(x+3)(x+5)(x-1)}$$
$$+ \frac{15(x+3)}{(x+3)(x+5)(x-1)}$$
$$= \frac{x(x-1)+15(x+3)}{(x+3)(x+5)(x-1)}$$
$$= \frac{x^2-x+15x+45}{(x+3)(x+5)(x-1)}$$
$$= \frac{x^2+14x+45}{(x+3)(x+5)(x-1)}$$
$$= \frac{(x+9)(x+5)}{(x+3)(x+5)(x-1)} = \frac{x+9}{(x+3)(x-1)}$$

**79.** $(x+y)^{-1} = \dfrac{1}{x+y}$ and

$$x^{-1} + y^{-1} = \frac{1}{x} + \frac{1}{y} = \frac{y}{xy} + \frac{x}{xy} = \frac{y+x}{xy}$$

The expressions are not equivalent.

**81.** $x^{-1} - y^{-1} = \dfrac{1}{x} - \dfrac{1}{y} = \dfrac{y}{xy} - \dfrac{x}{xy} = \dfrac{y-x}{xy}$

Section 8.3 Addition and Subtraction 271

**83.** $x(x+3)^{-1} + 3(x-2)^{-1}$

$= \dfrac{x}{x+3} + \dfrac{3}{x-2}$

$= \dfrac{x(x-2)}{(x+3)(x-2)} + \dfrac{3(x+3)}{(x+3)(x-2)}$

$= \dfrac{x^2 - 2x}{(x+3)(x-2)} + \dfrac{3x+9}{(x+3)(x-2)}$

$= \dfrac{x^2 - 2x + 3x + 9}{(x+3)(x-2)} = \dfrac{x^2 + x + 9}{(x+3)(x-2)}$

**85.** (a) The graph is shown below. It suggests that increasing the hours of training will reduce errors, but errors can never be reduced to zero.

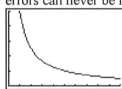

(b) $E_1(x) = \dfrac{5}{x} - \dfrac{5}{x+1}$

$= \dfrac{5(x+1)}{x(x+1)} - \dfrac{5x}{x(x+1)}$

$= \dfrac{5x+5}{x(x+1)} - \dfrac{5x}{x(x+1)} = \dfrac{5x+5-5x}{x(x+1)}$

$= \dfrac{5}{x(x+1)}$

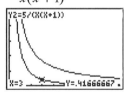

The graph suggests that the incentive plan will reduce errors even further.

**87.** To find the variable costs associated with driving 10,000 miles per year, we must multiply the variable cost function (which is given in cents per mile) by 10,000 miles and divide by 100 cents to have a function in terms of dollars.

$V_1(t) = 10,000 \left( \dfrac{28t + 157}{t + 25} \right) \div 100$

$= \dfrac{10,000}{100} \cdot \dfrac{28t + 157}{t + 25}$

$= \dfrac{100}{1} \cdot \dfrac{28t + 157}{t + 25}$

$= \dfrac{100(28t + 157)}{t + 25}$

**89.** $100 \left[ \dfrac{28t + 157}{t + 25} + \dfrac{10(9t + 73)}{t + 25} \right]$

$= 100 \left[ \dfrac{28t + 157}{t + 25} + \dfrac{90t + 730}{t + 25} \right]$

$= 100 \left[ \dfrac{28t + 157 + 90t + 730}{t + 25} \right]$

$= 100 \left[ \dfrac{118t + 887}{t + 25} \right]$

The expression represents the total annual fixed and variable costs (in dollars) for 10,000 miles of driving.

**91.** $\left( \dfrac{1}{x+h} - \dfrac{1}{x} \right) \div h$

$= \left( \dfrac{x}{x(x+h)} - \dfrac{(x+h)}{x(x+h)} \right) \div h$

$= \dfrac{x - (x+h)}{x(x+h)} \div h = \dfrac{x - x - h}{x(x+h)} \div h$

$= \dfrac{-h}{x(x+h)} \cdot \dfrac{1}{h} = \dfrac{-1}{x(x+h)}$

**93.** $\dfrac{\dfrac{1}{x} - \dfrac{2x}{x^2 + 1}}{\dfrac{1}{x}}$

$= \left( \dfrac{1}{x} - \dfrac{2x}{x^2 + 1} \right) \div \dfrac{1}{x}$

$= \left( \dfrac{x^2 + 1}{x(x^2 + 1)} - \dfrac{2x^2}{x(x^2 + 1)} \right) \div \dfrac{1}{x}$

$= \dfrac{x^2 + 1 - 2x^2}{x(x^2 + 1)} \div \dfrac{1}{x}$

$= \dfrac{1 - x^2}{x(x^2 + 1)} \cdot \dfrac{x}{1} = \dfrac{1 - x^2}{x^2 + 1}$

## Section 8.4 Complex Fractions

1. The expressions do not mean the same thing. The first expression means 2/3 divided by 4, and its value is 1/6. The second expression means 2 divided by 3/4, and its value is 8/3.

3. $\dfrac{\dfrac{3}{x}}{\dfrac{9}{x^3}} = \dfrac{3}{x} \div \dfrac{9}{x^3} = \dfrac{3}{x} \cdot \dfrac{x^3}{9} = \dfrac{x^2}{3}$

5. $\dfrac{\dfrac{2x}{x+3}}{\dfrac{6x^2}{x+3}} = \dfrac{2x}{x+3} \div \dfrac{6x^2}{x+3} = \dfrac{2x}{x+3} \cdot \dfrac{x+3}{6x^2} = \dfrac{1}{3x}$

7. $\dfrac{\dfrac{5}{x+5}}{\dfrac{10}{x^2-25}} = \dfrac{5}{x+5} \cdot \dfrac{x^2-25}{10}$
$= \dfrac{5}{x+5} \cdot \dfrac{(x+5)(x-5)}{10} = \dfrac{x-5}{2}$

9. $\dfrac{\dfrac{4x^2}{6x+54}}{\dfrac{32x^3}{x^2-81}} = \dfrac{4x^2}{6x+54} \cdot \dfrac{x^2-81}{32x^3}$
$= \dfrac{4x^2}{6(x+9)} \cdot \dfrac{(x-9)(x+9)}{32x^3}$
$= \dfrac{x-9}{48x}$

11. $\dfrac{\dfrac{8x}{x+5}}{\dfrac{16x^2}{x^2+x-20}} = \dfrac{8x}{x+5} \cdot \dfrac{x^2+x-20}{16x^2}$
$= \dfrac{8x}{x+5} \cdot \dfrac{(x+5)(x-4)}{16x^2}$
$= \dfrac{x-4}{2x}$

13. $\dfrac{\dfrac{5x}{x^2-4}}{\dfrac{35x^2}{x^2-x-6}}$
$= \dfrac{5x}{x^2-4} \cdot \dfrac{x^2-x-6}{35x^2}$
$= \dfrac{5x}{(x+2)(x-2)} \cdot \dfrac{(x-3)(x+2)}{35x^2}$
$= \dfrac{x-3}{7x(x-2)}$

15. $\dfrac{\dfrac{2x^2+3x-2}{x^2-x-6}}{\dfrac{1-x-2x^2}{x^2+3x+2}}$
$= \dfrac{2x^2+3x-2}{x^2-x-6} \cdot \dfrac{x^2+3x+2}{1-x-2x^2}$
$= \dfrac{(2x-1)(x+2)}{(x-3)(x+2)} \cdot \dfrac{(x+2)(x+1)}{(1-2x)(1+x)}$
$= \dfrac{(2x-1)(x+2)}{(x-3)(x+2)} \cdot \dfrac{(x+2)(x+1)}{-(2x-1)(1+x)}$
$= -\dfrac{x+2}{x-3}$

17. Add or subtract the fractions in the numerator and the denominator. Then multiply the numerator by the reciprocal of the denominator.

19. $\dfrac{\dfrac{2}{3}+\dfrac{4}{t}}{\dfrac{1}{t}+\dfrac{1}{2}} = \dfrac{\dfrac{2t}{3t}+\dfrac{12}{3t}}{\dfrac{2}{2t}+\dfrac{t}{2t}} = \dfrac{\dfrac{2t+12}{3t}}{\dfrac{t+2}{2t}}$
$= \dfrac{2t+12}{3t} \cdot \dfrac{2t}{t+2} = \dfrac{2(t+6)}{3t} \cdot \dfrac{2t}{t+2}$
$= \dfrac{4(t+6)}{3(t+2)}$

21. $\dfrac{\dfrac{2}{y^2}+\dfrac{1}{y}}{\dfrac{2}{y}+1} = \dfrac{\dfrac{2}{y^2}+\dfrac{y}{y^2}}{\dfrac{2}{y}+\dfrac{y}{y}} = \dfrac{\dfrac{2+y}{y^2}}{\dfrac{2+y}{y}}$
$= \dfrac{2+y}{y^2} \cdot \dfrac{y}{2+y} = \dfrac{1}{y}$

Section 8.4 Complex Fractions

**23.** 
$$\frac{4 - \frac{1}{x^2}}{2 + \frac{1}{x}} = \frac{\frac{4x^2}{x^2} - \frac{1}{x^2}}{\frac{2x}{x} + \frac{1}{x}} = \frac{\frac{4x^2 - 1}{x^2}}{\frac{2x + 1}{x}}$$
$$= \frac{4x^2 - 1}{x^2} \cdot \frac{x}{2x + 1}$$
$$= \frac{(2x+1)(2x-1)}{x^2} \cdot \frac{x}{2x+1}$$
$$= \frac{2x - 1}{x}$$

**25.** 
$$\frac{\frac{1}{c} - \frac{1}{cd}}{\frac{1}{cd} - \frac{1}{c}} = \frac{\frac{d}{cd} - \frac{1}{cd}}{\frac{1}{cd} - \frac{d}{cd}} = \frac{\frac{d-1}{cd}}{\frac{1-d}{cd}}$$
$$= \frac{d-1}{cd} \cdot \frac{cd}{1-d} = \frac{d-1}{1-d}$$
$$= \frac{d-1}{-(d-1)} = -1$$

**27.** 
$$\frac{\frac{5}{x^4} - \frac{2}{x^3}}{\frac{5}{x^3} - \frac{2}{x^2}} = \frac{\frac{5}{x^4} - \frac{2x}{x^4}}{\frac{5}{x^3} - \frac{2x}{x^3}} = \frac{\frac{5-2x}{x^4}}{\frac{5-2x}{x^3}}$$
$$= \frac{5 - 2x}{x^4} \cdot \frac{x^3}{5 - 2x} = \frac{1}{x}$$

**29.** 
$$\frac{x+y}{\frac{1}{x} + \frac{1}{y}} = \frac{x+y}{\frac{y}{xy} + \frac{x}{xy}} = \frac{x+y}{\frac{x+y}{xy}}$$
$$= \frac{x+y}{1} \cdot \frac{xy}{x+y} = xy$$

**31.** Multiply the numerator and the denominator by $(x+3)(x-3)$.

$$\frac{\frac{2}{x+3} - \frac{3}{x-3}}{2 + \frac{1}{x^2 - 9}} = \frac{\frac{2}{x+3} - \frac{3}{x-3}}{2 + \frac{1}{(x+3)(x-3)}}$$
$$= \frac{(x+3)(x-3)\left(\frac{2}{x+3} - \frac{3}{x-3}\right)}{(x+3)(x-3)\left(2 + \frac{1}{(x+3)(x-3)}\right)}$$
$$= \frac{2(x-3) - 3(x+3)}{2(x+3)(x-3) + 1}$$
$$= \frac{2x - 6 - 3x - 9}{2x^2 - 18 + 1} = \frac{-x - 15}{2x^2 - 17}$$

**33.** Multiply the numerator and the denominator by $2x+1$.

$$\frac{\frac{3}{2x+1} - (x+1)}{(x-3) - \frac{2}{2x+1}}$$
$$= \frac{(2x+1)\left(\frac{3}{2x+1} - (x+1)\right)}{(2x+1)\left((x-3) - \frac{2}{2x+1}\right)}$$
$$= \frac{3 - (2x+1)(x+1)}{(2x+1)(x-3) - 2} = \frac{3 - (2x^2 + 3x + 1)}{(2x^2 - 5x - 3) - 2}$$
$$= \frac{-2x^2 - 3x + 2}{2x^2 - 5x - 5} = -\frac{2x^2 + 3x - 2}{2x^2 - 5x - 5}$$

**35.** Multiply the numerator and the denominator by $(1-y)(1+y)$.

$$\dfrac{\dfrac{1}{1-y} - \dfrac{1}{1+y}}{\dfrac{1+y}{1-y} - \dfrac{1-y}{1+y}}$$

$$= \dfrac{(1-y)(1+y)\left(\dfrac{1}{1-y} - \dfrac{1}{1+y}\right)}{(1-y)(1+y)\left(\dfrac{1+y}{1-y} - \dfrac{1-y}{1+y}\right)}$$

$$= \dfrac{(1+y) - (1-y)}{(1+y)^2 - (1-y)^2}$$

$$= \dfrac{1+y-1+y}{(1+2y+y^2) - (1-2y+y^2)}$$

$$= \dfrac{2y}{1+2y+y^2 - 1+2y-y^2}$$

$$= \dfrac{2y}{4y} = \dfrac{1}{2}$$

**37.** Multiply the numerator and the denominator by $x^2$.

$$\dfrac{3 + \dfrac{1}{x} - \dfrac{14}{x^2}}{6 + \dfrac{11}{x} - \dfrac{7}{x^2}} = \dfrac{x^2\left(3 + \dfrac{1}{x} - \dfrac{14}{x^2}\right)}{x^2\left(6 + \dfrac{11}{x} - \dfrac{7}{x^2}\right)}$$

$$= \dfrac{3x^2 + x - 14}{6x^2 + 11x - 7}$$

$$= \dfrac{(3x+7)(x-2)}{(3x+7)(2x-1)} = \dfrac{x-2}{2x-1}$$

**39.** Multiply the numerator and the denominator by $xy$.

$$\dfrac{5 + \dfrac{2x}{y} + \dfrac{2y}{x}}{\dfrac{2x}{y} - \dfrac{2y}{x} - 3} = \dfrac{xy\left(5 + \dfrac{2x}{y} + \dfrac{2y}{x}\right)}{xy\left(\dfrac{2x}{y} - \dfrac{2y}{x} - 3\right)}$$

$$= \dfrac{5xy + 2x^2 + 2y^2}{2x^2 - 2y^2 - 3xy} = \dfrac{2x^2 + 5xy + 2y^2}{2x^2 - 3xy - 2y^2}$$

$$= \dfrac{(2x+y)(x+2y)}{(2x+y)(x-2y)} = \dfrac{x+2y}{x-2y}$$

**41.** Multiply the numerator and the denominator by $x^2$.

$$\dfrac{x + \dfrac{8}{x^2}}{1 - \dfrac{2}{x} + \dfrac{4}{x^2}} = \dfrac{x^2\left(x + \dfrac{8}{x^2}\right)}{x^2\left(1 - \dfrac{2}{x} + \dfrac{4}{x^2}\right)}$$

$$= \dfrac{x^3 + 8}{x^2 - 2x + 4}$$

$$= \dfrac{(x+2)(x^2 - 2x + 4)}{x^2 - 2x + 4}$$

$$= x + 2$$

**43.** To determine the domain, exclude values for which $Q$, $R$, or $S$ is 0.

**45.** Multiply the numerator and the denominator by $2y + x$.

$$\dfrac{\dfrac{4y^2}{2y+x} + (3x-y)}{\dfrac{5x^2}{2y+x} - (3y+2x)}$$

$$= \dfrac{(2y+x)\left(\dfrac{4y^2}{2y+x} + (3x-y)\right)}{(2y+x)\left(\dfrac{5x^2}{2y+x} - (3y+2x)\right)}$$

$$= \dfrac{4y^2 + (2y+x)(3x-y)}{5x^2 - (2y+x)(3y+2x)}$$

$$= \dfrac{4y^2 - 2y^2 + 5xy + 3x^2}{5x^2 - (6y^2 + 7xy + 2x^2)}$$

$$= \dfrac{3x^2 + 5xy + 2y^2}{5x^2 - 6y^2 - 7xy - 2x^2}$$

$$= \dfrac{3x^2 + 5xy + 2y^2}{3x^2 - 7xy - 6y^2}$$

$$= \dfrac{(3x+2y)(x+y)}{(3x+2y)(x-3y)} = \dfrac{x+y}{x-3y}$$

Section 8.4  Complex Fractions

**47.** Multiply the numerator and the denominator by $x(x+h)$.

$$\frac{\frac{1}{x+h}-\frac{1}{x}}{h} = \frac{x(x+h)\left(\frac{1}{x+h}-\frac{1}{x}\right)}{x(x+h)h}$$
$$= \frac{x-(x+h)}{hx(x+h)} = \frac{x-x-h}{hx(x+h)}$$
$$= \frac{-h}{hx(x+h)} = \frac{-1}{x(x+h)}$$

**49.** Multiply the numerator and the denominator by $a-b$.

$$\frac{a+b}{1+\frac{2b}{a-b}} = \frac{(a-b)(a+b)}{(a-b)\left(1+\frac{2b}{a-b}\right)}$$
$$= \frac{(a-b)(a+b)}{a-b+2b} = \frac{(a-b)(a+b)}{a+b} = a-b$$

**51.** Multiply the numerator and the denominator by $x-2$.

$$\frac{x+3+\frac{4}{x-2}}{x^2+3x+2} = \frac{(x-2)\left(x+3+\frac{4}{x-2}\right)}{(x-2)(x^2+3x+2)}$$
$$= \frac{x(x-2)+3(x-2)+4}{(x-2)(x^2+3x+2)}$$
$$= \frac{x^2-2x+3x-6+4}{(x-2)(x^2+3x+2)}$$
$$= \frac{x^2+x-2}{(x-2)(x^2+3x+2)}$$
$$= \frac{(x+2)(x-1)}{(x-2)(x+2)(x+1)} = \frac{x-1}{(x-2)(x+1)}$$

**53.** Multiply the numerator and the denominator by $a^3b^3$.

$$\frac{\frac{1}{a^3}+\frac{1}{b^3}}{b^2-ab+a^2} = \frac{a^3b^3\left(\frac{1}{a^3}+\frac{1}{b^3}\right)}{a^3b^3(b^2-ab+a^2)}$$
$$= \frac{b^3+a^3}{a^3b^3(b^2-ab+a^2)}$$
$$= \frac{(b+a)(b^2-ab+a^2)}{a^3b^3(b^2-ab+a^2)}$$
$$= \frac{a+b}{a^3b^3}$$

**55.** Multiplying the given expression by the LCD would change the value of the expression.

**57.** Using the method suggested in Exercise 56(a), multiply the numerator and denominator by $ab$.

$$\frac{a^{-1}b^{-1}}{a^{-1}+b^{-1}} = \frac{ab(a^{-1}b^{-1})}{ab(a^{-1}+b^{-1})}$$
$$= \frac{1}{b+a} = \frac{1}{a+b}$$

**59.** Using the method suggested in Exercise 56(a), multiply the numerator and denominator by $x^2$.

$$\frac{x^{-1}+x}{x^{-2}+x^2} = \frac{x^2(x^{-1}+x)}{x^2(x^{-2}+x^2)}$$
$$= \frac{x+x^3}{1+x^4} = \frac{x^3+x}{x^4+1}$$

**61.** Using the method suggested in Exercise 56(a), multiply the numerator and denominator by $x^2$.

$$\frac{1-x^{-2}}{3-x^{-1}-4x^{-2}} = \frac{x^2(1-x^{-2})}{x^2(3-x^{-1}-4x^{-2})}$$
$$= \frac{x^2-1}{3x^2-x-4} = \frac{(x+1)(x-1)}{(3x-4)(x+1)} = \frac{x-1}{3x-4}$$

**63.**

$$\frac{xy^{-3}+x^{-2}y}{x-y} = \frac{\frac{x}{y^3}+\frac{y}{x^2}}{x-y}$$
$$= \frac{\frac{x^3}{x^2y^3}+\frac{y^4}{x^2y^3}}{x-y} = \frac{\frac{x^3+y^4}{x^2y^3}}{x-y}$$
$$= \frac{x^3+y^4}{x^2y^3}\cdot\frac{1}{x-y} = \frac{x^3+y^4}{x^2y^3(x-y)}$$

**65.** The reciprocal of $x+\dfrac{5}{x+1}$ is

$\dfrac{1}{x+\dfrac{5}{x+1}}$. This expression is simplified as follows.

$$\frac{1}{x+\dfrac{5}{x+1}} = \frac{(x+1)\cdot 1}{(x+1)\left(x+\dfrac{5}{x+1}\right)}$$

$$= \frac{x+1}{x(x+1)+5} = \frac{x+1}{x^2+x+5}$$

**67.** (a) Let $x =$ the first integer, $x + 1 =$ the second integer, and $x + 2 =$ the third integer. The ratio of the first integer to the second integer is $\dfrac{x}{x+1}$. The ratio of the second integer to the third integer is $\dfrac{x+1}{x+2}$. The described complex fraction is

$$\frac{\dfrac{x}{x+1}}{\dfrac{x+1}{x+2}}.$$

(b) $\dfrac{\dfrac{x}{x+1}}{\dfrac{x+1}{x+2}} = \dfrac{x}{x+1} \cdot \dfrac{x+2}{x+1} = \dfrac{x^2+2x}{x^2+2x+1}$

Thus, the denominator of the result is always one more than the numerator of the result.

**69.** (a) The number of accepted male students who actually enrolled is $x - m$. The ratio of accepted male students who actually enrolled is $\dfrac{x-m}{x}$. Similarly the number of accepted female students who actually enrolled is $y - n$. Thus, the ratio we seek is $\dfrac{y-n}{y}$.

(b) The ratio of all accepted students who actually enrolled is $\dfrac{(x-m)+(y-n)}{x+y}$. This is not the sum of the fractions in part (a).

(c) $\dfrac{\dfrac{x-m}{x}}{\dfrac{y-n}{y}} = \dfrac{y(x-m)}{x(y-n)}$. This is a comparison of the ratio of accepted males who enrolled with the ratio of accepted females who enrolled.

**71.** If the length of the table is $x + 3$, then the width of the table is $(x + 3) - 2 = x + 1$.

The area of the red rectangle is:
$$(x+1)\left(\frac{1}{x^2+1}\right) = \frac{x+1}{x^2+1}$$

The area of the table is:
$(x+3)(x+1)$

The probability $P$ of the dime landing on the red rectangle is given by:

$$P = \frac{\dfrac{x+1}{x^2+1}}{(x+3)(x+1)} = \frac{x+1}{x^2+1} \cdot \frac{1}{(x+3)(x+1)}$$

$$= \frac{1}{(x^2+1)(x+3)}$$

**73.** The number of registered voters is found by multiplying the percent of registered voters (as a decimal) by the entire voting age population. Notice that the function for percent registered must be divided by 100 to obtain the percent as a decimal.

$$p(t) \cdot \frac{r(t)}{100}$$

$$= (-0.004t^2 + 0.27t - 0.48) \cdot \frac{65t}{100(0.7t+10)}$$

$$= (-0.004t^2 + 0.27t - 0.48) \cdot \frac{0.65t}{0.7t+10}$$

$$= \frac{-0.0026t^3 + 0.1755t^2 - 0.312t}{0.7t+10}$$

**75.** $\dfrac{650 \cdot S(t)}{A(t)}$

$$= \frac{\dfrac{650(34t^3 - 390t^2 + 2857t + 7)}{t+3}}{\dfrac{1000(102t-75)}{t+3}}$$

$$= \frac{650(34t^3 - 390t^2 + 2857t + 7)}{t+3} \cdot \frac{t+3}{1000(102t-75)}$$

$$= \frac{650(34t^3 - 390t^2 + 2857t + 7)}{1000(102t-75)}$$

$$= \frac{13(34t^3 - 390t^2 + 2857t + 7)}{20(102t-75)}$$

Section 8.5  Equations with Rational Expressions        277

**77.**

| X | Y1 |
|---|---|
| 15 | 31.21 |
| 16 | 35.545 |
| 17 | 40.32 |
| 18 | 45.534 |
| 19 | 51.188 |
| 20 | 57.275 |
| 21 | 63.801 |

Y1=51.185668277

From the table, we see that the price of a ticket first exceeds $50 in 2009 (when $x = 19$).

**79.** $f(x) = \dfrac{1}{2x-1}$

$f(x+h) = \dfrac{1}{2(x+h)-1} = \dfrac{1}{2x+2h-1}$

$\dfrac{f(x+h) - f(x)}{h} = \dfrac{\dfrac{1}{2x+2h-1} - \dfrac{1}{2x-1}}{h}$

$= \dfrac{(2x+2h-1)(2x-1)\left(\dfrac{1}{2x+2h-1} - \dfrac{1}{2x-1}\right)}{(2x+2h-1)(2x-1)h}$

$= \dfrac{2x-1-(2x+2h-1)}{(2x+2h-1)(2x-1)h}$

$= \dfrac{2x-1-2x-2h+1}{(2x+2h-1)(2x-1)h}$

$= \dfrac{-2h}{(2x+2h-1)(2x-1)h}$

$= \dfrac{-2}{(2x+2h-1)(2x-1)}$

**81.** $f(x) = \dfrac{x+1}{x-2}$

$f(x+h) = \dfrac{(x+h)+1}{(x+h)-2} = \dfrac{x+h+1}{x+h-2}$

$\dfrac{f(x+h) - f(x)}{h} = \dfrac{\dfrac{x+h+1}{x+h-2} - \dfrac{x+1}{x-2}}{h}$

$= \dfrac{(x+h-2)(x-2)\left(\dfrac{x+h+1}{x+h-2} - \dfrac{x+1}{x-2}\right)}{(x+h-2)(x-2)h}$

$= \dfrac{(x-2)(x+h+1) - (x+h-2)(x+1)}{(x+h-2)(x-2)h}$

$= \dfrac{x^2 + hx + x - 2x - 2h - 2 - (x^2 + hx - 2x + x + h - 2)}{(x+h-2)(x-2)h}$

$= \dfrac{x^2 + hx + x - 2x - 2h - 2 - x^2 - hx + 2x - x - h + 2}{(x+h-2)(x-2)h}$

$= \dfrac{-3h}{(x+h-2)(x-2)h} = \dfrac{-3}{(x+h-2)(x-2)}$

**83.** $2x - \dfrac{5 + \dfrac{3}{x}}{7 - \dfrac{2}{x}} = 2x - \dfrac{x \cdot \left(5 + \dfrac{3}{x}\right)}{x \cdot \left(7 - \dfrac{2}{x}\right)}$

$= 2x - \dfrac{5x+3}{7x-2} = \dfrac{2x(7x-2)}{7x-2} - \dfrac{5x+3}{7x-2}$

$= \dfrac{2x(7x-2) - (5x+3)}{7x-2}$

$= \dfrac{14x^2 - 4x - 5x - 3}{7x-2} = \dfrac{14x^2 - 9x - 3}{7x-2}$

**85.** $1 + \dfrac{1}{1 + \dfrac{1}{1 + \dfrac{1}{2}}} = 1 + \dfrac{1}{1 + \dfrac{1}{\dfrac{3}{2}}}$

$= 1 + \dfrac{1}{1 + \dfrac{2}{3}} = 1 + \dfrac{1}{\dfrac{5}{3}} = 1 + \dfrac{3}{5} = \dfrac{8}{5}$

## Section 8.5  Equations with Rational Expressions

**1.** We call $-3$ and $0$ restricted values because each one makes a denominator equal to $0$.

**3.** $\dfrac{x}{x+3} - \dfrac{2}{x-4} = \dfrac{1}{x}$

The restricted values are $-3$, $4$, and $0$.

**5.** $\dfrac{y}{y^2 - 9} + \dfrac{7}{y} = \dfrac{3y+2}{y+3}$

$\dfrac{y}{(y+3)(y-3)} + \dfrac{7}{y} = \dfrac{3y+2}{y+3}$

The restricted values are $-3$, $3$, and $0$.

**7.** No, the graphs will not intersect in two points because the apparent solution $2$ is a restricted value.

**9.** Graph $y_1 = \dfrac{3}{x}$ and $y_2 = x + 2$. The solutions appear to be $x = -3$ and $x = 1$.

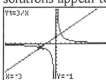
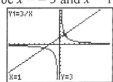

Verification with substitution:

$$\frac{3}{x} = x+2 \qquad \frac{3}{x} = x+2$$
$$\frac{3}{-3} = -3+2 \qquad \frac{3}{1} = 1+2$$
$$-1 = -1 \qquad 3 = 3$$

**11.** Graph $y_1 = \dfrac{x+3}{x+2}$ and $y_2 = x+3$. The solutions appear to be $x = -3, -1$.

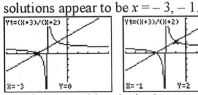

Verification with substitution:

$$\frac{x+3}{x+2} = x+3 \qquad \frac{x+3}{x+2} = x+3$$
$$\frac{-3+3}{-3+2} = -3+3 \qquad \frac{-1+3}{-1+2} = -1+3$$
$$0 = 0 \qquad 2 = 2$$

**13.** Graph $y_1 = \dfrac{12}{x}$ and $y_2 = x^2 + 3x - 4$.

The solutions appear to be $x = -3, -2,$ and $2$.

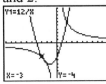 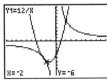

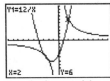

Verify these solutions with substitution.

**15.**
$$\frac{1}{8} = \frac{1}{2t} + \frac{2}{t}$$
$$8t \cdot \frac{1}{8} = 8t \cdot \frac{1}{2t} + 8t \cdot \frac{2}{t}$$
$$t = 4 + 16$$
$$t = 20$$

**17.**
$$\frac{5}{x-2} = 3 - \frac{1}{x-2}$$
$$(x-2) \cdot \frac{5}{x-2} = 3 \cdot (x-2) - (x-2) \cdot \frac{1}{x-2}$$
$$5 = 3x - 6 - 1$$
$$12 = 3x$$
$$4 = x$$

**19.**
$$\frac{7}{12} + \frac{1}{2y-10} = \frac{5}{3y-15}$$
$$\frac{7}{12} + \frac{1}{2(y-5)} = \frac{5}{3(y-5)}$$

The restricted value is 5. Multiply each term by the LCD of $12(y-5)$.

$$7(y-5) + 6 = 20$$
$$7y - 35 + 6 = 20$$
$$7y = 49$$
$$y = 7$$

**21.**
$$\frac{5}{x-2} = \frac{1}{x+2} + \frac{6}{x^2-4}$$
$$\frac{5}{x-2} = \frac{1}{x+2} + \frac{6}{(x+2)(x-2)}$$

The restricted values are $-2$ and $2$. Multiply each term by the LCD of $(x+2)(x-2)$.

$$5(x+2) = (x-2) + 6$$
$$5x + 10 = x - 2 + 6$$
$$4x = -6$$
$$x = -3/2$$

**23.**
$$\frac{t}{t+4} - \frac{2}{t-3} = 1$$

The restricted values are $-4$ and $3$. Multiply each term by the LCD of $(t+4)(t-3)$.

$$t(t-3) - 2(t+4) = (t-3)(t+4)$$
$$t^2 - 3t - 2t - 8 = t^2 + t - 12$$
$$-5t - 8 = t - 12$$
$$-6t = -4$$
$$t = 2/3$$

Section 8.5   Equations with Rational Expressions

**25.**
$$\frac{3}{x+3} = \frac{2}{x-4} - \frac{10}{x^2-x-12}$$
$$\frac{3}{x+3} = \frac{2}{x-4} - \frac{10}{(x-4)(x+3)}$$
The restricted values are 4 and $-3$. Multiply each term by the LCD $(x-4)(x+3)$.
$$3(x-4) = 2(x+3) - 10$$
$$3x - 12 = 2x + 6 - 10$$
$$x = 8$$

**27.** The restricted value is 0.
$$2x - \frac{15}{x} = 7$$
$$2x \cdot x - x \cdot \frac{15}{x} = 7 \cdot x$$
$$2x^2 - 15 = 7x$$
$$2x^2 - 7x - 15 = 0$$
$$(2x+3)(x-5) = 0$$
$$2x+3 = 0 \quad \text{or} \quad x-5 = 0$$
$$x = -1.5 \quad \text{or} \quad x = 5$$

**29.** The restricted value is 0.
$$\frac{x}{4} - \frac{5}{x} = \frac{1}{4}$$
$$4x \cdot \frac{x}{4} - 4x \cdot \frac{5}{x} = 4x \cdot \frac{1}{4}$$
$$x^2 - 20 = x$$
$$x^2 - x - 20 = 0$$
$$(x-5)(x+4) = 0$$
$$x - 5 = 0 \quad \text{or} \quad x + 4 = 0$$
$$x = 5 \quad \text{or} \quad x = -4$$

**31.**
$$\frac{t}{t-2} + 1 = \frac{8}{t-1}$$
The restricted values are 2 and 1. Multiply each term by the LCD $(t-2)(t-1)$.
$$t(t-1) + (t-2)(t-1) = 8(t-2)$$
$$t^2 - t + t^2 - 3t + 2 = 8t - 16$$
$$2t^2 - 12t + 18 = 0$$
$$2(t^2 - 6t + 9) = 0$$
$$2(t-3)^2 = 0$$
$$t - 3 = 0$$
$$t = 3$$

**33.**
$$\frac{3}{x+2} + 1 = \frac{6}{4-x^2}$$
$$\frac{3}{x+2} + 1 = \frac{6}{(2+x)(2-x)}$$
The restricted values are $-2$ and 2. Multiply each term by the LCD $(2+x)(2-x)$.
$$3(2-x) + (2+x)(2-x) = 6$$
$$6 - 3x + 4 - x^2 = 6$$
$$4 - 3x - x^2 = 0$$
$$(4+x)(1-x) = 0$$
$$4 + x = 0 \quad \text{or} \quad 1 - x = 0$$
$$x = -4 \quad \text{or} \quad x = 1$$

**35.** In (i), 9 is an extraneous solution. In (ii), clearing the fractions leads to $-4 = 4$, which is false, and the equation has no solution.

**37.**
$$\frac{1}{x+1} + \frac{2}{3x+3} = \frac{1}{3}$$
$$\frac{1}{x+1} + \frac{2}{3(x+1)} = \frac{1}{3}$$
The restricted value is $-1$. Multiply each term by the LCD $3(x+1)$.
$$3 + 2 = x + 1$$
$$4 = x$$

**39.**
$$\frac{x+2}{x+3} = \frac{x-1}{x+1}$$
The restricted values are $-3$ and $-1$. Multiply each term by the LCD $(x+3)(x+1)$.
$$(x+2)(x+1) = (x-1)(x+3)$$
$$x^2 + 3x + 2 = x^2 + 2x - 3$$
$$x = -5$$

**41.**
$$\frac{5}{x+5} = 4 - \frac{x}{x+5}$$
The restricted value is $-5$. Multiply each term by the LCD $(x+5)$.
$$5 = 4(x+5) - x$$
$$5 = 4x + 20 - x$$
$$-15 = 3x$$
$$-5 = x$$
This is a restricted value and is not a solution. There is no solution.

**43.** $\dfrac{10}{(2x-1)^2} = 4 + \dfrac{3}{2x-1}$

The restricted value is 0.5. Multiply each term by the LCD $(2x-1)^2$.

$$10 = 4(2x-1)^2 + 3(2x-1)$$
$$10 = 4(4x^2 - 4x + 1) + 3(2x-1)$$
$$10 = 16x^2 - 16x + 4 + 6x - 3$$
$$0 = 16x^2 - 10x - 9$$
$$0 = (8x-9)(2x+1)$$
$$8x - 9 = 0 \quad \text{or} \quad 2x + 1 = 0$$
$$x = 9/8 \quad \text{or} \quad x = -1/2$$

**45.** $\dfrac{2}{x^2+x-2} + \dfrac{3x}{x^2+5x+6} = \dfrac{5x}{x^2+2x-3}$

$\dfrac{2}{(x+2)(x-1)} + \dfrac{3x}{(x+3)(x+2)} = \dfrac{5x}{(x+3)(x-1)}$

The restricted values are $-3, -2,$ and $1$. Multiply each term by the LCD $(x+3)(x+2)(x-1)$.

$$2(x+3) + 3x(x-1) = 5x(x+2)$$
$$2x + 6 + 3x^2 - 3x = 5x^2 + 10x$$
$$-2x^2 - 11x + 6 = 0$$
$$2x^2 + 11x - 6 = 0$$
$$(2x-1)(x+6) = 0$$
$$2x - 1 = 0 \quad \text{or} \quad x + 6 = 0$$
$$x = 1/2 \quad \text{or} \quad x = -6$$

**47.** The restricted value is 0.
$$2x + 15 = \dfrac{8}{x}$$
$$2x \cdot x + 15 \cdot x = \dfrac{8}{x} \cdot x$$
$$2x^2 + 15x = 8$$
$$2x^2 + 15x - 8 = 0$$
$$(2x-1)(x+8) = 0$$
$$2x - 1 = 0 \quad \text{or} \quad x + 8 = 0$$
$$x = 1/2 \quad \text{or} \quad x = -8$$

**49.** $\dfrac{x}{x-3} + \dfrac{2}{x} = \dfrac{3}{x-3}$

The restricted values are 3 and 0. Multiply each term by the LCD $x(x-3)$.

$$x^2 + 2(x-3) = 3x$$
$$x^2 + 2x - 6 = 3x$$
$$x^2 - x - 6 = 0$$
$$(x-3)(x+2) = 0$$
$$x - 3 = 0 \quad \text{or} \quad x + 2 = 0$$
$$x = 3 \quad \text{or} \quad x = -2$$

Because 3 is a restricted value, the only solution is $-2$.

**51.** $\dfrac{8r}{r-4} - \dfrac{7}{r^2-16} = 8$

$\dfrac{8r}{r-4} - \dfrac{7}{(r+4)(r-4)} = 8$

The restricted values are $-4$ and $4$. Multiply each term by the LCD $(r+4)(r-4)$.

$$8r(r+4) - 7 = 8(r+4)(r-4)$$
$$8r^2 + 32r - 7 = 8r^2 - 128$$
$$32r = -121$$
$$r = -121/32$$

**53.** $\dfrac{5}{x-7} + \dfrac{2}{x+5} = \dfrac{1}{x^2-2x-35}$

$\dfrac{5}{x-7} + \dfrac{2}{x+5} = \dfrac{1}{(x-7)(x+5)}$

The restricted values are $-5$ and $7$. Multiply each term by the LCD $(x-7)(x+5)$.

$$5(x+5) + 2(x-7) = 1$$
$$5x + 25 + 2x - 14 = 1$$
$$7x + 11 = 1$$
$$7x = -10$$
$$x = -10/7$$

**55.** $\dfrac{x+4}{6x^2+5x-6} + \dfrac{x}{2x+3} = \dfrac{x}{3x-2}$

$\dfrac{x+4}{(2x+3)(3x-2)} + \dfrac{x}{2x+3} = \dfrac{x}{3x-2}$

The restricted values are $-3/2$ and $2/3$. Multiply each term by the LCD $(2x+3)(3x-2)$.

$$(x+4) + x(3x-2) = x(2x+3)$$
$$x + 4 + 3x^2 - 2x = 2x^2 + 3x$$
$$x^2 - 4x + 4 = 0$$
$$(x-2)^2 = 0$$
$$x - 2 = 0$$
$$x = 2$$

**57.**
$$\frac{2}{x+1} - \frac{1}{x-3} = \frac{-8}{x^2 - 2x - 3}$$
$$\frac{2}{x+1} - \frac{1}{x-3} = \frac{-8}{(x-3)(x+1)}$$

The restricted values are $-1$ and $3$. Multiply each term by the LCD $(x-3)(x+1)$.
$$2(x-3) - (x+1) = -8$$
$$2x - 6 - x - 1 = -8$$
$$x = -1$$

This is a restricted value and is not a solution. There is no solution.

**59.**
$$\frac{x}{x+2} + \frac{2}{x+3} + \frac{2}{x^2 + 5x + 6} = 0$$
$$\frac{x}{x+2} + \frac{2}{x+3} + \frac{2}{(x+2)(x+3)} = 0$$

The restricted values are $-3$ and $-2$. Multiply each term by the LCD $(x+2)(x+3)$.
$$x(x+3) + 2(x+2) + 2 = 0$$
$$x^2 + 3x + 2x + 4 + 2 = 0$$
$$x^2 + 5x + 6 = 0$$
$$(x+3)(x+2) = 0$$
$$x+3 = 0 \quad \text{or} \quad x+2 = 0$$
$$x = -3 \quad \text{or} \quad x = -2$$

These are both restricted values and are not solutions. There is no solution.

**61.** To solve the equation, clear the fractions by multiplying both sides by the LCD. To perform the addition, rewrite each expression with the LCD as the denominator.

**63.** The LCD is $(x-8)(x-3)$.
$$\frac{3}{x-8} + \frac{x}{x-3}$$
$$= \frac{3(x-3)}{(x-8)(x-3)} + \frac{x(x-8)}{(x-8)(x-3)}$$
$$= \frac{3(x-3) + x(x-8)}{(x-8)(x-3)}$$
$$= \frac{3x - 9 + x^2 - 8x}{(x-8)(x-3)} = \frac{x^2 - 5x - 9}{(x-8)(x-3)}$$

**65.**
$$4 + \frac{19}{2t-3} = \frac{t+2}{3-2t}$$
$$4 + \frac{19}{2t-3} = -\frac{t+2}{2t-3}$$

The restricted value is $3/2$. Multiply each term by the LCD $(2t-3)$.
$$4(2t-3) + 19 = -(t+2)$$
$$8t - 12 + 19 = -t - 2$$
$$9t = -9$$
$$t = -1$$

**67.**
$$\frac{y^2 - 49}{8y + 56} \cdot \frac{y^2 - 7y}{(y-7)^2}$$
$$= \frac{(y+7)(y-7)}{8(y+7)} \cdot \frac{y(y-7)}{(y-7)^2}$$
$$= \frac{y}{8}$$

**69.**
$$1 - 2t - \frac{2t}{t+3} = \frac{t+3}{t+3} - \frac{2t(t+3)}{t+3} - \frac{2t}{t+3}$$
$$= \frac{t+3 - 2t(t+3) - 2t}{t+3}$$
$$= \frac{t + 3 - 2t^2 - 6t - 2t}{t+3}$$
$$= \frac{-2t^2 - 7t + 3}{t+3}$$

**71.**
$$\frac{x+5}{x-1} = \frac{8}{x-3} - \frac{7}{x^2 - 4x + 3}$$
$$\frac{x+5}{x-1} = \frac{8}{x-3} - \frac{7}{(x-3)(x-1)}$$

The restricted values are $1$ and $3$. Multiply each term by the LCD $(x-1)(x-3)$.
$$(x+5)(x-3) = 8(x-1) - 7$$
$$x^2 + 2x - 15 = 8x - 8 - 7$$
$$x^2 - 6x = 0$$
$$x(x-6) = 0$$
$$x = 0 \quad \text{or} \quad x - 6 = 0$$
$$x = 6$$

**73.** 
$$\frac{x^2+x-6}{x^2+9x+18} \div \frac{x^2-4}{4x+24}$$
$$= \frac{x^2+x-6}{x^2+9x+18} \cdot \frac{4x+24}{x^2-4}$$
$$= \frac{(x+3)(x-2)}{(x+6)(x+3)} \cdot \frac{4(x+6)}{(x+2)(x-2)}$$
$$= \frac{4}{x+2}$$

**75.** 
$$I = \frac{E}{R}$$
$$IR = E$$
$$R = \frac{E}{I}$$

**77.** 
$$\frac{P_1V_1}{T_1} = \frac{P_2V_2}{T_2}$$
$$P_1V_1T_2 = P_2V_2T_1$$
$$\frac{P_1V_1T_2}{P_2V_2} = T_1$$

**79.** 
$$P = \frac{I}{rt}$$
$$Prt = I$$
$$r = \frac{I}{Pt}$$

**81.** 
$$F = \frac{m_1 m_2}{\mu r^2}$$
$$F\mu r^2 = m_1 m_2$$
$$\frac{F\mu r^2}{m_1} = m_2$$

**83.** 
$$S = \frac{a(1-r^n)}{1-r}$$
$$S(1-r) = a(1-r^n)$$
$$\frac{S(1-r)}{1-r^n} = a$$

**85.** 
$$\frac{x}{a} + \frac{y}{b} = 1$$
$$ab \cdot \frac{x}{a} + ab \cdot \frac{y}{b} = 1 \cdot ab$$
$$bx + ay = ab$$
$$bx = ab - ay$$
$$bx = a(b-y)$$
$$\frac{bx}{b-y} = a$$

**87.** 
$$\frac{y-4}{x+3} = -2$$
$$y - 4 = -2(x+3)$$
$$y - 4 = -2x - 6$$
$$y = -2x - 2$$

**89.** 
$$\frac{y}{x+4} = -\frac{4}{5}$$
$$y = -\frac{4}{5}(x+4)$$

**91.** The greatest frequency of DUI arrests per 100,000 drivers occurs among younger drivers.

**93.** The highest frequency of DUI arrests per 100,000 drivers is 356, which occurs at age 25.6.

**95.** 
$$\frac{\dfrac{x+3}{x-2}}{\dfrac{2}{x-2}} = 5$$
$$\frac{x+3}{x-2} \cdot \frac{x-2}{2} = 5$$
$$\frac{x+3}{2} = 5$$
$$x + 3 = 10$$
$$x = 7$$

Section 8.6   Applications

**97.**
$$\frac{1}{18} + \frac{1}{9x} = \frac{1}{2x^2} + \frac{1}{x^3}$$
$$\frac{18x^3}{18} + \frac{18x^3}{9x} = \frac{18x^3}{2x^2} + \frac{18x^3}{x^3}$$
$$x^3 + 2x^2 = 9x + 18$$
$$x^3 + 2x^2 - 9x - 18 = 0$$
$$(x^3 + 2x^2) - (9x + 18) = 0$$
$$x^2(x+2) - 9(x+2) = 0$$
$$(x+2)(x^2 - 9) = 0$$
$$(x+2)(x+3)(x-3) = 0$$
$x+2 = 0$  or  $x+3 = 0$  or  $x-3 = 0$
$x = -2$  or  $x = -3$  or  $x = 3$

**99.**
$$\frac{y - y_1}{x - x_1} = m$$
$$y - y_1 = m(x - x_1)$$
$$y = m(x - x_1) + y_1$$

**101.**
$$\left(\frac{1}{x+1}\right)^2 + 3\left(\frac{1}{x+1}\right) - 4 = 0$$
$$u^2 + 3u - 4 = 0$$
$$(u+4)(u-1) = 0$$
$u + 4 = 0$
$u = -4$
$\frac{1}{x+1} = -4$       $u - 1 = 0$
$1 = -4(x+1)$   or   $u = 1$
$-\frac{1}{4} = x + 1$        $\frac{1}{x+1} = 1$
$-\frac{5}{4} = x$           $1 = x + 1$
                             $0 = x$

## Section 8.6   Applications

**1.** Let $x$ = the number.
$$2x + \frac{3}{x} = 7$$
$$2x \cdot x + \frac{3}{x} \cdot x = 7 \cdot x$$
$$2x^2 + 3 = 7x$$
$$2x^2 - 7x + 3 = 0$$
$$(2x - 1)(x - 3) = 0$$
$2x - 1 = 0$  or  $x - 3 = 0$
$x = 1/2$  or  $x = 3$
The number is either 1/2 or 3.

**3.** Let $x$ = the smaller number and $x + 4$ = the larger number.
$$\frac{1}{x+4} + \frac{3}{x} = \frac{3}{5}$$
The restricted values are 0 and $-4$.
Multiply each term by the LCD $5x(x+4)$.
$$5x + 3(5)(x+4) = 3x(x+4)$$
$$5x + 15x + 60 = 3x^2 + 12x$$
$$0 = 3x^2 - 8x - 60$$
$$0 = (3x + 10)(x - 6)$$
$3x + 10 = 0$     or   $x - 6 = 0$
$x = -10/3$  or   $x = 6$
The numbers are either 6 and 10 or $-10/3$ and 2/3

**5.** In (i) we are asked to find the sum of two fractions. Write the fractions with the LCD and add. In (ii) we are asked to solve an equation. Clear the fractions by multiplying both sides by the LCD.

**7.** Let $b$ = the length of the base of the blue piece. Then the length of the side of the blue piece is $b + 2$.
$$\frac{b+2}{b} = \frac{3}{2}$$
$$3b = 2(b+2)$$
$$3b = 2b + 4$$
$$b = 4$$
The length of the base of the blue triangle is 4 inches.

**9.** Let $x$ = the number of nontraditional-age students that attend the college.
If the ratio of traditional-age students to nontraditional-age students is 4 to 3, then the ratio of the nontraditional-age

then the ratio of the nontraditional-age students to the total enrollment is 3 to 7.

$$\frac{3}{7} = \frac{x}{1260}$$
$$7x = 3(1260)$$
$$7x = 3780$$
$$x = 540$$

There are 540 nontraditional-age students attending the college.

11. Let $x$ = the number of hits in a row needed.

$$\frac{\text{Number of hits}}{\text{Number of at bats}} = \frac{112 + x}{444 + x}$$

The equation is
$$\frac{112 + x}{444 + x} = 0.267$$
$$112 + x = 0.267(444 + x)$$
$$112 + x = 118.548 + 0.267x$$
$$0.733x = 6.548$$
$$x \approx 8.9$$

So, 9 hits in a row are needed to raise the average to 0.267.

13. Let $x$ = the number of females in 1980.

| | 1980 | 1990 |
|---|---|---|
| Male | 360 | 660 |
| Female | $x$ | $x + 700$ |
| Totals | $360 + x$ | $1360 + x$ |

The percentage of males in 1980 is $\frac{360}{360 + x}$. The percentage males in 1990 is $\frac{660}{1360 + x}$. Because the percentage in 1980 and 1990 are the same, we have

$$\frac{360}{360 + x} = \frac{660}{1360 + x}$$
$$\frac{6}{360 + x} = \frac{11}{1360 + x}$$
$$6(1360 + x) = 11(360 + x)$$
$$8160 + 6x = 3960 + 11x$$
$$4200 = 5x$$
$$840 = x$$

The number of females in 1980 was 840. The total enrollment in 1980 was $360 + 840 = 1200$ students.

15. 
$$y = \frac{k}{x}$$
$$9 = \frac{k}{5}$$
$$45 = k$$

The constant of variation is 45.
$$y = \frac{45}{x}$$
$$y = \frac{45}{3} = 15$$

When $x = 3$, $y = 15$.

17. 
$$w = \frac{k}{z^2}$$
$$24 = \frac{k}{2^2}$$
$$96 = k$$

The constant of variation is 96.
$$w = \frac{96}{z^2}$$
$$w = \frac{96}{4^2} = 6$$

When $z = 4$, $w = 6$.

19. Let $t$ = the time required to do the job and $p$ = the number of people working.

$$t = \frac{k}{p}$$
$$9 = \frac{k}{4}$$
$$36 = k$$

The constant of variation is 36.
$$t = \frac{36}{p}$$
$$t = \frac{36}{5}$$
$$t = 7.2$$

It will take 7.2 hours for five workers to finish a roof of the same size.

21. Let $p$ = the price per barrel of oil and $s$ = the number millions of barrels supplied.

$$p = \frac{k}{s}$$
$$26 = \frac{k}{3}$$
$$78 = k$$

The constant of variation is 78.

Section 8.6   Applications

$$p = \frac{78}{s}$$
$$p = \frac{78}{4}$$
$$p = 19.5$$

When 4 million barrels are supplied, the price is $19.50 per barrel.

**23.** Let $x =$ the number of minutes for the other person working alone.

|  | Number of minutes | Work rate |
|---|---|---|
| One person | 30 | 1/30 |
| Other person | $x$ | $1/x$ |
| Together | 10 | 1/10 |

The combined work rate is the sum of the individual work rates.

$$\frac{1}{30} + \frac{1}{x} = \frac{1}{10}$$
$$\frac{30x}{30} + \frac{30x}{x} = \frac{30x}{10}$$
$$x + 30 = 3x$$
$$30 = 2x$$
$$15 = x$$

It takes 15 minutes for the other person to wash the car.

**25.** Let $x =$ the number of hours for the professional painter working alone.

|  | Number of hours | Work rate |
|---|---|---|
| Apprentice 1 | 15 | 1/15 |
| Apprentice 2 | 10 | 1/10 |
| Professional | $x$ | $1/x$ |

The professional's work rate is equal to the sum of the apprentices' work rates.

$$\frac{1}{15} + \frac{1}{10} = \frac{1}{x}$$
$$\frac{30x}{15} + \frac{30x}{10} = \frac{30x}{x}$$
$$2x + 3x = 30$$
$$5x = 30$$
$$x = 6$$

The professional can paint the trim in 6 hours without help.

**27.** Let $x =$ the rate of the train and $x + 90 =$ the rate of the plane.

|  | $d$ | $r$ | $t = d/r$ |
|---|---|---|---|
| Train | 210 | $x$ | $210/x$ |
| Plane | 525 | $x + 90$ | $525/(x + 90)$ |

Because the times are equal, the equation is:

$$\frac{210}{x} = \frac{525}{x + 90}$$
$$210(x + 90) = 525x$$
$$210x + 18{,}900 = 525x$$
$$18{,}900 = 315x$$
$$60 = x$$

The rate of the train is 60 mph. The rate of the plane is 150 mph.

**29.** Let $x =$ the rate of the novice and $x + 3 =$ the rate of the experienced cyclist.

|  | $d$ | $r$ | $t = d/r$ |
|---|---|---|---|
| Novice | 4.4 | $x$ | $4.4/x$ |
| Experienced | 5 | $x + 3$ | $5/(x + 3)$ |

The difference in times was 2 minutes, or 1/30 of an hour. The equation is:

$$\frac{4.4}{x} - \frac{5}{x + 3} = \frac{1}{30}$$
$$4.4(30)(x + 3) - 5x(30) = x(x + 3)$$
$$132x + 396 - 150x = x^2 + 3x$$
$$x^2 + 21x - 396 = 0$$
$$(x + 33)(x - 12) = 0$$
$$x + 33 = 0 \quad \text{or} \quad x - 12 = 0$$
$$x = -33 \quad \text{or} \quad x = 12$$

The rate of the novice cyclist was 12 mph.

**31.** Choices (c) and (e) illustrate inverse relations because an increase in the illiteracy rate would most likely correspond to a decrease in library circulation, and an increase in unemployment would most likely correspond to a decrease in home ownership.

**33.** Let $x =$ the height of the tree.
Using similar triangles, the equation is:

$$\frac{36}{15} = \frac{x}{25}$$
$$15x = 900$$
$$x = 60$$

The tree is 60 feet tall.

**35.** Let $x$ = the number of minutes for the person using the snow blower alone.

| | Number of minutes | Work rate |
|---|---|---|
| Shovel | $2x$ | $1/(2x)$ |
| Snow blower | $x$ | $1/x$ |
| Together | 8 | $1/8$ |

The combined work rate is the sum of the individual work rates.
$$\frac{1}{2x} + \frac{1}{x} = \frac{1}{8}$$
$$\frac{8x}{2x} + \frac{8x}{x} = \frac{8x}{8}$$
$$4 + 8 = x$$
$$12 = x$$

It takes 12 minutes for person with the snow blower and 24 minutes for the person with the shovel.

**37.** Let $h$ = the number of hairs last year and $t$ = the average winter temperature last year.
$$h = \frac{k}{t}$$
$$ht = k$$
The constant of variation is $ht$.
$$1.5h = \frac{ht}{t-10}$$
$$1.5h(t-10) = ht$$
$$1.5(t-10) = t$$
$$1.5t - 15 = t$$
$$0.5t = 15$$
$$t = 30$$

The average winter temperature last year was 30 degrees.

**39.** Let $x$ = the number of seconds for the first 50 yards and $x + 1$ = the number of seconds for the second 50 yards. The average speed is equal to the total distance divided by the total time.

$$\frac{100}{x+(x+1)} = 9\frac{1}{11}$$
$$\frac{100}{x+x+1} = \frac{100}{11}$$
$$\frac{100}{2x+1} = \frac{100}{11}$$
$$100(2x+1) = 100(11)$$
$$2x+1 = 11$$
$$2x = 10$$
$$x = 5$$

It took 5 seconds to run the first 50 yards and 6 seconds to run the second 50 yards.

**41.** Let $x$ = the speed of the plane in still air.

| | $d$ | $r$ | $t = d/r$ |
|---|---|---|---|
| With wind | 300 | $x+40$ | $300/(x+40)$ |
| Against wind | 200 | $x-40$ | $200/(x-40)$ |

The times are equal. The equation is:
$$\frac{300}{x+40} = \frac{200}{x-40}$$
$$300(x-40) = 200(x+40)$$
$$3(x-40) = 2(x+40)$$
$$3x - 120 = 2x + 80$$
$$x = 200$$

The rate of the plane in still air is 200 mph.

**43.** Let $x$ = the number of hours for the tank to drain under these conditions.

| | Number of hours | Work rate |
|---|---|---|
| Commercial outtake | 7 | $1/7$ |
| Residential outtake | 9 | $1/9$ |
| Intake | 14 | $1/14$ |
| Together | $x$ | $1/x$ |

Because we are interested in how long it takes to drain the tank, we must consider the intake valve to be working to do the opposite: to fill the tank. Therefore, the work rate of the intake valve is subtracted from the sum of the work rates of the outtake valves.

Section 8.6   Applications

$$\frac{1}{7}+\frac{1}{9}-\frac{1}{14}=\frac{1}{x}$$
$$\frac{126x}{7}+\frac{126x}{9}-\frac{126x}{14}=\frac{126x}{x}$$
$$18x+14x-9x=126$$
$$23x=126$$
$$x \approx 5.5$$

The tank will drain in approximately 5.5 hours under these conditions.

45. Two cats can eat 7-pounds in 24 days, for a rate of 7/24 per day. The first cat can eat 3.5 pounds in 3 weeks (21 days), for a rate of 3.5/21 per day. Let $x$ = the number of days it takes the second cat to eat 3.5 pounds. The rate of the second cat is 3.5/$x$.

$$\frac{3.5}{21}+\frac{3.5}{x}=\frac{7}{24}$$
$$168x \cdot \frac{3.5}{21}+168x \cdot \frac{3.5}{x}=168x \cdot \frac{7}{24}$$
$$28x+588=49x$$
$$588=21x$$
$$28=x$$

It takes 28 days for the second cat to eat 3.5 pounds of food.

47. (a) The percentage of aluminum that was recovered is the quotient of the amount of aluminum recovered $A_2(x)$ and the amount of aluminum generated $A_1(x)$, multiplied by 100 to get a percent:
$$\frac{100 A_2(x)}{A_1(x)}=\frac{100(0.04x+0.4)}{0.07x+1.9}$$

(b) The percentage of glass that was recovered is found similarly:
$$\frac{100 G_2(x)}{G_1(x)}=\frac{100(0.12x+1.02)}{-0.16x+14.9}$$

49. Let $D$ = the number of Democrats and $R$ = the number of Republicans. We are told that in a recent election, 5/6 of Democrats, 7/9 of Republicans, and all 11 Independents voted. This can be translated as
$$\frac{5}{6}D+\frac{7}{9}R+11$$

We are also told that all of the women voted, which is 1/2 of Democrats, 2/3 of Republicans, and 9 Independents. This can be translated as
$$\frac{1}{2}D+\frac{2}{3}R+9$$

If 75% of the those who voted were women, we have the equation
$$0.75\left(\frac{5}{6}D+\frac{7}{9}R+11\right)=\frac{1}{2}D+\frac{2}{3}R+9$$

We are also told in the problem that there is an equal number of Democrats and Republicans. Using the substitution $R = D$, we have
$$0.75\left(\frac{5}{6}D+\frac{7}{9}R+I\right)=\frac{1}{2}D+\frac{2}{3}R+9$$
$$0.75\left(\frac{5}{6}D+\frac{7}{9}D+11\right)=\frac{1}{2}D+\frac{2}{3}D+9$$
$$0.75\left(\frac{15}{18}D+\frac{14}{18}D+11\right)=\frac{3}{6}D+\frac{4}{6}D+9$$
$$0.75\left(\frac{29}{18}D+11\right)=\frac{7}{6}D+9$$
$$\frac{29}{24}D+8.25=\frac{7}{6}D+9$$
$$29D+198=28D+216$$
$$29D-28D=216-198$$
$$D=18$$

There are 18 Democrats, 18 Republicans, and 11 Independents, for a total of 47 registered voters in Woodville Notch.

51. Let $x$ = the number of pounds of small bolts, then $x + 4$ = the number of pounds of medium bolts, and $x - 3$ = the number of pound of large bolts. The unit prices are 3.20/$x$ for small bolts, 8.40/($x + 4$) for medium bolts, and 5.50/($x - 3$) for large bolts.

$$\frac{3.20}{x}+\frac{8.40}{x+4}=\frac{5.50}{x-3}$$
$$\frac{32}{x}+\frac{84}{x+4}=\frac{55}{x-3}$$
$$32(x+4)(x-3)+84x(x-3)=55x(x+4)$$
$$32(x^2+x-12)+84(x^2-3x)=55(x^2+4x)$$
$$32x^2+32x-384+84x^2-252x=55x^2+220x$$
$$61x^2-440x-384=0$$
$$(61x+48)(x-8)=0$$
$$61x+48=0 \quad \text{or} \quad x-8=0$$
$$x=-48/61 \quad \text{or} \quad x=8$$

The builder bought 8 pounds of small bolts, 12 pounds of medium bolts, and 5 pounds of large bolts.

**53.**
$$(AQ)^2 = BQ \cdot QC$$
$$\left(\frac{8}{x}\right)^2 = \frac{x}{1} \cdot \frac{32}{x+2}$$
$$\frac{64}{x^2} = \frac{32x}{x+2}$$
$$64(x+2) = 32x^3$$
$$2(x+2) = x^3$$
$$2x + 4 = x^3$$
$$0 = x^3 - 2x - 4$$

We can use a graph to estimate the solution of this equation.

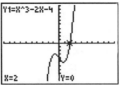

The solution appears to be $x = 2$.
Verification:
$$0 = x^3 - 2x - 4$$
$$0 = (2)^3 - 2(2) - 4$$
$$0 = 8 - 4 - 4$$
$$0 = 0$$

The length of $AQ$ is $8/x = 8/2 = 4$.

## Chapter 8 Project

**1.** The curve should rise from 1990 to 1994, fall from 1994 to 1995, and then rise again from 1995 onward.

**3.** $x \approx 3$ and $x \approx 8$; The model estimates that there were 262 million-dollar players in 1993 and 1998.

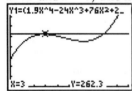 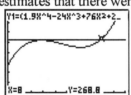

**5.** $P(x) = 1.9x^3 - 25.9x^2 + 101.9x + 138.4$; The domain of $P$ is **R**, whereas the domain of $N$ excludes $x = -1$.

## Chapter 8 Review Exercises

**1.**
$$f(x) = \frac{x-7}{x^2 - 2x - 15} = \frac{x-7}{(x-5)(x+3)}$$
The numbers are 5 and $-3$, which are called restricted values.

**3.**
$$y = \frac{5x-2}{2x-5}$$
To find the restricted values, set the denominator equal to 0 and solve for $x$.

$$2x - 5 = 0$$
$$x = 2.5$$
The domain is $\{x \mid x \neq 2.5\}$.

**5.** In the first expression, 3 is a factor in both the numerator and the denominator. There is no common factor in the numerator and the denominator of the second expression.

7. $\dfrac{x^2+x-12}{2x^2-9x+9} = \dfrac{(x+4)(x-3)}{(2x-3)(x-3)}$
$\phantom{\dfrac{x^2+x-12}{2x^2-9x+9}} = \dfrac{x+4}{2x-3}$

9. $\dfrac{ab+4a-3b-12}{ab+2a-3b-6} = \dfrac{(ab+4a)-(3b+12)}{(ab+2a)-(3b+6)}$
$\phantom{\dfrac{ab+4a-3b-12}{ab+2a-3b-6}} = \dfrac{a(b+4)-3(b+4)}{a(b+2)-3(b+2)}$
$\phantom{\dfrac{ab+4a-3b-12}{ab+2a-3b-6}} = \dfrac{(b+4)(a-3)}{(b+2)(a-3)}$
$\phantom{\dfrac{ab+4a-3b-12}{ab+2a-3b-6}} = \dfrac{b+4}{b+2}$

11. $\dfrac{5}{3x+12} = \dfrac{?}{3x^2-48} = \dfrac{?}{3(x+4)(x-4)}$
Multiply both the numerator and the denominator of the original expression by $x-4$. The unknown numerator is $5(x-4)$.

13. $\dfrac{8x-40}{4x^3} \cdot \dfrac{4x^2+20x}{x^2-25}$
$= \dfrac{8(x-5)}{4x^3} \cdot \dfrac{4x(x+5)}{(x+5)(x-5)}$
$= \dfrac{8}{x^2}$

15. $\dfrac{x^2+3x+2}{x^2+8x+12} \cdot \dfrac{x^2+7x+6}{x^2-x-2}$
$= \dfrac{(x+2)(x+1)}{(x+6)(x+2)} \cdot \dfrac{(x+6)(x+1)}{(x-2)(x+1)}$
$= \dfrac{x+1}{x-2}$

17. (a) Their product is $xy$. The reciprocal of their product is $\dfrac{1}{xy}$.
(b) The product of their reciprocals is $\dfrac{1}{x} \cdot \dfrac{1}{y} = \dfrac{1}{xy}$.
(c) Yes, the expressions are equivalent.

19. $\dfrac{x^2-5x+6}{x^2+3x-18} \div \dfrac{x^2-4}{6x+36}$
$= \dfrac{x^2-5x+6}{x^2+3x-18} \cdot \dfrac{6x+36}{x^2-4}$
$= \dfrac{(x-3)(x-2)}{(x+6)(x-3)} \cdot \dfrac{6(x+6)}{(x+2)(x-2)}$
$= \dfrac{6}{x+2}$

21. $\dfrac{x^2+3x-4}{x^2-2x-24} \div \dfrac{x^2-5x+4}{x^2-7x+6}$
$= \dfrac{x^2+3x-4}{x^2-2x-24} \cdot \dfrac{x^2-7x+6}{x^2-5x+4}$
$= \dfrac{(x+4)(x-1)}{(x-6)(x+4)} \cdot \dfrac{(x-6)(x-1)}{(x-4)(x-1)}$
$= \dfrac{x-1}{x-4}$

23. $\dfrac{x^2+5x+6}{x^2-3x-18} \cdot \dfrac{2x-12}{x^2-4}$
$= \dfrac{(x+3)(x+2)}{(x-6)(x+3)} \cdot \dfrac{2(x-6)}{(x+2)(x-2)}$
$= \dfrac{2}{x-2}$

25. $x^2-49 = (x+7)(x-7)$
$9x-63 = 9(x-7)$
The LCD is $9(x+7)(x-7)$.

27. If the denominators were not the same, the terms would not have a common factor, and the Distributive Property would not apply.

29. $\dfrac{2x-1}{5x-4} + \dfrac{6x-6}{4-5x} = \dfrac{2x-1}{5x-4} - \dfrac{6x-6}{5x-4}$
$= \dfrac{2x-1-(6x-6)}{5x-4}$
$= \dfrac{2x-1-6x+6}{5x-4}$
$= \dfrac{-4x+5}{5x-4}$

**31.**
$$\frac{7}{x^2+x-12} + \frac{3}{x^2-16}$$
$$= \frac{7}{(x+4)(x-3)} + \frac{3}{(x+4)(x-4)}$$
$$= \frac{7(x-4)}{(x-4)(x+4)(x-3)}$$
$$\quad + \frac{3(x-3)}{(x-4)(x+4)(x-3)}$$
$$= \frac{7(x-4) + 3(x-3)}{(x-4)(x+4)(x-3)}$$
$$= \frac{7x - 28 + 3x - 9}{(x-4)(x+4)(x-3)}$$
$$= \frac{10x - 37}{(x-4)(x+4)(x-3)}$$

**33.**
$$2x^{-1} + 3(x+2)^{-1} = \frac{2}{x} + \frac{3}{x+2}$$
$$= \frac{2(x+2)}{x(x+2)} + \frac{3x}{x(x+2)}$$
$$= \frac{2x+4}{x(x+2)} + \frac{3x}{x(x+2)}$$
$$= \frac{2x+4+3x}{x(x+2)}$$
$$= \frac{5x+4}{x(x+2)}$$

**35.**
$$\frac{7}{x^2+2x-48} - \frac{1}{x^2-x-30}$$
$$= \frac{7}{(x-6)(x+8)} - \frac{1}{(x+5)(x-6)}$$
$$= \frac{7(x+5)}{(x+5)(x-6)(x+8)}$$
$$\quad - \frac{1(x+8)}{(x+5)(x-6)(x+8)}$$
$$= \frac{7(x+5) - (x+8)}{(x+5)(x-6)(x+8)}$$
$$= \frac{7x + 35 - x - 8}{(x+5)(x-6)(x+8)}$$
$$= \frac{6x+27}{(x+5)(x-6)(x+8)}$$

**37.** Multiply the numerator by the reciprocal of the denominator or multiply the numerator and the denominator by the LCD of all the fractions.

**39.**
$$\frac{\dfrac{a}{9} - \dfrac{1}{a}}{\dfrac{1}{3} + \dfrac{a+4}{a}} = \frac{9a \cdot \left(\dfrac{a}{9} - \dfrac{1}{a}\right)}{9a \cdot \left(\dfrac{1}{3} + \dfrac{a+4}{a}\right)}$$
$$= \frac{a^2 - 9}{3a + 9(a+4)} = \frac{a^2 - 9}{3a + 9a + 36}$$
$$= \frac{a^2 - 9}{12a + 36} = \frac{(a-3)(a+3)}{12(a+3)} = \frac{a-3}{12}$$

**41.**
$$\frac{\dfrac{m^{-1}}{n^{-1}}}{\dfrac{n}{m}} = \frac{m^{-1}}{n^{-1}} \cdot \frac{m}{n} = \frac{m^{-1+1}}{n^{-1+1}} = \frac{m^0}{n^0} = \frac{1}{1} = 1$$

**43.**
$$\frac{\dfrac{x+4}{x-4} + \dfrac{x-4}{x+4}}{\dfrac{1}{x+4} + \dfrac{1}{x-4}}$$
$$= \frac{(x-4)(x+4) \cdot \left(\dfrac{x+4}{x-4} + \dfrac{x-4}{x+4}\right)}{(x-4)(x+4) \cdot \left(\dfrac{1}{x+4} + \dfrac{1}{x-4}\right)}$$
$$= \frac{(x+4)^2 + (x-4)^2}{x-4+x+4}$$
$$= \frac{x^2 + 8x + 16 + x^2 - 8x + 16}{2x}$$
$$= \frac{2x^2 + 32}{2x} = \frac{2(x^2 + 16)}{2x} = \frac{x^2 + 16}{x}$$

**45.** An apparent solution does not check if it makes a denominator zero. Such a number is called an extraneous solution.

**47.** The restricted value is 0. Multiply each term by the LCD $x$.
$$10x - \frac{2}{x} = 1$$
$$10x^2 - 2 = x$$
$$10x^2 - x - 2 = 0$$
$$(5x+2)(2x-1) = 0$$
$$5x + 2 = 0 \quad \text{or} \quad 2x - 1 = 0$$
$$x = -2/5 \quad \text{or} \quad x = 1/2$$

**49.** The restricted value is 5. Multiply each term by the LCD $x-5$.
$$\frac{x+3}{x-5} = \frac{8}{x-5}$$
$$x+3 = 8$$
$$x = 5$$
This is a restricted value and is not a solution. Therefore, there is no solution.

**51.**
$$\frac{7}{x+1} - \frac{2}{x-1} = \frac{1}{x^2-1}$$
$$\frac{7}{x+1} - \frac{2}{x-1} = \frac{1}{(x+1)(x-1)}$$
The restricted values are $-1$ and $1$. Multiply each term by the LCD $(x+1)(x-1)$.
$$7(x-1) - 2(x+1) = 1$$
$$7x - 7 - 2x - 2 = 1$$
$$5x - 9 = 1$$
$$5x = 10$$
$$x = 2$$

**53.**
$$\frac{2x}{x-2} - \frac{16}{x^2-4} = 2$$
$$\frac{2x}{x-2} - \frac{16}{(x+2)(x-2)} = 2$$
The restricted values are $-2$ and $2$. Multiply each term by $(x+2)(x-2)$.
$$2x(x+2) - 16 = 2(x+2)(x-2)$$
$$2x^2 + 4x - 16 = 2x^2 - 8$$
$$4x = 8$$
$$x = 2$$
This is a restricted value and is not a solution. Therefore, there is no solution.

**55.** Graph $y_1 = x^2 + 1$ and $y_2 = \dfrac{3}{x-2}$.

The graphs intersect in one point (remember that the vertical portion of the rational graph is not actually part of the graph—it corresponds to the restricted value of the rational function.) There will be one solution to the given equation.

**57.** Let $x$ = the number of hours for the experienced welder working alone.

| | Number of hours | Work rate |
|---|---|---|
| Helper | $2x$ | $1/(2x)$ |
| Welder | $x$ | $1/x$ |
| Together | 8 | $1/8$ |

The combined work rate is the sum of the individual work rates.
$$\frac{1}{2x} + \frac{1}{x} = \frac{1}{8}$$
$$\frac{8x}{2x} + \frac{8x}{x} = \frac{8x}{8}$$
$$4 + 8 = x$$
$$12 = x$$
It takes 12 hours for the experienced welder to do the job alone and 24 hours for the helper to do the job alone.

**59.** Let $x$ = the number of hours for the tank to fill under these conditions.

| | Number of hours | Work rate |
|---|---|---|
| Outtake | 3 | $1/3$ |
| Intake | 2.5 | $1/2.5$ |
| Together | $x$ | $1/x$ |

Because we are interested in how long it takes to fill the tank, we must consider the outtake pipe to be working to do the opposite: to drain the tank. Therefore, the work rate of the outtake pipe is subtracted from the rate of the intake pipe.
$$\frac{1}{2.5} - \frac{1}{3} = \frac{1}{x}$$
$$\frac{15x}{2.5} - \frac{15x}{3} = \frac{15x}{x}$$
$$6x - 5x = 15$$
$$x = 15$$
The tank will fill in 15 hours under these conditions.

**61.** Let $x$ = the average rate hiking and $x + 50$ = the average rate driving.

| | $d$ | $r$ | $t = d/r$ |
|---|---|---|---|
| Hiking | 2 | $x$ | $2/x$ |
| Driving | 81 | $x + 50$ | $81/(x+50)$ |

The total time was 2 hours.

$$\frac{2}{x} + \frac{81}{x+50} = 2$$
$$2(x+50) + 81x = 2x(x+50)$$
$$2x + 100 + 81x = 2x^2 + 100x$$
$$0 = 2x^2 + 17x - 100$$
$$0 = (2x+25)(x-4)$$
$$2x + 25 = 0 \quad \text{or} \quad x - 4 = 0$$
$$x = -25/2 \quad \text{or} \quad x = 4$$
The hiking rate was 4 mph and the driving rate was 54 mph.

## Chapter 8 Looking Back

1. (a) $(x^2)^4 = x^8$

   (b) $(a^{-3})^5 = a^{-15} = \dfrac{1}{a^{15}}$

3. (a) $(a^2 b^3)^4 = a^8 b^{12}$

   (b) $\left(\dfrac{3}{x^5}\right)^2 = \dfrac{3^2}{(x^5)^2} = \dfrac{9}{x^{10}}$

5. (a) $5^{-2} = \dfrac{1}{5^2} = \dfrac{1}{25}$

   (b) $-3 \cdot 6^0 = -3 \cdot 1 = -3$

   (c) $|-8| = 8$

7. $a^4 b^6 + a^2 b^8 = a^2 b^6 (a^2 + b^2)$

9. 
$$x^2 = 16$$
$$x^2 - 16 = 0$$
$$(x+4)(x-4) = 0$$
$$x + 4 = 0 \quad \text{or} \quad x - 4 = 0$$
$$x = -4 \quad \text{or} \quad x = 4$$

11. 
$$x^2 - 3x = 2^2$$
$$x^2 - 3x = 4$$
$$x^2 - 3x - 4 = 0$$
$$(x-4)(x+1) = 0$$
$$x - 4 = 0 \quad \text{or} \quad x + 1 = 0$$
$$x = 4 \quad \text{or} \quad x = -1$$

## Chapter 8 Test

1. $\dfrac{x-3}{x^2 + 5x - 24} = \dfrac{x-3}{(x+8)(x-3)}$

   When $x = 3$, the denominator is zero.

3. $\dfrac{6x^2 + 12x}{3x^3 - 6x^2} = \dfrac{6x(x+2)}{3x^2(x-2)} = \dfrac{2(x+2)}{x(x-2)}$

5. $\dfrac{x^2 + xy - 2y^2}{y^2 - x^2} = \dfrac{(x+2y)(x-y)}{(y+x)(y-x)}$
   $= -\dfrac{x+2y}{x+y}$

Chapter 8 Test

**7.** 
$$\frac{y^2-25}{2y+10} \div (y^2-10y+25)$$
$$= \frac{y^2-25}{2y+10} \cdot \frac{1}{y^2-10y+25}$$
$$= \frac{(y+5)(y-5)}{2(y+5)} \cdot \frac{1}{(y-5)^2}$$
$$= \frac{1}{2(y-5)}$$

**9.** The expression can be written as $\dfrac{r}{m+n} - \dfrac{r-1}{m+n}$. Because the denominators are the same, the subtraction is easy to perform. The result is $\dfrac{1}{m+n}$.

**11.** $\dfrac{3}{t^2} - \dfrac{5}{t} = \dfrac{3}{t^2} - \dfrac{5t}{t^2} = \dfrac{3-5t}{t^2}$

**13.**
(i) $\dfrac{\frac{m}{n}}{p} = \dfrac{m}{n} \cdot \dfrac{1}{p} = \dfrac{m}{np}$

(ii) $\dfrac{m}{\frac{n}{p}} = \dfrac{m}{1} \cdot \dfrac{p}{n} = \dfrac{mp}{n}$

(iii) $\dfrac{\frac{m}{n}}{\frac{1}{p}} = \dfrac{m}{n} \cdot \dfrac{p}{1} = \dfrac{mp}{n}$

All three are complex fractions. The last two are equivalent.

**15.**
$$\dfrac{x-3+\frac{x-3}{x^2}}{x+2+\frac{1}{x}+\frac{2}{x^2}} = \dfrac{x^2\left(x-3+\frac{x-3}{x^2}\right)}{x^2\left(x+2+\frac{1}{x}+\frac{2}{x^2}\right)}$$
$$= \dfrac{x^3-3x^2+x-3}{x^3+2x^2+x+2}$$
$$= \dfrac{(x^3-3x^2)+(x-3)}{(x^3+2x^2)+(x+2)}$$
$$= \dfrac{x^2(x-3)+(x-3)}{x^2(x+2)+(x+2)}$$
$$= \dfrac{(x-3)(x^2+1)}{(x+2)(x^2+1)}$$
$$= \dfrac{x-3}{x+2}$$

**17.** The restricted values are $-3$ and $2$.
$$\dfrac{3}{a+3} = \dfrac{2}{a-2}$$
$$3(a-2) = 2(a+3)$$
$$3a-6 = 2a+6$$
$$a = 12$$

**19.** The restricted value is 2. Multiply each term by the LCD $3(x-2)$.
$$\dfrac{x}{x-2} + \dfrac{2}{3} = \dfrac{2}{x-2}$$
$$3x + 2(x-2) = 6$$
$$3x + 2x - 4 = 6$$
$$5x = 10$$
$$x = 2$$
This is a restricted value and is not a solution. Therefore, there is no solution.

**21.**
(a) $\dfrac{x-2}{x-1} = \dfrac{1}{1-x}$
$\dfrac{x-2}{x-1} = \dfrac{-1}{x-1}$
$x - 2 = -1$
$x = 1$

(b) $\dfrac{x-2}{x-1} = \dfrac{x+3}{x-1}$
$x - 2 = x + 3$
$-2 = 3$

In (a) the apparent solution is 1, but it is an extraneous solution. In (b) the equation leads to $-2 = 3$, so the equation has no solution.

**23.** Let $s$ = success rate and $w$ = wind speed.
$$s = \frac{k}{w}$$
$$0.6 = \frac{k}{20}$$
$$12 = k$$
The constant of variation is 12.
$$s = \frac{12}{w}$$
$$s = \frac{12}{30}$$
$$s = 0.4$$
The success rate is 40%.

**25.** Let $x$ = the paddling rate in still water.

|  | $d$ | $r$ | $t = d/r$ |
|---|---|---|---|
| Upstream | 0.5 | $x - 3$ | $0.5/(x-3)$ |
| Downstream | 2 | $x + 3$ | $2/(x+3)$ |

$$\frac{0.5}{x-3} = \frac{2}{x+3}$$
$$0.5(x+3) = 2(x-3)$$
$$0.5x + 1.5 = 2x - 6$$
$$7.5 = 1.5x$$
$$5 = x$$
The paddling rate in still water is 5 mph.

## Cumulative Test for Chapters 6-8

**1.** $(x-3)^0 = 1$

**3.** $(2x^2)^{-3} = \frac{1}{(2x^2)^3} = \frac{1}{8x^6}$

**5.** $\frac{18x^{-3}}{27z^{-2}} = \frac{2z^2}{3x^3}$

**7.** (a) $2.7 \cdot 10^{-5} = 0.000027$
   (b) $0.0000000083 = 8.3 \cdot 10^{-9}$

**9.** $(3x^2 - 6x + 8) - (x^3 - 2x^2 - 7) +$
   $(2x^3 - 4x^2 + 7x - 16)$
   $= 3x^2 - 6x + 8 - x^3 + 2x^2 + 7 + 2x^3$
   $\quad - 4x^2 + 7x - 16$
   $= x^3 + x^2 + x - 1$

**11.** (a) $x^4y^3 - x^3y^4 = x^3y^3(x - y)$
   (b) $ax + 3ay - 7bx - 21by$
   $= (ax + 3ay) - (7bx + 21by)$
   $= a(x + 3y) - 7b(x + 3y)$
   $= (x + 3y)(a - 7b)$

**13.** (a) $z^2 + 2z - 63 = (z + 9)(z - 7)$
   (b) $a^2 - 4ab - 45b^2 = (a + 5b)(a - 9b)$
   (c) $12 + 9x - 3x^2 = 3(4 - x)(1 + x)$

**15.** (a) $(x-4)(x-3) = 12$
   $x^2 - 7x + 12 = 12$
   $x^2 - 7x = 0$
   $x(x - 7) = 0$
   $x = 0$ or $x - 7 = 0$
   $\qquad\qquad x = 7$

   (b) $a^4 + 25 = 26a^2$
   $a^4 - 26a^2 + 25 = 0$
   $(a^2 - 1)(a^2 - 25) = 0$
   $(a+1)(a-1)(a+5)(a-5) = 0$
   $a + 1 = 0$ or $a - 1 = 0$ or
   $a = -1$ or $\;a = 1\;$ or
   $a + 5 = 0$ or $a - 5 = 0$
   $a = -5$ or $\;a = 5$

**17.**

$$\begin{array}{r}2c^3+1c^2+3c-2\phantom{)}\\2c-1{\overline{\smash{\big)}\,4c^4+0c^3+5c^2-7c+0\phantom{)}}}\\\underline{4c^4-2c^3\phantom{+5c^2-7c+0)}}\\2c^3+5c^2\phantom{-7c+0)}\\\underline{2c^3-1c^2\phantom{-7c+0)}}\\6c^2-7c\phantom{+0)}\\\underline{6c^2-3c\phantom{+0)}}\\-4c+0\phantom{)}\\\underline{-4c+2\phantom{)}}\\-2\phantom{)}\end{array}$$

$Q(c) = 2c^3 + c^2 + 3c - 2, \ R(c) = -2$

**19.** $\dfrac{x^2 - x - 12}{4x^2 - 13x - 12} = \dfrac{(x-4)(x+3)}{(4x+3)(x-4)}$
$= \dfrac{x+3}{4x+3}$

**21.** $\dfrac{3}{y^2+11y+10} + \dfrac{y}{y^2+5y+4}$

$= \dfrac{3}{(y+10)(y+1)} + \dfrac{y}{(y+1)(y+4)}$

$= \dfrac{3(y+4)}{(y+10)(y+1)(y+4)}$
$\quad + \dfrac{y(y+10)}{(y+10)(y+1)(y+4)}$

$= \dfrac{3(y+4) + y(y+10)}{(y+10)(y+1)(y+4)}$

$= \dfrac{3y+12+y^2+10y}{(y+10)(y+1)(y+4)}$

$= \dfrac{y^2+13y+12}{(y+10)(y+1)(y+4)}$

$= \dfrac{(y+12)(y+1)}{(y+10)(y+1)(y+4)}$

$= \dfrac{y+12}{(y+10)(y+4)}$

**23.** $\dfrac{2}{x} - \dfrac{x+8}{x+2} = \dfrac{12}{x^2+2x}$

$\dfrac{2}{x} - \dfrac{x+8}{x+2} = \dfrac{12}{x(x+2)}$

The restricted values are 0 and $-2$.
Multiply each term by $x(x+2)$.

$2(x+2) - x(x+8) = 12$
$2x + 4 - x^2 - 8x = 12$
$x^2 + 6x + 8 = 0$
$(x+4)(x+2) = 0$
$x+4 = 0 \quad \text{or} \quad x+2 = 0$
$x = -4 \quad \text{or} \quad x = -2$

Because $-2$ is a restricted value, the only solution is $-4$.

**25.** Let $x =$ the rate during the first 220 miles of the trip. Then $x + 5 =$ the rate during the last 120 miles of the trip.

$\dfrac{220}{x} + \dfrac{120}{x+5} = 6$

$220(x+5) + 120x = 6x(x+5)$
$220x + 1100 + 120x = 6x^2 + 30x$
$6x^2 - 310x - 1100 = 0$
$(x-55)(6x+20) = 0$
$x - 55 = 0 \quad \text{or} \quad 6x + 20 = 0$
$x = 55 \quad \text{or} \quad x = -10/3$

The average speed during the first 220 miles was 55 mph.

# Chapter 9

# Radical Expressions

## Section 9.1  Radicals

1. The number 9 has two square roots, $-3$ and 3, because squaring either results in 9. The expression $\sqrt{9}$ means the principal or positive square root of 9, which is 3.

3. The square roots of 36 are 6 and $-6$ because $6^2 = 36$ and $(-6)^2 = 36$.

5. The square root of $-4$ is not a real number because there is no real number $a$ such that $a^2 = -4$.

7. The third root of 27 is 3 because $3^3 = 27$.

9. The fourth roots of 625 are $-5$ and 5 because $5^4 = 625$ and $(-5)^4 = 625$.

11. The fifth root of $-243$ is $-3$ because $(-3)^5 = -243$.

13. The sixth root of $-64$ is not a real number because there is no real number $a$ such that $a^6 = -64$.

15. $\sqrt{16} = 4$ because $4^2 = 16$.

17. $\sqrt[3]{-64} = -4$ because $(-4)^3 = -64$.

19. $\sqrt{\dfrac{4}{9}} = \dfrac{2}{3}$ because $\left(\dfrac{2}{3}\right)^2 = \dfrac{4}{9}$.

21. $\sqrt[4]{-16}$ is not a real number because this is an even root of a negative number.

23. $-\sqrt{16} = -4$ because
$-\sqrt{16} = -1 \cdot \sqrt{16} = -1 \cdot 4 = -4$.

25. $4\sqrt{25} = 20$ because $4 \cdot \sqrt{25} = 4 \cdot 5 = 20$.

27. $\sqrt[3]{64} = 4$ because $4^3 = 64$.

29. $-\sqrt[5]{-32} = -1 \cdot \sqrt[5]{-32} = -1 \cdot (-2) = 2$ because $(-2)^5 = -32$.

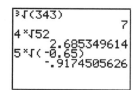

**Figure for Exercises 31, 33, and 35**

31. $\sqrt[3]{343} = 7$   (see first entry)

33. $\sqrt[4]{52} \approx 2.69$   (see second entry)

35. $\sqrt[5]{-0.65} \approx -0.92$   (see third entry)

37. The expression $x^3$ could be positive or negative depending on the value of $x$. The expression $x^2$ is nonnegative for all $x$.

39. $\sqrt{(-4)^2} = |-4| = 4$

41. $\left(\sqrt[4]{7}\right)^4 = 7$

43. $\sqrt[3]{10^3} = 10$

45. $\sqrt{4^2 - 4(3)(1)} = \sqrt{16-12} = \sqrt{4} = 2$

Section 9.1   Radicals

**47.** $\sqrt{9^2 - 4(10)(-9)} = \sqrt{81-(-360)}$
$= \sqrt{441} = 21$

**49.** "Twice the square root of $x$" is translated as $2\sqrt{x}$.

**51.** "The difference of the fifth root of $a$ and 9" is translated as $\sqrt[5]{a} - 9$.

**53.** "The square root of the quantity of $c + b$" is translated as $\sqrt{c+b}$.

**55.** $\sqrt{x^{10}} = \sqrt{(x^5)^2} = x^5$

**57.** $4\sqrt{y^6} = 4\sqrt{(y^3)^2} = 4y^3$

**59.** $\sqrt[3]{x^{21}} = \sqrt[3]{(x^7)^3} = x^7$

**61.** $-\sqrt[3]{y^{15}} = -\sqrt[3]{(y^5)^3} = -y^5$

**63.** $\sqrt[5]{x^{35}} = \sqrt[5]{(x^7)^5} = x^7$

**65.** $-\sqrt[4]{x^8} = -\sqrt[4]{(x^2)^4} = -x^2$

**67.** $\sqrt{(3x)^6} = \sqrt{[(3x)^3]^2} = (3x)^3$

**69.** $\sqrt{(x+3)^{10}} = \sqrt{[(x+3)^5]^2} = (x+3)^5$

**71.**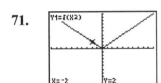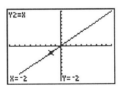

(a) The graph of $y_1$ is the same as the graph of $|x|$, which is not the same as the graph of $y_2$.

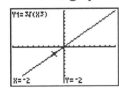 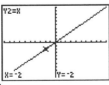

(b) The graphs are the same.

**73.** $\sqrt{x^6} = \sqrt{(x^3)^2} = |x^3|$

**75.** $-\sqrt{x^8} = -\sqrt{(x^4)^2} = -|x^4| = -x^4$

**77.** $\sqrt[5]{x^{15}} = \sqrt[5]{(x^3)^5} = x^3$

**79.** $\sqrt[4]{a^4} = |a|$

**81.** $\sqrt{(x+2)^2} = |x+2|$

**83.** $-2\sqrt{(x-1)^2} = -2|x-1|$

**85.** $\sqrt[3]{(1-x)^3} = 1-x$

**87.** $\sqrt{7-5} > \sqrt{7} - \sqrt{5}$

```
√(7-5)
        1.414213562
√(7)-√(5)
        .4096833336
```

**89.** $\sqrt{25-9} > \sqrt{25} - \sqrt{9}$

```
√(25-9)
              4
√(25)-√(9)
              2
```

**91.** Because $\sqrt{x^4} = \sqrt{(x^2)^2} = |x^2| = x^2$, the missing number is 4.

**93.** Because $\sqrt[8]{y^{24}} = \sqrt[8]{(y^3)^8} = |y^3|$, the missing number is 8.

**95.** Because $-3\sqrt[4]{16} = -3(2) = -6$, the missing number is $-3$.

**97.** $P(x) = 148\sqrt[5]{x}$

(a) For the year 2004, $x = 14$.
$P(14) = 148\sqrt[5]{14} \approx \$251$ million.

(b) From the table, it appears that during the 2012, the PAC contributions will reach $275 million.

| X | Y1 |
|---|---|
| 18 | 263.83 |
| 19 | 266.69 |
| 20 | 269.44 |
| 21 | 272.09 |
| 22 | **274.63** |
| 23 | 277.08 |
| 24 | 279.45 |

Y1=274.628908966

**99.** Cheyenne:
```
26.5→T
          26.5
Y1(15.3)
     -10.05570084
```

Boston:
```
28.6→T
          28.6
Y1(13.8)
     -3.622012586
```

Seattle:
```
40.1→T
          40.1
Y1(9.7)
      21.78255665
```

Norfolk:
```
39.1→T
          39.1
Y1(8.1)
      24.14748135
```

**101.**
$91.4 - (91.4 - T)[0.478 + 0.301(\sqrt{x} - 0.02x)] = y$
$91.4 - (91.4 - T)[0.478 + 0.301(\sqrt{9} - 0.02(9))] = 20$
$91.4 - (91.4 - T)[0.478 + 0.301(3 - 0.18)] = 20$
$91.4 - (91.4 - T)[0.478 + 0.301(2.82)] = 20$
$91.4 - (91.4 - T)(1.32682) = 20$
$91.4 - (121.271348 - 1.32682T) = 20$
$-121.271348 + 1.32682T = -71.4$
$1.32682T = 49.87134$
$T \approx 37.5°$

**103.** $\sqrt[3]{-\sqrt[4]{1}} = \sqrt[3]{-1} = -1$

**105.** $\sqrt[3]{\sqrt{64}} = \sqrt[3]{8} = 2$

**107.** $\sqrt{\sqrt[3]{64}} = \sqrt{4} = 2$

**109.** (a) If we did not exclude $n = 1$, then $\sqrt[1]{b} = a$ would imply that $a^1 = b$, which would mean that $\sqrt[1]{b} = b$. Thus, nothing is gained by including $n = 1$ in the definition.

(b) For $b = 1$, $\sqrt[0]{b} = a$ implies that $a^0 = b$ or $a^0 = 1$. But this statement is true for all $a \neq 0$. Thus, $\sqrt[0]{b}$ would not have a unique value if $b = 1$.
For $b \neq 1$, $\sqrt[0]{b} = a$ implies that $a^0 = b$, which implies that $a^0 \neq 1$. Because this statement is false, $\sqrt[0]{b}$ has no value.

## Section 9.2 Rational Exponents

**1.** Choice (iii) is false because $16^{1/2}$ is the principal square root 4.

**3.** $49^{1/2} = \sqrt{49} = 7$

**5.** $(-64)^{1/3} = \sqrt[3]{-64} = -4$

**7.** $(-16)^{1/4} = \sqrt[4]{-16}$ is not a real number because the $n$ is even and $b < 0$.

**9.** $\left(\dfrac{8}{27}\right)^{1/3} = \sqrt[3]{\dfrac{8}{27}} = \dfrac{2}{3}$

**11.** $(-27)^{1/3} = \sqrt[3]{-27} = -3$

**13.** $-32^{1/5} = -\sqrt[5]{32} = -2$

Section 9.2  Rational Exponents

15. $-(-625)^{1/4} = -\sqrt[4]{-625}$ is not a real number because the $n$ is even and $b < 0$.

17. $-625^{1/4} = -\sqrt[4]{625} = -5$

**Figure for Exercises 19 and 23**

19. $15^{1/4} \approx 1.97$ (see first entry)

21. $(-12)^{1/2} = \sqrt[2]{-12}$ is not a real number because the $n$ is even and $b < 0$. The calculator gives a nonreal answer error.

23. $-4^{1/4} \approx -1.41$ (see second entry)

25. The numerator is the exponent and the denominator is the index.

27. $3x^{2/3} = 3\sqrt[3]{x^2}$

29. $(3x)^{2/3} = \sqrt[3]{(3x)^2}$

31. $(2x+3)^{2/3} = \sqrt[3]{(2x+3)^2}$

33. $3x^{1/2} + (2y)^{1/2} = 3\sqrt{x} + \sqrt{2y}$

35. $\sqrt[3]{y^2} = y^{2/3}$

37. $\sqrt[4]{3x^3} = (3x^3)^{1/4}$

39. $\sqrt{x^2+4} = (x^2+4)^{1/2}$

41. $\sqrt{t} - \sqrt{5} = t^{1/2} - 5^{1/2}$

43. $\sqrt[3]{(2x-1)^2} = (2x-1)^{2/3}$

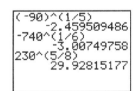

**Figure for Exercises 45, 47, and 49**

45. $\sqrt[5]{-90} = (-90)^{1/5} \approx -2.46$ (see first entry)

47. $-\sqrt[6]{740} = -740^{1/6} \approx -3.01$ (see second entry)

49. $\sqrt[8]{230^5} = 230^{5/8} \approx 29.93$ (see third entry)

51. Use the $\sqrt[x]{\phantom{a}}$ key with $x = 5$ or evaluate $470^{1/5}$.

53. $9^{-1/2} = \dfrac{1}{9^{1/2}} = \dfrac{1}{\sqrt{9}} = \dfrac{1}{3}$

55. $\left(\dfrac{1}{9}\right)^{-1/2} = 9^{1/2} = \sqrt{9} = 3$

57. $27^{2/3} = \left(\sqrt[3]{27}\right)^2 = 3^2 = 9$

59. $(-27)^{-2/3} = \dfrac{1}{(-27)^{2/3}} = \dfrac{1}{\left(\sqrt[3]{-27}\right)^2}$
$= \dfrac{1}{(-3)^2} = \dfrac{1}{9}$

61. $32^{3/5} = \left(\sqrt[5]{32}\right)^3 = 2^3 = 8$

63. $(-16)^{-1/2} = \dfrac{1}{(-16)^{1/2}} = \dfrac{1}{\sqrt{-16}}$ is not a real number because the $n$ is even and $b < 0$.

65. $8^{-2/3} = \dfrac{1}{8^{2/3}} = \dfrac{1}{\left(\sqrt[3]{8}\right)^2} = \dfrac{1}{2^2} = \dfrac{1}{4}$

67. $\left(\dfrac{8}{27}\right)^{-2/3} = \left(\dfrac{27}{8}\right)^{2/3} = \left(\sqrt[3]{\dfrac{27}{8}}\right)^2$
$= \left(\dfrac{3}{2}\right)^2 = \dfrac{9}{4}$

**69.** $-(-8)^{-4/3} = -\dfrac{1}{(-8)^{4/3}} = -\dfrac{1}{\left(\sqrt[3]{-8}\right)^4}$

$= -\dfrac{1}{(-2)^4} = -\dfrac{1}{16}$

**71.** $(-36)^{5/2} = \left(\sqrt[2]{-36}\right)^5$ is not a real number because the $n$ is even and $b < 0$.

**73.** $-4^{-5/2} = -\dfrac{1}{4^{5/2}} = -\dfrac{1}{\left(\sqrt[2]{4}\right)^5} = -\dfrac{1}{2^5}$

$= -\dfrac{1}{32}$

**75.** $16^{1/2} + 49^{1/2} = \sqrt{16} + \sqrt{49} = 4 + 7 = 11$

**77.** $27^{1/3} + 4^{-1/2} = \sqrt[3]{27} + \dfrac{1}{4^{1/2}} = 3 + \dfrac{1}{\sqrt{4}}$

$= 3 + \dfrac{1}{2} = \dfrac{6}{2} + \dfrac{1}{2} = \dfrac{7}{2}$

**79.** $(-8)^{2/3}\left(16^{1/4}\right) = \left(\sqrt[3]{-8}\right)^2 \left(\sqrt[4]{16}\right)$

$= (-2)^2(2) = (4)(2) = 8$

**81.** $16^{0.25} = 16^{1/4} = \sqrt[4]{16} = 2$

**83.** $9^{-0.5} = 9^{-1/2} = \dfrac{1}{9^{1/2}} = \dfrac{1}{\sqrt{9}} = \dfrac{1}{3}$

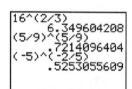

**Figure for Exercises 85, 87, and 89**

**85.** $16^{2/3} \approx 6.35$  (see first entry)

**87.** $\left(\dfrac{5}{9}\right)^{5/9} \approx 0.72$  (see second entry)

**89.** $(-5)^{-2/5} \approx 0.53$  (see third entry)

**91.** Because $16^{-1/2} = \dfrac{1}{16^{1/2}} = \dfrac{1}{\sqrt{16}} = \dfrac{1}{4}$, the missing number is 16.

**93.** Because

$(-8)^{5/3} = \left(\sqrt[3]{-8}\right)^5 = (-2)^5 = -32$,

the missing number is $-8$.

**95.** Because

$(-27)^{2/3} = \left(\sqrt[3]{-27}\right)^2 = (-3)^2 = 9$,

the missing number is 2/3.

**97.**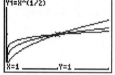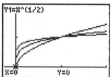

(a) From the graphs, we can see that all three graphs intersect at $(0, 0)$ and $(1, 1)$. If $x = 0$, then $x^{1/n} = 0^{1/n} = 0$ for all $n > 1$. If $x = 1$, then $x^{1/n} = 1^{1/n} = 1$ for all $n > 1$.

(b) By tracing the graphs, we see that $x^{1/2} < x^{1/4} < x^{1/8}$ when $0 < x < 1$.

(c) By tracing the graphs, we see that $x^{1/2} > x^{1/4} > x^{1/8}$ when $x > 1$.

**99.** (a) Graph $y_1 = 60x^{2/9}$ and $y_2 = 70$. By finding the point of intersection of these two graphs, we find that if the passing grade is 70, approximately 2 hours of study are required to pass.

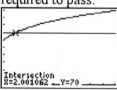

(b) Add the graph of $y_3 = 90$. To earn a 90 on the exam, a total of approximately 6.2 hours of study, or 4.2 additional hours, are needed.

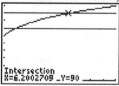

(c) $G(0) = 60(0)^{2/9} = 0$

Yes, from the graph we can see that it is possible to get a 100 on the exam, according to the model. Even if a student does not study for the final, it is unlikely that the exam

Section 9.3    Properties of Rational Exponents

final, it is unlikely that the exam grade would be 0. The model indicates that a perfect score can be earned with about 10 hours of study. This seems realistic.

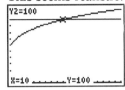

101. (a) Let $x = 25$:
$$f(x) = 154x^{-1/5}$$
$$f(25) = 154(25)^{-1/5}$$
$$\approx 81$$
About 81% of those of age 25 are expected to file early.

(b) From the graph, we see that at about age 37, 75% are expected to file early.

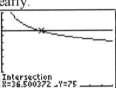

103. $\sqrt{16^{-1/2}} = \sqrt{\dfrac{1}{16^{1/2}}} = \sqrt{\dfrac{1}{\sqrt{16}}} = \sqrt{\dfrac{1}{4}} = \dfrac{1}{2}$

105. $\left(\sqrt[3]{64}\right)^{-1/2} = 4^{-1/2} = \dfrac{1}{4^{1/2}} = \dfrac{1}{\sqrt{4}} = \dfrac{1}{2}$

107. The graph is displayed as a set of distinguishable points rather than as a smooth curve because the function is not defined if $x$ is an even number.

## Section 9.3  Properties of Rational Exponents

1. Choice (i) is true according to the Power of a Product and the Power to a Power Rule for Exponents. Choice (ii) is untrue for $a = 3$ and $b = 4$, for example:
$$(3^2 + 4^2)^{1/2} = (9 + 16)^{1/2}$$
$$= 25^{1/2} = 5 \neq 3 + 4 = 7$$

3. $(6^3)^{2/3} = 6^{3 \cdot (2/3)} = 6^2 = 36$

5. $(2^{2/7})^{14} = 2^{(2/7) \cdot 14} = 2^4 = 16$

7. $12^{1/2} \cdot 3^{1/2} = (12 \cdot 3)^{1/2} = 36^{1/2} = \sqrt{36} = 6$

9. $\left(\dfrac{7^3}{2^6}\right)^{-2/3} = \left(\dfrac{2^6}{7^3}\right)^{2/3} = \dfrac{2^{6 \cdot (2/3)}}{7^{3 \cdot (2/3)}} = \dfrac{2^4}{7^2}$
$$= \dfrac{16}{49}$$

11. $65^{1/3} \cdot 65^{2/3} = 65^{(1/3)+(2/3)} = 65^1 = 65$

13. $49^{7/10} \cdot 49^{-1/5} = 49^{(7/10)-(1/5)}$
$$= 49^{(7/10)-(2/10)}$$
$$= 49^{5/10} = 49^{1/2}$$
$$= \sqrt{49} = 7$$

15. $8^{2/9} \cdot 8^{1/9} = 8^{(2/9)+(1/9)} = 8^{3/9} = 8^{1/3}$
$$= \sqrt[3]{8} = 2$$

17. $6^{0.6} \cdot 6^{0.4} = 6^{0.6+0.4} = 6^1 = 6$

19. $\dfrac{27^{1/3}}{27^{2/3}} = 27^{(1/3)-(2/3)} = 27^{-1/3}$
$$= \dfrac{1}{27^{1/3}} = \dfrac{1}{\sqrt[3]{27}} = \dfrac{1}{3}$$

**21.** $8^{1/6} \div 8^{1/2} = 8^{1/6} \cdot 8^{-1/2} = 8^{(1/6)-(1/2)}$
$= 8^{(1/6)-(3/6)} = 8^{-2/6} = 8^{-1/3}$
$= \dfrac{1}{8^{1/3}} = \dfrac{1}{\sqrt[3]{8}} = \dfrac{1}{2}$

**23.** Choice (i) is false because the exponents were multiplied when they should have been added. Choice (ii) is true because the Product Rule for Exponents was applied properly.

**25.** $x^{2/3} \cdot x^{1/3} = x^{(2/3)+(1/3)} = x^1 = x$

**27.** $t \cdot t^{5/7} = t^{1+(5/7)} = t^{(7/7)+(5/7)} = t^{12/7}$

**29.** $(a^{-1/2}b)(a^{3/4}b^{1/2})$
$= a^{(3/4)-(1/2)}b^{1+(1/2)}$
$= a^{(3/4)-(2/4)}b^{(2/2)+(1/2)}$
$= a^{1/4}b^{3/2}$

**31.** $\dfrac{z^{1/2}}{z^{1/3}} = z^{(1/2)-(1/3)} = z^{(3/6)-(2/6)} = z^{1/6}$

**33.** $\dfrac{a^{1/3}b}{b^{-1/3}} = a^{1/3}b^{1-(-1/3)} = a^{1/3}b^{4/3}$

**35.** $(a^{-2})^{1/2} = a^{-2 \cdot (1/2)} = a^{-1} = \dfrac{1}{a}$

**37.** $(y^{2/3})^{3/2} = y^{(2/3) \cdot (3/2)} = y^1 = y$

**39.** $(a^6 b^4)^{3/2} = (a^6)^{3/2}(b^4)^{3/2}$
$= a^{6(3/2)}b^{4(3/2)} = a^9 b^6$

**41.** Because $(9x^4)^{1/2} = 9^{1/2}x^{4 \cdot (1/2)} = 3x^2$, the missing number is 1/2.

**43.** Because $(x^{2/3})^{3/2} = x^{(2/3) \cdot (3/2)} = x^1 = x$, the missing number is 3/2.

**45.** In the first expression, the index and the exponent have a common factor. Thus, when the expression is written with a rational exponent, the exponent can be reduced. None of this is true for the second expression.

**47.** $\left[(-2)^6\right]^{1/2} = \left\{\left[(-2)^3\right]^2\right\}^{1/2} = \left|(-2)^3\right|$
$= |-8| = 8$

**49.** $(a^{15})^{1/5} = ((a^3)^5)^{1/5} = a^3$

**51.** $(a^4 b^2)^{1/2} = (a^4)^{1/2}(b^2)^{1/2}$
$= ((a^2)^2)^{1/2}(b^2)^{1/2}$
$= |a^2| \cdot |b|$
$= a^2 |b|$

**53.** $\left[(x+y)^2\right]^{1/2} = |x+y|$

**55.** $(a^{-4}b^{1/3})^{1/2} = a^{-4 \cdot (1/2)}b^{(1/3) \cdot (1/2)}$
$= a^{-2}b^{1/6} = \dfrac{b^{1/6}}{a^2}$

**57.** $(a^6 b^{-9})^{2/3} = a^{6(2/3)}b^{-9(2/3)}$
$= a^4 b^{-6} = \dfrac{a^4}{b^6}$

**59.** $(-3^{2/3}x^{1/2}y^{-1/3})^3$
$= (-1 \cdot 3^{2/3}x^{1/2}y^{-1/3})^3$
$= (-1)^3 \cdot 3^{(2/3) \cdot 3}x^{(1/2) \cdot 3}y^{(-1/3) \cdot 3}$
$= -1 \cdot 3^2 \cdot x^{3/2} y^{-1}$
$= \dfrac{-9x^{3/2}}{y}$

**61.** $\left(\dfrac{a^{4/5}}{b^{8/9}}\right)^{-15/4} = \left(\dfrac{b^{8/9}}{a^{4/5}}\right)^{15/4}$
$= \dfrac{b^{(8/9) \cdot (15/4)}}{a^{(4/5) \cdot (15/4)}} = \dfrac{b^{120/36}}{a^{60/20}}$
$= \dfrac{b^{10/3}}{a^3}$

**63.** $\left(\dfrac{x^{-1/3}}{y^{-1/2}}\right)^{-6} = \left(\dfrac{y^{-1/2}}{x^{-1/3}}\right)^6 = \left(\dfrac{x^{1/3}}{y^{1/2}}\right)^6$
$= \dfrac{x^{(1/3) \cdot 6}}{y^{(1/2) \cdot 6}} = \dfrac{x^2}{y^3}$

**65.** $\left(\dfrac{16x^{-12}y^8}{z^4}\right)^{1/4} = \left(\dfrac{16y^8}{x^{12}z^4}\right)^{1/4}$

$= \dfrac{16^{1/4}y^{8 \cdot (1/4)}}{x^{12 \cdot (1/4)}z^{4 \cdot (1/4)}} = \dfrac{2y^2}{x^3 z}$

**67.** $(a^{-2/3}b^3)^6 (a^{5/7}b^{-2})^7$

$= a^{(-2/3) \cdot 6} b^{3 \cdot 6} a^{(5/7) \cdot 7} b^{-2 \cdot 7}$

$= a^{-4} b^{18} a^5 b^{-14}$

$= a^{5-4} b^{18-14} = ab^4$

**69.** $\dfrac{(a^{5/6} b^{-2/5} c^{7/60})^{30}}{a^{21} b^{-5}}$

$= \dfrac{a^{(5/6) \cdot 30} b^{(-2/5) \cdot 30} c^{(7/60) \cdot 30}}{a^{21} b^{-5}}$

$= \dfrac{a^{25} b^{-12} c^{7/2}}{a^{21} b^{-5}} = \dfrac{a^4 c^{7/2}}{b^7}$

**71.** $\dfrac{(9x^4 y^{-2})^{-1/2}}{(x^6 y^{-3})^{1/3}} = \dfrac{9^{-1/2} x^{4 \cdot (-1/2)} y^{(-2) \cdot (-1/2)}}{x^{6 \cdot (1/3)} y^{(-3) \cdot (1/3)}}$

$= \dfrac{x^{-2} y}{9^{1/2} x^2 y^{-1}} = \dfrac{y^2}{3x^4}$

**73.** $a^{3/4}(a^{5/4} + a^{-3/4})$

$= a^{3/4} \cdot a^{5/4} + a^{3/4} \cdot a^{-3/4}$

$= a^{(3/4)+(5/4)} + a^{(3/4)-(3/4)}$

$= a^{8/4} + a^0 = a^2 + 1$

**75.** $x^{-1/4}(x^{9/4} - x^{2/3})$

$= x^{-1/4} \cdot x^{9/4} - x^{-1/4} \cdot x^{2/3}$

$= x^{(-1/4)+(9/4)} - x^{(-1/4)+(2/3)}$

$= x^{8/4} - x^{(-3/12)+(8/12)}$

$= x^2 - x^{5/12}$

**77.** $(x^{1/2} + y^{1/2})(x^{1/2} - y^{1/2})$

$= (x^{1/2})^2 - (y^{1/2})^2 = x - y$

**79.** $(x^{1/2} - 3^{1/2})^2$

$= (x^{1/2})^2 - 2x^{1/2} 3^{1/2} + (3^{1/2})^2$

$= x - 2x^{1/2} 3^{1/2} + 3$

$= x + 3 - 2(3x)^{1/2}$

**81.** $\sqrt[6]{x^3} = x^{3/6} = x^{1/2} = \sqrt{x}$

**83.** $\sqrt[9]{8x^3} = \sqrt[9]{2^3 x^3} = \sqrt[9]{(2x)^3} = (2x)^{3/9}$

$= (2x)^{1/3} = \sqrt[3]{2x}$

**85.** $\sqrt[10]{x^4 y^6} = \sqrt[10]{(x^2 y^3)^2} = (x^2 y^3)^{2/10}$

$= (x^2 y^3)^{1/5} = \sqrt[5]{x^2 y^3}$

**87.** $\left(\sqrt[4]{a}\right)^2 = a^{2/4} = a^{1/2} = \sqrt{a}$

**89.** $\sqrt[3]{x^2} \sqrt[4]{x} = x^{2/3} \cdot x^{1/4} = x^{(2/3)+(1/4)}$

$= x^{(8/12)+(3/12)} = x^{11/12} = \sqrt[12]{x^{11}}$

**91.** $\dfrac{\sqrt[4]{x}}{\sqrt[3]{x^2}} = \dfrac{x^{1/4}}{x^{2/3}} = x^{(1/4)-(2/3)} = x^{(3/12)-(8/12)}$

$= x^{-5/12} = \dfrac{1}{x^{5/12}} = \dfrac{1}{\sqrt[12]{x^5}}$

**93.** $\sqrt[4]{\sqrt[3]{7}} = (7^{1/3})^{1/4} = 7^{1/12} = \sqrt[12]{7}$

**95.** $\sqrt{2} \sqrt[3]{5} = 2^{1/2} \cdot 5^{1/3} = 2^{3/6} \cdot 5^{2/6}$

$= (2^3)^{1/6} \cdot (5^2)^{1/6} = (8)^{1/6} \cdot (25)^{1/6}$

$= (8 \cdot 25)^{1/6} = (200)^{1/6} = \sqrt[6]{200}$

**97.** (a) For $p = 90$, $16p^{1/4} = 16(90)^{1/4} \approx 49$

In a city of 90,000, there would be approximately 49 crime victims per 1000 citizens.

(b) The total number of crimes is found by multiplying the number of crimes per thousand by the population of the city (in thousands).

$p \cdot 16p^{1/4} = 16p^{(1/4)+1} = 16p^{5/4}$

(c) For $p = 1000$ thousand people,

$16p^{5/4} = 16(1000)^{5/4} \approx 89{,}975$

There would be a total of approximately 89,975 crimes.

**99.** Let $t = 11$:

$E(t) = 1150t^{1/4}$

$E(11) = 1150(11)^{1/4}$

$\approx 2094$

The expenditures in 2005 will be $2094 million.

$V(t) = 250t^{1/12}$
$V(11) = 250(11)^{1/12}$
$\approx 305$
The number of visitors in 2005 will be 305 million.

**101.** The function $C(t)$ models the average expenditure per visitor.

**103.** $3x^{-1/2} + 5x^{3/2} = x^{-1/2}(3 + 5x^2)$

**105.** $5a^{1/2} - 10a^{-1/2} = 5a^{-1/2}(a - 2)$

**107.** $t^{2/5} + 4t^{1/5} - 5 = (t^{1/5} + 5)(t^{1/5} - 1)$

**109.** $2x^{1/3} - 7x^{1/6} - 15$
$= 2x^{2/6} - 7x^{1/6} - 15$
$= (2x^{1/6} + 3)(x^{1/6} - 5)$

**111.** $\sqrt{\sqrt{\sqrt{x}}} = \left[(x^{1/2})^{1/2}\right]^{1/2} = \left[x^{1/4}\right]^{1/2}$
$= x^{1/8} = \sqrt[8]{x}$

## Section 9.4 The Product Rule for Radicals

**1.** The Product Rule for Radicals applies to products of real numbers. The factors $\sqrt{-3}$ and $\sqrt{-12}$ are not real numbers.

**3.** $\sqrt{2}\sqrt{3} = \sqrt{2 \cdot 3} = \sqrt{6}$

**5.** $\sqrt{3x}\sqrt{5y} = \sqrt{3 \cdot 5 \cdot xy} = \sqrt{15xy}$

**7.** $(-4\sqrt{2})(3\sqrt{7}) = -4 \cdot 3\sqrt{2 \cdot 7} = -12\sqrt{14}$

**9.** $\sqrt[3]{x-2}\sqrt[3]{x+3} = \sqrt[3]{(x-2)(x+3)}$
$= \sqrt[3]{x^2 + x - 6}$

**11.** The first step is to factor the trinomial: $\sqrt{(x+5)^2} = x+5$. This step is necessary because there is no rule for simplifying a radical whose radicand is a sum. Although $x^2 + 10x + 21$ can be factored, the result is not a perfect square. Therefore, the radical cannot be simplified.

**13.** $\sqrt{25y^2} = \sqrt{25} \cdot \sqrt{y^2} = 5y$

**15.** $\sqrt{49w^8 t^4} = \sqrt{49} \cdot \sqrt{w^8} \cdot \sqrt{t^4} = 7w^4 t^2$

**17.** $\sqrt[3]{8t^6} = \sqrt[3]{8} \cdot \sqrt[3]{t^6} = 2t^2$

**19.** $\sqrt[4]{81x^{12} y^{20}} = \sqrt[4]{81} \cdot \sqrt[4]{x^{12}} \cdot \sqrt[4]{y^{20}}$
$= 3x^3 y^5$

**21.** $\sqrt{x^2(y-3)^4} = \sqrt{x^2} \cdot \sqrt{(y-3)^4}$
$= x(y-3)^2$

**23.** $\sqrt{x^2 + 12x + 36} = \sqrt{(x+6)^2} = x+6$

**25.** $\sqrt{4y^2 + 4y + 1} = \sqrt{(2y+1)^2} = 2y+1$

**27.** The radicand 32 has a perfect square factor, 16. The radicand 30 does not have a perfect square factor.

**29.** $\sqrt{12} = \sqrt{4 \cdot 3} = \sqrt{4} \cdot \sqrt{3} = 2\sqrt{3}$

**31.** $\sqrt{50} = \sqrt{25 \cdot 2} = \sqrt{25} \cdot \sqrt{2} = 5\sqrt{2}$

**33.** $\sqrt[3]{54} = \sqrt[3]{27 \cdot 2} = \sqrt[3]{27} \cdot \sqrt[3]{2} = 3\sqrt[3]{2}$

**35.** $\sqrt[3]{-48} = \sqrt[3]{-8 \cdot 6} = \sqrt[3]{-8} \cdot \sqrt[3]{6} = -2\sqrt[3]{6}$

**37.** $\sqrt[4]{32} = \sqrt[4]{16 \cdot 2} = \sqrt[4]{16} \cdot \sqrt[4]{2} = 2\sqrt[4]{2}$

**39.** $\sqrt[5]{256} = \sqrt[5]{32 \cdot 8} = \sqrt[5]{32} \cdot \sqrt[5]{8} = 2\sqrt[5]{8}$

**41.** $\sqrt{a^5 b^3} = \sqrt{a \cdot a^4 \cdot b \cdot b^2}$
$= \sqrt{a^4 b^2} \cdot \sqrt{ab}$
$= a^2 b \sqrt{ab}$

**43.** $\sqrt[3]{a^9 b^8} = \sqrt[3]{a^9 \cdot b^6 \cdot b^2}$
$= \sqrt[3]{a^9 b^6} \cdot \sqrt[3]{b^2}$
$= a^3 b^2 \sqrt[3]{b^2}$

**45.** $\sqrt[4]{x^{16} y^9} = \sqrt[4]{x^{16} \cdot y^8 \cdot y}$
$= \sqrt[4]{x^{16} y^8} \cdot \sqrt[4]{y}$
$= x^4 y^2 \sqrt[4]{y}$

**47.** $\sqrt{60t} = \sqrt{4 \cdot 15 \cdot t}$
$= \sqrt{4} \cdot \sqrt{15t}$
$= 2\sqrt{15t}$

**49.** $\sqrt{25 x^{25}} = \sqrt{25 \cdot x^{24} \cdot x}$
$= \sqrt{25 x^{24}} \cdot \sqrt{x}$
$= 5 x^{12} \sqrt{x}$

**51.** $\sqrt{500 a^7 b^{14}} = \sqrt{100 \cdot a^6 \cdot b^{14} \cdot 5 \cdot a}$
$= \sqrt{100 a^6 b^{14}} \cdot \sqrt{5a}$
$= 10 a^3 b^7 \sqrt{5a}$

**53.** $-\sqrt{49 x^6 y^7} = -\sqrt{49 \cdot x^6 \cdot y^6 \cdot y}$
$= -\sqrt{49 x^6 y^6} \cdot \sqrt{y}$
$= -7 x^3 y^3 \sqrt{y}$

**55.** $\sqrt{28 w^6 t^9} = \sqrt{4 w^6 t^8} \cdot \sqrt{7t}$
$= 2 w^3 t^4 \sqrt{7t}$

**57.** $\sqrt[3]{81 x^9 y^{11}} = \sqrt[3]{27 x^9 y^9} \cdot \sqrt[3]{3 y^2}$
$= 3 x^3 y^3 \sqrt[3]{3 y^2}$

**59.** $\sqrt[5]{-32 x^7 y^9} = \sqrt[5]{-32 x^5 y^5} \cdot \sqrt[5]{x^2 y^4}$
$= -2xy \sqrt[5]{x^2 y^4}$

**61.** Factor out the common factor 4 from the radicand. Then apply the Product Rule for Radicals.

**63.** $\sqrt{12x + 8y} = \sqrt{4(3x + 2y)}$
$= \sqrt{4} \cdot \sqrt{3x + 2y}$
$= 2\sqrt{3x + 2y}$

**65.** $\sqrt{4x^2 + 16} = \sqrt{4(x^2 + 4)}$
$= \sqrt{4} \cdot \sqrt{x^2 + 4}$
$= 2\sqrt{x^2 + 4}$

**67.** $\sqrt{9 x^7 y^6 + 9 x^6 y^7} = \sqrt{9 x^6 y^6 (x + y)}$
$= \sqrt{9 x^6 y^6} \cdot \sqrt{x + y}$
$= 3 x^3 y^3 \sqrt{x + y}$

**69.** $\sqrt{9 y^2 + 9 y^4} = \sqrt{9 y^2 (1 + y^2)}$
$= \sqrt{9 y^2} \cdot \sqrt{1 + y^2}$
$= 3|y| \sqrt{1 + y^2}$

**71.** $\sqrt{r^4 t^2 + r^6 t^4} = \sqrt{r^4 t^2 (1 + r^2 t^2)}$
$= \sqrt{r^4 t^2} \cdot \sqrt{1 + r^2 t^2}$
$= r^2 |t| \sqrt{1 + r^2 t^2}$

**73.** $\sqrt{y^2 - 10y + 25} = \sqrt{(y-5)^2} = |y - 5|$

**75.** $\left(-2\sqrt{5}\right)\left(2\sqrt{5}\right) = -4\sqrt{25} = -4 \cdot 5 = -20$

**77.** $\sqrt[3]{3} \sqrt[3]{54} = \sqrt[3]{162} = \sqrt[3]{27} \cdot \sqrt[3]{6} = 3\sqrt[3]{6}$

**79.** $\sqrt{7x} \sqrt{14x} = \sqrt{98 x^2}$
$= \sqrt{49 x^2} \cdot \sqrt{2}$
$= 7x \sqrt{2}$

**81.** $\sqrt{3 x^3} \sqrt{2 x^6} = \sqrt{6 x^9}$
$= \sqrt{x^8} \cdot \sqrt{6x}$
$= x^4 \sqrt{6x}$

**83.** $\left(2\sqrt{5}\right)^2 = 2^2 \left(\sqrt{5}\right)^2 = 4 \cdot 5 = 20$

**85.** $\sqrt{6 x y^3} \sqrt{12 x^3 y^2} = \sqrt{72 x^4 y^5}$
$= \sqrt{36 x^4 y^4} \cdot \sqrt{2y}$
$= 6 x^2 y^2 \sqrt{2y}$

87. $\sqrt[3]{25x^2y^2}\ \sqrt[3]{5x^3y^4} = \sqrt[3]{125x^5y^6}$
$= \sqrt[3]{125x^3y^6} \cdot \sqrt[3]{x^2}$
$= 5xy^2\ \sqrt[3]{x^2}$

89. $\sqrt[4]{54x^3y}\ \sqrt[4]{3x^2y^4} = \sqrt[4]{162x^5y^5}$
$= \sqrt[4]{81x^4y^4} \cdot \sqrt[4]{2xy}$
$= 3xy\ \sqrt[4]{2xy}$

91. Because
$\sqrt{16a^4b^3} = \sqrt{16a^4b^2} \cdot \sqrt{b} = 4a^2b\sqrt{b}$,
the missing numbers are 16, 4, and 3.

93. Because $\sqrt{3x^2y^3}\ \sqrt{3x^2y^3} = 3x^2y^3$, the missing numbers are 2, 3, and 3.

95. Because
$\sqrt[3]{16x^4y^3} = \sqrt[3]{8x^3y^3} \cdot \sqrt[3]{2x} = 2xy\sqrt[3]{2x}$,
the missing numbers are 16, 4, and 3.

97. (a) The function we seek, $N_1(x)$, is the product of the percent of African-American households with computers (as a decimal, rather than a percent) and the number of African-American households:
$N_1(x) = \dfrac{C_1(x)}{100} \cdot H_1(x)$
$= \dfrac{(0.7x\sqrt[5]{x})}{100} \cdot (6.9\sqrt[5]{x})$
$= 0.0483x\sqrt[5]{x^2}$

(b) Similarly, we have:
$N_2(x) = \dfrac{C_2(x)}{100} \cdot H_2(x)$
$= \dfrac{0.31x\sqrt{x}}{100} \cdot 2.1\sqrt{x}$
$= 0.00651x(\sqrt{x})^2$
$= 0.00651x(x)$
$= 0.00651x^2$

99.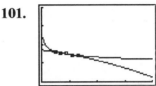
The function is
$Q_1(t) = -0.004t^2 - 0.2t + 17.2$

101.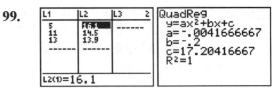
Function $M$ projects a continued slight decline in union membership, whereas function $Q_1$ projects a sharp decline. In the long run, function $M$ is probably the more realistic model.

103. The radical is not simplified because the radicand has a perfect cube factor.

105. $\sqrt[8]{16x^{12}y^{20}} = \sqrt[8]{2^4x^{12}y^{20}}$
$= 2^{4/8}x^{12/8}y^{20/8}$
$= 2^{1/2}x^{3/2}y^{5/2}$
$= 2^{1/2} \cdot x \cdot x^{1/2} \cdot y^2 \cdot y^{1/2}$
$= xy^2\ \sqrt{2xy}$

107. $\sqrt[21]{64x^{18}y^{15}} = \sqrt[21]{2^6x^{18}y^{15}}$
$= 2^{6/21}x^{18/21}y^{15/21}$
$= 2^{2/7}x^{6/7}y^{5/7}$
$= \sqrt[7]{2^2x^6y^5}$
$= \sqrt[7]{4x^6y^5}$

109. If a number larger than $\sqrt{n}$ divided $n$ evenly, then the quotient would be a number less than $\sqrt{n}$, and that number would have already been tested. Thus, only numbers that do not exceed $\sqrt{n}$ need to be tested.

## Section 9.5  The Quotient Rule for Radicals

**1.** The rule does not apply because the indices are different.

**3.** $\dfrac{\sqrt{30}}{\sqrt{5}} = \sqrt{\dfrac{30}{5}} = \sqrt{6}$

**5.** $\dfrac{\sqrt{48}}{\sqrt{3}} = \sqrt{\dfrac{48}{3}} = \sqrt{16} = 4$

**7.** $\dfrac{\sqrt{x^7}}{\sqrt{x}} = \sqrt{\dfrac{x^7}{x}} = \sqrt{x^6} = x^3$

**9.** $\dfrac{\sqrt{49a^7b^4}}{\sqrt{4a}} = \dfrac{\sqrt{49}\sqrt{a^7b^4}}{\sqrt{4}\sqrt{a}}$
$= \dfrac{7}{2}\sqrt{\dfrac{a^7b^4}{a}}$
$= \dfrac{7}{2}\sqrt{a^6b^4}$
$= \dfrac{7a^3b^2}{2}$

**11.** $\dfrac{\sqrt[3]{x^{11}}}{\sqrt[3]{x^8}} = \sqrt[3]{\dfrac{x^{11}}{x^8}} = \sqrt[3]{x^3} = x$

**13.** $\dfrac{\sqrt{x^5y^{12}}}{\sqrt{xy^3}} = \sqrt{\dfrac{x^5y^{12}}{xy^3}}$
$= \sqrt{x^4y^9}$
$= \sqrt{x^4y^8} \cdot \sqrt{y}$
$= x^2y^4\sqrt{y}$

**15.** $\dfrac{\sqrt{45w^9t^8}}{\sqrt{w^2t^3}} = \sqrt{\dfrac{45w^9t^8}{w^2t^3}}$
$= \sqrt{45w^7t^5}$
$= \sqrt{9w^6t^4} \cdot \sqrt{5wt}$
$= 3w^3t^2\sqrt{5wt}$

**17.** $\dfrac{\sqrt{18x^9d^{13}}}{\sqrt{2x^4d^{10}}} = \sqrt{\dfrac{18x^9d^{13}}{2x^4d^{10}}}$
$= \sqrt{9x^5d^3}$
$= \sqrt{9x^4d^2} \cdot \sqrt{xd}$
$= 3x^2d\sqrt{xd}$

**19.** $\dfrac{\sqrt[3]{54x^{13}}}{\sqrt[3]{x^2}} = \sqrt[3]{\dfrac{54x^{13}}{x^2}}$
$= \sqrt[3]{54x^{11}}$
$= \sqrt[3]{27x^9} \cdot \sqrt[3]{2x^2}$
$= 3x^3\sqrt[3]{2x^2}$

**21.** $\dfrac{\sqrt{10}}{\sqrt{360}} = \sqrt{\dfrac{10}{360}} = \sqrt{\dfrac{1}{36}} = \dfrac{\sqrt{1}}{\sqrt{36}} = \dfrac{1}{6}$

**23.** $\dfrac{\sqrt{t^3}}{\sqrt{t^9}} = \sqrt{\dfrac{t^3}{t^9}} = \sqrt{\dfrac{1}{t^6}} = \dfrac{\sqrt{1}}{\sqrt{t^6}} = \dfrac{1}{t^3}$

**25.** $\dfrac{\sqrt{3a^3}}{\sqrt{75a^5}} = \sqrt{\dfrac{3a^3}{75a^5}}$
$= \sqrt{\dfrac{1}{25a^2}}$
$= \dfrac{\sqrt{1}}{\sqrt{25a^2}}$
$= \dfrac{1}{5a}$

**27.** $\dfrac{\sqrt{16xy}}{\sqrt{9x^3y^5}} = \sqrt{\dfrac{16xy}{9x^3y^5}}$
$= \sqrt{\dfrac{16}{9x^2y^4}}$
$= \dfrac{\sqrt{16}}{\sqrt{9x^2y^4}}$
$= \dfrac{4}{3xy^2}$

**29.** $\dfrac{\sqrt[4]{x^6}}{\sqrt[4]{16x^2}} = \sqrt[4]{\dfrac{x^6}{16x^2}} = \sqrt[4]{\dfrac{x^4}{16}} = \dfrac{\sqrt[4]{x^4}}{\sqrt[4]{16}} = \dfrac{x}{2}$

**31.** The radical could be simplified if $Q$ is a perfect square or if $Q$ divides $P$.

**33.** $\sqrt{\dfrac{5}{49}} = \dfrac{\sqrt{5}}{\sqrt{49}} = \dfrac{\sqrt{5}}{7}$

**35.** $\sqrt[3]{\dfrac{16x^2}{2x^{14}}} = \sqrt[3]{\dfrac{8}{x^{12}}} = \dfrac{\sqrt[3]{8}}{\sqrt[3]{x^{12}}} = \dfrac{2}{x^4}$

**37.** $\sqrt{\dfrac{28}{9}} = \dfrac{\sqrt{28}}{\sqrt{9}} = \dfrac{\sqrt{4}\sqrt{7}}{\sqrt{9}} = \dfrac{2\sqrt{7}}{3}$

**39.** $\sqrt[3]{\dfrac{-16a}{1000}} = \dfrac{\sqrt[3]{-16a}}{\sqrt[3]{1000}} = \dfrac{\sqrt[3]{-8}\sqrt[3]{2a}}{\sqrt[3]{1000}} = -\dfrac{\sqrt[3]{2a}}{5}$

**41.** $\sqrt{\dfrac{60x^{11}y^3}{3y}} = \sqrt{20x^{11}y^2}$
$= \sqrt{4x^{10}y^2} \cdot \sqrt{5x}$
$= 2x^5 y \sqrt{5x}$

**43.** $\sqrt{\dfrac{75x^7 y^{13}}{x^3 y^5}} = \sqrt{75x^4 y^8}$
$= \sqrt{25x^4 y^8} \cdot \sqrt{3}$
$= 5x^2 y^4 \sqrt{3}$

**45.** $\sqrt{\dfrac{245x^7 y^9 z^{12}}{5x^3 y^2 z^5}} = \sqrt{49x^4 y^7 z^7}$
$= \sqrt{49x^4 y^6 z^6} \cdot \sqrt{yz}$
$= 7x^2 y^3 z^3 \sqrt{yz}$

**47.** $\sqrt[3]{\dfrac{72a^9 b^{10}}{3a^2 b}} = \sqrt[3]{24a^7 b^9}$
$= \sqrt[3]{8a^6 b^9} \cdot \sqrt[3]{3a}$
$= 2a^2 b^3 \sqrt[3]{3a}$

**49.** $\sqrt{\dfrac{x^3 y^5}{x^5 y}} = \sqrt{\dfrac{y^4}{x^2}} = \dfrac{\sqrt{y^4}}{\sqrt{x^2}} = \dfrac{y^2}{x}$

**51.** $\sqrt{\dfrac{2a^{12}b^3}{18a^9 b^9}} = \sqrt{\dfrac{a^3}{9b^6}}$
$= \dfrac{\sqrt{a^3}}{\sqrt{9b^6}}$
$= \dfrac{\sqrt{a^2} \cdot \sqrt{a}}{\sqrt{9b^6}}$
$= \dfrac{a\sqrt{a}}{3b^3}$

**53.** $\sqrt[4]{\dfrac{x^7}{16x^2}} = \sqrt[4]{\dfrac{x^5}{16}}$
$= \dfrac{\sqrt[4]{x^5}}{\sqrt[4]{16}}$
$= \dfrac{\sqrt[4]{x^4} \cdot \sqrt[4]{x}}{\sqrt[4]{16}}$
$= \dfrac{x\sqrt[4]{x}}{2}$

**55.** Either method is correct, but multiplying by $\sqrt{3}$ requires fewer steps.

**57.** $\dfrac{5}{\sqrt{5}} = \dfrac{5 \cdot \sqrt{5}}{\sqrt{5} \cdot \sqrt{5}} = \dfrac{5\sqrt{5}}{5} = \sqrt{5}$

**59.** $\dfrac{-\sqrt{5}}{\sqrt{3}} = \dfrac{-\sqrt{5} \cdot \sqrt{3}}{\sqrt{3} \cdot \sqrt{3}} = -\dfrac{\sqrt{15}}{3}$

**61.** $\dfrac{15}{\sqrt{10}} = \dfrac{15 \cdot \sqrt{10}}{\sqrt{10} \cdot \sqrt{10}} = \dfrac{15\sqrt{10}}{10} = \dfrac{3\sqrt{10}}{2}$

**63.** $\dfrac{7}{\sqrt[3]{3}} = \dfrac{7 \cdot \sqrt[3]{9}}{\sqrt[3]{3} \cdot \sqrt[3]{9}} = \dfrac{7\sqrt[3]{9}}{\sqrt[3]{27}} = \dfrac{7\sqrt[3]{9}}{3}$

**65.** $\dfrac{9}{\sqrt[4]{8}} = \dfrac{9 \cdot \sqrt[4]{2}}{\sqrt[4]{8} \cdot \sqrt[4]{2}} = \dfrac{9\sqrt[4]{2}}{\sqrt[4]{16}} = \dfrac{9\sqrt[4]{2}}{2}$

**67.** $\dfrac{6}{\sqrt{3x}} = \dfrac{6 \cdot \sqrt{3x}}{\sqrt{3x} \cdot \sqrt{3x}} = \dfrac{6\sqrt{3x}}{3x} = \dfrac{2\sqrt{3x}}{x}$

**69.** $\dfrac{2}{\sqrt{x+2}} = \dfrac{2 \cdot \sqrt{x+2}}{\sqrt{x+2} \cdot \sqrt{x+2}} = \dfrac{2\sqrt{x+2}}{x+2}$

Section 9.5  The Quotient Rule for Radicals

**71.**
$$\sqrt{\frac{a^3}{8}} = \frac{\sqrt{a^3}}{\sqrt{8}} = \frac{\sqrt{a^3} \cdot \sqrt{2}}{\sqrt{8} \cdot \sqrt{2}}$$
$$= \frac{\sqrt{a^3} \cdot \sqrt{2}}{\sqrt{16}}$$
$$= \frac{\sqrt{a^2} \cdot \sqrt{a} \cdot \sqrt{2}}{4}$$
$$= \frac{a\sqrt{2a}}{4}$$

**73.**
$$\sqrt{\frac{5}{y^5}} = \frac{\sqrt{5}}{\sqrt{y^5}} = \frac{\sqrt{5} \cdot \sqrt{y}}{\sqrt{y^5} \cdot \sqrt{y}} = \frac{\sqrt{5y}}{\sqrt{y^6}} = \frac{\sqrt{5y}}{y^3}$$

**75.**
$$\frac{\sqrt{5x^2}}{\sqrt{12x^7}} = \sqrt{\frac{5x^2}{12x^7}} = \sqrt{\frac{5x}{12x^6}}$$
$$= \frac{\sqrt{5x} \cdot \sqrt{3}}{\sqrt{12x^6} \cdot \sqrt{3}}$$
$$= \frac{\sqrt{15x}}{\sqrt{36x^6}}$$
$$= \frac{\sqrt{15x}}{6x^3}$$

**77.**
$$\frac{18x^2y^3}{\sqrt{3x^2y^5}} = \frac{18x^2y^3 \cdot \sqrt{3y}}{\sqrt{3x^2y^5} \cdot \sqrt{3y}}$$
$$= \frac{18x^2y^3 \sqrt{3y}}{\sqrt{9x^2y^6}}$$
$$= \frac{18x^2y^3 \sqrt{3y}}{3xy^3}$$
$$= 6x\sqrt{3y}$$

**79.**
$$\frac{9x}{\sqrt{12x^5}} = \frac{9x \cdot \sqrt{3x}}{\sqrt{12x^5} \cdot \sqrt{3x}}$$
$$= \frac{9x \cdot \sqrt{3x}}{\sqrt{36x^6}}$$
$$= \frac{9x\sqrt{3x}}{6x^3}$$
$$= \frac{3\sqrt{3x}}{2x^2}$$

**81.**
$$\sqrt{\frac{x^5y^2}{20y^3}} = \sqrt{\frac{x^5y}{20y^2}} = \frac{\sqrt{x^5y}}{\sqrt{20y^2}}$$
$$= \frac{\sqrt{x^5y} \cdot \sqrt{5}}{\sqrt{20y^2} \cdot \sqrt{5}} = \frac{\sqrt{5x^5y}}{\sqrt{100y^2}}$$
$$= \frac{\sqrt{x^4} \cdot \sqrt{5xy}}{10y} = \frac{x^2\sqrt{5xy}}{10y}$$

**83.** Because $\frac{\sqrt{3}}{\sqrt{5}} = \frac{\sqrt{3} \cdot \sqrt{5}}{\sqrt{5} \cdot \sqrt{5}} = \frac{\sqrt{15}}{5}$, the missing number is 5.

**85.** Because $\sqrt{\frac{2}{7}} = \frac{\sqrt{2}}{\sqrt{7}} = \frac{\sqrt{2} \cdot \sqrt{7}}{\sqrt{7} \cdot \sqrt{7}} = \frac{\sqrt{14}}{7}$, the missing number is $\sqrt{14}$.

**87.** Because $\frac{3}{\sqrt{3}} = \frac{3 \cdot \sqrt{3}}{\sqrt{3} \cdot \sqrt{3}} = \frac{3\sqrt{3}}{3} = \sqrt{3}$, the missing number is $\sqrt{3}$.

**89.**
$$\frac{2b}{\sqrt[3]{b^2}} = \frac{2b \cdot \sqrt[3]{b}}{\sqrt[3]{b^2} \cdot \sqrt[3]{b}}$$
$$= \frac{2b\sqrt[3]{b}}{\sqrt[3]{b^3}}$$
$$= \frac{2b\sqrt[3]{b}}{b}$$
$$= 2\sqrt[3]{b}$$

**91.**
$$\sqrt[3]{\frac{5}{4}} = \frac{\sqrt[3]{5}}{\sqrt[3]{4}} = \frac{\sqrt[3]{5} \cdot \sqrt[3]{2}}{\sqrt[3]{4} \cdot \sqrt[3]{2}} = \frac{\sqrt[3]{10}}{\sqrt[3]{8}} = \frac{\sqrt[3]{10}}{2}$$

**93.**
$$\frac{x}{\sqrt[4]{xy^3}} = \frac{x \cdot \sqrt[4]{x^3y}}{\sqrt[4]{xy^3} \cdot \sqrt[4]{x^3y}}$$
$$= \frac{x \cdot \sqrt[4]{x^3y}}{\sqrt[4]{x^4y^4}}$$
$$= \frac{x\sqrt[4]{x^3y}}{xy}$$
$$= \frac{\sqrt[4]{x^3y}}{y}$$

**95.** 
$$\sqrt[4]{\frac{4x^5}{64xy^3}} = \sqrt[4]{\frac{x^4}{16y^3}}$$
$$= \frac{\sqrt[4]{x^4}}{\sqrt[4]{16y^3}}$$
$$= \frac{\sqrt[4]{x^4} \cdot \sqrt[4]{y}}{\sqrt[4]{16y^3} \cdot \sqrt[4]{y}}$$
$$= \frac{x\sqrt[4]{y}}{\sqrt[4]{16y^4}}$$
$$= \frac{x\sqrt[4]{y}}{2y}$$

**97.** $4\sqrt{x} = \sqrt{ax}$
Because $\sqrt{16x} = \sqrt{16} \cdot \sqrt{x} = 4\sqrt{x}$, the value of $a$ is 16.

**99.** $\frac{6}{\sqrt{x}} = \sqrt{\frac{a}{x}}$
Because $\sqrt{\frac{36}{x}} = \frac{\sqrt{36}}{\sqrt{x}} = \frac{6}{\sqrt{x}}$, the value of $a$ is 36.

**101.** $\sqrt{x^3} = \sqrt[a]{x^6}$
Because $\sqrt[4]{x^6} = x^{6/4} = x^{3/2} = \sqrt{x^3}$, the value of $a$ is 4.

**103.** (a) We can find the average price of each car sold by dividing the total revenue by the number of sales.
$$\frac{90{,}000d^2}{30\sqrt{d^3}} = \frac{90{,}000d^2 \cdot \sqrt{d}}{30\sqrt{d^3} \cdot \sqrt{d}}$$
$$= \frac{90{,}000d^2 \sqrt{d}}{30\sqrt{d^4}}$$
$$= \frac{3000d^2 \sqrt{d}}{d^2}$$
$$= 3000\sqrt{d}$$

(b) If $d = 4$, then the average price of a car is $3000\sqrt{d} = 3000\sqrt{4} = \$6000$.

**105.**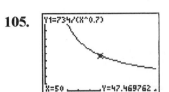
The graph suggests that the percentage of taxpayers who expect a refund decreases with the age of the taxpayer.

**107.**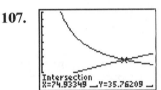
From the graph, we see that the graphs of the two models intersect at approximately (75, 36). This means that the age at which the percentage of people expecting refunds equals the percentage of people expecting to pay more taxes is 75 years old.

**109.**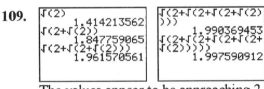

The values appear to be approaching 2. Repeating the experiment for $n = 3$:

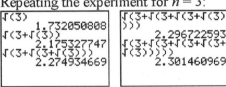

The pattern does not hold for $n = 3$.

**111.** 
$$\frac{1}{\sqrt{\sqrt[3]{x}}} = \frac{1}{(x^{1/3})^{1/2}}$$
$$= \frac{1}{x^{1/6}}$$
$$= \frac{\sqrt[6]{x^5}}{\sqrt[6]{x} \cdot \sqrt[6]{x^5}}$$
$$= \frac{\sqrt[6]{x^5}}{x}$$

**113.**

```
7/√(7)
         2.645751311
√(7)
         2.645751311
```

This shows that 7 divided by $\sqrt{7}$ is $\sqrt{7}$.

Let $x$ be any positive number.

$$\frac{x}{\sqrt{x}} = \frac{x}{x^{1/2}} = x^{1-(1/2)} = x^{1/2} = \sqrt{x}$$

Thus, it is true that dividing a positive number by its square root always results in the square root of the number.

## Section 9.6  Operations with Radicals

**1.** The *product* $\sqrt{7} \cdot \sqrt{7} = 7$, whereas the *sum* $\sqrt{7} + \sqrt{7} = 2\sqrt{7}$.

**3.**
$$5\sqrt{13} + \sqrt{13} = 5\sqrt{13} + 1\sqrt{13}$$
$$= (5+1)\sqrt{13}$$
$$= 6\sqrt{13}$$

**5.** $6\sqrt{x} + 10\sqrt{x} = (6+10)\sqrt{x} = 16\sqrt{x}$

**7.** $9\sqrt[3]{y} - 12\sqrt[3]{y} = (9-12)\sqrt[3]{y} = -3\sqrt[3]{y}$

**9.**
$$\sqrt{2x} + 4\sqrt{3} - 5\sqrt{2x} - 3\sqrt{3}$$
$$= 1\sqrt{2x} + 4\sqrt{3} - 5\sqrt{2x} - 3\sqrt{3}$$
$$= (1-5)\sqrt{2x} + (4-3)\sqrt{3}$$
$$= -4\sqrt{2x} + 1\sqrt{3}$$
$$= -4\sqrt{2x} + \sqrt{3}$$

**11.** The terms must have a common factor for the Distributive Property to apply. The radicals should be simplified before we can decide whether the terms can be combined.

**13.**
$$\sqrt{27} - \sqrt{12} = \sqrt{9} \cdot \sqrt{3} - \sqrt{4} \cdot \sqrt{3}$$
$$= 3\sqrt{3} - 2\sqrt{3}$$
$$= (3-2)\sqrt{3}$$
$$= 1\sqrt{3} = \sqrt{3}$$

**15.**
$$\sqrt[3]{250} + \sqrt[3]{54} = \sqrt[3]{125} \cdot \sqrt[3]{2} + \sqrt[3]{27} \cdot \sqrt[3]{2}$$
$$= 5\sqrt[3]{2} + 3\sqrt[3]{2}$$
$$= (5+3)\sqrt[3]{2}$$
$$= 8\sqrt[3]{2}$$

**17.**
$$\sqrt[4]{162} - \sqrt[4]{32} = \sqrt[4]{81} \cdot \sqrt[4]{2} - \sqrt[4]{16} \cdot \sqrt[4]{2}$$
$$= 3\sqrt[4]{2} - 2\sqrt[4]{2}$$
$$= \sqrt[4]{2}$$

**19.**
$$5\sqrt{32} + 7\sqrt{72} = 5\sqrt{16} \cdot \sqrt{2} + 7\sqrt{36} \cdot \sqrt{2}$$
$$= 5 \cdot 4\sqrt{2} + 7 \cdot 6\sqrt{2}$$
$$= 20\sqrt{2} + 42\sqrt{2}$$
$$= 62\sqrt{2}$$

**21.**
$$4\sqrt{12} - 2\sqrt{27} + 5\sqrt{8}$$
$$= 4\sqrt{4} \cdot \sqrt{3} - 2\sqrt{9} \cdot \sqrt{3} + 5\sqrt{4} \cdot \sqrt{2}$$
$$= 4 \cdot 2 \cdot \sqrt{3} - 2 \cdot 3 \cdot \sqrt{3} + 5 \cdot 2 \cdot \sqrt{2}$$
$$= 8\sqrt{3} - 6\sqrt{3} + 10\sqrt{2}$$
$$= 2\sqrt{3} + 10\sqrt{2}$$

**23.**
$$\sqrt{72x} - \sqrt{50x} = \sqrt{36} \cdot \sqrt{2x} - \sqrt{25} \cdot \sqrt{2x}$$
$$= 6\sqrt{2x} - 5\sqrt{2x}$$
$$= \sqrt{2x}$$

25. $x\sqrt{48} + 3\sqrt{27x^2}$
$= x\sqrt{16} \cdot \sqrt{3} + 3\sqrt{9x^2} \cdot \sqrt{3}$
$= 4x\sqrt{3} + 3 \cdot 3x \cdot \sqrt{3}$
$= 4x\sqrt{3} + 9x\sqrt{3}$
$= 13x\sqrt{3}$

27. $4a\sqrt{20a^2 b} + a^2\sqrt{45b}$
$= 4a\sqrt{4a^2} \cdot \sqrt{5b} + a^2\sqrt{9} \cdot \sqrt{5b}$
$= 4a \cdot 2a\sqrt{5b} + 3a^2\sqrt{5b}$
$= 8a^2\sqrt{5b} + 3a^2\sqrt{5b}$
$= 11a^2\sqrt{5b}$

29. $x\sqrt{2x} - 3\sqrt{8x^2} + \sqrt{50x^3}$
$= x\sqrt{2x} - 3\sqrt{4x^2} \cdot \sqrt{2} + \sqrt{25x^2}\sqrt{2x}$
$= x\sqrt{2x} - 3 \cdot 2x\sqrt{2} + 5x\sqrt{2x}$
$= x\sqrt{2x} - 6x\sqrt{2} + 5x\sqrt{2x}$
$= 6x\sqrt{2x} - 6x\sqrt{2}$

31. $\sqrt{9xy^3} + 4\sqrt{x^3 y} - 5y\sqrt{4xy}$
$= \sqrt{9y^2} \cdot \sqrt{xy} + 4\sqrt{x^2} \cdot \sqrt{xy} - 5y\sqrt{4} \cdot \sqrt{xy}$
$= 3y\sqrt{xy} + 4x\sqrt{xy} - 5y \cdot 2\sqrt{xy}$
$= 3y\sqrt{xy} + 4x\sqrt{xy} - 10y\sqrt{xy}$
$= 4x\sqrt{xy} - 7y\sqrt{xy}$

33. $2b\sqrt[3]{54b^5} - 3\sqrt[3]{16b^8}$
$= 2b\sqrt[3]{27b^3} \cdot \sqrt[3]{2b^2} - 3\sqrt[3]{8b^6} \cdot \sqrt[3]{2b^2}$
$= 2b \cdot 3b \cdot \sqrt[3]{2b^2} - 3 \cdot 2b^2 \cdot \sqrt[3]{2b^2}$
$= 6b^2\sqrt[3]{2b^2} - 6b^2\sqrt[3]{2b^2} = 0$

35. $\sqrt[3]{-54x^3} - \sqrt[3]{-16x^3}$
$= \sqrt[3]{-27x^3} \cdot \sqrt[3]{2} - \sqrt[3]{-8x^3} \cdot \sqrt[3]{2}$
$= -3x\sqrt[3]{2} - (-2x)\sqrt[3]{2}$
$= -3x\sqrt[3]{2} + 2x\sqrt[3]{2}$
$= -x\sqrt[3]{2}$

37. $4\sqrt{3}(2\sqrt{3} - \sqrt{7})$
$= (4\sqrt{3})(2\sqrt{3}) - (4\sqrt{3})(\sqrt{7})$
$= 8 \cdot 3 - 4\sqrt{21}$
$= 24 - 4\sqrt{21}$

39. $5\sqrt{7}(4\sqrt{3} + 7)$
$= (5\sqrt{7})(4\sqrt{3}) + (5\sqrt{7})(7)$
$= 20\sqrt{21} + 35\sqrt{7}$

41. $\sqrt{5}(\sqrt{35} - \sqrt{15})$
$= (\sqrt{5})(\sqrt{35}) - (\sqrt{5})(\sqrt{15})$
$= \sqrt{175} - \sqrt{75}$
$= \sqrt{25} \cdot \sqrt{7} - \sqrt{25} \cdot \sqrt{3}$
$= 5\sqrt{7} - 5\sqrt{3}$

43. $\sqrt{3}(\sqrt{y} - \sqrt{3y})$
$= (\sqrt{3})(\sqrt{y}) - (\sqrt{3})(\sqrt{3y})$
$= \sqrt{3y} - \sqrt{9y}$
$= \sqrt{3y} - \sqrt{9} \cdot \sqrt{y}$
$= \sqrt{3y} - 3\sqrt{y}$

45. Because
$\sqrt{7}(3 - \sqrt{2}) = (\sqrt{7})(3) - (\sqrt{7})(\sqrt{2})$
$= 3\sqrt{7} - \sqrt{14}$
the missing number is $\sqrt{7}$.

47. Because
$4(x + 2\sqrt{3}) = 4(x) + 4(2\sqrt{3})$
$= 4x + 8\sqrt{3}$
the missing number is $x + 2\sqrt{3}$.

49.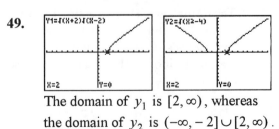

The domain of $y_1$ is $[2, \infty)$, whereas the domain of $y_2$ is $(-\infty, -2] \cup [2, \infty)$.

Section 9.6 Operations with Radicals

**51.** $(\sqrt{3}-2)(\sqrt{2}+\sqrt{3})$
$= \sqrt{3}\sqrt{2}+\sqrt{3}\sqrt{3}-2\sqrt{2}-2\sqrt{3}$
$= \sqrt{6}+3-2\sqrt{2}-2\sqrt{3}$

**53.** $(\sqrt{7x}-3)(\sqrt{3}+\sqrt{7x})$
$= \sqrt{3}\sqrt{7x}+\sqrt{7x}\sqrt{7x}-3\sqrt{3}-3\sqrt{7x}$
$= \sqrt{21x}+7x-3\sqrt{3}-3\sqrt{7x}$

**55.** $(2\sqrt{10}+\sqrt{5})(2\sqrt{2}-5)$
$= 2\sqrt{2}\cdot 2\sqrt{10}-5\cdot 2\sqrt{10}+2\sqrt{2}\sqrt{5}-5\sqrt{5}$
$= 4\sqrt{20}-10\sqrt{10}+2\sqrt{10}-5\sqrt{5}$
$= 4\sqrt{4}\sqrt{5}-10\sqrt{10}+2\sqrt{10}-5\sqrt{5}$
$= 8\sqrt{5}-10\sqrt{10}+2\sqrt{10}-5\sqrt{5}$
$= 3\sqrt{5}-8\sqrt{10}$

**57.** $(\sqrt{2}-\sqrt{6})^2 = (\sqrt{2})^2 - 2\cdot\sqrt{2}\sqrt{6}+(\sqrt{6})^2$
$= 2-2\sqrt{12}+6$
$= 8-2\sqrt{12}$
$= 8-2\sqrt{4}\sqrt{3}$
$= 8-2\cdot 2\sqrt{3}$
$= 8-4\sqrt{3}$

**59.** $(\sqrt{5}+\sqrt{10})^2 = (\sqrt{5})^2 + 2\cdot\sqrt{5}\sqrt{10}+(\sqrt{10})^2$
$= 5+2\sqrt{50}+10$
$= 15+2\sqrt{25}\sqrt{2}$
$= 15+2\cdot 5\sqrt{2}$
$= 15+10\sqrt{2}$

**61.** $(2\sqrt{2}-3)(2\sqrt{2}+3) = (2\sqrt{2})^2 - 3^2$
$= 4\cdot 2 - 9$
$= 8-9$
$= -1$

**63.** $(3\sqrt{2}+4\sqrt{6})(3\sqrt{2}-4\sqrt{6})$
$= (3\sqrt{2})^2 - (4\sqrt{6})^2$
$= 9\cdot 2 - 16\cdot 6$
$= 18-96$
$= -78$

**65.** $(10-5t\sqrt{2})(10+5t\sqrt{2})$
$= (10)^2 - (5t\sqrt{2})^2$
$= 100-25t^2\cdot 2$
$= 100-50t^2$

**67.** $\dfrac{\sqrt{3}+2}{\sqrt{3}-2}\cdot\dfrac{\sqrt{3}+2}{\sqrt{3}+2} = \dfrac{(\sqrt{3}+2)^2}{(\sqrt{3})^2-2^2}$
$= \dfrac{(\sqrt{3})^2+2\cdot 2\cdot\sqrt{3}+2^2}{3-4}$
$= \dfrac{3+4\sqrt{3}+4}{-1}$
$= \dfrac{7+4\sqrt{3}}{-1}$
$= -7-4\sqrt{3}$

**69.** $\dfrac{\sqrt{7}}{\sqrt{3}+\sqrt{7}}\cdot\dfrac{\sqrt{3}-\sqrt{7}}{\sqrt{3}-\sqrt{7}} = \dfrac{\sqrt{7}\cdot\sqrt{3}-\sqrt{7}\cdot\sqrt{7}}{(\sqrt{3})^2-(\sqrt{7})^2}$
$= \dfrac{\sqrt{21}-7}{3-7}$
$= \dfrac{\sqrt{21}-7}{-4}$
$= \dfrac{7-\sqrt{21}}{4}$

**71.** $x^2-4x+1=0,\ 2+\sqrt{3},\ 2-\sqrt{3}$
$(2+\sqrt{3})^2 - 4(2+\sqrt{3})+1=0$
$2^2+2\cdot 2\sqrt{3}+(\sqrt{3})^2-8-4\sqrt{3}+1=0$
$4+4\sqrt{3}+3-8-4\sqrt{3}+1=0$
$0=0$

$(2-\sqrt{3})^2 - 4(2-\sqrt{3})+1=0$
$2^2-2\cdot 2\sqrt{3}+(\sqrt{3})^2-8+4\sqrt{3}+1=0$
$4-4\sqrt{3}+3-8+4\sqrt{3}+1=0$
$0=0$

Both the given number and its conjugate are solutions of the equation.

**73.** $4x^2 - 12x + 7 = 0$, $\dfrac{3-\sqrt{2}}{2}$, $\dfrac{3+\sqrt{2}}{2}$

$4\left(\dfrac{3-\sqrt{2}}{2}\right)^2 - 12\left(\dfrac{3-\sqrt{2}}{2}\right) + 7 = 0$

$4 \cdot \dfrac{(3-\sqrt{2})^2}{2^2} - 6(3-\sqrt{2}) + 7 = 0$

$(3-\sqrt{2})^2 - 18 + 6\sqrt{2} + 7 = 0$

$3^2 - 2 \cdot 3\sqrt{2} + (\sqrt{2})^2 - 18 + 6\sqrt{2} + 7 = 0$

$9 - 6\sqrt{2} + 2 - 18 + 6\sqrt{2} + 7 = 0$

$0 = 0$

$4\left(\dfrac{3+\sqrt{2}}{2}\right)^2 - 12\left(\dfrac{3+\sqrt{2}}{2}\right) + 7 = 0$

$4 \cdot \dfrac{(3+\sqrt{2})^2}{2^2} - 6(3+\sqrt{2}) + 7 = 0$

$(3+\sqrt{2})^2 - 18 - 6\sqrt{2} + 7 = 0$

$3^2 + 2 \cdot 3\sqrt{2} + (\sqrt{2})^2 - 18 - 6\sqrt{2} + 7 = 0$

$9 + 6\sqrt{2} + 2 - 18 - 6\sqrt{2} + 7 = 0$

$0 = 0$

Both the given number and its conjugate are solutions of the equation.

**75.** $\dfrac{3}{3-\sqrt{6}} = \dfrac{3 \cdot (3+\sqrt{6})}{(3-\sqrt{6}) \cdot (3+\sqrt{6})}$

$= \dfrac{9 + 3\sqrt{6}}{3^2 - (\sqrt{6})^2}$

$= \dfrac{9 + 3\sqrt{6}}{9 - 6}$

$= \dfrac{9 + 3\sqrt{6}}{3} = 3 + \sqrt{6}$

**77.** $\dfrac{\sqrt{5}-\sqrt{3}}{\sqrt{5}+\sqrt{3}} = \dfrac{(\sqrt{5}-\sqrt{3})(\sqrt{5}-\sqrt{3})}{(\sqrt{5}+\sqrt{3})(\sqrt{5}-\sqrt{3})}$

$= \dfrac{(\sqrt{5})^2 - 2\sqrt{5} \cdot \sqrt{3} + (\sqrt{3})^2}{(\sqrt{5})^2 - (\sqrt{3})^2}$

$= \dfrac{5 - 2\sqrt{15} + 3}{5 - 3}$

$= \dfrac{8 - 2\sqrt{15}}{2} = 4 - \sqrt{15}$

**79.** $\dfrac{\sqrt{y}+5}{\sqrt{y}-4} = \dfrac{(\sqrt{y}+5)(\sqrt{y}+4)}{(\sqrt{y}-4)(\sqrt{y}+4)}$

$= \dfrac{\sqrt{y^2} + 4\sqrt{y} + 5\sqrt{y} + 20}{(\sqrt{y})^2 - 4^2}$

$= \dfrac{y + 9\sqrt{y} + 20}{y - 16}$

**81.** $\dfrac{\sqrt{x}}{\sqrt{x}+2} = \dfrac{(\sqrt{x})(\sqrt{x}-2)}{(\sqrt{x}+2)(\sqrt{x}-2)}$

$= \dfrac{\sqrt{x^2} - 2\sqrt{x}}{(\sqrt{x})^2 - 2^2}$

$= \dfrac{x - 2\sqrt{x}}{x - 4}$

**83.** $\dfrac{1}{\sqrt{x+1}-2} = \dfrac{(\sqrt{x+1}+2)}{(\sqrt{x+1}-2)(\sqrt{x+1}+2)}$

$= \dfrac{\sqrt{x+1}+2}{(\sqrt{x+1})^2 - 2^2}$

$= \dfrac{\sqrt{x+1}+2}{x+1-4}$

$= \dfrac{\sqrt{x+1}+2}{x-3}$

Section 9.6   Operations with Radicals

**85.**
$$\frac{\sqrt{x}-\sqrt{y}}{\sqrt{x}+\sqrt{y}} = \frac{(\sqrt{x}-\sqrt{y})(\sqrt{x}-\sqrt{y})}{(\sqrt{x}+\sqrt{y})(\sqrt{x}-\sqrt{y})}$$
$$= \frac{(\sqrt{x})^2 - 2\sqrt{x}\cdot\sqrt{y} + (\sqrt{y})^2}{(\sqrt{x})^2 - (\sqrt{y})^2}$$
$$= \frac{x - 2\sqrt{xy} + y}{x - y}$$

**87.** The operation cannot be performed because the denominators are not the same. It is not necessary to rationalize the denominators in order to add. It is only necessary that the fractions have the same denominator.

**89.**
$$\frac{\sqrt{x}-3}{\sqrt{x}} = \frac{\sqrt{x}(\sqrt{x}-3)}{\sqrt{x}\cdot\sqrt{x}}$$
$$= \frac{\sqrt{x}\cdot\sqrt{x} - 3\sqrt{x}}{x}$$
$$= \frac{x - 3\sqrt{x}}{x}$$

**91.**
$$\frac{1}{\sqrt{5}} + \sqrt{45} = \frac{1\cdot\sqrt{5}}{\sqrt{5}\cdot\sqrt{5}} + \sqrt{45}$$
$$= \frac{\sqrt{5}}{5} + \frac{5\sqrt{45}}{5}$$
$$= \frac{\sqrt{5} + 5\sqrt{9}\cdot\sqrt{5}}{5}$$
$$= \frac{\sqrt{5} + 15\sqrt{5}}{5}$$
$$= \frac{16\sqrt{5}}{5}$$

**93.**
$$1 - \frac{3}{\sqrt{2}} = 1 - \frac{3\sqrt{2}}{\sqrt{2}\cdot\sqrt{2}}$$
$$= \frac{2}{2} - \frac{3\sqrt{2}}{2}$$
$$= \frac{2 - 3\sqrt{2}}{2}$$

**95.**
$$2\sqrt{x+1} - \frac{2x}{\sqrt{x+1}}$$
$$= \frac{2\sqrt{x+1}\cdot\sqrt{x+1}}{\sqrt{x+1}} - \frac{2x}{\sqrt{x+1}}$$
$$= \frac{2(x+1) - 2x}{\sqrt{x+1}}$$
$$= \frac{2}{\sqrt{x+1}}$$
$$= \frac{2\sqrt{x+1}}{\sqrt{x+1}\cdot\sqrt{x+1}}$$
$$= \frac{2\sqrt{x+1}}{x+1}$$

**97.**
$$\frac{2}{\sqrt{3}+2} - \frac{1}{\sqrt{3}-2}$$
$$= \frac{2(\sqrt{3}-2)}{(\sqrt{3}-2)(\sqrt{3}+2)} - \frac{1(\sqrt{3}+2)}{(\sqrt{3}-2)(\sqrt{3}+2)}$$
$$= \frac{2(\sqrt{3}-2)}{(\sqrt{3})^2 - (2)^2} - \frac{1(\sqrt{3}+2)}{(\sqrt{3})^2 - (2)^2}$$
$$= \frac{2(\sqrt{3}-2) - (\sqrt{3}+2)}{3-4}$$
$$= \frac{2\sqrt{3} - 4 - \sqrt{3} - 2}{3-4}$$
$$= \frac{\sqrt{3}-6}{-1} = 6 - \sqrt{3}$$

**99.**
$$\frac{\sqrt{5}+\sqrt{7}}{4} = \frac{(\sqrt{5}+\sqrt{7})(\sqrt{5}-\sqrt{7})}{4(\sqrt{5}-\sqrt{7})}$$
$$= \frac{(\sqrt{5})^2 - (\sqrt{7})^2}{4(\sqrt{5}-\sqrt{7})}$$
$$= \frac{5-7}{4(\sqrt{5}-\sqrt{7})}$$
$$= \frac{-2}{4(\sqrt{5}-\sqrt{7})}$$
$$= \frac{-1}{2(\sqrt{5}-\sqrt{7})}$$

**101.**
$$\frac{\sqrt{x}+3}{\sqrt{x}-3} = \frac{(\sqrt{x}+3)(\sqrt{x}-3)}{(\sqrt{x}-3)(\sqrt{x}-3)}$$
$$= \frac{(\sqrt{x})^2 - 3^2}{(\sqrt{x})^2 - 2 \cdot 3\sqrt{x} + 3^2}$$
$$= \frac{x-9}{x - 6\sqrt{x} + 9}$$

**103.**
$$\frac{\sqrt{x+2} - \sqrt{x}}{2}$$
$$= \frac{(\sqrt{x+2} - \sqrt{x})(\sqrt{x+2} + \sqrt{x})}{2(\sqrt{x+2} + \sqrt{x})}$$
$$= \frac{(\sqrt{x+2})^2 - (\sqrt{x})^2}{2(\sqrt{x+2} + \sqrt{x})}$$
$$= \frac{x+2-x}{2(\sqrt{x+2} + \sqrt{x})}$$
$$= \frac{2}{2(\sqrt{x+2} + \sqrt{x})}$$
$$= \frac{1}{\sqrt{x+2} + \sqrt{x}}$$

**105.** Because 0 and 1 are consecutive integers and $\sqrt{0} = 0$ and $\sqrt{1} = 1$ are consecutive integers, two consecutive integers whose square roots are also consecutive integers are 0 and 1.

**107.** (a) Using the Pythagorean Theorem, we have
$$x^2 = d^2 + 50^2$$
$$x^2 - 50^2 = d^2$$
$$d^2 = x^2 - 2500$$
$$d = \sqrt{x^2 - 2500}$$

(b) The shorter guy wire is 10 feet shorter than the longer wire. Thus, the length of the shorter wire is $x - 10$. Using the Pythagorean Theorem, we have

$$d^2 + h^2 = (x - 10)^2$$
$$x^2 - 2500 + h^2 = (x - 10)^2$$
$$x^2 - 2500 + h^2 = x^2 - 20x + 100$$
$$-2500 + h^2 = -20x + 100$$
$$h^2 = -20x + 100 + 2500$$
$$h^2 = 2600 - 20x$$
$$h = \sqrt{2600 - 20x}$$

**109.** (a) Begin by calculating the percentage of the catch that is not marketable by dividing the pounds that are not marketable by the total number of pounds: $\frac{\sqrt{T}}{100} = \frac{\sqrt{60}}{100} \approx 0.077$ or about 8 % is unmarketable. That means that $100\% - 8\% = 92\%$ are marketable when the water temperature is 60°F.

(b) The pounds of fish that are not marketable are given by $\sqrt{T}$. The pounds of fish that are marketable are given by $100 - \sqrt{T}$. The ratio is
$$\frac{\sqrt{T}}{100 - \sqrt{T}} = \frac{\sqrt{T}}{100 - \sqrt{T}} \cdot \frac{(100 + \sqrt{T})}{(100 + \sqrt{T})}$$
$$= \frac{100\sqrt{T} + T}{10{,}000 - T}$$

**111.**
$$F(t) = \frac{1252t}{\sqrt[5]{t^3}} \cdot \frac{\sqrt[5]{t^2}}{\sqrt[5]{t^2}}$$
$$= \frac{1252t \sqrt[5]{t^2}}{\sqrt[5]{t^5}}$$
$$= \frac{1252t \sqrt[5]{t^2}}{t}$$
$$= 1252 \sqrt[5]{t^2}$$

**113.** The model $T(t)$ is the sum of the functions $F(t)$ and $B(t)$, using the forms from Exercises 111 and 112 with the rationalized denominators:
$$T(t) = F(t) + B(t)$$
$$= 1252\sqrt[5]{t^2} + 15.3t\sqrt[5]{t^2} + 135\sqrt[5]{t^2}$$
$$= 15.3t\sqrt[5]{t^2} + 1387\sqrt[5]{t^2}$$

Section 9.6   Operations with Radicals

**115.** $\sqrt[3]{3\sqrt[3]{3}} = \left(3^1 \cdot 3^{1/3}\right)^{1/3} = (3^{4/3})^{1/3} = 3^{4/9}$
$= 81^{1/9} \approx 1.63$

**117.** $\sqrt[4]{x^2} + \sqrt{x} = x^{2/4} + x^{1/2} = x^{1/2} + x^{1/2}$
$= 2x^{1/2} = 2\sqrt{x}$

**119.** $\sqrt[6]{8} - 2\sqrt[4]{4} = \sqrt[6]{2^3} - 2\sqrt[4]{2^2}$
$= 2^{3/6} - 2 \cdot 2^{2/4}$
$= 2^{1/2} - 2 \cdot 2^{1/2}$
$= \sqrt{2} - 2\sqrt{2}$
$= -\sqrt{2}$

**121.** $\dfrac{\sqrt{2} - \sqrt{3}}{\sqrt{2} + \sqrt{3} + \sqrt{5}} = \dfrac{\sqrt{2} - \sqrt{3}}{\left[\sqrt{2} + \left(\sqrt{3} + \sqrt{5}\right)\right]}$

Multiply the numerator and the denominator by the conjugate of the denominator: $\left[\sqrt{2} - \left(\sqrt{3} + \sqrt{5}\right)\right]$.

$\dfrac{\sqrt{2} - \sqrt{3}}{\sqrt{2} + \sqrt{3} + \sqrt{5}} = \dfrac{\sqrt{2} - \sqrt{3}}{\left[\sqrt{2} + \left(\sqrt{3} + \sqrt{5}\right)\right]}$

$= \dfrac{\left(\sqrt{2} - \sqrt{3}\right)\left[\sqrt{2} - \left(\sqrt{3} + \sqrt{5}\right)\right]}{\left[\sqrt{2} + \left(\sqrt{3} + \sqrt{5}\right)\right]\left[\sqrt{2} - \left(\sqrt{3} + \sqrt{5}\right)\right]}$

$= \dfrac{2 - \sqrt{2}\left(\sqrt{3} + \sqrt{5}\right) - \sqrt{6} + \sqrt{3}\left(\sqrt{3} + \sqrt{5}\right)}{2 - \left(\sqrt{3} + \sqrt{5}\right)^2}$

$= \dfrac{5 - 2\sqrt{6} - \sqrt{10} + \sqrt{15}}{-6 - 2\sqrt{15}}$

$= \dfrac{\left(5 - 2\sqrt{6} - \sqrt{10} + \sqrt{15}\right)\left(-6 + 2\sqrt{15}\right)}{\left(-6 - 2\sqrt{15}\right)\left(-6 + 2\sqrt{15}\right)}$

$= \dfrac{-30 + 10\sqrt{15} + 12\sqrt{6} - 4\sqrt{90} + 6\sqrt{10} - 2\sqrt{150} - 6\sqrt{15} + 30}{36 - 60}$

$= \dfrac{-30 + 10\sqrt{15} + 12\sqrt{6} - 12\sqrt{10} + 6\sqrt{10} - 10\sqrt{6} - 6\sqrt{15} + 30}{36 - 60}$

$= \dfrac{4\sqrt{15} + 2\sqrt{6} - 6\sqrt{10}}{-24}$

$= \dfrac{3\sqrt{10} - \sqrt{6} - 2\sqrt{15}}{12}$

**123.** Recall the special factoring pattern:
$x^3 - c^3 = (x-c)(x^2 + cx + c^2)$

$\dfrac{3}{\sqrt[3]{x} - \sqrt[3]{2}}$

$= \dfrac{3 \cdot \left(\sqrt[3]{x^2} + \sqrt[3]{2x} + \sqrt[3]{4}\right)}{\left(\sqrt[3]{x} - \sqrt[3]{2}\right) \cdot \left(\sqrt[3]{x^2} + \sqrt[3]{2x} + \sqrt[3]{4}\right)}$

$= \dfrac{3\left(\sqrt[3]{x^2} + \sqrt[3]{2x} + \sqrt[3]{4}\right)}{x - 2}$

**125.** $\dfrac{\dfrac{1}{\sqrt{3}} + \dfrac{\sqrt{3}}{\sqrt{x}}}{\dfrac{\sqrt{x}}{\sqrt{3}} + \dfrac{3}{\sqrt{3x}}} = \dfrac{\dfrac{\sqrt{x} + 3}{\sqrt{3x}}}{\dfrac{x+3}{\sqrt{3x}}}$

$= \dfrac{\sqrt{x}+3}{\sqrt{3x}} \cdot \dfrac{\sqrt{3x}}{x+3}$

$= \dfrac{\sqrt{x}+3}{x+3}$

## Section 9.7 Equations with Radicals and Exponents

**1.** To determine the domain of the function, solve the inequality $R \geq 0$.

**3.** The expression $\sqrt{x-2}$ is defined for all numbers for which $x - 2 \geq 0$. To determine the domain, we solve the following inequality.
$$x - 2 \geq 0$$
$$x \geq 2$$
The domain of the expression is $\{x \mid x \geq 2\}$.

**5.** The expression $\sqrt{2x-3}$ is defined for all numbers for which $2x - 3 \geq 0$. To determine the domain, we solve the following inequality.
$$2x - 3 \geq 0$$
$$2x \geq 3$$
$$x \geq 1.5$$
The domain of the expression is $\{x \mid x \geq 1.5\}$.

**7.** The expression $\sqrt{2-x}$ is defined for all numbers for which $2 - x \geq 0$. To determine the domain, we solve the following inequality.
$$2 - x \geq 0$$
$$-x \geq -2$$
$$x \leq 2$$
The domain of the expression is $\{x \mid x \leq 2\}$.

**9.** The expression $\sqrt{1+3x^2}$ is defined for all numbers for which $1 + 3x^2 \geq 0$. Because this inequality is true for any real value of $x$, the domain of the expression is $\mathbf{R}$.

**11.** Because it is a cube root, the expression $\sqrt[3]{3-2x}$ is defined for any real value of $x$. The domain of the expression is $\mathbf{R}$.

**13.** The expression $\sqrt[4]{x+3}$ is defined for all numbers for which $x + 3 \geq 0$. To determine the domain, we solve the following inequality.
$$x + 3 \geq 0$$
$$x \geq -3$$
The domain of the expression is $\{x \mid x \geq -3\}$.

**15.** The equations are not equivalent because the first has no solution, whereas the second has the solution 25. The methods used to solve a radical equation may introduce an extraneous solution.

**17.**
$$\sqrt{5x+1} = 4$$
$$\left(\sqrt{5x+1}\right)^2 = 4^2$$
$$5x + 1 = 16$$
$$5x = 15$$
$$x = 3$$
This value checks as a valid solution.

**19.**
$$\sqrt[3]{3x-1} = 2$$
$$\left(\sqrt[3]{3x-1}\right)^3 = 2^3$$
$$3x - 1 = 8$$
$$3x = 9$$
$$x = 3$$
This value checks as a valid solution.

**21.**
$$\sqrt{1-5x} + 6 = 3$$
$$\sqrt{1-5x} = -3$$
Because it is not possible for a square root expression to be negative, this equation has no real number solution.

**23.**
$$\sqrt[4]{x+1} = 2$$
$$\left(\sqrt[4]{x+1}\right)^4 = 2^4$$
$$x + 1 = 16$$
$$x = 15$$
This value checks as a valid solution.

Section 9.7   Equations with Radicals and Exponents

**25.**
$$\sqrt{x^2 - 3x} = 2$$
$$\left(\sqrt{x^2 - 3x}\right)^2 = 2^2$$
$$x^2 - 3x = 4$$
$$x^2 - 3x - 4 = 0$$
$$(x - 4)(x + 1) = 0$$
$x - 4 = 0$ or $x + 1 = 0$
$x = 4$ or $x = -1$
Both values check as valid solutions.

**27.**
$$\sqrt{x}\,\sqrt{x - 5} = 6$$
$$\left(\sqrt{x}\,\sqrt{x - 5}\right)^2 = 6^2$$
$$x(x - 5) = 36$$
$$x^2 - 5x - 36 = 0$$
$$(x - 9)(x + 4) = 0$$
$x - 9 = 0$ or $x + 4 = 0$
$x = 9$ or $x = -4$
A check shows that only 9 is a valid solution. The value of $-4$ is an extraneous solution.

**29.**
$$x\sqrt{2} = \sqrt{5x - 2}$$
$$\left(x\sqrt{2}\right)^2 = \left(\sqrt{5x - 2}\right)^2$$
$$2x^2 = 5x - 2$$
$$2x^2 - 5x + 2 = 0$$
$$(2x - 1)(x - 2) = 0$$
$2x - 1 = 0$ or $x - 2 = 0$
$x = 1/2$ or $x = 2$
Both values check as valid solutions.

**31.**
$$\sqrt{4x + 13} = 2x - 1$$
$$\left(\sqrt{4x + 13}\right)^2 = (2x - 1)^2$$
$$4x + 13 = 4x^2 - 4x + 1$$
$$0 = 4x^2 - 8x - 12$$
$$0 = 4(x^2 - 2x - 3)$$
$$0 = 4(x - 3)(x + 1)$$
$x - 3 = 0$ or $x + 1 = 0$
$x = 3$ or $x = -1$
A check shows that only 3 is a valid solution. The value of $-1$ is an extraneous solution.

**33.**
$$\sqrt[3]{x^2 + x + 2} = 2$$
$$\left(\sqrt[3]{x^2 + x + 2}\right)^3 = 2^3$$
$$x^2 + x + 2 = 8$$
$$x^2 + x - 6 = 0$$
$$(x - 2)(x + 3) = 0$$
$x - 2 = 0$ or $x + 3 = 0$
$x = 2$ or $x = -3$
Both values check as valid solutions.

**35.**   $\sqrt[4]{x + 1} + \sqrt[4]{2x - 3} = 0$
$$\sqrt[4]{x + 1} = -\sqrt[4]{2x - 3}$$
Because it is not possible for a fourth root expression to be negative, this equation has no real number solution.

**37.**
$$\sqrt{x(x - 2)} - x = -10$$
$$\sqrt{x(x - 2)} = x - 10$$
$$\left(\sqrt{x(x - 2)}\right)^2 = (x - 10)^2$$
$$x(x - 2) = x^2 - 20x + 100$$
$$x^2 - 2x = x^2 - 20x + 100$$
$$18x = 100$$
$$x = \frac{50}{9}$$
A check shows that this value is an extraneous solution. There is no real number solution to this equation.

**39.**
$$x - 9 = \sqrt{x^2 - x - 4}$$
$$(x - 9)^2 = \left(\sqrt{x^2 - x - 4}\right)^2$$
$$x^2 - 18x + 81 = x^2 - x - 4$$
$$-17x = -85$$
$$x = 15$$
A check shows that this value is an extraneous solution. There is no real number solution to this equation.

**41.**
$$\sqrt{\frac{t}{3}} = \sqrt{\frac{t+4}{2+t}}$$
$$\left(\sqrt{\frac{t}{3}}\right)^2 = \left(\sqrt{\frac{t+4}{2+t}}\right)^2$$
$$\frac{t}{3} = \frac{t+4}{2+t}$$
$$t(2+t) = 3(t+4)$$
$$2t + t^2 = 3t + 12$$
$$t^2 - t - 12 = 0$$
$$(t-4)(t+3) = 0$$
$$t - 4 = 0 \quad \text{or} \quad t + 3 = 0$$
$$t = 4 \quad \text{or} \quad t = -3$$
A check shows that only 4 is a valid solution. The value of −3 is an extraneous solution.

**43.** The radical in each equation can be eliminated by squaring once. The first two resulting equations are second degree.

**45.**
$$\sqrt{x-3} + 1 = \sqrt{x}$$
$$\sqrt{x-3} = \sqrt{x} - 1$$
$$\left(\sqrt{x-3}\right)^2 = \left(\sqrt{x} - 1\right)^2$$
$$x - 3 = x - 2\sqrt{x} + 1$$
$$-4 = -2\sqrt{x}$$
$$(-4)^2 = \left(-2\sqrt{x}\right)^2$$
$$16 = 4x$$
$$4 = x$$
This value checks as a valid solution.

**47.**
$$\sqrt{4x-3} - \sqrt{3x-5} = 1$$
$$\sqrt{4x-3} = 1 + \sqrt{3x-5}$$
$$\left(\sqrt{4x-3}\right)^2 = \left(1 + \sqrt{3x-5}\right)^2$$
$$4x - 3 = 1 + 2\sqrt{3x-5} + 3x - 5$$
$$x + 1 = 2\sqrt{3x-5}$$
$$(x+1)^2 = \left(2\sqrt{3x-5}\right)^2$$
$$x^2 + 2x + 1 = 4(3x-5)$$
$$x^2 + 2x + 1 = 12x - 20$$
$$x^2 - 10x + 21 = 0$$
$$(x-3)(x-7) = 0$$
$$x - 3 = 0 \quad \text{or} \quad x - 7 = 0$$
$$x = 3 \quad \text{or} \quad x = 7$$
Both values check as valid solutions.

**49.**
$$\sqrt{5x-9} + 3 = \sqrt{x}$$
$$\sqrt{5x-9} = \sqrt{x} - 3$$
$$\left(\sqrt{5x-9}\right)^2 = \left(\sqrt{x} - 3\right)^2$$
$$5x - 9 = x - 6\sqrt{x} + 9$$
$$4x - 18 = -6\sqrt{x}$$
$$(4x-18)^2 = \left(-6\sqrt{x}\right)^2$$
$$16x^2 - 144x + 324 = 36x$$
$$16x^2 - 180x + 324 = 0$$
$$4(4x^2 - 45x + 81) = 0$$
$$4(4x - 9)(x - 9) = 0$$
$$4x - 9 = 0 \quad \text{or} \quad x - 9 = 0$$
$$x = 9/4 \quad \text{or} \quad x = 9$$
A check shows that these values are extraneous solutions. There is no real number solution to this equation.

**51.**
$$\sqrt{7x+4} - \sqrt{x+1} = 3$$
$$\sqrt{7x+4} = 3 + \sqrt{x+1}$$
$$\left(\sqrt{7x+4}\right)^2 = \left(3 + \sqrt{x+1}\right)^2$$
$$7x + 4 = 9 + 6\sqrt{x+1} + x + 1$$
$$6x - 6 = 6\sqrt{x+1}$$
$$6(x-1) = 6\sqrt{x+1}$$
$$x - 1 = \sqrt{x+1}$$
$$(x-1)^2 = \left(\sqrt{x+1}\right)^2$$
$$x^2 - 2x + 1 = x + 1$$
$$x^2 - 3x = 0$$
$$x(x-3) = 0$$
$$x = 0 \quad \text{or} \quad x - 3 = 0$$
$$x = 3$$
A check shows that only 3 is a valid solution. The value of 0 is extraneous.

**53.**
$$\sqrt{x+3} = 1 + \sqrt{x-2}$$
$$\left(\sqrt{x+3}\right)^2 = \left(1 + \sqrt{x-2}\right)^2$$
$$x + 3 = 1 + 2\sqrt{x-2} + x - 2$$
$$4 = 2\sqrt{x-2}$$
$$2 = \sqrt{x-2}$$
$$2^2 = \left(\sqrt{x-2}\right)^2$$
$$4 = x - 2$$
$$6 = x$$
This value checks as a valid solution.

## Section 9.7 Equations with Radicals and Exponents

**55.**
$\sqrt{5x-6} = 1 + \sqrt{3x-5}$
$(\sqrt{5x-6})^2 = (1+\sqrt{3x-5})^2$
$5x - 6 = 1 + 2\sqrt{3x-5} + 3x - 5$
$2x - 2 = 2\sqrt{3x-5}$
$2(x-1) = 2\sqrt{3x-5}$
$x - 1 = \sqrt{3x-5}$
$(x-1)^2 = (\sqrt{3x-5})^2$
$x^2 - 2x + 1 = 3x - 5$
$x^2 - 5x + 6 = 0$
$(x-3)(x-2) = 0$
$x - 3 = 0$ or $x - 2 = 0$
$x = 3$ or $x = 2$
Both values check as valid solutions.

**57.**
$\sqrt{4x+1} - \sqrt{3x-2} = 1$
$\sqrt{4x+1} = 1 + \sqrt{3x-2}$
$(\sqrt{4x+1})^2 = (1+\sqrt{3x-2})^2$
$4x + 1 = 1 + 2\sqrt{3x-2} + 3x - 2$
$x + 2 = 2\sqrt{3x-2}$
$(x+2)^2 = (2\sqrt{3x-2})^2$
$x^2 + 4x + 4 = 4(3x-2)$
$x^2 + 4x + 4 = 12x - 8$
$x^2 - 8x + 12 = 0$
$(x-2)(x-6) = 0$
$x - 2 = 0$ or $x - 6 = 0$
$x = 2$ or $x = 6$
Both values check as valid solutions.

**59.**
$\sqrt{7x-3} - \sqrt{2x+1} = 2$
$\sqrt{7x-3} = 2 + \sqrt{2x+1}$
$(\sqrt{7x-3})^2 = (2+\sqrt{2x+1})^2$
$7x - 3 = 4 + 4\sqrt{2x+1} + 2x + 1$
$5x - 8 = 4\sqrt{2x+1}$
$(5x-8)^2 = (4\sqrt{2x+1})^2$
$25x^2 - 80x + 64 = 16(2x+1)$
$25x^2 - 80x + 64 = 32x + 16$
$25x^2 - 112x + 48 = 0$
$(25x - 12)(x-4) = 0$
$25x - 12 = 0$ or $x - 4 = 0$
$x = 0.48$ or $x = 4$

A check shows that only 4 is a valid solution. The value of 0.48 is an extraneous solution.

**61.** Raise both sides to the $b$th power to obtain an integer exponent on $x$.

**63.** $x^2 = 6$
$x = \pm\sqrt{6}$

**65.** $t^2 + 9 = 0$
$t^2 = -9$
There are no real solutions.

**67.** $(2x-5)^2 = 16$
$2x - 5 = \pm\sqrt{16}$
$2x - 5 = \pm 4$
$2x = 5 \pm 4$
$x = \dfrac{5 \pm 4}{2}$
$x = 4.5$ or $x = 0.5$

**69.** $(3x-1)^2 = 3$
$3x - 1 = \pm\sqrt{3}$
$3x = 1 \pm \sqrt{3}$
$x = \dfrac{1 \pm \sqrt{3}}{3}$

**71.** $(x-5)^2 = 0$
$x - 5 = \sqrt{0}$
$x - 5 = 0$
$x = 5$

**73.** $(2x-3)^5 = -32$
$2x - 3 = \sqrt[5]{-32}$
$2x - 3 = -2$
$2x = 1$
$x = \dfrac{1}{2}$

**75.** $x^{2/3} = 9$
$(x^{2/3})^3 = 9^3$
$x^2 = 729$
$x = \pm\sqrt{729}$
$x = \pm 27$

**77.**
$$x^{-3/4} = 27$$
$$\left(x^{-3/4}\right)^4 = 27^4$$
$$x^{-3} = 27^4$$
$$\frac{1}{x^3} = 27^4$$
$$x^3 = \frac{1}{27^4}$$
$$x^3 = \frac{1}{(3^3)^4}$$
$$x = \sqrt[3]{\frac{1}{(3^4)^3}}$$
$$x = \frac{1}{3^4} = \frac{1}{81}$$

**79.**
$$(x-3)^{-2/3} = 4$$
$$\left((x-3)^{-2/3}\right)^3 = 4^3$$
$$(x-3)^{-2} = 64$$
$$\frac{1}{(x-3)^2} = 64$$
$$(x-3)^2 = \frac{1}{64}$$
$$x - 3 = \pm\sqrt{\frac{1}{64}}$$
$$x = 3 \pm \frac{1}{8}$$
$$x = \frac{25}{8} \quad \text{or} \quad x = \frac{23}{8}$$

**81.**
$$(3x-2)^{3/4} = 8$$
$$\left((3x-2)^{3/4}\right)^4 = 8^4$$
$$(3x-2)^3 = (2^3)^4$$
$$3x - 2 = \sqrt[3]{(2^3)^4}$$
$$3x = 2 + \sqrt[3]{(2^4)^3}$$
$$x = \frac{2 + 2^4}{3} = 6$$

**83.** Let $x$ = the positive number.
$$\sqrt{x \cdot x^2} = 2x$$
$$\sqrt{x^3} = 2x$$
$$x^{3/2} = 2x$$
$$(x^{3/2})^2 = (2x)^2$$
$$x^3 = 4x^2$$
$$x^3 - 4x^2 = 0$$
$$x^2(x-4) = 0$$
$$x^2 = 0 \quad \text{or} \quad x - 4 = 0$$
$$x = 0 \quad \text{or} \quad x = 4$$
Because 0 is not considered positive, the only answer is 4.

**85.** By the Pythagorean Theorem, we know that
$$x^2 + 8^2 = 11^2$$
$$x^2 + 64 = 121$$
$$x^2 = 57$$
$$x = \pm\sqrt{57}$$
Because we consider length to be positive, the value of $x$ is $\sqrt{57} \approx 7.55$.

**87.** By the Pythagorean Theorem, we know that
$$x^2 + \left(\sqrt{7}\right)^2 = \left(2\sqrt{3}\right)^2$$
$$x^2 + 7 = 12$$
$$x^2 = 5$$
$$x = \pm\sqrt{5}$$
Because we consider length to be positive, the value of $x$ is $\sqrt{5} \approx 2.24$.

**89.** Let $x$ = the distance up the wall at which the ladder rests.
$$x^2 + 7^2 = 26^2$$
$$x^2 + 49 = 676$$
$$x^2 = 627$$
$$x = \pm\sqrt{627}$$
Because we consider length to be positive, the ladder rests $\sqrt{627} \approx 25.04$ feet up the wall.

Section 9.7   Equations with Radicals and Exponents

**91.** Let $x$ = the length of the brace. The brace forms the hypotenuse of a right triangle.
$$30^2 + 10^2 = x^2$$
$$900 + 100 = x^2$$
$$1000 = x^2$$
$$x = \pm\sqrt{1000}$$
Because we consider length to be positive, the length of the brace is $\sqrt{1000} \approx 31.62$ feet long.

**93.** Let $x$ = the distance between the stake and the tree at point $B$.
$$x^2 + 125^2 = 150^2$$
$$x^2 + 15{,}625 = 22{,}500$$
$$x^2 = 6875$$
$$x = \pm\sqrt{6875}$$
Because we consider length to be positive, the distance between the stake and the tree at point $B$ is $\sqrt{6875} \approx 82.92$ feet.

**95.** The area of a circle is given by $\pi r^2 = A$. From the given information, we can find the radius of the circle.
$$\pi r^2 = 95$$
$$r^2 = \frac{95}{\pi}$$
$$r = \pm\sqrt{\frac{95}{\pi}} \approx \pm 5.499 \text{ feet}$$
The rug will fit in a square room 11 feet on each side because the diameter of the rug (twice the radius of the rug) is slightly less than 11 feet.

**97.**
$$C(w) = w - \sqrt{w}$$
$$2 = w - \sqrt{w}$$
$$\sqrt{w} = w - 2$$
$$(\sqrt{w})^2 = (w-2)^2$$
$$w = w^2 - 4w + 4$$
$$0 = w^2 - 5w + 4$$
$$0 = (w-1)(w-4)$$
$w - 1 = 0$  or  $w - 4 = 0$
$w = 1$  or  $w = 4$
Only the value of 4 checks as a valid solution. After 4 weeks, the dolphin understands two commands.

**99.**
$$9.75\sqrt{x} + 3.45 = 40$$
$$9.75\sqrt{x} = 36.55$$
$$\sqrt{x} = 3.748717949$$
$$(\sqrt{x})^2 = (3.748717949)^2$$
$$x \approx 14$$
The amount from loans is approximately $40 billion in 2004.

**101.** $9.75\sqrt{x} + 3.45 - \sqrt{3.2x + 37.4} = 35$

**103.**

From the graph, it appears the domain of the function is $\{x \mid x \leq -4 \text{ or } x \geq 3\}$.

**105.**
$$\sqrt[6]{3 + t - t^2} = \sqrt[3]{2t + 3}$$
$$(3 + t - t^2)^{1/6} = (2t + 3)^{1/3}$$
$$\left[(3 + t - t^2)^{1/6}\right]^6 = \left[(2t + 3)^{1/3}\right]^6$$
$$3 + t - t^2 = (2t + 3)^2$$
$$3 + t - t^2 = 4t^2 + 12t + 9$$
$$0 = 5t^2 + 11t + 6$$
$$0 = (5t + 6)(t + 1)$$
$5t + 6 = 0$   or   $t + 1 = 0$
$t = -6/5$   or   $t = -1$
Both values check as valid solutions.

**107.**
$$\sqrt{x+2} - \sqrt{x-3} = \sqrt{4x-1}$$
$$(\sqrt{x+2} - \sqrt{x-3})^2 = (\sqrt{4x-1})^2$$
$$x + 2 - 2\sqrt{(x+2)(x-3)} + x - 3 = 4x - 1$$
$$-2\sqrt{(x+2)(x-3)} = 2x$$
$$-\sqrt{(x+2)(x-3)} = x$$
$$\left(-\sqrt{(x+2)(x-3)}\right)^2 = x^2$$
$$(x+2)(x-3) = x^2$$
$$x^2 - x - 6 = x^2$$
$$-x - 6 = 0$$
$$x = -6$$
A check shows that this value is an extraneous solution. The equation has no real number solution.

**109.**
(a) For $\sqrt{x^2} = x$, negative values do not check. The solution set is $\{x \mid x \geq 0\}$.

(b) For $\sqrt{x^2} = -x$, positive values do not check. The solution set is $\{x \mid x \leq 0\}$.

(c) For $\sqrt{x^4} = x^2$, every real number checks. The solution set is **R**.

## Section 9.8 Complex Numbers

**1.** The number is $i$, and the properties are $i^2 = -1$ and $i = \sqrt{-1}$.

**3.** $\sqrt{-9} = i\sqrt{9} = 3i$

**5.** $\sqrt{-18} = i\sqrt{9}\sqrt{2} = 3i\sqrt{2}$

**7.** $\sqrt{25} + \sqrt{-36} = \sqrt{25} + i\sqrt{36} = 5 + 6i$

**9.** $\dfrac{-4 - \sqrt{-12}}{2} = \dfrac{-4 - i\sqrt{4}\sqrt{3}}{2}$
$= \dfrac{-4 - 2i\sqrt{3}}{2}$
$= -2 - i\sqrt{3}$

**11.** Write $i^{34}$ as a power of $i^2$:
$i^{34} = (i^2)^{17} = (-1)^{17} = -1$.

**13.** $(a + 1) + 2bi = 3 - 2i$
Equate the real parts and equate the imaginary parts and solve.
$a + 1 = 3$ and $2b = -2$
$a = 2$ and $b = -1$

**15.** $2a + bi = \sqrt{-25} - 2$
$2a + bi = i\sqrt{25} - 2$
$2a + bi = -2 + 5i$
Equate the real parts and equate the imaginary parts and solve.
$2a = -2$ and $b = 5$
$a = -1$

**17.** (a) By definition, the imaginary numbers are of the form $a + bi$, where $a$ and $b$ are real numbers and $b \neq 0$. The real numbers when written in this form have $b = 0$. So, the statement that the set of real numbers is a subset of the set of imaginary numbers is false.

(b) The statement that if $a$ and $b$ are nonzero real numbers, then $a + bi$ is an imaginary number is true. The statement would still be true if $a = 0$ but not if $b = 0$.

(c) The statement that 0 is both a real number and an imaginary number is false. For a number to be an imaginary number, you must be able to write it in the form $a + bi$ with $b \neq 0$.

**19.** $(5 + 6i) + (-4 - 5i) = (5 - 4) + (6 - 5)i$
$= 1 + i$

**21.** $(7 + 3i) - (-8 - 7i) = 7 + 3i + 8 + 7i$
$= 15 + 10i$

**23.** $7i - 2(-3 - 7i) = 7i + 6 + 14i$
$= 6 + 21i$

**25.** $(7 - 2i) - (3 - 4i) + 2(5 - i)$
$= 7 - 2i - 3 + 4i + 10 - 2i$
$= 14$

**27.** $\sqrt{-36} + \sqrt{-49} = i\sqrt{36} + i\sqrt{49}$
$= 6i + 7i$
$= 13i$

Section 9.8   Complex Numbers

**29.** $3 - i - \sqrt{4} - \sqrt{-9} = 3 - i - 2 - i\sqrt{9}$
$= 3 - i - 2 - 3i$
$= 1 - 4i$

**31.** $(a + 3i) + (2 + bi) = i$
$(a + 2) + (3 + b)i = 0 + 1i$
Equate the real parts and equate the imaginary parts and solve.
$a + 2 = 0$ and $3 + b = 1$
$a = -2$ and $b = -2$

**33.** $\left(5 - \sqrt{-4}\right) - (a + bi) = 6$
$\left(5 - i\sqrt{4}\right) - (a + bi) = 6$
$(5 - 2i) - (a + bi) = 6$
$5 - 2i - a - bi = 6$
$(5 - a) + (-2 - b)i = 6 + 0i$
Equate the real parts and equate the imaginary parts and solve.
$5 - a = 6$ and $-2 - b = 0$
$a = -1$ and $b = -2$

**35.** $(2i)(5i) = 10i^2 = 10(-1) = -10$

**37.** $7i(3 - 4i) = 21i - 28i^2$
$= 21i - 28(-1)$
$= 28 + 21i$

**39.** $(2 - 3i)(3 - 4i) = 6 - 8i - 9i + 12i^2$
$= 6 - 8i - 9i + 12(-1)$
$= 6 - 8i - 9i - 12$
$= -6 - 17i$

**41.** $(3 + i)^2 = 3^2 + 6i + i^2$
$= 9 + 6i - 1$
$= 8 + 6i$

**43.** $\sqrt{-8}\sqrt{-2} = i\sqrt{8} \cdot i\sqrt{2}$
$= i^2 \cdot \sqrt{16}$
$= -1 \cdot 4$
$= -4$

**45.** $\sqrt{-3}\left(\sqrt{-6} + \sqrt{-27}\right)$
$= i\sqrt{3}\left(i\sqrt{6} + i\sqrt{27}\right)$
$= i^2\sqrt{18} + i^2\sqrt{81}$
$= (-1)\sqrt{9}\sqrt{2} + (-1)\sqrt{81}$
$= -3\sqrt{2} - 9 = -9 - 3\sqrt{2}$

**47.** $i^{15} = (i^2)^7 \cdot i = (-1)^7 \cdot i = -1 \cdot i = -i$

**49.** $i^{52} = (i^2)^{26} = (-1)^{26} = 1$

**51.** $i^{30} = (i^2)^{15} = (-1)^{15} = -1$

**53.** $i^{13} = (i^2)^6 \cdot i = (-1)^6 \cdot i = 1 \cdot i = i$

**55.** In part (a) the result is $a^2 - b^2$, but in part (b) the result is $a^2 + b^2$.

**57.** The complex conjugate of $7i$ is $-7i$.
Product:
$(7i)(-7i) = -49i^2 = -49(-1) = 49$

**59.** The complex conjugate of $2 + 3i$ is $2 - 3i$. Product:
$(2 + 3i)(2 - 3i) = 2^2 + 3^2 = 4 + 9 = 13$

**61.** The complex conjugate of $\sqrt{5} - 4i$ is $\sqrt{5} + 4i$. Product:
$\left(\sqrt{5} - 4i\right)\left(\sqrt{5} + 4i\right) = \left(\sqrt{5}\right)^2 + 4^2$
$= 5 + 16 = 21$

**63.** $\dfrac{-3}{2i} = \dfrac{(-3)(i)}{(2i)(i)} = \dfrac{-3i}{2i^2} = \dfrac{-3i}{-2} = \dfrac{3}{2}i$

**65.** $\dfrac{9 + 6i}{3i} = \dfrac{(9 + 6i)(i)}{(3i)(i)}$
$= \dfrac{9i + 6i^2}{3i^2}$
$= \dfrac{9i - 6}{-3}$
$= -3i + 2 = 2 - 3i$

**67.** $\dfrac{3 - i}{1 + 2i} = \dfrac{(3 - i)(1 - 2i)}{(1 + 2i)(1 - 2i)}$
$= \dfrac{3 - 7i + 2i^2}{1 + 4}$
$= \dfrac{3 - 7i - 2}{5}$
$= \dfrac{1 - 7i}{5}$

69. $\dfrac{3-4i}{-7-5i} = \dfrac{(3-4i)(-7+5i)}{(-7-5i)(-7+5i)}$

$= \dfrac{-21+43i-20i^2}{49+25}$

$= \dfrac{-21+43i+20}{74}$

$= \dfrac{-1+43i}{74}$

71. $\dfrac{3}{2+\sqrt{-3}} = \dfrac{3}{2+i\sqrt{3}}$

$= \dfrac{3(2-i\sqrt{3})}{(2+i\sqrt{3})(2-i\sqrt{3})}$

$= \dfrac{6-3i\sqrt{3}}{4+3}$

$= \dfrac{6-3i\sqrt{3}}{7}$

73. $\dfrac{\sqrt{-5}-\sqrt{3}}{\sqrt{-3}-\sqrt{9}} = \dfrac{i\sqrt{5}-\sqrt{3}}{i\sqrt{3}-3} = \dfrac{-\sqrt{3}+i\sqrt{5}}{-3+i\sqrt{3}}$

$= \dfrac{(-\sqrt{3}+i\sqrt{5})(-3-i\sqrt{3})}{(-3+i\sqrt{3})(-3-i\sqrt{3})}$

$= \dfrac{3\sqrt{3}+3i-3i\sqrt{5}-i^2\sqrt{15}}{9+3}$

$= \dfrac{3\sqrt{3}+3i-3i\sqrt{5}+\sqrt{15}}{12}$

$= \dfrac{\sqrt{15}+3\sqrt{3}+3i-3i\sqrt{5}}{12}$

75. $(3-4i)+(5+i) = (3+5)+(-4+1)i$
$= 8-3i$

77. $(4i)(-7i) = -28i^2 = -28(-1) = 28$

79. $\dfrac{1+2i}{3i} = \dfrac{(1+2i)(-3i)}{(3i)(-3i)}$

$= \dfrac{-3i-6i^2}{9}$

$= \dfrac{-3i+6}{9}$

$= \dfrac{2-i}{3}$

81. $3(2+i)+3i(1-i) = 6+3i+3i-3i^2$
$= 6+3i+3i-3(-1)$
$= 6+3i+3i+3$
$= 9+6i$

83. $(-6+5i)^2 = (-6)^2 - 60i + (5i)^2$
$= 36-60i-25$
$= 11-60i$

85. $\dfrac{3+4i}{3-4i} = \dfrac{(3+4i)(3+4i)}{(3-4i)(3+4i)}$

$= \dfrac{3^2+24i+(4i)^2}{9+16}$

$= \dfrac{9+24i-16}{25}$

$= \dfrac{-7+24i}{25}$

87. $7\sqrt{-25}-4\sqrt{-49} = 7i\sqrt{25}-4i\sqrt{49}$
$= (7i)(5)-(4i)(7)$
$= 35i-28i$
$= 7i$

89. $(1+\sqrt{-3})(-1-\sqrt{-12})$
$= (1+i\sqrt{3})(-1-i\sqrt{4}\sqrt{3})$
$= (1+i\sqrt{3})(-1-2i\sqrt{3})$
$= -1-2i\sqrt{3}-i\sqrt{3}-2i^2(3)$
$= 5-3i\sqrt{3}$

91. $\dfrac{\sqrt{-75}}{\sqrt{-3}} = \dfrac{i\sqrt{25}\sqrt{3}}{i\sqrt{3}} = \sqrt{25} = 5$

93. $x^2-4x+13=0, \quad 2+3i, \quad 2-3i$

$(2+3i)^2 = 4+12i+9i^2 = -5+12i$
$(2-3i)^2 = 4-12i+9i^2 = -5-12i$

$x^2-4x+13=0$
$(2+3i)^2-4(2+3i)+13=0$
$-5+12i-8-12i+13=0$
$\qquad\qquad\qquad 0=0 \quad \text{True}$

Section 9.8  Complex Numbers

$$x^2 - 4x + 13 = 0$$
$$(2-3i)^2 - 4(2-3i) + 13 = 0$$
$$-5 - 12i - 8 + 12i + 13 = 0$$
$$0 = 0 \quad \text{True}$$

Both the given number and its conjugate are solutions of the given equation.

**95.** $x^2 - 4x + 5 = 0, \quad 2-i, \quad 2+i$

$$(2-i)^2 = 4 - 4i + i^2 = 3 - 4i$$
$$(2+i)^2 = 4 + 4i + i^2 = 3 + 4i$$

$$x^2 - 4x + 5 = 0$$
$$(2-i)^2 - 4(2-i) + 5 = 0$$
$$3 - 4i - 8 + 4i + 5 = 0$$
$$0 = 0 \quad \text{True}$$

$$x^2 - 4x + 5 = 0$$
$$(2+i)^2 - 4(2+i) + 5 = 0$$
$$3 + 4i - 8 - 4i + 5 = 0$$
$$0 = 0 \quad \text{True}$$

Both the given number and its conjugate are solutions of the given equation.

**97.** $x^2 + 9 = (x + 3i)(x - 3i)$

**99.** $4x^2 + 1 = (2x + i)(2x - i)$

**101.**

(a) $i^{-2} = \dfrac{1}{i^2} = \dfrac{1}{(-1)} = -1$

(b) $i^{-3} = \dfrac{1}{i^3} = \dfrac{1}{-i} = \dfrac{1 \cdot i}{-i \cdot i} = \dfrac{i}{-i^2} = i$

(c) $(2+3i)^{-1} = \dfrac{1}{2+3i} = \dfrac{2-3i}{(2+3i)(2-3i)}$
$$= \dfrac{2-3i}{4+9} = \dfrac{2-3i}{13}$$

(d) $(1-2i)^{-2} = \dfrac{1}{(1-2i)^2} = \dfrac{1}{1 - 4i + 4i^2}$
$$= \dfrac{1}{-3 - 4i}$$
$$= \dfrac{-3 + 4i}{(-3 - 4i)(-3 + 4i)}$$
$$= \dfrac{-3 + 4i}{9 + 16} = \dfrac{-3 + 4i}{25}$$

**103.** Conjecture: there are three cube roots of $-8$.

(a) $\left(1 + i\sqrt{3}\right)^3 = \left(1 + i\sqrt{3}\right)\left(1 + 2i\sqrt{3} - 3\right)$
$$= 1 + 2i\sqrt{3} - 3 + i\sqrt{3}$$
$$\quad - 6 - 3i\sqrt{3}$$
$$= -8$$

(b) $\left(1 - i\sqrt{3}\right)^3 = \left(1 - i\sqrt{3}\right)\left(1 - 2i\sqrt{3} - 3\right)$
$$= 1 - 2i\sqrt{3} - 3 - i\sqrt{3}$$
$$\quad - 6 + 3i\sqrt{3}$$
$$= -8$$

**105.** $\dfrac{2+3i}{2-3i} + \dfrac{i}{1-i}$
$$= \dfrac{(2+3i)(1-i)}{(2-3i)(1-i)} + \dfrac{i(2-3i)}{(2-3i)(1-i)}$$
$$= \dfrac{5 + i + 3 + 2i}{(2-3i)(1-i)} = \dfrac{5 + i + 3 + 2i}{-1 - 5i}$$
$$= \dfrac{8 + 3i}{-1 - 5i} = \dfrac{(8+3i)(-1+5i)}{(-1-5i)(-1+5i)}$$
$$= \dfrac{-23 + 37i}{1 + 25} = \dfrac{-23 + 37i}{26}$$

## Chapter 9 Project

1. By tracing a graph of the function, we can see that the maximum of 70 million is reached in approximately 2005 (when $x = 22$).

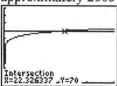

3. $\dfrac{P(x)}{M(x)} = \dfrac{54.6x^{2/25}}{30.4x^{-3/25}} = \dfrac{54.6}{30.4}x^{2/25+3/25} = \dfrac{273}{152}x^{5/25} = \dfrac{273x^{1/5}}{152}$; In 2005 ($x = 22$) this ratio is projected to be $\dfrac{273(22)^{1/5}}{152} \approx 3$, or about 3 to 1.

5. If the land were square, the area would be given by $A = x^2$, where $x$ is the length of the side of the square. By solving the equation at the right, we find that the square would be approximately 30 feet long on each side.

$x^2 = A$
$x^2 = 0.02(43{,}560)$
$x^2 = 871.2$
$x = \pm\sqrt{871.2} \approx \pm 29.52$

## Chapter 9 Review Exercises

1. The number $x^2$ has two square roots, $x$ and $-x$, but $\sqrt{x^2}$ is the principal (nonnegative) square root. Thus, we use absolute value symbols to assure that the result is nonnegative. The number $x^3$ has only one cube root, and so absolute value symbols are not used.

3. $-\sqrt{64} = -\sqrt{8^2} = -8$

5. $-\sqrt[3]{64} = -\sqrt[3]{4^3} = -4$

7. $\sqrt[4]{x^8} = \sqrt[4]{(x^2)^4} = x^2$

9. $\sqrt{(x+4)^2} = x+4$

11. (a) "The square root of 10 minus $a$" is translated as $\sqrt{10} - a$.
    (b) "The square root of the quantity 10 minus $a$" is translated as $\sqrt{10-a}$.

13. The expression $(-4)^{1/2}$ means $\sqrt{-4}$, but there is no real number whose square is $-4$.

15. $16^{3/2} = \left(\sqrt{16}\right)^3 = 4^3 = 64$

17. $(-27)^{-2/3} = \dfrac{1}{(-27)^{2/3}}$
$= \dfrac{1}{\left(\sqrt[3]{-27}\right)^2}$
$= \dfrac{1}{(-3)^2} = \dfrac{1}{9}$

19. $(-9)^{-1/2} = \dfrac{1}{(-9)^{1/2}} = \dfrac{1}{\sqrt{-9}}$ is not a real number

21. $25^{1.5} = 25^{3/2} = \left(\sqrt{25}\right)^3 = 5^3 = 125$

Chapter 9 Review Exercises

**23.** When $n$ is even, the initial result is always $|x|$. However, if $x$ is nonnegative, then $|x| = x$.

**25.** (a) $18^{1/2} \cdot 2^{1/2} = (18 \cdot 2)^{1/2} = 36^{1/2} = 6$

(b) $8^{1/6} \cdot 8^{1/2} = 8^{(1/6)+(1/2)} = 8^{(1/6)+(3/6)}$
$= 8^{4/6} = 8^{2/3} = \left(\sqrt[3]{8}\right)^2$
$= 2^2 = 4$

**27.** $(b^{-4/3} c^2)^9 = (b^{-4/3})^9 (c^2)^9$
$= b^{-36/3} c^{18}$
$= b^{-12} c^{18}$
$= \dfrac{c^{18}}{b^{12}}$

**29.** $\left(\dfrac{27 x^6 z^{-9}}{y^{-3}}\right)^{1/3} = \dfrac{27^{1/3} (x^6)^{1/3} (z^{-9})^{1/3}}{(y^{-3})^{1/3}}$
$= \dfrac{3 x^{6/3} z^{-9/3}}{y^{-1}}$
$= \dfrac{3 x^2 z^{-3}}{y^{-1}}$
$= \dfrac{3 x^2 y}{z^3}$

**31.** $\dfrac{(3x^{-1})^{-2}}{x^{1/2}} = \dfrac{3^{-2} x^{(-1)\cdot(-2)}}{x^{1/2}}$
$= \dfrac{9^{-1} x^2}{x^{1/2}}$
$= \dfrac{x^2}{9 x^{1/2}}$
$= \dfrac{x^{3/2}}{9}$

**33.** $x^{2/3}(x^{-5/3} - x^{7/3})$
$= x^{(2/3)-(5/3)} - x^{(2/3)+(7/3)}$
$= x^{-3/3} - x^{9/3}$
$= x^{-1} - x^3$
$= \dfrac{1}{x} - x^3, \quad x \neq 0$

**35.** The Product Rule cannot be used because the indices are different. We can use rational exponents to write
$\sqrt[3]{5} \cdot \sqrt{5} = 5^{1/3} \cdot 5^{1/2}$
$= 5^{2/6} \cdot 5^{3/6}$
$= 5^{5/6}$
$= \sqrt[6]{5^5}$

**37.** $\left(3\sqrt{5x}\right)^2 = 3^2 \left(\sqrt{5x}\right)^2 = 9 \cdot 5x = 45x$

**39.** $\sqrt{6} \cdot \sqrt{7x} = \sqrt{6 \cdot 7 \cdot x} = \sqrt{42x}$

**41.** $\sqrt{x^5} = \sqrt{x^4} \cdot \sqrt{x} = x^2 \sqrt{x}$

**43.** $\sqrt{c^{17} d^{20}} = \sqrt{c^{16} d^{20}} \cdot \sqrt{c} = c^8 d^{10} \sqrt{c}$

**45.** If $x$ represents any real number,
$\sqrt{x^2 - 6x + 9} = \sqrt{(x-3)^2} = |x-3|$
which is the definition of the distance between $x$ and 3 on a number line.

**47.** The radicand must contain no perfect $n$th power factors. The radicand must not contain a fraction. There must be no radical in the denominator of a fraction.

**49.** $\dfrac{\sqrt[3]{x^5}}{\sqrt[3]{x^2}} = \sqrt[3]{\dfrac{x^5}{x^2}} = \sqrt[3]{x^3} = x$

**51.** $\sqrt{\dfrac{5a^7}{9b^4}} = \dfrac{\sqrt{5a^7}}{\sqrt{9b^4}} = \dfrac{\sqrt{a^6}\sqrt{5a}}{3b^2} = \dfrac{a^3 \sqrt{5a}}{3b^2}$

**53.** $\dfrac{6}{\sqrt{6}} = \dfrac{6 \cdot \sqrt{6}}{\sqrt{6} \cdot \sqrt{6}} = \dfrac{6\sqrt{6}}{6} = \sqrt{6}$

**55.** $\dfrac{3x}{\sqrt{x^5}} = \dfrac{3x \sqrt{x}}{\sqrt{x^5}\sqrt{x}} = \dfrac{3x\sqrt{x}}{\sqrt{x^6}}$
$= \dfrac{3x\sqrt{x}}{x^3} = \dfrac{3\sqrt{x}}{x^2}$

**57.** $\dfrac{\sqrt{18xy^5}}{\sqrt{2x^2y^3}} = \sqrt{\dfrac{18xy^5}{2x^2y^3}} = \sqrt{\dfrac{9y^2}{x}} = \dfrac{3y}{\sqrt{x}}$

$= \dfrac{3y \cdot \sqrt{x}}{\sqrt{x} \cdot \sqrt{x}} = \dfrac{3y\sqrt{x}}{x}$

**59.** Let $x = $ a positive number.

$\sqrt{\dfrac{1}{x}} = \dfrac{\sqrt{1}}{\sqrt{x}} = \dfrac{1}{\sqrt{x}}$,

which shows that the square root of the reciprocal of a positive number is the reciprocal of the square root of the number.

**61.** (a) The indices are the same, but the radicands are different.
(b) The radicands are the same, but the indices are different.

**63.** $3\sqrt{45x} + 2\sqrt{20x} = 3\sqrt{9}\sqrt{5x} + 2\sqrt{4}\sqrt{5x}$
$= 3 \cdot 3\sqrt{5x} + 2 \cdot 2\sqrt{5x}$
$= 9\sqrt{5x} + 4\sqrt{5x}$
$= 13\sqrt{5x}$

**65.** $(\sqrt{3} + \sqrt{5})^2 = (\sqrt{3})^2 + 2(\sqrt{3})(\sqrt{5}) + (\sqrt{5})^2$
$= 3 + 2\sqrt{15} + 5$
$= 8 + 2\sqrt{15}$

**67.** Because
$\sqrt{2}(\sqrt{3} + \sqrt{5x}) = \sqrt{2}\sqrt{3} + \sqrt{2}\sqrt{5x}$
$= \sqrt{6} + \sqrt{10x}$
The missing number is $\sqrt{2}$.

**69.** $\dfrac{5}{5 - \sqrt{3}} = \dfrac{5(5 + \sqrt{3})}{(5 - \sqrt{3})(5 + \sqrt{3})}$

$= \dfrac{5(5 + \sqrt{3})}{5^2 - 3}$

$= \dfrac{5(5 + \sqrt{3})}{22}$

**71.** $\dfrac{1}{\sqrt{3}} + \dfrac{5\sqrt{3}}{3} = \dfrac{\sqrt{3}}{\sqrt{3} \cdot \sqrt{3}} + \dfrac{5\sqrt{3}}{3}$

$= \dfrac{\sqrt{3}}{3} + \dfrac{5\sqrt{3}}{3}$

$= \dfrac{\sqrt{3} + 5\sqrt{3}}{3}$

$= \dfrac{6\sqrt{3}}{3} = 2\sqrt{3}$

**73.** The perimeter is
$\sqrt{20} + \sqrt{125} + \sqrt{180}$
$= \sqrt{4} \cdot \sqrt{5} + \sqrt{25} \cdot \sqrt{5} + \sqrt{36} \cdot \sqrt{5}$
$= 2\sqrt{5} + 5\sqrt{5} + 6\sqrt{5}$
$= 13\sqrt{5}$

**75.** When an algebraic method is correctly used to solve an equation, an apparent solution may not satisfy the equation. Such a number is an extraneous solution.

**77.** $(2x - 3)^3 = 27$
$2x - 3 = \sqrt[3]{27}$
$2x - 3 = 3$
$2x = 6$
$x = 3$
This value checks as a valid solution.

**79.** $\sqrt{5x - 4} - \sqrt{x + 3} = -1$
$\sqrt{5x - 4} = \sqrt{x + 3} - 1$
$(\sqrt{5x - 4})^2 = (\sqrt{x + 3} - 1)^2$
$5x - 4 = x + 3 - 2\sqrt{x + 3} + 1$
$4x - 8 = -2\sqrt{x + 3}$
$4(x - 2) = -2\sqrt{x + 3}$
$2x - 4 = -\sqrt{x + 3}$
$(2x - 4)^2 = (-\sqrt{x + 3})^2$
$4x^2 - 16x + 16 = x + 3$
$4x^2 - 17x + 13 = 0$
$(4x - 13)(x - 1) = 0$
$4x - 13 = 0$ or $x - 1 = 0$
$x = 13/4$ or $x = 1$
A check shows that only 1 is a valid solution. The value of 13/4 is an extraneous solution.

**81.** From the given information, we can identify the 20-foot wire as the hypotenuse of a right triangle with one leg 13 feet long. The height $x$ of the totem pole is the length of the other leg of the triangle

$$x^2 + 13^2 = 20^2$$
$$x^2 + 169 = 400$$
$$x^2 = 231$$
$$x = \pm\sqrt{231}$$
$$x \approx \pm 15.20$$

The totem pole is approximately 15.20 feet tall.

**83.** Let $x$ = the smaller number and $x + 5$ = the larger number.

$$\sqrt{x} \cdot \sqrt{x+5} = 6$$
$$\left(\sqrt{x} \cdot \sqrt{x+5}\right)^2 = 6^2$$
$$x(x+5) = 36$$
$$x^2 + 5x - 36 = 0$$
$$(x+9)(x-4) = 0$$
$$x + 9 = 0 \quad \text{or} \quad x - 4 = 0$$
$$x = -9 \quad \text{or} \quad x = 4$$

The smaller number is 4 and the larger number is 9.

**85.** The square root of a negative number is not a real number.
$$\sqrt{-4}\sqrt{-9} = (2i)(3i) = 6i^2 = -6$$

**87.** $(3 - 5i) - (5 - 7i) = 3 - 5i - 5 + 7i$
$$= -2 + 2i$$

**89.** $(6 - 7i)(2 + 3i) = 12 + 18i - 14i - 21i^2$
$$= 12 + 18i - 14i - 21(-1)$$
$$= 33 + 4i$$

**91.** $\dfrac{8 - 4i}{2} = \dfrac{8}{2} - \dfrac{4i}{2} = 4 - 2i$

**93.** $i^{57} = (i^2)^{28} \cdot i = (-1)^{28} \cdot i = i$

**95.** $i(a + bi) = ai + bi^2 = -b + ai = 6 + 0i$

Equate the real parts and equate the imaginary parts and solve.
$$-b = 6 \quad \text{and} \quad a = 0$$
$$b = -6$$

## Chapter 9 Looking Back

**1.**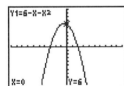

The $x$-intercepts appear to be $(-3, 0)$ and $(2, 0)$ and the $y$-intercept appears to be $(0, 6)$.

**3.**
$$9x^2 + 4 = 12x$$
$$9x^2 - 12x + 4 = 0$$
$$(3x - 2)^2 = 0$$
$$3x - 2 = 0$$
$$3x = 2$$
$$x = 2/3$$

**5.**
$$x^3 + 3x^2 - 4x - 12 = 0$$
$$(x^3 + 3x^2) - (4x + 12) = 0$$
$$x^2(x + 3) - 4(x + 3) = 0$$
$$(x + 3)(x^2 - 4) = 0$$
$$(x + 3)(x + 2)(x - 2) = 0$$
$$x + 3 = 0 \quad \text{or} \quad x + 2 = 0 \quad \text{or} \quad x - 2 = 0$$
$$x = -3 \quad \text{or} \quad x = -2 \quad \text{or} \quad x = 2$$

7.  $a = 2, b = -7, c = 3$
    $$\frac{-b + \sqrt{b^2 - 4ac}}{2a}$$
    $$= \frac{-(-7) + \sqrt{(-7)^2 - 4(2)(3)}}{2(2)}$$
    $$= \frac{7 + \sqrt{49 - 24}}{4}$$
    $$= \frac{7 + \sqrt{25}}{4} = \frac{7 + 5}{4} = 3$$

9.  The restricted values are $-3$ and $3$.
    $$\frac{12}{t^2 - 9} = \frac{2}{t - 3} - 1$$
    $$\frac{12}{t^2 - 9} - \frac{2}{t - 3} + 1 = 0$$
    Multiply each term by the LCD $t^2 - 9$.
    $$12 - 2(t + 3) + t^2 - 9 = 0$$
    $$t^2 - 2t - 3 = 0$$
    $$(t - 3)(t + 1) = 0$$
    $t - 3 = 0$ or $t + 1 = 0$
    $t = 3$ $\quad$ $t = -1$
    Because 3 is a restricted value, only $-1$ is a solution.

11. Graph the solid boundary line $y = x - 6$. Because the inequality is in the form $y \le mx + b$, shade below the line.

## Chapter 9 Test

1.  To find the domain of $f(x) = \sqrt{3 - x}$, find the values for which $3 - x \ge 0$.
    $$3 - x \ge 0$$
    $$-x \ge -3$$
    $$x \le 3$$
    The domain is $\{x \mid x \le 3\}$.

3.  $-25^{1/2} = -\sqrt{25} = -5$.
    $(-25)^{1/2} = \sqrt{-25}$, which is not a real number.

5.  $\sqrt[3]{\sqrt[4]{x^7}} = \sqrt[3]{x^{7/4}} = \left(x^{7/4}\right)^{1/3} = x^{7/12} = \sqrt[12]{x^7}$

7.  $$\frac{\sqrt{12}}{4} + \frac{\sqrt{27}}{6} = \frac{\sqrt{4}\sqrt{3}}{4} + \frac{\sqrt{9}\sqrt{3}}{6}$$
    $$= \frac{2\sqrt{3}}{4} + \frac{3\sqrt{3}}{6}$$
    $$= \frac{\sqrt{3}}{2} + \frac{\sqrt{3}}{2} = \sqrt{3}$$

9.  The conditions are (i) $b$ is divisible by $a$, or (ii) $a$ is a perfect square.

11. $c^2 = \left(\sqrt{2x + 3}\right)^2 + (x + 1)^2$
    $c^2 = 2x + 3 + x^2 + 2x + 1$
    $c^2 = x^2 + 4x + 4$
    $c^2 = (x + 2)^2$
    $c = \pm\sqrt{(x + 2)^2}$
    $c = \pm(x + 2)$
    Because length is considered to be positive, the length of the hypotenuse is $x + 2$.

13. $$\left(\frac{16}{25}\right)^{-3/2} = \left(\frac{25}{16}\right)^{3/2} = \left(\sqrt{\frac{25}{16}}\right)^3$$
    $$= \left(\frac{5}{4}\right)^3 = \frac{125}{64}$$

15. Extraneous solutions can occur.

**17.**
$$\sqrt{1-3x} = \sqrt{3x}+1$$
$$\left(\sqrt{1-3x}\right)^2 = \left(\sqrt{3x}+1\right)^2$$
$$1-3x = 3x + 2\sqrt{3x}+1$$
$$-6x = 2\sqrt{3x}$$
$$-3x = \sqrt{3x}$$
$$(-3x)^2 = \left(\sqrt{3x}\right)^2$$
$$9x^2 = 3x$$
$$9x^2 - 3x = 0$$
$$3x(3x-1) = 0$$
$3x = 0$ or $3x - 1 = 0$
$x = 0$ or $x = 1/3$

A check shows that only 0 is a valid solution. The value of 1/3 is an extraneous solution. The length of each side is 1.

**19.**
$$(3+2i)^2 = 3^2 + 12i + 4i^2$$
$$= 9 + 12i - 4$$
$$= 5 + 12i$$

**21.** Before they can be multiplied, rewrite the numbers as $\sqrt{-7} = i\sqrt{7}$ and $\sqrt{-3} = i\sqrt{3}$.

**23.**
$$V = s^3$$
$$V^{1/3} = (s^3)^{1/3}$$
$$V^{1/3} = s$$
For a cube with a volume of 200 cubic inches, $s = V^{1/3} = 200^{1/3} \approx 5.85$ inches.

**25.** (a) If $A^n = B$ and $n$ is an even positive integer, then the equation has no real number solution if $B < 0$.

(b) For a positive integer $n$, $\sqrt[n]{a^n} = a$ $a > 0$ or if $n$ is odd.

# Chapter 10

# Quadratic Equations

## Section 10.1 Special Methods

1. The $x$-intercepts correspond to the solutions of the equation.

3. $2x^2 + 5x - 3 = 0$
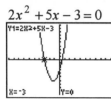
The estimated solutions are $-3$ and $0.5$.

5. $3 - 8x = 3x^2$
$0 = 3x^2 + 8x - 3$
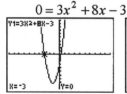
The estimated solutions are $-3$ and $1/3$.

7. $x^2 - 6x + 9 = 0$
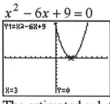
The estimated solution is $3$.

9. $10x = 25 + x^2$
$0 = 25 - 10x + x^2$
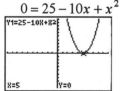
The estimated solution is $5$.

11. $x^2 + 3x + 5 = 0$
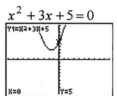
Because the graph has no $x$-intercept, the equation has no real number solution.

13. $10x - 27 - x^2 = 0$

Because the graph has no $x$-intercept, the equation has no real number solution.

15. $x^2 + 7x + 6 = 0$
$(x+6)(x+1) = 0$
$x + 6 = 0$  or  $x + 1 = 0$
$x = -6$  or  $x = -1$

17. $x^2 = x$
$x^2 - x = 0$
$x(x-1) = 0$
$x = 0$  or  $x - 1 = 0$
$\quad\quad\quad\quad\quad\; x = 1$

19. $\dfrac{3x}{4} = \dfrac{1}{2} + \dfrac{2}{x}$
$8x \cdot \dfrac{3x}{4} = 8x \cdot \dfrac{1}{2} + 8x \cdot \dfrac{2}{x}$
$6x^2 = 4x + 16$
$6x^2 - 4x - 16 = 0$
$2(3x^2 - 2x - 8) = 0$
$2(3x + 4)(x - 2) = 0$

Section 10.1  Special Methods

$$3x+4=0 \quad \text{or} \quad x-2=0$$
$$x=-4/3 \quad \text{or} \quad x=2$$

**21.**
$$x(x-3)=x-3$$
$$x^2-3x=x-3$$
$$x^2-4x+3=0$$
$$(x-3)(x-1)=0$$
$$x-3=0 \quad \text{or} \quad x-1=0$$
$$x=3 \quad \text{or} \quad x=1$$

**23.**
$$8x^2+x-1=3x^2-2x+1$$
$$5x^2+3x-2=0$$
$$(5x-2)(x+1)=0$$
$$5x-2=0 \quad \text{or} \quad x+1=0$$
$$x=2/5 \quad \text{or} \quad x=-1$$

**25.** If $A^2 = c$ and $c > 0$, then $A = \pm\sqrt{c}$. Thus the equation $x^2 = 6$ has two solutions, $\pm\sqrt{6}$. If $A^2 = c$ and $c < 0$, then there are no real number solutions. Thus $x^2 = -6$ has no real number solutions.

**27.**
$$(5x-4)^2=9$$
$$5x-4=\pm\sqrt{9}$$
$$5x=4\pm 3$$
$$x=\frac{4\pm 3}{5}$$
$$x=0.2,\ 1.4$$

**29.**
$$16(x+5)^2-5=0$$
$$16(x+5)^2=5$$
$$(x+5)^2=\frac{5}{16}$$
$$x+5=\pm\sqrt{\frac{5}{16}}$$
$$x=-5\pm\frac{\sqrt{5}}{4}$$
$$x\approx -4.44,\ -5.56$$

**31.**
$$\left(x-\frac{3}{2}\right)^2=\frac{9}{4}$$
$$x-\frac{3}{2}=\pm\sqrt{\frac{9}{4}}$$
$$x-\frac{3}{2}=\pm\frac{3}{2}$$
$$x=\frac{3}{2}\pm\frac{3}{2}$$
$$x=0,\ 3$$

**33.**
$$x(x-6)=3(3-2x)$$
$$x^2-6x=9-6x$$
$$x^2=9$$
$$x=\pm\sqrt{9}$$
$$x=\pm 3$$
$$x=3,\ -3$$

**35.**
$$t^2+16=0$$
$$t^2=-16$$
$$t=\pm\sqrt{-16}=\pm i\sqrt{16}=\pm 4i$$

**37.**
$$24+y^2=0$$
$$y^2=-24$$
$$y=\pm\sqrt{-24}=\pm 2i\sqrt{6}\approx \pm 4.90i$$

**39.**
$$(x-3)^2=-4$$
$$x-3=\pm\sqrt{-4}$$
$$x-3=\pm i\sqrt{4}$$
$$x-3=\pm 2i$$
$$x=3\pm 2i$$

**41.**
$$(t-1)^2+169=144$$
$$(t-1)^2=-25$$
$$t-1=\pm\sqrt{-25}$$
$$t-1=\pm i\sqrt{25}$$
$$t-1=\pm 5i$$
$$t=1\pm 5i$$

**43.**
$$(2x+5)^2+6=0$$
$$(2x+5)^2=-6$$
$$2x+5=\pm\sqrt{-6}$$
$$2x+5=\pm i\sqrt{6}$$
$$2x=-5\pm i\sqrt{6}$$
$$x=\frac{-5\pm i\sqrt{6}}{2}\approx -2.5\pm 1.22i$$

**45.** 
$$8(x+1)^2 + 21 = 3$$
$$8(x+1)^2 = -18$$
$$(x+1)^2 = -\frac{18}{8}$$
$$x+1 = \pm\sqrt{-\frac{9}{4}}$$
$$x+1 = \pm i\sqrt{\frac{9}{4}}$$
$$x = -1 \pm \frac{3}{2}i = -1 \pm 1.5i$$

**47.**
$$10 + \frac{4}{y^2} = 1$$
$$\frac{4}{y^2} = -9$$
$$4 = -9y^2$$
$$y^2 = -\frac{4}{9}$$
$$y = \pm\sqrt{-\frac{4}{9}}$$
$$y = \pm i\sqrt{\frac{4}{9}} = \pm\frac{2}{3}i$$

**49.**
$$4 = \frac{9}{(1-x)^2}$$
$$4(1-x)^2 = 9$$
$$(1-x)^2 = \frac{9}{4}$$
$$1-x = \pm\sqrt{\frac{9}{4}}$$
$$1-x = \pm\frac{3}{2}$$
$$x = 1 \pm 1.5 = -0.5, 2.5$$

**51.**
$$x(x+4) = -4(1-x)$$
$$x^2 + 4x = -4 + 4x$$
$$x^2 = -4$$
$$x = \pm\sqrt{-4}$$
$$x = \pm i\sqrt{4}$$
$$x = \pm 2i$$

**53.** Because there are no x-intercepts, the equation has no real number solution. There are two imaginary number solutions.

**55.**
$$3x = x^2$$
$$0 = x^2 - 3x$$
$$0 = x(x-3)$$
$$x = 0 \quad \text{or} \quad x - 3 = 0$$
$$x = 3$$

**57.**
$$t^2 + 45 = 0$$
$$t^2 = -45$$
$$t = \pm\sqrt{-45}$$
$$t = \pm i\sqrt{45}$$
$$t = \pm 3i\sqrt{5} \approx \pm 6.71i$$

**59.**
$$36x^2 + 12x + 1 = 0$$
$$(6x+1)^2 = 0$$
$$6x + 1 = 0$$
$$x = -\frac{1}{6}$$

**61.**
$$y^2 = -1.44$$
$$y = \pm\sqrt{-1.44}$$
$$y = \pm i\sqrt{1.44}$$
$$y = \pm 1.2i$$

**63.**
$$x(x-5) = 5(2-x)$$
$$x^2 - 5x = 10 - 5x$$
$$x^2 = 10$$
$$x = \pm\sqrt{10} \approx \pm 3.16$$

**65.**
$$x - 2 = \frac{35}{x}$$
$$x(x-2) = 35$$
$$x^2 - 2x - 35 = 0$$
$$(x-7)(x+5) = 0$$
$$x - 7 = 0 \quad \text{or} \quad x + 5 = 0$$
$$x = 7 \quad \text{or} \quad x = -5$$

**67.**
$$(4x+5)^2 + 49 = 0$$
$$(4x+5)^2 = -49$$
$$4x + 5 = \pm\sqrt{-49}$$
$$4x + 5 = \pm i\sqrt{49}$$
$$4x + 5 = \pm 7i$$
$$4x = -5 \pm 7i$$
$$x = \frac{-5 \pm 7i}{4} = -1.25 \pm 1.75i$$

**69.** 
$$\left(\frac{3x+2}{4}\right)^2 = 12$$
$$\frac{3x+2}{4} = \pm\sqrt{12}$$
$$\frac{3x+2}{4} = \pm 2\sqrt{3}$$
$$3x+2 = \pm 8\sqrt{3}$$
$$3x = -2 \pm 8\sqrt{3}$$
$$x = \frac{-2 \pm 8\sqrt{3}}{3} \approx 3.95, \ -5.29$$

**71.** 
$$t(t+3) = 3(t-2)$$
$$t^2 + 3t = 3t - 6$$
$$t^2 = -6$$
$$t = \pm\sqrt{-6}$$
$$t = \pm i\sqrt{6} \approx \pm 2.45i$$

**73.** 
$$x(x-4) = x-4$$
$$x^2 - 4x = x - 4$$
$$x^2 - 5x + 4 = 0$$
$$(x-4)(x-1) = 0$$
$$x-4 = 0 \quad \text{or} \quad x-1 = 0$$
$$x = 4 \quad \text{or} \quad x = 1$$

**75.** 
$$(4x-1)(x-1) = 7x - 8$$
$$4x^2 - 5x + 1 = 7x - 8$$
$$4x^2 - 12x + 9 = 0$$
$$(2x-3)^2 = 0$$
$$2x - 3 = 0$$
$$x = 3/2$$

**77.** 
$$x = 3 \quad \text{or} \quad x = 5$$
$$x - 3 = 0 \quad \text{or} \quad x - 5 = 0$$
$$(x-3)(x-5) = 0$$
$$x^2 - 8x + 15 = 0$$

**79.** 
$$x = 3 \quad \text{or} \quad x = 3$$
$$x - 3 = 0 \quad \text{or} \quad x - 3 = 0$$
$$(x-3)(x-3) = 0$$
$$x^2 - 6x + 9 = 0$$

**81.** 
$$x = 3 \quad \text{or} \quad x = 4/3$$
$$x - 3 = 0 \quad \text{or} \quad 3x - 4 = 0$$
$$(x-3)(3x-4) = 0$$
$$3x^2 - 13x + 12 = 0$$

**83.** 
$$x = -4i \quad \text{or} \quad x = 4i$$
$$x + 4i = 0 \quad \text{or} \quad x - 4i = 0$$
$$(x+4i)(x-4i) = 0$$
$$x^2 + 16 = 0$$

**85.** Let $x$ = the first negative integer and $x + 1$ = the second consecutive negative integer.
$$x(x+1) = 72$$
$$x^2 + x - 72 = 0$$
$$(x+9)(x-8) = 0$$
$$x + 9 = 0 \quad \text{or} \quad x - 8 = 0$$
$$x = -9 \quad \text{or} \quad x = 8$$
The numbers are $-9$ and $-8$.

**87.** Let $x$ = the first even integer and $x + 2$ = the second consecutive even integer.
$$x(x+2) = 360$$
$$x^2 + 2x - 360 = 0$$
$$(x+20)(x-18) = 0$$
$$x + 20 = 0 \quad \text{or} \quad x - 18 = 0$$
$$x = -20 \quad \text{or} \quad x = 18$$
Because both numbers are even, the numbers are 18 and 20.

**89.** Solve $18t^2 - 90t = 0$ to find how long it takes to complete one round trip.
$$18t^2 - 90t = 0$$
$$18t(t-5) = 0$$
$$t = 0 \quad \text{or} \quad t - 5 = 0$$
$$t = 5$$
It takes 5 minutes for one round trip.

**91.** (a) The equation needed to estimate the year in which 212 million subscribers are predicted is
$$0.82t^2 + 1.5t + 5 = 212.$$

(b) From the table, we see that 212 million subscribers are predicted for $t = 15$, which is equivalent to the year 2005.

| X | Y1 |
|---|---|
| 13 | 163.08 |
| 14 | 186.72 |
| 15 | 212 |
| 16 | 238.92 |
| 17 | 267.48 |
| 18 | 297.68 |
| 19 | 329.52 |

Y1=212

93. $(x+2)^2 = -y$
$x+2 = \pm\sqrt{-y}$
$x+2 = \pm i\sqrt{y}$
$x = -2 \pm i\sqrt{y}$

95. $(x+c)^2 = -16$
$x+c = \pm\sqrt{-16}$
$x+c = \pm i\sqrt{16}$
$x+c = \pm 4i$
$x = -c \pm 4i$

97. $x-1+i$ and $x-1-i$ are both factors of $x^2 - 2x + 2$ if we obtain $x^2 - 2x + 2$ when the factors are multiplied.
$(x-1+i)(x-1-i)$
$= x^2 - x - ix - x + 1 + i + ix - i + 1$
$= x^2 - 2x + 2$

Using the Zero Factor Property:
$x^2 - 2x + 2 = 0$
$(x-1+i)(x-1-i) = 0$
$x-1+i = 0$ or $x-1-i = 0$
$x = 1-i$ or $x = 1+i$
The solutions are $1 \pm i$.

99. $(2x+5)^2 = (x-1)^2$
$2x+5 = \pm\sqrt{(x-1)^2}$
$2x+5 = \pm(x-1)$
$2x+5 = x-1$ or $2x+5 = -x+1$
$x = -6$ or $3x = -4$
$x = -4/3$

## Section 10.2 Completing the Square

1. The expression $A$ is a first-degree binomial.

3. $x^2 + 6x + k$
Half the coefficient of $x$ is 3 and $3^2 = 9$. So, the value of $k = 9$ makes $x^2 + 6x + k$ a perfect square trinomial.

5. $x^2 + 7x + k$
Half the coefficient of $x$ is $7/2$ and $\left(\frac{7}{2}\right)^2 = \frac{49}{4}$. So, the value of $k = \frac{49}{4}$ makes $x^2 + 7x + k$ a perfect square trinomial.

7. $x^2 + \frac{4}{3}x + k$
Half the coefficient of $x$ is $4/6 = 2/3$ and $\left(\frac{2}{3}\right)^2 = \frac{4}{9}$. So, the value of $k = \frac{4}{9}$ makes $x^2 + \frac{4}{3}x + k$ a perfect square trinomial.

9. Both methods are correct. The factoring method can be used because $x^2 - 8x + 12$ is factorable. The method of completing the square can be used for any quadratic equation.

11. Factoring:
$x^2 + 4x + 3 = 0$
$(x+3)(x+1) = 0$
$x+3 = 0$ or $x+1 = 0$
$x = -3$ or $x = -1$

Section 10.2   Completing the Square

Completing the square:
$$x^2 + 4x + 3 = 0$$
$$x^2 + 4x = -3$$
$$x^2 + 4x + 4 = -3 + 4$$
$$(x+2)^2 = 1$$
$$x + 2 = \pm\sqrt{1}$$
$$x = -2 \pm 1 = -3, -1$$

13. Factoring:
$$t^2 + 2t = 15$$
$$t^2 + 2t - 15 = 0$$
$$(t+5)(t-3) = 0$$
$$t + 5 = 0 \quad \text{or} \quad t - 3 = 0$$
$$t = -5 \quad \text{or} \quad t = 3$$
Completing the square:
$$t^2 + 2t = 15$$
$$t^2 + 2t + 1 = 15 + 1$$
$$(t+1)^2 = 16$$
$$t + 1 = \pm\sqrt{16}$$
$$t = -1 \pm 4 = -5, 3$$

15. Factoring:
$$x^2 - x - 12 = 0$$
$$(x-4)(x+3) = 0$$
$$x - 4 = 0 \quad \text{or} \quad x + 3 = 0$$
$$x = 4 \quad \text{or} \quad x = -3$$
Completing the square:
$$x^2 - x - 12 = 0$$
$$x^2 - x = 12$$
$$x^2 - x + \frac{1}{4} = 12 + \frac{1}{4}$$
$$\left(x - \frac{1}{2}\right)^2 = \frac{49}{4}$$
$$x - \frac{1}{2} = \pm\sqrt{\frac{49}{4}}$$
$$x = \frac{1}{2} \pm \frac{7}{2} = 4, -3$$

17. Factoring:
$$6 = x^2 + x$$
$$0 = x^2 + x - 6$$
$$0 = (x+3)(x-2)$$
$$x + 3 = 0 \quad \text{or} \quad x - 2 = 0$$
$$x = -3 \quad \text{or} \quad x = 2$$

Completing the square:
$$x^2 + x = 6$$
$$x^2 + x + \frac{1}{4} = 6 + \frac{1}{4}$$
$$\left(x + \frac{1}{2}\right)^2 = \frac{25}{4}$$
$$x + \frac{1}{2} = \pm\sqrt{\frac{25}{4}}$$
$$x = -\frac{1}{2} \pm \frac{5}{2} = -3, 2$$

19. $$x^2 - 6x + 4 = 0$$
$$x^2 - 6x = -4$$
$$x^2 - 6x + 9 = -4 + 9$$
$$(x-3)^2 = 5$$
$$x - 3 = \pm\sqrt{5}$$
$$x = 3 \pm \sqrt{5} \approx 0.76, 5.24$$

21. $$x^2 + 2x = 24$$
$$x^2 + 2x + 1 = 24 + 1$$
$$(x+1)^2 = 25$$
$$x + 1 = \pm\sqrt{25}$$
$$x + 1 = \pm 5$$
$$x = -1 \pm 5 = -6, 4$$

23. $$x^2 = 4x + 4$$
$$x^2 - 4x = 4$$
$$x^2 - 4x + 4 = 4 + 4$$
$$(x-2)^2 = 8$$
$$x - 2 = \pm\sqrt{8}$$
$$x = 2 \pm 2\sqrt{2} \approx -0.83, 4.83$$

25. Divide both sides by 3 to make the leading coefficient 1.

27. Factoring:
$$3y^2 + 4y + 1 = 0$$
$$(3y+1)(y+1) = 0$$
$$3y + 1 = 0 \quad \text{or} \quad y + 1 = 0$$
$$y = -1/3 \quad \text{or} \quad y = -1$$

Completing the square:
$$3y^2 + 4y + 1 = 0$$
$$y^2 + \frac{4}{3}y = -\frac{1}{3}$$
$$y^2 + \frac{4}{3}y + \frac{4}{9} = -\frac{1}{3} + \frac{4}{9}$$
$$\left(y + \frac{2}{3}\right)^2 = \frac{1}{9}$$
$$y + \frac{2}{3} = \pm\sqrt{\frac{1}{9}}$$
$$y = -\frac{2}{3} \pm \frac{1}{3} = -1, \ -\frac{1}{3}$$

**29.** Factoring:
$$15 = 2x^2 + 7x$$
$$0 = 2x^2 + 7x - 15$$
$$0 = (2x - 3)(x + 5)$$
$$2x - 3 = 0 \quad \text{or} \quad x + 5 = 0$$
$$x = 3/2 \quad \text{or} \quad x = -5$$
Completing the Square:
$$2x^2 + 7x = 15$$
$$x^2 + \frac{7}{2}x = \frac{15}{2}$$
$$x^2 + \frac{7}{2}x + \frac{49}{16} = \frac{15}{2} + \frac{49}{16}$$
$$\left(x + \frac{7}{4}\right)^2 = \frac{169}{16}$$
$$x + \frac{7}{4} = \pm\sqrt{\frac{169}{16}}$$
$$x = -\frac{7}{4} \pm \frac{13}{4} = -5, \ \frac{3}{2}$$

**31.** 
$$4x^2 = 8x + 21$$
$$x^2 - 2x = \frac{21}{4}$$
$$x^2 - 2x + 1 = \frac{21}{4} + 1$$
$$(x - 1)^2 = \frac{25}{4}$$
$$x - 1 = \pm\sqrt{\frac{25}{4}}$$
$$x = 1 \pm \frac{5}{2} = -1.5, \ 3.5$$

**33.**
$$4x^2 + 25 = 20x$$
$$4x^2 - 20x = -25$$
$$x^2 - 5x = -\frac{25}{4}$$
$$x^2 - 5x + \frac{25}{4} = -\frac{25}{4} + \frac{25}{4}$$
$$\left(x - \frac{5}{2}\right)^2 = 0$$
$$x - \frac{5}{2} = 0$$
$$x = 2.5$$

**35.**
$$x(2x - 3) = 1$$
$$2x^2 - 3x = 1$$
$$x^2 - \frac{3}{2}x = \frac{1}{2}$$
$$x^2 - \frac{3}{2}x + \frac{9}{16} = \frac{1}{2} + \frac{9}{16}$$
$$\left(x - \frac{3}{4}\right)^2 = \frac{17}{16}$$
$$x - \frac{3}{4} = \pm\sqrt{\frac{17}{16}}$$
$$x = \frac{3}{4} \pm \frac{\sqrt{17}}{4} \approx -0.28, \ 1.78$$

**37.**
$$2x^2 + 7x - 1 = 0$$
$$2x^2 + 7x = 1$$
$$x^2 + \frac{7}{2}x = \frac{1}{2}$$
$$x^2 + \frac{7}{2}x + \frac{49}{16} = \frac{1}{2} + \frac{49}{16}$$
$$\left(x + \frac{7}{4}\right)^2 = \frac{57}{16}$$
$$x + \frac{7}{4} = \pm\sqrt{\frac{57}{16}}$$
$$x = -\frac{7}{4} \pm \frac{\sqrt{57}}{4} \approx -3.64, \ 0.14$$

**39.** Because $9x^2 + 6x + 1$ is a perfect square, the value of $k$ that makes $kx^2 + 6x + 1$ a perfect square is 9.

Section 10.2 Completing the Square

**41.**
$x^2 - 4x + 8 = 0$
$x^2 - 4x = -8$
$x^2 - 4x + 4 = -8 + 4$
$(x-2)^2 = -4$
$x - 2 = \pm\sqrt{-4}$
$x - 2 = \pm 2i$
$x = 2 \pm 2i$

**43.**
$x^2 + 6 = 4x$
$x^2 - 4x = -6$
$x^2 - 4x + 4 = -6 + 4$
$(x-2)^2 = -2$
$x - 2 = \pm\sqrt{-2}$
$x - 2 = \pm i\sqrt{2}$
$x = 2 \pm i\sqrt{2} \approx 2 \pm 1.41i$

**45.**
$x^2 + 3x + 5 = 0$
$x^2 + 3x = -5$
$x^2 + 3x + \dfrac{9}{4} = -5 + \dfrac{9}{4}$
$\left(x + \dfrac{3}{2}\right)^2 = -\dfrac{11}{4}$
$x + \dfrac{3}{2} = \pm\sqrt{-\dfrac{11}{4}}$
$x + \dfrac{3}{2} = \pm\dfrac{i\sqrt{11}}{2}$
$x = -\dfrac{3}{2} \pm \dfrac{i\sqrt{11}}{2} = -1.5 \pm 1.66i$

**47.**
$2x^2 + x + 3 = 0$
$2x^2 + x = -3$
$x^2 + \dfrac{1}{2}x = -\dfrac{3}{2}$
$x^2 + \dfrac{1}{2}x + \dfrac{1}{16} = -\dfrac{3}{2} + \dfrac{1}{16}$
$\left(x + \dfrac{1}{4}\right)^2 = -\dfrac{23}{16}$
$x + \dfrac{1}{4} = \pm\sqrt{-\dfrac{23}{16}}$
$x + \dfrac{1}{4} = \pm\dfrac{i\sqrt{23}}{4}$
$x = -\dfrac{1}{4} \pm \dfrac{i\sqrt{23}}{4} = -0.25 \pm 1.20i$

**49.**
$x^2 + bx = k$
$x^2 + bx + \left(\dfrac{b}{2}\right)^2 = k + \left(\dfrac{b}{2}\right)^2$
$\left(x + \dfrac{b}{2}\right)^2 = k + \left(\dfrac{b}{2}\right)^2$

The next step in the solution is to take the square root of both sides. If the right side of the equation is negative, the solutions will be imaginary numbers. The right side is negative if

$k + \left(\dfrac{b}{2}\right)^2 < 0$
$k < -\left(\dfrac{b}{2}\right)^2$

**51.**
$x^2 + 16 = 10x$
$x^2 - 10x + 16 = 0$
$(x-8)(x-2) = 0$
$x - 8 = 0$ or $x - 2 = 0$
$x = 8$ or $x = 2$

**53.**
$36x^2 - 49 = 0$
$(6x + 7)(6x - 7) = 0$
$6x + 7 = 0$ or $6x - 7 = 0$
$x = -7/6$ or $x = 7/6$

**55.**
$x(x - 3) = 10$
$x^2 - 3x - 10 = 0$
$(x - 5)(x + 2) = 0$
$x - 5 = 0$ or $x + 2 = 0$
$x = 5$ or $x = -2$

**57.**
$4(x + 3)^2 - 7 = 0$
$(x + 3)^2 = \dfrac{7}{4}$
$x + 3 = \pm\sqrt{\dfrac{7}{4}}$
$x = -3 \pm \dfrac{\sqrt{7}}{2} \approx -4.32, -1.68$

**59.** 
$$x(x+1)+9=0$$
$$x^2+x=-9$$
$$x^2+x+\frac{1}{4}=-9+\frac{1}{4}$$
$$\left(x+\frac{1}{2}\right)^2=-\frac{35}{4}$$
$$x+\frac{1}{2}=\pm\sqrt{-\frac{35}{4}}$$
$$x=-\frac{1}{2}\pm\frac{i\sqrt{35}}{2}\approx -0.5\pm 2.96i$$

**61.**
$$(x-3)(x-1)=5x-6$$
$$x^2-4x+3=5x-6$$
$$x^2-9x=-9$$
$$x^2-9x+\frac{81}{4}=-9+\frac{81}{4}$$
$$\left(x-\frac{9}{2}\right)^2=\frac{45}{4}$$
$$x-\frac{9}{2}=\pm\sqrt{\frac{45}{4}}$$
$$x=\frac{9}{2}\pm\frac{3\sqrt{5}}{2}\approx 1.15,\ 7.85$$

**63.**
$$x^2+\frac{3}{2}x=\frac{3}{4}$$
$$x^2+\frac{3}{2}x+\frac{9}{16}=\frac{3}{4}+\frac{9}{16}$$
$$\left(x+\frac{3}{4}\right)^2=\frac{21}{16}$$
$$x+\frac{3}{4}=\pm\sqrt{\frac{21}{16}}$$
$$x=-\frac{3}{4}\pm\frac{\sqrt{21}}{4}\approx -1.90,\ 0.40$$

**65.**
$$x^2=x-9$$
$$x^2-x=-9$$
$$x^2-x+\frac{1}{4}=-9+\frac{1}{4}$$
$$\left(x-\frac{1}{2}\right)^2=-\frac{35}{4}$$
$$x-\frac{1}{2}=\pm\sqrt{-\frac{35}{4}}$$
$$x=\frac{1}{2}\pm\frac{i\sqrt{35}}{2}\approx 0.5\pm 2.96i$$

**67.**
$$4x^2+7x=3$$
$$x^2+\frac{7}{4}x=\frac{3}{4}$$
$$x^2+\frac{7}{4}x+\frac{49}{64}=\frac{3}{4}+\frac{49}{64}$$
$$\left(x+\frac{7}{8}\right)^2=\frac{97}{64}$$
$$x+\frac{7}{8}=\pm\sqrt{\frac{97}{64}}$$
$$x=-\frac{7}{8}\pm\frac{\sqrt{97}}{8}\approx -2.11,\ 0.36$$

**69.**
$$3x^2+x=\frac{2}{3}$$
$$x^2+\frac{x}{3}=\frac{2}{9}$$
$$x^2+\frac{x}{3}+\frac{1}{36}=\frac{2}{9}+\frac{1}{36}$$
$$\left(x+\frac{1}{6}\right)^2=\frac{1}{4}$$
$$x+\frac{1}{6}=\pm\sqrt{\frac{1}{4}}$$
$$x=-\frac{1}{6}\pm\frac{1}{2}=-\frac{2}{3},\ \frac{1}{3}$$

**71.** $(x-2)^2=-3(x-2)$
Let $u=(x-2)$.
$$u^2=-3u$$
$$u^2+3u=0$$
$$u(u+3)=0$$

$u=0$    or    $u+3=0$
$x-2=0$      $u=-3$
$x=2$      $x-2=-3$
     $x=-1$

Section 10.2 Completing the Square

**73.**
$$4x^2 + 9 = 2x$$
$$4x^2 - 2x = -9$$
$$x^2 - \frac{1}{2}x = -\frac{9}{4}$$
$$x^2 - \frac{1}{2}x + \frac{1}{16} = -\frac{9}{4} + \frac{1}{16}$$
$$\left(x - \frac{1}{4}\right)^2 = -\frac{35}{16}$$
$$x - \frac{1}{4} = \pm\sqrt{-\frac{35}{16}}$$
$$x = \frac{1}{4} \pm \frac{i\sqrt{35}}{4} \approx 0.25 \pm 1.48i$$

**75.** $ax^2 + x - 5 = 0 \quad x = 1$
$$a(1)^2 + 1 - 5 = 0$$
$$a - 4 = 0$$
$$a = 4$$
For $x = 1$ to be a solution, $a = 4$.
$$4x^2 + x - 5 = 0$$
$$(4x + 5)(x - 1) = 0$$
Another solution is $-5/4 = -1.25$.

**77.** $6x^2 + bx - 3 = 0 \quad x = \frac{1}{3}$
$$6\left(\frac{1}{3}\right)^2 + b\left(\frac{1}{3}\right) - 3 = 0$$
$$\frac{2}{3} + \frac{b}{3} - 3 = 0$$
$$\frac{b}{3} = \frac{7}{3}$$
$$b = 7$$
For $x = 1/3$ to be a solution, $b = 7$.
$$6x^2 + 7x - 3 = 0$$
$$(3x - 1)(2x + 3) = 0$$
Another solution is $-3/2 = -1.5$.

**79.** $x^2 + 2x + c = 0 \quad x = 6$
$$6^2 + 2(6) + c = 0$$
$$c = -48$$
For $x = 6$ to be a solution, $c = -48$.
$$x^2 + 2x - 48 = 0$$
$$(x + 8)(x - 6) = 0$$
Another solution is $-8$.

**81.** Let $x$ = the integer.
$$\frac{1}{x} + \frac{1}{x^2} = \frac{6}{25}$$
$$25x + 25 = 6x^2$$
$$6x^2 - 25x - 25 = 0$$
$$(6x + 5)(x - 5) = 0$$
$6x + 5 = 0 \quad$ or $\quad x - 5 = 0$
$x = -5/6 \quad$ or $\quad x = 5$
Because $-5/6$ is not an integer, the only valid solution is 5. The number is 5.

**83.** Let $x$ = the positive number.
$$x^2 = 5x + 3$$
$$x^2 - 5x = 3$$
$$x^2 - 5x + \frac{25}{4} = 3 + \frac{25}{4}$$
$$\left(x - \frac{5}{2}\right)^2 = \frac{37}{4}$$
$$x - \frac{5}{2} = \pm\sqrt{\frac{37}{4}}$$
$$x = \frac{5}{2} \pm \frac{\sqrt{37}}{2} \approx -0.54, \; 5.54$$
Disregard the negative solution. The solution is 5.54.

**85.** Let $x$ = the length of the shortest side and $x + 8$ = the length of the other leg.
$$x^2 + (x + 8)^2 = 50^2$$
$$x^2 + x^2 + 16x + 64 = 2500$$
$$2x^2 + 16x = 2436$$
$$x^2 + 8x = 1218$$
$$x^2 + 8x + 16 = 1218 + 16$$
$$(x + 4)^2 = 1234$$
$$x + 4 = \pm\sqrt{1234}$$
$$x = -4 \pm \sqrt{1234}$$
$$x \approx -39, \; 31$$
The length of the shortest side is approximately 31 feet.

**87.** Let $x =$ the number of hours it takes the woman to till her garden and $x - 0.5 =$ the number of hours it takes her neighbor to till the garden.

$$\frac{1}{x} + \frac{1}{x - 0.5} = \frac{1}{1}$$
$$x - 0.5 + x = x(x - 0.5)$$
$$2x - 0.5 = x^2 - 0.5x$$
$$x^2 - \frac{5}{2}x = -\frac{1}{2}$$
$$x^2 - \frac{5}{2}x + \frac{25}{16} = -\frac{1}{2} + \frac{25}{16}$$
$$\left(x - \frac{5}{4}\right)^2 = \frac{17}{16}$$
$$x - \frac{5}{4} = \pm\sqrt{\frac{17}{16}}$$
$$x = \frac{5}{4} \pm \frac{\sqrt{17}}{4} = 0.22,\ 2.28$$

It takes the woman approximately 2.28 hours, or 2 hours and 17 minutes, to till her garden by herself.

**89.**
$$A(t) = 3000$$
$$13.4t^2 - 72t + 1759 = 3000$$

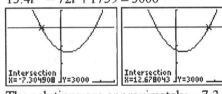

The solutions are approximately $-7.3$ and $12.7$.

**91.**
$$A(t) = 3000$$
$$13.4t^2 - 72t + 1759 = 3000$$
$$13.4t^2 - 72t = 1241$$
$$\frac{13.4t^2}{13.4} - \frac{72t}{13.4} = \frac{1241}{13.4}$$
$$t^2 - \frac{72}{13.4}t = \frac{1241}{13.4}$$
$$t^2 - \frac{72}{13.4}t + \frac{32{,}400}{4489} = \frac{1241}{13.4} + \frac{32{,}400}{4489}$$
$$\left(t - \frac{180}{67}\right)^2 = \frac{448{,}135}{4489}$$
$$t - \frac{180}{67} = \pm\sqrt{\frac{448{,}135}{4489}}$$
$$t = \frac{180}{67} \pm \sqrt{\frac{448{,}135}{4489}}$$
$$t \approx -7.3,\ 12.7$$

**93.**
$$x^2 - 2cx + 4c^2 = 0$$
$$x^2 - 2cx = -4c^2$$
$$x^2 - 2cx + c^2 = -4c^2 + c^2$$
$$(x - c)^2 = -3c^2$$
$$x - c = \pm\sqrt{-3c^2}$$
$$x = c \pm ci\sqrt{3}$$

**95.**
$$x^2 + 2\sqrt{5}\,x - 4 = 0$$
$$x^2 + 2\sqrt{5}\,x = 4$$
$$x^2 + 2\sqrt{5}\,x + 5 = 4 + 5$$
$$\left(x + \sqrt{5}\right)^2 = 9$$
$$x + \sqrt{5} = \pm\sqrt{9}$$
$$x = -\sqrt{5} \pm 3 \approx -5.24,\ 0.76$$

**97.**
$$x^2 + kx + 1 = 0$$
$$x^2 + kx = -1$$
$$x^2 + kx + \left(\frac{k}{2}\right)^2 = -1 + \left(\frac{k}{2}\right)^2$$
$$\left(x + \frac{k}{2}\right)^2 = \left(\frac{k}{2}\right)^2 - 1$$
$$x + \frac{k}{2} = \pm\sqrt{\left(\frac{k}{2}\right)^2 - 1}$$
$$x = -\frac{k}{2} \pm \sqrt{\left(\frac{k}{2}\right)^2 - 1}$$

(a) The solutions will be real if $\left(\frac{k}{2}\right)^2 - 1 \geq 0$. From the graph, we see that this will occur when $k \leq -2$ or $k \geq 2$.

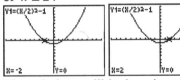

(b) The solutions will be imaginary if $\left(\frac{k}{2}\right)^2 - 1 < 0$. From the graph, we see that this will occur when $-2 < k < 2$.

**99.**
$$x^2 + 4ix - 8 = 0$$
$$x^2 + 4ix = 8$$
$$x^2 + 4ix + (2i)^2 = 8 + (2i)^2$$
$$x^2 + 4ix - 4 = 8 - 4$$
$$(x + 2i)^2 = 4$$
$$x + 2i = \pm\sqrt{4}$$
$$x = -2i \pm 2$$

**101.** (a) The area of the top square is $4x$. The area of the lower left square is $x^2$. The area of the lower right square is $4x$. The total area is $4x + x^2 + 4x = x^2 + 8x$.

(b) The area of the missing corner in the figure is $4(4) = 16$.

(c) The area of the complete square is $x^2 + 8x + 16$.

(d) The area of the square is $(x+4)(x+4) = (x+4)^2$.

(e) Because the area of the complete square in part (c) and in part (d) represent the same thing, we have:
$$(x+4)^2 = x^2 + 8x + 16.$$

## Section 10.3 The Quadratic Formula

**1.** The equation must be in standard form.

**3.**
$$x(3x - 2) = 7x + 6$$
$$3x^2 - 2x - 7x - 6 = 0$$
$$3x^2 - 9x - 6 = 0$$
$$a = 3, b = -9, c = -6$$

**5.**
$$(x+1)^2 = 3x - 1$$
$$x^2 + 2x + 1 - 3x + 1 = 0$$
$$x^2 - x + 2 = 0$$
$$a = 1, b = -1, c = 2$$

**7.** Factoring:
$$x^2 - x - 2 = 0$$
$$(x - 2)(x + 1) = 0$$
$$x - 2 = 0 \quad \text{or} \quad x + 1 = 0$$
$$x = 2 \quad \text{or} \quad x = -1$$
Quadratic Equation:
$$a = 1, b = -1, c = -2$$
$$x = \frac{-b \pm \sqrt{b^2 - 4ac}}{2a}$$
$$= \frac{-(-1) \pm \sqrt{(-1)^2 - 4(1)(-2)}}{2(1)}$$
$$= \frac{1 \pm \sqrt{1 + 8}}{2} = \frac{1 \pm \sqrt{9}}{2} = \frac{1 \pm 3}{2}$$
$$= -1, 2$$

**9.** Factoring:
$$x^2 + 16 = 8x$$
$$x^2 - 8x + 16 = 0$$
$$(x-4)^2 = 0$$
$$x - 4 = 0$$
$$x = 4$$
Quadratic Equation:
$a = 1$, $b = -8$, $c = 16$
$$x = \frac{-b \pm \sqrt{b^2 - 4ac}}{2a}$$
$$= \frac{-(-8) \pm \sqrt{(-8)^2 - 4(1)(16)}}{2(1)}$$
$$= \frac{8 \pm \sqrt{64 - 64}}{2} = \frac{8 \pm 0}{2} = 4$$

**11.** Factoring:
$$2x^2 + 7x = 4$$
$$2x^2 + 7x - 4 = 0$$
$$(2x - 1)(x + 4) = 0$$
$$2x - 1 = 0 \quad \text{or} \quad x + 4 = 0$$
$$x = 1/2 \quad \text{or} \quad x = -4$$
Quadratic Equation:
$a = 2$, $b = 7$, $c = -4$
$$x = \frac{-b \pm \sqrt{b^2 - 4ac}}{2a}$$
$$= \frac{-7 \pm \sqrt{7^2 - 4(2)(-4)}}{2(2)}$$
$$= \frac{-7 \pm \sqrt{49 + 32}}{4} = \frac{-7 \pm \sqrt{81}}{4}$$
$$= \frac{-7 \pm 9}{4} = -4, \ 1/2$$

**13.** Factoring:
$$x^2 = 5x$$
$$x^2 - 5x = 0$$
$$x(x - 5) = 0$$
$$x = 0 \quad \text{or} \quad x - 5 = 0$$
$$x = 5$$
Quadratic Equation:
$a = 1$, $b = -5$, $c = 0$
$$x = \frac{-b \pm \sqrt{b^2 - 4ac}}{2a}$$
$$= \frac{-(-5) \pm \sqrt{(-5)^2 - 4(1)(0)}}{2(1)}$$
$$= \frac{5 \pm \sqrt{25 - 0}}{2} = \frac{5 \pm 5}{2} = 0, 5$$

**15.**
$$4x^2 = 3 + 9x$$
$$4x^2 - 9x - 3 = 0$$
$a = 4$, $b = -9$, $c = -3$
$$x = \frac{-b \pm \sqrt{b^2 - 4ac}}{2a}$$
$$= \frac{-(-9) \pm \sqrt{(-9)^2 - 4(4)(-3)}}{2(4)}$$
$$= \frac{9 \pm \sqrt{81 + 48}}{8} = \frac{9 \pm \sqrt{129}}{8}$$
$$\approx -0.29, \ 2.54$$

**17.** $x^2 - 2x + 9 = 0$
$a = 1$, $b = -2$, $c = 9$
$$x = \frac{-b \pm \sqrt{b^2 - 4ac}}{2a}$$
$$= \frac{-(-2) \pm \sqrt{(-2)^2 - 4(1)(9)}}{2(1)}$$
$$= \frac{2 \pm \sqrt{4 - 36}}{2} = \frac{2 \pm \sqrt{-32}}{2}$$
$$\approx 1 \pm 2.83i$$

**19.** $x(x + 2) = 5$
$$x^2 + 2x - 5 = 0$$
$a = 1$, $b = 2$, $c = -5$
$$x = \frac{-b \pm \sqrt{b^2 - 4ac}}{2a}$$
$$= \frac{-2 \pm \sqrt{2^2 - 4(1)(-5)}}{2(1)}$$
$$= \frac{-2 \pm \sqrt{4 + 20}}{2} = \frac{-2 \pm \sqrt{24}}{2}$$
$$= -1 \pm \sqrt{6} \approx -3.45, \ 1.45$$

## Section 10.3 The Quadratic Formula

**21.** 
$$x + 2x^2 + 4 = 0$$
$$2x^2 + x + 4 = 0$$
$$a = 2, \ b = 1, \ c = 4$$
$$x = \frac{-b \pm \sqrt{b^2 - 4ac}}{2a}$$
$$= \frac{-1 \pm \sqrt{1^2 - 4(2)(4)}}{2(2)}$$
$$= \frac{-1 \pm \sqrt{1 - 32}}{4} = \frac{-1 \pm \sqrt{-31}}{4}$$
$$= \frac{-1 \pm i\sqrt{31}}{4} \approx -0.25 \pm 1.39i$$

**23.** The keystrokes are:
(− B + √ (B^2 − 4AC)) / (2A) and
(− B − √ (B^2 − 4AC)) / (2A)
If parentheses in the numerator are omitted, only the radical term will be divided by $2A$.

**25.** 
$$6x - 9x^2 = 0$$
$$3x(2 - 3x) = 0$$
$$3x = 0 \quad \text{or} \quad 2 - 3x = 0$$
$$x = 0 \quad \text{or} \quad x = 2/3$$

**27.** 
$$9 - 49x^2 = 0$$
$$(3 - 7x)(3 + 7x) = 0$$
$$3 - 7x = 0 \quad \text{or} \quad 3 + 7x = 0$$
$$x = 3/7 \quad \text{or} \quad x = -3/7$$

**29.** 
$$49 = 14x - x^2$$
$$x^2 - 14x + 49 = 0$$
$$(x - 7)^2 = 0$$
$$x - 7 = 0$$
$$x = 7$$

**31.** 
$$(5x - 4)^2 = 10$$
$$5x - 4 = \pm\sqrt{10}$$
$$5x = 4 \pm \sqrt{10}$$
$$x = \frac{4 \pm \sqrt{10}}{5} \approx 0.17, \ 1.43$$

**33.** 
$$0.25x^2 - 0.5x - 2 = 0$$
$$a = 0.25, \ b = -0.5, \ c = -2$$
$$x = \frac{-b \pm \sqrt{b^2 - 4ac}}{2a}$$
$$= \frac{-(-0.5) \pm \sqrt{(-0.5)^2 - 4(0.25)(-2)}}{2(0.25)}$$
$$= \frac{0.5 \pm \sqrt{0.25 + 2}}{0.5} = \frac{0.5 \pm \sqrt{2.25}}{0.5}$$
$$= \frac{0.5 \pm 1.5}{0.5} = -2, \ 4$$

**35.** 
$$x(x + 4) = 5$$
$$x^2 + 4x - 5 = 0$$
$$(x + 5)(x - 1) = 0$$
$$x + 5 = 0 \quad \text{or} \quad x - 1 = 0$$
$$x = -5 \quad \text{or} \quad x = 1$$

**37.** 
$$5x^2 + 2x + 7 = 0$$
$$a = 5, \ b = 2, \ c = 7$$
$$x = \frac{-b \pm \sqrt{b^2 - 4ac}}{2a}$$
$$= \frac{-2 \pm \sqrt{2^2 - 4(5)(7)}}{2(5)}$$
$$= \frac{-2 \pm \sqrt{4 - 140}}{10} = \frac{-2 \pm \sqrt{-136}}{10}$$
$$\approx -0.2 \pm 1.17i$$

**39.** 
$$x^2 - 3x + 8 = 0$$
$$a = 1, \ b = -3, \ c = 8$$
$$x = \frac{-b \pm \sqrt{b^2 - 4ac}}{2a}$$
$$= \frac{-(-3) \pm \sqrt{(-3)^2 - 4(1)(8)}}{2(1)}$$
$$= \frac{3 \pm \sqrt{9 - 32}}{2} = \frac{3 \pm \sqrt{-23}}{2}$$
$$= \frac{3 \pm i\sqrt{23}}{2} \approx 1.5 \pm 2.40i$$

**41.**
$6x - x^2 - 4 = 0$
$-x^2 + 6x - 4 = 0$
$a = -1, b = 6, c = -4$
$x = \dfrac{-b \pm \sqrt{b^2 - 4ac}}{2a}$
$= \dfrac{-6 \pm \sqrt{6^2 - 4(-1)(-4)}}{2(-1)}$
$= \dfrac{-6 \pm \sqrt{36 - 16}}{-2} = \dfrac{-6 \pm \sqrt{20}}{-2}$
$= 3 \pm \sqrt{5} \approx 0.76,\ 5.24$

**43.**
$x(5 + x) + 9 = 0$
$x^2 + 5x + 9 = 0$
$a = 1, b = 5, c = 9$
$x = \dfrac{-b \pm \sqrt{b^2 - 4ac}}{2a}$
$= \dfrac{-5 \pm \sqrt{5^2 - 4(1)(9)}}{2(1)}$
$= \dfrac{-5 \pm \sqrt{25 - 36}}{2} = \dfrac{-5 \pm \sqrt{-11}}{2}$
$= \dfrac{-5 \pm i\sqrt{11}}{2} \approx -2.5 \pm 1.66i$

**45.**
$\dfrac{1}{2}x^2 - \dfrac{2}{3}x + \dfrac{4}{3} = 0$
$3x^2 - 4x + 8 = 0$
$a = 3, b = -4, c = 8$
$x = \dfrac{-b \pm \sqrt{b^2 - 4ac}}{2a}$
$= \dfrac{-(-4) \pm \sqrt{(-4)^2 - 4(3)(8)}}{2(3)}$
$= \dfrac{4 \pm \sqrt{16 - 96}}{6} = \dfrac{4 \pm \sqrt{-80}}{6}$
$= \dfrac{4 \pm 4i\sqrt{5}}{6} \approx 0.67 \pm 1.49i$

**47.**
$x^2 = \dfrac{5x}{2} + \dfrac{1}{2}$
$2x^2 - 5x - 1 = 0$
$a = 2, b = -5, c = -1$
$x = \dfrac{-b \pm \sqrt{b^2 - 4ac}}{2a}$
$= \dfrac{-(-5) \pm \sqrt{(-5)^2 - 4(2)(-1)}}{2(2)}$
$= \dfrac{5 \pm \sqrt{25 + 8}}{4} = \dfrac{5 \pm \sqrt{33}}{4} \approx -0.19,\ 2.69$

**49.**
$(x - 3)(2x + 1) = x(x - 4)$
$2x^2 - 5x - 3 = x^2 - 4x$
$x^2 - x - 3 = 0$
$a = 1, b = -1, c = -3$
$x = \dfrac{-b \pm \sqrt{b^2 - 4ac}}{2a}$
$= \dfrac{-(-1) \pm \sqrt{(-1)^2 - 4(1)(-3)}}{2(1)}$
$= \dfrac{1 \pm \sqrt{1 + 12}}{2} = \dfrac{1 \pm \sqrt{13}}{2} \approx -1.30,\ 2.30$

**51.**
$4x^2 + 12x = 7$
$4x^2 + 12x - 7 = 0$
$a = 4, b = 12, c = -7$
Discriminant:
$b^2 - 4ac = 12^2 - 4(4)(-7) = 256 > 0$
Because 256 is positive and a perfect square ($16^2 = 256$), there are two rational solutions.

**53.**
$2x^2 + 5x + 7 = 0$
$a = 2, b = 5, c = 7$
Discriminant:
$b^2 - 4ac = 5^2 - 4(2)(7) = -31 < 0$
Because the discriminant is negative, there are two imaginary solutions.

**55.**
$36x^2 + 1 = 12x$
$36x^2 - 12x + 1 = 0$
$a = 36, b = -12, c = 1$
Discriminant:
$b^2 - 4ac = (-12)^2 - 4(36)(1) = 0$
Because the discriminant is zero, there is one rational solution (double root).

**57.** $3x^2 + 4x - 8 = 0$
$a = 3$, $b = 4$, $c = -8$
Discriminant:
$b^2 - 4ac = 4^2 - 4(3)(-8) = 112 > 0$

Because the discriminant is positive and not a perfect square, there are two irrational solutions.

**59.** $x^2 + 8x + k = 0$
$a = 1$, $b = 8$, $c = k$
Discriminant:
$b^2 - 4ac = 8^2 - 4(1)(k) = 64 - 4k$

For the equation to have two unequal real roots, the discriminant must be greater than 0.
$64 - 4k > 0$
$-4k > -64$
$k < 16$

**61.** $kx^2 - 4x + 1 = 0$
$a = k$, $b = -4$, $c = 1$
Discriminant:
$b^2 - 4ac = (-4)^2 - 4(k)(1) = 16 - 4k$

For the equation to have two unequal real roots, the discriminant must be greater than 0.
$16 - 4k > 0$
$-4k > -16$
$k < 4$

**63.** $x^2 + 6x + k = 0$
$a = 1$, $b = 6$, $c = k$
Discriminant:
$b^2 - 4ac = 6^2 - 4(1)(k) = 36 - 4k$

For the equation to have two imaginary solutions, the discriminant must be less than 0.
$36 - 4k < 0$
$-4k < -36$
$k > 9$

**65.** $kx^2 - 3x + 1 = 0$
$a = k$, $b = -3$, $c = 1$
Discriminant:
$b^2 - 4ac = (-3)^2 - 4(k)(1) = 9 - 4k$

For the equation to have two imaginary solutions, the discriminant must be less than 0.
$9 - 4k < 0$
$-4k < -9$
$k > \dfrac{9}{4}$

**67.** $x^2 + kx + 9 = 0$
$a = 1$, $b = k$, $c = 9$
Discriminant:
$b^2 - 4ac = k^2 - 4(1)(9) = k^2 - 36$

For the equation to have a double root, the discriminant must equal 0.
$k^2 - 36 = 0$
$k^2 = 36$
$k = \pm\sqrt{36} = \pm 6$

**69.** $x^2 + kx + 2k = 0$
$a = 1$, $b = k$, $c = 2k$
Discriminant:
$b^2 - 4ac = k^2 - 4(1)(2k) = k^2 - 8k$

For the equation to have a double root, the discriminant must equal 0.
$k^2 - 8k = 0$
$k(k - 8) = 0$
$k = 0$ or $k - 8 = 0$
$k = 8$

**71.** (a) The discriminant is $a^2 + 20$, which is positive for all values of $a$. Therefore, the solutions are always real numbers.

(b) The discriminant is $a^2 + 4k$. Because $k$ is positive, the discriminant is always positive, and the solutions are always real numbers.

**73.** $x^2 + 4x - 480$
$a = 1$, $b = 4$, $c = -480$
$$x = \frac{-b \pm \sqrt{b^2 - 4ac}}{2a}$$
$$= \frac{-4 \pm \sqrt{4^2 - 4(1)(-480)}}{2(1)}$$
$$= \frac{-4 \pm \sqrt{16 + 1920}}{2} = \frac{-4 \pm \sqrt{1936}}{2}$$
$$= \frac{-4 \pm 44}{2} = -24, 20$$
$x^2 + 4x - 480 = (x + 24)(x - 20)$

**75.** $x^2 + 7x - 450$
$a = 1$, $b = 7$, $c = -450$
$$x = \frac{-b \pm \sqrt{b^2 - 4ac}}{2a}$$
$$= \frac{-7 \pm \sqrt{7^2 - 4(1)(-450)}}{2(1)}$$
$$= \frac{-7 \pm \sqrt{49 + 1800}}{2} = \frac{-7 \pm \sqrt{1849}}{2}$$
$$= \frac{-7 \pm 43}{2} = -25, 18$$
$x^2 + 7x - 450 = (x + 25)(x - 18)$

**77.** By the Quadratic Formula, the roots of $ax^2 + bx + c = 0$ are $\dfrac{-b - \sqrt{b^2 - 4ac}}{2a}$ and $\dfrac{-b + \sqrt{b^2 - 4ac}}{2a}$. Adding these, we obtain:
$$\frac{-b - \sqrt{b^2 - 4ac}}{2a} + \frac{-b + \sqrt{b^2 - 4ac}}{2a}$$
$$= \frac{-b - \sqrt{b^2 - 4ac} - b + \sqrt{b^2 - 4ac}}{2a}$$
$$= \frac{-2b}{2a} = -\frac{b}{a}$$
Test:
$2x^2 - 7x - 15 = 0$
$(2x + 3)(x - 5) = 0$
$2x + 3 = 0$ or $x - 5 = 0$
$x = -1.5$ or $x = 5$
The sum of the roots is: $-1.5 + 5 = 3.5$
For this equation, $a = 2$ and $b = -7$.

$-\dfrac{b}{a} = -\dfrac{-7}{2} = 3.5$

Thus, the conclusion tests true that the sum of the roots is equal to $-\dfrac{b}{a}$.

**79.** By the Quadratic Formula, the roots of $ax^2 + bx + c = 0$ are $\dfrac{-b - \sqrt{b^2 - 4ac}}{2a}$ and $\dfrac{-b + \sqrt{b^2 - 4ac}}{2a}$. Multiplying these, we obtain:
$$\frac{-b - \sqrt{b^2 - 4ac}}{2a} \cdot \frac{-b + \sqrt{b^2 - 4ac}}{2a}$$
$$= \frac{(-b)^2 - \left(\sqrt{b^2 - 4ac}\right)^2}{4a^2}$$
$$= \frac{b^2 - (b^2 - 4ac)}{4a^2}$$
$$= \frac{b^2 - b^2 + 4ac}{4a^2} = \frac{4ac}{4a^2} = \frac{c}{a}$$
Thus, the product of the roots of $ax^2 + bx + c = 0$ is $\dfrac{c}{a}$.

**81.** $3x^2 - 2x - 1 = 0$, $a = 3$, $b = -2$, $c = -1$
Using the results of Exercises 77 and 79, the sum of the roots is $-\dfrac{b}{a}$ and the product of the roots is $\dfrac{c}{a}$.

Sum: $-\dfrac{b}{a} = -\dfrac{-2}{3} = \dfrac{2}{3}$

Product: $\dfrac{c}{a} = \dfrac{-1}{3} = -\dfrac{1}{3}$

**83.** $x^2 + x + 1 = 0$, $a = 1$, $b = 1$, $c = 1$
Using the results of Exercises 77 and 79, the sum of the roots is $-\dfrac{b}{a}$ and the product of the roots is $\dfrac{c}{a}$.

Sum: $-\dfrac{b}{a} = -\dfrac{1}{1} = -1$; Product: $\dfrac{c}{a} = \dfrac{1}{1} = 1$

**85.** $x^2 + kx + (k-1) = 0$

Discriminant:
$b^2 - 4ac = k^2 - 4(1)(k-1) = k^2 - 4k + 4$

The equation will have two real number solutions when the discriminant is greater than zero.
$$k^2 - 4k + 4 > 0$$
$$(k-2)^2 > 0$$

The expression $(k-2)^2$ is always positive except for $k = 2$, when it is equal to 0. Thus, there will be two real number solutions for any value of $k$ except 2.

**87.** Let $x =$ the distance between consecutive bases. Then the distance between third and first base is the length of the hypotenuse of a right triangle with legs represented by the distance between first and second base and between second and third base. Using the Pythagorean Theorem, we obtain
$$x^2 + x^2 = 84.9^2$$
$$2x^2 = 7208.01$$
$$x^2 = 3604.005$$
$$x = \pm\sqrt{3604.005} \approx \pm 60$$

Because the distances between consecutive bases is approximately 60 feet, this is a regulation softball field.

**89.** The revenue is the product of the number of desks sold and the price per desk. Let $x =$ the number of $2 reductions in price. Because the number of desks sold increases by 10 for each $2 reduction in price, the number of desks sold is given by $200 + 10x$. The price is given by $90 - 2x$. A function for the total revenue is $R(x) = (200 + 10x)(90 - 2x)$. We must solve the equation
$$(200 + 10x)(90 - 2x) = 20,000$$
$$18,000 + 500x - 20x^2 = 20,000$$
$$20x^2 - 500x + 2000 = 0$$
$$20(x^2 - 25x + 100) = 0$$
$$20(x - 20)(x - 5) = 0$$
$$x - 20 = 0 \quad \text{or} \quad x - 5 = 0$$
$$x = 20 \qquad\qquad x = 5$$

The total revenue is $20,000 when there are 5 $2-reductions in price or when there are 20 $2-reductions in price. Therefore, the total revenue is $20,000 when the price is $80 or $50.

**91.** Let $x = 50$:
$$p(x) = 0.01(0.96x^2 + 63x + 2950)$$
$$p(50) = 0.01(0.96(50)^2 + 63(50) + 2950)$$
$$= 85$$

The percentage who will pursue more schooling in 2010 is 85%.

**93.** From Exercise 92, the point is (30, 57). This means that in 1990, about 57% of those who received a GED planned additional schooling.

**95.** $x^2 + ax - 2a = 0$
$$x = \frac{-a \pm \sqrt{a^2 - 4(1)(-2a)}}{2(1)}$$
$$= \frac{-a \pm \sqrt{a^2 + 8a}}{2}$$

**97.** $x = y^2 + 2y - 8$
$$0 = y^2 + 2y - 8 - x$$
$$y = \frac{-2 \pm \sqrt{2^2 - 4(1)(-8-x)}}{2(1)}$$
$$= \frac{-2 \pm \sqrt{4 + 32 + 4x}}{2}$$
$$= \frac{-2 \pm \sqrt{36 + 4x}}{2}$$
$$= \frac{-2 \pm \sqrt{4(9+x)}}{2}$$
$$= \frac{-2 \pm 2\sqrt{9+x}}{2}$$
$$= -1 \pm \sqrt{9+x}$$

**99.** If $-5 \pm 2\sqrt{5}$ are solutions, then the equation is:
$$[x-(-5+2\sqrt{5})][x-(-5-2\sqrt{5})] = 0$$
$$[x+5-2\sqrt{5}][x+5+2\sqrt{5}] = 0$$
$$[(x+5)-2\sqrt{5}][(x+5)+2\sqrt{5}] = 0$$
$$(x+5)^2 - (2\sqrt{5})^2 = 0$$
$$x^2 + 10x + 25 - 20 = 0$$
$$x^2 + 10x + 5 = 0$$

**101.** $x^2 + \sqrt{3}x - 1 = 0$
$a = 1$, $b = \sqrt{3}$, $c = -1$
$$x = \frac{-b \pm \sqrt{b^2 - 4ac}}{2a}$$
$$= \frac{-\sqrt{3} \pm \sqrt{(\sqrt{3})^2 - 4(1)(-1)}}{2(1)}$$
$$= \frac{-\sqrt{3} \pm \sqrt{3+4}}{2}$$
$$= \frac{-\sqrt{3} \pm \sqrt{7}}{2} \approx -2.19, \ 0.46$$

**103.** $x^2 - 2ix - 3 = 0$
$a = 1$, $b = -2i$, $c = -3$
$$x = \frac{-b \pm \sqrt{b^2 - 4ac}}{2a}$$
$$= \frac{-(-2i) \pm \sqrt{(-2i)^2 - 4(1)(-3)}}{2(1)}$$
$$= \frac{2i \pm \sqrt{-4 + 12}}{2}$$
$$= \frac{2i \pm \sqrt{8}}{2} = \frac{2i \pm 2\sqrt{2}}{2}$$
$$= i \pm \sqrt{2} \approx 1.41 + i, \ -1.41 + i$$

## Section 10.4 Equations in Quadratic Form

**1.** First, clear the fractions by multiplying each term by the LCD.
$$x + \frac{1}{x-1} - \frac{5-4x}{x-1} = 0$$
$$x(x-1) + 1 - (5-4x) = 0$$
$$x^2 - x + 1 - 5 + 4x = 0$$
$$x^2 + 3x - 4 = 0$$
The result is the quadratic equation $x^2 + 3x - 4 = 0$. Multiplying both sides of the original equation by $x - 1$ introduces an extraneous solution.

**3.** $\frac{3}{x} = 2 + \frac{4}{x^2}$
$$x^2 \cdot \frac{3}{x} = 2x^2 + x^2 \cdot \frac{4}{x^2}$$
$$3x = 2x^2 + 4$$
$$0 = 2x^2 - 3x + 4$$
$a = 2$, $b = -3$, $c = 4$
$$x = \frac{-b \pm \sqrt{b^2 - 4ac}}{2a}$$
$$= \frac{-(-3) \pm \sqrt{(-3)^2 - 4(2)(4)}}{2(2)}$$
$$= \frac{3 \pm \sqrt{9 - 32}}{4} = \frac{3 \pm \sqrt{-23}}{4}$$
$$= \frac{3 \pm i\sqrt{23}}{4} \approx 0.75 \pm 1.20i$$

Section 10.4  Equations in Quadratic Form

**5.**
$$\frac{t-3}{t} = \frac{2}{t-3}$$
$$(t-3)^2 = 2t$$
$$t^2 - 6t + 9 - 2t = 0$$
$$t^2 - 8t + 9 = 0$$
$$a = 1,\ b = -8,\ c = 9$$
$$t = \frac{-b \pm \sqrt{b^2 - 4ac}}{2a}$$
$$= \frac{-(-8) \pm \sqrt{(-8)^2 - 4(1)(9)}}{2(1)}$$
$$= \frac{8 \pm \sqrt{64 - 36}}{2} = \frac{8 \pm \sqrt{28}}{2}$$
$$= \frac{8 \pm 2\sqrt{7}}{2} = 4 \pm \sqrt{7} \approx 1.35,\ 6.65$$

**7.**
$$x + 2 = \frac{6}{1-x}$$
$$(x+2)(1-x) = 6$$
$$0 = x^2 + x - 2 + 6$$
$$0 = x^2 + x + 4$$
$$a = 1,\ b = 1,\ c = 4$$
$$x = \frac{-b \pm \sqrt{b^2 - 4ac}}{2a}$$
$$= \frac{-1 \pm \sqrt{1^2 - 4(1)(4)}}{2(1)}$$
$$= \frac{-1 \pm \sqrt{1 - 16}}{2} = \frac{-1 \pm \sqrt{-15}}{2}$$
$$= \frac{-1 \pm i\sqrt{15}}{2} \approx -0.5 \pm 1.94i$$

**9.**
$$\frac{x}{x+3} + \frac{3}{x} = 4$$
$$x(x+3) \cdot \frac{x}{x+3} + x(x+3) \cdot \frac{3}{x} = 4x(x+3)$$
$$x^2 + 3x + 9 = 4x^2 + 12x$$
$$3x^2 + 9x - 9 = 0$$
$$x^2 + 3x - 3 = 0$$
$$a = 1,\ b = 3,\ c = -3$$

$$x = \frac{-b \pm \sqrt{b^2 - 4ac}}{2a}$$
$$= \frac{-3 \pm \sqrt{3^2 - 4(1)(-3)}}{2(1)}$$
$$= \frac{-3 \pm \sqrt{9 + 12}}{2} = \frac{-3 \pm \sqrt{21}}{2}$$
$$\approx -3.79,\ 0.79$$

**11.**
$$\frac{x}{x-2} - \frac{5}{x+2} = \frac{12}{x^2 - 4}$$
$$x(x+2) - 5(x-2) = 12$$
$$x^2 + 2x - 5x + 10 = 12$$
$$x^2 - 3x - 2 = 0$$
$$a = 1,\ b = -3,\ c = -2$$
$$x = \frac{-b \pm \sqrt{b^2 - 4ac}}{2a}$$
$$= \frac{-(-3) \pm \sqrt{(-3)^2 - 4(1)(-2)}}{2(1)}$$
$$= \frac{3 \pm \sqrt{9 + 8}}{2} = \frac{3 \pm \sqrt{17}}{2} \approx -0.56,\ 3.56$$

**13.**
$$\frac{x+3}{x-2} - \frac{5}{2x^2 - 3x - 2} = \frac{2}{2x+1}$$
$$\frac{x+3}{x-2} - \frac{5}{(2x+1)(x-2)} = \frac{2}{2x+1}$$
The restricted values are 2 and $-0.5$.
Multiply each term by the LCD.
$$(x+3)(2x+1) - 5 = 2(x-2)$$
$$2x^2 + 7x + 3 - 5 = 2x - 4$$
$$2x^2 + 5x + 2 = 0$$
$$(2x+1)(x+2) = 0$$
$2x + 1 = 0$    or    $x + 2 = 0$
$x = -0.5$    or    $x = -2$
Because $-0.5$ is a restricted value, the only valid solution is $-2$.

**15.** First, isolate the radicals and square both sides.
$$\sqrt{x^2 - 5} + 2\sqrt{x} = 0$$
$$\sqrt{x^2 - 5} = -2\sqrt{x}$$
$$\left(\sqrt{x^2 - 5}\right)^2 = \left(-2\sqrt{x}\right)^2$$
$$x^2 - 5 = 4x$$
$$x^2 - 4x - 5 = 0$$

Squaring both sides of the original equation introduces extraneous solutions.

**17.**
$$5 = 3\sqrt{x} + \frac{2}{\sqrt{x}}$$
$$5\sqrt{x} = 3x + 2$$
$$(5\sqrt{x})^2 = (3x+2)^2$$
$$25x = 9x^2 + 12x + 4$$
$$0 = 9x^2 - 13x + 4$$
$$0 = (9x-4)(x-1)$$
$$9x - 4 = 0 \quad \text{or} \quad x - 1 = 0$$
$$x = 4/9 \quad \text{or} \quad x = 1$$

**19.**
$$2y + 1 = \sqrt{11y - 1}$$
$$(2y+1)^2 = \left(\sqrt{11y-1}\right)^2$$
$$4y^2 + 4y + 1 = 11y - 1$$
$$4y^2 - 7y + 2 = 0$$
$$y = \frac{-(-7) \pm \sqrt{(-7)^2 - 4(4)(2)}}{2(4)}$$
$$= \frac{7 \pm \sqrt{49 - 32}}{8} = \frac{7 \pm \sqrt{17}}{8}$$
$$\approx 0.36,\ 1.39$$

**21.**
$$x - 3 = \sqrt{x - 1}$$
$$(x-3)^2 = \left(\sqrt{x-1}\right)^2$$
$$x^2 - 6x + 9 = x - 1$$
$$x^2 - 7x + 10 = 0$$
$$(x-5)(x-2) = 0$$
$$x - 5 = 0 \quad \text{or} \quad x - 2 = 0$$
$$x = 5 \quad \text{or} \quad x = 2$$
A check shows that 2 is an extraneous solution. The only solution is 5.

**23.**
$$\sqrt{3x - 2} = \sqrt{x + 2} - 1$$
$$\left(\sqrt{3x-2}\right)^2 = \left(\sqrt{x+2} - 1\right)^2$$
$$3x - 2 = x + 2 - 2\sqrt{x+2} + 1$$
$$2x - 5 = -2\sqrt{x+2}$$
$$(2x-5)^2 = \left(-2\sqrt{x+2}\right)^2$$
$$4x^2 - 20x + 25 = 4x + 8$$
$$4x^2 - 24x + 17 = 0$$
$$x = \frac{-(-24) \pm \sqrt{(-24)^2 - 4(4)(17)}}{2(4)}$$
$$= \frac{24 \pm \sqrt{304}}{8} \approx 0.82,\ 5.18$$
A check shows that 5.18 is an extraneous solution. The only solution is 0.82.

**25.**
$$\sqrt{x}\sqrt{x+3} = 1$$
$$\left(\sqrt{x}\sqrt{x+3}\right)^2 = 1^2$$
$$x(x+3) = 1$$
$$x^2 + 3x - 1 = 0$$
$$x = \frac{-3 \pm \sqrt{3^2 - 4(1)(-1)}}{2(1)}$$
$$= \frac{-3 \pm \sqrt{13}}{2} \approx -3.30,\ 0.30$$
A check shows that $-3.30$ is an extraneous solution. The only solution is 0.30.

**27.** By making an appropriate substitution, we write the given equation in the form $au^2 + bu + c = 0$, where $u$ represents an expression. After solving this equation for $u$, we replace $u$ with the expression that it represents and solve for the original variable.

**29.**
$$b^4 - 7b^2 + 12 = 0$$
$$u^2 - 7u + 12 = 0 \quad \text{Let } u = b^2.$$
$$(u-4)(u-3) = 0$$
$$u - 4 = 0 \quad \text{or} \quad u - 3 = 0$$
$$u = 4 \quad \text{or} \quad u = 3$$
Now, replace $u$ with $b^2$.
$$b^2 = 4 \quad \text{or} \quad b^2 = 3$$
$$b = \pm 2 \quad \text{or} \quad b = \pm\sqrt{3} \approx \pm 1.73$$

**31.**
$$y^4 + 2y^2 = 63$$
$$u^2 + 2u - 63 = 0 \quad \text{Let } u = y^2.$$
$$(u-7)(u+9) = 0$$
$$u - 7 = 0 \quad \text{or} \quad u + 9 = 0$$
$$u = 7 \quad \text{or} \quad u = -9$$
Now, replace $u$ with $y^2$.

Section 10.4    Equations in Quadratic Form

$$y^2 = 7 \qquad y^2 = -9$$
$$y = \pm\sqrt{7} \quad \text{or} \quad y = \pm\sqrt{-9}$$
$$\approx \pm 2.65 \qquad = \pm 3i$$

**33.**
$$x - 6x^{1/2} + 8 = 0$$
$$u^2 - 6u + 8 = 0 \qquad \text{Let } u = x^{1/2}.$$
$$(u - 4)(u - 2) = 0$$
$$u - 4 = 0 \quad \text{or} \quad u - 2 = 0$$
$$u = 4 \quad \text{or} \quad u = 2$$
Now, replace $u$ with $x^{1/2}$.
$$x^{1/2} = 4 \quad \text{or} \quad x^{1/2} = 2$$
$$x = 4^2 = 16 \qquad x = 2^2 = 4$$

**35.**
$$y^{-2} + 2 = y^{-1}$$
$$u^2 - u + 2 = 0 \qquad \text{Let } u = y^{-1}.$$
$$u = \frac{-(-1) \pm \sqrt{(-1)^2 - 4(1)(2)}}{2(1)}$$
$$= \frac{1 \pm \sqrt{-7}}{2} \approx 0.5 \pm 1.32i$$
Now, replace $u$ with $y^{-1}$.
$$y^{-1} = \frac{1 \pm i\sqrt{7}}{2}$$
$$y = \frac{2}{1 \pm i\sqrt{7}} \approx 0.25 \pm 0.66i$$

**37.**
$$y^{2/3} + 12 = 7y^{1/3}$$
$$u^2 - 7u + 12 = 0 \qquad \text{Let } u = y^{1/3}.$$
$$(u - 4)(u - 3) = 0$$
$$u - 4 = 0 \quad \text{or} \quad u - 3 = 0$$
$$u = 4 \quad \text{or} \quad u = 3$$
Now, replace $u$ with $y^{1/3}$.
$$y^{1/3} = 4 \quad \text{or} \quad y^{1/3} = 3$$
$$y = 4^3 = 64 \qquad y = 3^3 = 27$$

**39.**
$$(x^2 - 5x)^2 - 8(x^2 - 5x) = 84$$
$$u^2 - 8u - 84 = 0$$
$$(u - 14)(u + 6) = 0$$
$$u - 14 = 0 \quad \text{or} \quad u + 6 = 0$$
$$u = 14 \quad \text{or} \quad u = -6$$
Now, replace $u$ with $x^2 - 5x$.

$$x^2 - 5x = 14$$
$$x^2 - 5x - 14 = 0$$
$$(x - 7)(x + 2) = 0$$
$$x = -2, 7$$
or
$$x^2 - 5x = -6$$
$$x^2 - 5x + 6 = 0$$
$$(x - 3)(x - 2) = 0$$
$$x = 2, 3$$

**41.**
$$(x + 1)^2 + 2(x + 1) = 2$$
$$u^2 + 2u - 2 = 0$$
$$u = \frac{-2 \pm \sqrt{2^2 - 4(1)(-2)}}{2(1)}$$
$$= \frac{-2 \pm \sqrt{12}}{2} \approx -2.73, \ 0.73$$
Now, replace $u$ with $x + 1$.
$$x + 1 = -2.73 \quad \text{or} \quad x + 1 = 0.73$$
$$x = -3.73 \qquad x = -0.27$$

**43.**
$$(x^2 - 3)^2 - 5(x^2 - 3) - 6 = 0$$
$$u^2 - 5u - 6 = 0$$
$$(u - 6)(u + 1) = 0$$
$$u - 6 = 0 \quad \text{or} \quad u + 1 = 0$$
$$u = 6 \quad \text{or} \quad u = -1$$
Now, replace $u$ with $x^2 - 3$.
$$x^2 - 3 = 6 \qquad x^2 - 3 = -1$$
$$x^2 = 9 \quad \text{or} \quad x^2 = 2$$
$$x = \pm 3 \qquad x = \pm\sqrt{2} \approx \pm 1.41$$

**45.**
$$3(\sqrt{x} - 2)^2 - 7(\sqrt{x} - 2) + 2 = 0$$
$$3u^2 - 7u + 2 = 0$$
$$(3u - 1)(u - 2) = 0$$
$$3u - 1 = 0 \quad \text{or} \quad u - 2 = 0$$
$$u = 1/3 \quad \text{or} \quad u = 2$$
Now, replace $u$ with $\sqrt{x} - 2$.
$$\sqrt{x} - 2 = \frac{1}{3} \qquad \sqrt{x} - 2 = 2$$
$$\sqrt{x} = \frac{7}{3} \quad \text{or} \quad \sqrt{x} = 4$$
$$x = \frac{49}{9} \qquad x = 16$$

**47.**
$$x - \sqrt{x} = 6$$
$$u^2 - u - 6 = 0 \quad \text{Let } u = \sqrt{x}.$$
$$(u-3)(u+2) = 0$$
$$u - 3 = 0 \quad \text{or} \quad u + 2 = 0$$
$$u = 3 \quad \text{or} \quad u = -2$$
Now, replace $u$ with $\sqrt{x}$.
$$\sqrt{x} = 3 \quad \text{or} \quad \sqrt{x} = -2$$
$$x = 9 \quad\quad\quad x = (-2)^2 = 4$$
A check shows that 4 is an extraneous solution. The only solution is 9.

**49.**
$$x^4 - 7x^2 + 12 = 0$$
$$u^2 - 7u + 12 = 0 \quad \text{Let } u = x^2.$$
$$(u-4)(u-3) = 0$$
$$u - 4 = 0 \quad \text{or} \quad u - 3 = 0$$
$$u = 4 \quad \text{or} \quad u = 3$$
Now, replace $u$ with $x^2$.
$$x^2 = 4 \quad \text{or} \quad x^2 = 3$$
$$x = \pm 2 \quad\quad x = \pm\sqrt{3} \approx \pm 1.73$$

**51.**
$$\frac{x}{x+2} + \frac{4}{x} = 5$$
$$\frac{x \cdot x(x+2)}{x+2} + \frac{4 \cdot x(x+2)}{x} = 5x(x+2)$$
$$x^2 + 4(x+2) = 5x(x+2)$$
$$x^2 + 4x + 8 = 5x^2 + 10x$$
$$4x^2 + 6x - 8 = 0$$
$$2(2x^2 + 3x - 4) = 0$$
$$x = \frac{-3 \pm \sqrt{3^2 - 4(2)(-4)}}{2(2)}$$
$$= \frac{-3 \pm \sqrt{41}}{4} \approx -2.35, \ 0.85$$

**53.**
$$2x + 5 = 2\sqrt{x+3}$$
$$(2x+5)^2 = \left(2\sqrt{x+3}\right)^2$$
$$4x^2 + 20x + 25 = 4x + 12$$
$$4x^2 + 16x + 13 = 0$$
$$x = \frac{-16 \pm \sqrt{16^2 - 4(4)(13)}}{2(4)}$$
$$= \frac{-16 \pm \sqrt{48}}{8} \approx -2.87, \ -1.13$$
A check shows that $-2.87$ is extraneous. The only solution is $-1.13$.

**55.**
$$a^4 + a^2 = 20$$
$$a^4 + a^2 - 20 = 0$$
$$u^2 + u - 20 = 0 \quad \text{Let } u = a^2.$$
$$(u-4)(u+5) = 0$$
$$u - 4 = 0 \quad \text{or} \quad u + 5 = 0$$
$$u = 4 \quad \text{or} \quad u = -5$$
Now, replace $u$ with $a^2$.
$$a^2 = 4 \quad \text{or} \quad a^2 = -5$$
$$a = \pm 2 \quad\quad a = \pm\sqrt{-5}$$
$$\quad\quad\quad\quad\quad a = \pm i\sqrt{5} \approx \pm 2.24i$$

**57.**
$$x - 13x^{1/2} + 42 = 0$$
$$u^2 - 13u + 42 = 0 \quad \text{Let } u = x^{1/2}.$$
$$(u-6)(u-7) = 0$$
$$u - 6 = 0 \quad \text{or} \quad u - 7 = 0$$
$$u = 6 \quad \text{or} \quad u = 7$$
Now, replace $u$ with $x^{1/2}$.
$$x^{1/2} = 6 \quad \text{or} \quad x^{1/2} = 7$$
$$x = 6^2 = 36 \quad\quad x = 7^2 = 49$$

**59.**
$$\frac{x+3}{x-1} = \frac{5}{x}$$
$$x(x+3) = 5(x-1)$$
$$x^2 + 3x = 5x - 5$$
$$x^2 - 2x + 5 = 0$$
$$x = \frac{-(-2) \pm \sqrt{(-2)^2 - 4(1)(5)}}{2(1)}$$
$$= \frac{2 \pm \sqrt{-16}}{2} \approx 1 \pm 2i$$

**61.**
$$\sqrt{3x+12} - \sqrt{x+8} = 2$$
$$\sqrt{3x+12} = 2 + \sqrt{x+8}$$
$$\left(\sqrt{3x+12}\right)^2 = \left(2 + \sqrt{x+8}\right)^2$$
$$3x + 12 = 4 + 4\sqrt{x+8} + x + 8$$
$$2x = 4\sqrt{x+8}$$
$$x = 2\sqrt{x+8}$$
$$x^2 = 4x + 32$$
$$x^2 - 4x - 32 = 0$$
$$(x-8)(x+4) = 0$$
$$x = -4, \ 8$$
A check shows that $-4$ is extraneous. The only solution is 8.

Section 10.4  Equations in Quadratic Form

**63.**
$$a^{2/3} = 2a^{1/3} + 15$$
$$u^2 - 2u - 15 = 0 \quad \text{Let } u = a^{1/3}.$$
$$(u-5)(u+3) = 0$$
$$u = -3, 5$$

Now, replace $u$ with $a^{1/3}$.

$$\begin{array}{ll} a^{1/3} = -3 & a^{1/3} = 5 \\ a = (-3)^3 \quad \text{or} & a = 5^3 \\ \quad = -27 & \quad = 125 \end{array}$$

**65.**
$$\sqrt{3}\,x = \sqrt{2-7x}$$
$$\left(\sqrt{3}\,x\right)^2 = \left(\sqrt{2-7x}\right)^2$$
$$3x^2 = 2 - 7x$$
$$3x^2 + 7x - 2 = 0$$
$$x = \frac{-7 \pm \sqrt{7^2 - 4(3)(-2)}}{2(3)}$$
$$= \frac{-7 \pm \sqrt{73}}{6} \approx -2.59,\ 0.26$$

A check shows that $-2.59$ is an extraneous solution. The only solution is $0.26$.

**67.**
$$\frac{x+1}{3-x} + \frac{12}{x^2 - 3x} + \frac{5}{x} = 0$$
$$-\frac{x+1}{x-3} + \frac{12}{x(x-3)} + \frac{5}{x} = 0$$

The restricted values are $0$ and $3$.
Multiply each term by the LCD.
$$-x(x+1) + 12 + 5(x-3) = 0$$
$$-x^2 - x + 12 + 5x - 15 = 0$$
$$-x^2 + 4x - 3 = 0$$
$$x^2 - 4x + 3 = 0$$
$$(x-3)(x-1) = 0$$
$$x = 1, 3$$

Because 3 is a restricted value, the only solution is 1.

**69.**
$$(3-y)^2 - 3(3-y) = 6$$
$$u^2 - 3u - 6 = 0 \quad \text{Let } u = 3 - y.$$
$$u = \frac{-(-3) \pm \sqrt{(-3)^2 - 4(1)(-6)}}{2(1)}$$
$$= \frac{3 \pm \sqrt{33}}{2} \approx -1.37,\ 4.37$$

Now, replace $u$ with $3 - y$.

$$\begin{array}{ll} 3 - y = -1.37 & 3 - y = 4.37 \\ \quad y = 4.37 \quad \text{or} & \quad y = -1.37 \end{array}$$

**71.**
$$\sqrt{2x-3} = 4 - \sqrt{x+2}$$
$$\left(\sqrt{2x-3}\right)^2 = \left(4 - \sqrt{x+2}\right)^2$$
$$2x - 3 = 16 - 8\sqrt{x+2} + x + 2$$
$$x - 21 = -8\sqrt{x+2}$$
$$(x-21)^2 = \left(-8\sqrt{x+2}\right)^2$$
$$x^2 - 42x + 441 = 64x + 128$$
$$x^2 - 106x + 313 = 0$$
$$x = \frac{-(-106) \pm \sqrt{(-106)^2 - 4(1)(313)}}{2(1)}$$
$$= \frac{106 \pm \sqrt{9984}}{2} \approx 3.04,\ 102.96$$

A check shows that 102.96 is an extraneous solution. The only solution is 3.04.

**73.**
$$(3-x)^{-2} - 3(3-x)^{-1} = 4$$
$$u^2 - 3u - 4 = 0$$
$$(u-4)(u+1) = 0$$
$$u = -1, 4$$

Now, replace $u$ with $(3-x)^{-1}$.

$$\begin{array}{ll} (3-x)^{-1} = -1 & (3-x)^{-1} = 4 \\ 3 - x = -1 \quad \text{or} & 3 - x = \dfrac{1}{4} \\ x = 4 & x = 2.75 \end{array}$$

**75.**
$$\frac{x}{4(x-4)} = \frac{1}{x+2} + \frac{6}{x^2 - 2x - 8}$$
$$\frac{x}{4(x-4)} = \frac{1}{x+2} + \frac{6}{(x-4)(x+2)}$$

The restricted values are $-2$ and $4$.
Multiply each term by the LCD.
$$x(x+2) = 4(x-4) + 24$$
$$x^2 + 2x = 4x - 16 + 24$$
$$x^2 - 2x - 8 = 0$$
$$(x-4)(x+2) = 0$$
$$x = -2, 4$$

Because both of these are restricted values, there is no solution.

**77.** $(x^2 - x)^2 + 3(x^2 - x) + 2 = 0$
$$u^2 + 3u + 2 = 0$$
$$(u + 2)(u + 1) = 0$$
$$u = -2, -1$$

Now, replace $u$ with $x^2 - x$.

$x^2 - x = -2$ or $x^2 - x = -1$
$x^2 - x + 2 = 0 \qquad x^2 - x + 1 = 0$

$$x = \frac{-(-1) \pm \sqrt{(-1)^2 - 4(1)(2)}}{2(1)}$$
$$= \frac{1 \pm \sqrt{-7}}{2} \approx 0.5 \pm 1.32i$$

$$x = \frac{-(-1) \pm \sqrt{(-1)^2 - 4(1)(1)}}{2(1)}$$
$$= \frac{1 \pm \sqrt{-3}}{2} \approx 0.5 \pm 0.87i$$

**79.** The break-even point occurs when the cost is equal to the total revenue. Set the expressions equal and solve for $x$.
$$8000 + 8000x^2 - 80x^4 = 1680x^2 - 4960$$
$$12{,}960 + 6320x^2 - 80x^4 = 0$$
$$162 + 79x^2 - x^4 = 0$$
$$162 + 79u - u^2 = 0$$
$$(81 - u)(2 + u) = 0$$
$$u = -2, 81$$

Now, replace $u$ with $x^2$. Because only real number solutions apply to this situation, we will only consider positive values of $u$.
$$x^2 = 81$$
$$x = \pm 9$$

The company must make and sell 9 hundred, or 900, motors to break even.

**81.** Let $x =$ the total number of students that can be accommodated and $x - 40 =$ the number of students enrolled after the initial registration. The description in the problem tells us that the difference between the average cost per student for the initial enrollment and the total enrollment is $100.

$$\frac{195{,}000}{x - 40} - \frac{195{,}000}{x} = 100$$
$$\frac{1950}{x - 40} - \frac{1950}{x} = 1$$

The restricted values are 0 and 40. Multiply each term by the LCD.
$$1950x - 1950(x - 40) = x(x - 40)$$
$$1950x - 1950x + 78{,}000 = x^2 - 40x$$
$$x^2 - 40x - 78{,}000 = 0$$
$$(x + 260)(x - 300) = 0$$
$$x = -260, 300$$

The program can accommodate 300 students. If all spaces are filled, the cost per person is $650.

**83.** (a) The equation is
$$-0.3(x + 5)^2 + 5.14(x + 5) + 2.34 = 15.5$$

(b) Let $u = x + 5$. Then we have
$$-0.3u^2 + 5.14u + 2.34 = 15.5$$
where $u \approx 3.13$ or $u = 14$. Replacing $u$ with $x + 5$, we obtain $x = 9$ (2004).

**85.** For 2006, $t = 16$:
$$C(t) = 6.98(t + 10)^2 + 38.4(t + 10) + 2246$$
$$C(16) = 6.98(16 + 10)^2 + 38.4(16 + 10) + 2246$$
$$= 6.98(26)^2 + 38.4(26) + 2246$$
$$= 7962.88$$

If the population is 285 million, then the per capita cost for the school lunch program in 2006, is
$7962.88 \div 285 \approx \$27.94$.

**87.** $C(-10) = 2246$, which is the cost (in millions of dollars) of the school lunch program in 1980.

**89.**
$$t = \sqrt{t + 3}$$
$$t + 3 = \sqrt{t + 3} + 3$$
$$t + 3 - \sqrt{t + 3} - 3 = 0$$
$$u^2 - u - 3 = 0$$
$$u = \frac{-(-1) \pm \sqrt{(-1)^2 - 4(1)(-3)}}{2(1)}$$
$$= \frac{1 \pm \sqrt{13}}{2} \approx -1.30, 2.30$$

Now, replace $u$ with $\sqrt{t + 3}$.

Section 10.4   Equations in Quadratic Form

$\sqrt{t+3} = 2.30278$     $\sqrt{t+3} = -1.30278$
$t+3 \approx 5.30$   or   $t+3 \approx 1.69722$
$t \approx 2.30$              $t \approx -1.30$

A check shows that $-1.30$ is extraneous. The only solution is 2.30.

**91.**
$$x^6 - 7x^3 = 8$$
$$x^6 - 7x^3 - 8 = 0$$
$$u^2 - 7u - 8 = 0 \quad \text{Let } u = x^3.$$
$$(u-8)(u+1) = 0$$
$$u = -1, 8$$

Now, replace $u$ with $x^3$.
$$x^3 = -1$$
$$x^3 + 1 = 0$$
$$(x+1)(x^2 - x + 1) = 0$$
$x^2 - x + 1 = 0$    or    $x + 1 = 0$
$\phantom{x^2 - x + 1 = 0}$        $x = -1$

$$x = \frac{-(-1) \pm \sqrt{(-1)^2 - 4(1)(1)}}{2(1)}$$
$$= \frac{1 \pm \sqrt{-3}}{2} \approx 0.5 \pm 0.87i$$

or
$$x^3 = 8$$
$$x^3 - 8 = 0$$
$$(x-2)(x^2 + 2x + 4) = 0$$
$x^2 + 2x + 4 = 0$    or    $x - 2 = 0$
$\phantom{x^2 + 2x + 4 = 0}$        $x = 2$

$$x = \frac{-2 \pm \sqrt{2^2 - 4(1)(4)}}{2(1)}$$
$$= \frac{-2 \pm \sqrt{-12}}{2} \approx -1 \pm 1.73i$$

Solution: $-1, 2, -1 \pm 1.73i, 0.5 \pm 0.87i$.

**93.** $x^2 + 2x = |2x + 1|$ can be written as
$x^2 + 2x = -2x - 1$ or $x^2 + 2x = 2x + 1$

$$x^2 + 2x = -2x - 1$$
$$x^2 + 4x + 1 = 0$$
$$x = \frac{-4 \pm \sqrt{4^2 - 4(1)(1)}}{2(1)}$$
$$= \frac{-4 \pm \sqrt{12}}{2} \approx -3.73, -0.27$$

or
$$x^2 + 2x = 2x + 1$$
$$x^2 - 1 = 0$$
$$(x+1)(x-1) = 0$$
$$x = -1, 1$$

A check shows that $-1$ and $-0.27$ are extraneous solutions. The only solutions are $-3.73$ and $1$.

**95.**
$$y^{-2} + 1 = y^{-1}$$
$$y^{-2} - y^{-1} + 1 = 0$$
$$u^2 - u + 1 = 0 \quad \text{Let } u = y^{-1}.$$
$$u = \frac{-(-1) \pm \sqrt{(-1)^2 - 4(1)(1)}}{2(1)}$$
$$= \frac{1 \pm \sqrt{-3}}{2} \approx 0.5 \pm 0.87i$$

Now, replace $u$ with $y^{-1}$.
$$y^{-1} = \frac{1 \pm \sqrt{-3}}{2}$$
$$y = \left(\frac{1 \pm \sqrt{-3}}{2}\right)^{-1} \approx 0.5 \pm 0.87i$$

## Section 10.5 Applications

1. Let $x$ = the width of the border in feet.

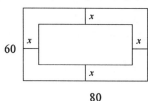

Then the dimensions of the smaller rectangle are $60 - 2x$ and $80 - 2x$. The area of this smaller rectangle is to be one-fourth of the area of the larger rectangle. The equation is
$$(60 - 2x)(80 - 2x) = \frac{1}{4}(60)(80)$$
$$4800 - 280x + 4x^2 = 1200$$
$$4x^2 - 280x + 3600 = 0$$
$$x^2 - 70x + 900 = 0$$
$$x = \frac{-(-70) \pm \sqrt{(-70)^2 - 4(1)(900)}}{2(1)}$$
$$= \frac{70 \pm \sqrt{1300}}{2} \approx 16.97,\ 53.03$$

The second value has no meaning in this context. To complete three-fourths of the job, the border should be about 16.97 feet wide.

3. Let $x$ = the length of the shorter side and $x + 30$ = the length of the side perpendicular to the shorter side. The sides with the higher wall correspond to the legs of a right triangle.

By the Pythagorean Theorem, we have
$$x^2 + (x+30)^2 = 160^2$$
$$x^2 + x^2 + 60x + 900 = 25{,}600$$
$$2x^2 + 60x - 24{,}700 = 0$$
$$x^2 + 30x - 12{,}350 = 0$$
$$x = \frac{-30 \pm \sqrt{30^2 - 4(1)(-12{,}350)}}{2(1)}$$
$$= \frac{-30 \pm \sqrt{50{,}300}}{2} \approx -127.14,\ 97.14$$

Disregard the negative value. The shorter leg is 97.14 feet long and the longer leg is 127.14 feet long.

5. Use the Golden Ratio and the information given in the figure.

$$\frac{W}{L} = \frac{L}{W+L}$$
$$\frac{x}{28-x} = \frac{28-x}{28}$$
$$28x = (28-x)^2$$
$$28x = 784 - 56x + x^2$$
$$0 = 784 - 84x + x^2$$
$$x = \frac{-(-84) \pm \sqrt{(-84)^2 - 4(1)(784)}}{2(1)}$$
$$= \frac{84 \pm \sqrt{3920}}{2} \approx 10.70,\ 73.30$$

The value 73.30 has no meaning in this context. The dimensions are 10.70 inches and 17.3 inches.

7. Use the fact that corresponding sides of similar triangles are proportional.
$$\frac{10}{x} = \frac{x+9}{4}$$
$$40 = x(x+9)$$
$$0 = x^2 + 9x - 40$$
$$x = \frac{-9 \pm \sqrt{9^2 - 4(1)(-40)}}{2(1)}$$
$$= \frac{-9 \pm \sqrt{241}}{2} \approx -12.26,\ 3.26$$

The negative value has no meaning in this context. The side $DE$ has a length of approximately 3.26 and the side $AC$ has a length of approximately 12.26.

**Section 10.5   Applications**

**9.** Let $x$ = the rate of the boat in still water.

| | d | r | t = d/r |
|---|---|---|---|
| Up | 46 | $x - 4$ | $46/(x-4)$ |
| Down | 46 | $x + 4$ | $46/(x+4)$ |

The sum of the times is 3.5 or 7/2 hours.
$$\frac{46}{x-4} + \frac{46}{x+4} = \frac{7}{2}$$
$$92(x+4) + 92(x-4) = 7(x^2 - 16)$$
$$92x + 368 + 92x - 368 = 7x^2 - 112$$
$$7x^2 - 184x - 112 = 0$$
$$x = \frac{-(-184) \pm \sqrt{(-184)^2 - 4(7)(-112)}}{2(7)}$$
$$= \frac{184 \pm \sqrt{36{,}992}}{14} \approx -0.60,\ 26.88$$

Using the positive value, the boat would have traveled approximately 26.88 mph in still water.

**11.** Let $x$ = the hiking rate and $x + 55$ = the driving rate.

| | d | r | t = d/r |
|---|---|---|---|
| Hiking | 1 | $x$ | $1/x$ |
| Driving | 45 | $x + 55$ | $45/(x+55)$ |

The sum of the times is 1 hour and 15 minutes, or 5/4 hours.
$$\frac{1}{x} + \frac{45}{x+55} = \frac{5}{4}$$
$$4(x+55) + 180x = 5x(x+55)$$
$$4x + 220 + 180x = 5x^2 + 275x$$
$$0 = 5x^2 + 91x - 220$$
$$x = \frac{-91 \pm \sqrt{91^2 - 4(5)(-220)}}{2(5)}$$
$$= \frac{-91 \pm \sqrt{12{,}681}}{10} \approx -20.36,\ 2.16$$

Using the positive value, the hiking rate is approximately 2.16 mph and the driving rate is approximately 57.16 mph.

**13.** Let $x$ = the number of hours for the slower painter to complete the job alone.

| | Time | Work rate |
|---|---|---|
| Slower | $x$ | $1/x$ |
| Faster | $x - 2$ | $1/(x-2)$ |
| Together | 6 | 1/6 |

$$\frac{1}{x} + \frac{1}{x-2} = \frac{1}{6}$$
$$6(x-2) + 6x = x(x-2)$$
$$6x - 12 + 6x = x^2 - 2x$$
$$0 = x^2 - 14x + 12$$
$$x = \frac{-(-14) \pm \sqrt{(-14)^2 - 4(1)(12)}}{2(1)}$$
$$= \frac{14 \pm \sqrt{148}}{2} \approx 0.92,\ 13.08$$

The smaller value has no meaning in this context. The slower painter could paint the apartment in 13.08 hours and the faster painter could do it in 11.08 hours.

**15.** Let $x$ = the number of hours for the second fill pipe alone to fill the pool.

| | Time | Work rate |
|---|---|---|
| 1st fill pipe | 9 | 1/9 |
| 2nd fill pipe | $x$ | $1/x$ |
| Drain pipe | $x + 1$ | $1/(x+1)$ |
| Together | 8 | 1/8 |

$$\frac{1}{9} + \frac{1}{x} - \frac{1}{x+1} = \frac{1}{8}$$
$$8x(x+1) + 72(x+1) - 72x = 9x(x+1)$$
$$8x^2 + 8x + 72x + 72 - 72x = 9x^2 + 9x$$
$$x^2 + x - 72 = 0$$
$$(x+9)(x-8) = 0$$
$$x = -9,\ 8$$

Using the positive value, the drain pipe could empty the pool in $x + 1 = 9$ hours with the two fill pipes closed.

**17.** Let $x$ = the number of shares owned ten years ago ($x < 400$). Then $x + 405$ represents the number of shares owned today. The price per share ten years ago was $\dfrac{16{,}095}{x}$. The price per share today is $\dfrac{34{,}875}{x + 405}$. The price per share today is $1.50 more than it was ten years ago.

$$\frac{34{,}875}{x+405} - \frac{16{,}095}{x} = 1.50$$
$$34{,}875x - 16{,}095(x+405) = 1.5x(x+405)$$
$$34{,}875x - 16{,}095x - 6{,}518{,}475 = 1.5x^2 + 607.5x$$
$$1.5x^2 - 18{,}172.5x + 6{,}518{,}475 = 0$$
$$x^2 - 12{,}115x + 4{,}345{,}650 = 0$$
$$(x - 370)(x - 11{,}745) = 0$$
$$x = 370,\ 11{,}745$$

The value 11,745 does not meet the conditions of the problem. So, the man originally had 370 shares.

**19.** Let $x =$ the number of boxes sold last year.

|  | Boxes sold | Revenue | Price per box |
|---|---|---|---|
| Last year | $x$ | 8568 | $8568/x$ |
| This year | $x + 102$ | 11,594 | $11{,}594/(x+102)$ |

This year's price is $2 per box more than last year's price.

$$\frac{11{,}594}{x+102} - \frac{8568}{x} = 2$$
$$11{,}594x - 8568(x+102) = 2x(x+102)$$
$$11{,}594x - 8568x - 873{,}936 = 2x^2 + 204x$$
$$2x^2 - 2822x + 873{,}936 = 0$$
$$x^2 - 1411 + 436{,}968 = 0$$
$$(x - 459)(x - 952) = 0$$
$$x = 459,\ 952$$

Because fewer than 500 boxes were sold last year, there were 459 boxes sold last year at $18.67 per box and 561 boxes sold this year at $20.67 per box.

**21.** Let $x =$ the number of occupied units before the rent increase.

|  | Occupied Units | Rent collected | Rent per unit |
|---|---|---|---|
| Before | $x$ | 24,000 | $24{,}000/x$ |
| After | $x - 10$ | 24,000 | $24{,}000/(x-10)$ |

The difference in rent per unit is $80.

$$\frac{24{,}000}{x-10} - \frac{24{,}000}{x} = 80$$
$$24{,}000x - 24{,}000(x-10) = 80x(x-10)$$
$$24{,}000x - 24{,}000x + 240{,}000 = 80x^2 - 800x$$
$$80x^2 - 800x - 240{,}000 = 0$$
$$x^2 - 10x - 3000 = 0$$
$$(x + 50)(x - 60) = 0$$
$$x = -50,\ 60$$

Disregard the negative solution. All the units are occupied at the lower rent, so there are 60 apartments in the complex.

**23.** Let $x =$ the smaller of two consecutive odd integers and $x + 2 =$ the larger.

$$(x+2)^3 - x^3 = 866$$
$$x^3 + 6x^2 + 12x + 8 - x^3 = 866$$
$$6x^2 + 12x - 858 = 0$$
$$x^2 + 2x - 143 = 0$$
$$(x + 13)(x - 11) = 0$$
$$x = -13,\ 11$$

The numbers are either $-13$ and $-11$ or 11 and 13.

**25.** Let $x =$ the distance from the river to the parallel side (minimum 70 feet) and $300 - 2x =$ the feet of fencing remaining to be used for the parallel side. Because a gate takes up 8 feet, the length of the parallel side is $(300 - 2x) + 8 = 308 - 2x$. The area is to be 11,500 square feet.

$$x(308 - 2x) = 11{,}500$$
$$308x - 2x^2 = 11{,}500$$
$$0 = 2x^2 - 308x + 11{,}500$$
$$x = \frac{-(-308) \pm \sqrt{(-308)^2 - 4(2)(11{,}500)}}{2(2)}$$
$$= \frac{308 \pm \sqrt{2864}}{4} \approx 63.62,\ 90.38$$

Because the distance from the river to the parallel side is to be at least 70 feet, the distance should be approximately 90.38 feet.

**27.** Use the Golden Ratio and the information given in the figure.

$$\frac{W}{L} = \frac{L}{W+L}$$
$$\frac{5}{x} = \frac{x}{x+5}$$
$$x^2 = 5(x+5)$$
$$x^2 = 5x + 25$$
$$x^2 - 5x - 25 = 0$$
$$x = \frac{-(-5) \pm \sqrt{(-5)^2 - 4(1)(-25)}}{2(1)}$$
$$= \frac{5 \pm \sqrt{125}}{2} \approx -3.09, \, 8.09$$

The negative value has no meaning in this context. The length of the sign is approximately 8 feet.

**29.**
$$\frac{1}{2}n(n-3) = 27$$
$$n(n-3) = 54$$
$$n^2 - 3n - 54 = 0$$
$$(n+6)(n-9) = 0$$
$$n = -6, \, 9$$

The negative value has no meaning in this context. A polygon with 27 diagonals has 9 sides.

**31.** We are given that $b = a + 2.5$ and $h = 3$.
$$\frac{a}{h} = \frac{h}{b}$$
$$\frac{a}{3} = \frac{3}{a+2.5}$$
$$a^2 + 2.5a = 9$$
$$a^2 + 2.5a - 9 = 0$$
$$2a^2 + 5a - 18 = 0$$
$$(2a+9)(a-2) = 0$$
$$a = -4.5, \, 2$$

The negative value has no meaning in this context. Therefore, $a = 2$ and $b = 4.5$. Use the Pythagorean Theorem with $h$ and $b$ to find $c$.

$$h^2 + b^2 = c^2$$
$$3^2 + 4.5^2 = c^2$$
$$9 + 20.25 = c^2$$
$$29.25 = c^2$$
$$\pm\sqrt{29.25} = c$$
$$\pm 5.41 \approx c$$

The negative value has no meaning in this context. Therefore, the length of panel $c$ is approximately 5.41 feet.

**33.**

$$C_S = 2\pi r \quad C_L = 2\pi R$$
$$A_S = \pi r^2 \quad A_L = \pi R^2$$

First, use the information about the difference in circumferences to obtain a relationship between the two radii.
$$C_L - C_S = 4\pi$$
$$2\pi R - 2\pi r = 4\pi$$
$$R - r = 2$$
$$R = r + 2$$

The area of the larger circle in terms of $r$ is $\pi(r+2)^2$.

$$\pi(r+2)^2 + \pi r^2 = 164\pi$$
$$r^2 + 4r + 4 + r^2 = 164$$
$$2r^2 + 4r - 160 = 0$$
$$r^2 + 2r - 80 = 0$$
$$(r+10)(r-8) = 0$$
$$r = -10, \, 8$$

The radius of the smaller circular area is 8 yards.

**35.**

The area of a triangle is given by $A = \frac{1}{2}bh$.

$$\frac{1}{2}x(x-10) = 440$$
$$x(x-10) = 880$$
$$x^2 - 10x - 880 = 0$$

$$x = \frac{-(-10) \pm \sqrt{(-10)^2 - 4(1)(-880)}}{2(1)}$$

$$= \frac{10 \pm \sqrt{3620}}{2} \approx -25.08, \ 35.08$$

The negative value has no meaning in this context. The base of the sail is approximately 35.08 feet and the height is approximately 25.08 feet.

**37.**

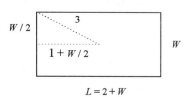

Using the Pythagorean Theorem with the information in the figure, we obtain

$$\left(\frac{W}{2}\right)^2 + \left(\frac{W}{2} + 1\right)^2 = 3^2$$

$$\frac{W^2}{4} + \frac{W^2}{4} + W + 1 = 9$$

$$\frac{W^2}{2} + W + 1 = 9$$

$$W^2 + 2W - 16 = 0$$

$$W = \frac{-2 \pm \sqrt{2^2 - 4(1)(-16)}}{2(1)}$$

$$= \frac{-2 \pm \sqrt{68}}{2} \approx -5.12, \ 3.12$$

The negative value has no meaning in this context. The width of the table should be approximately 3.12 feet.

**39.** Let $x =$ the number.

$$x + 1 = \sqrt{x + 1} + 2$$
$$x - 1 = \sqrt{x + 1}$$
$$(x - 1)^2 = \left(\sqrt{x + 1}\right)^2$$
$$x^2 - 2x + 1 = x + 1$$
$$x^2 - 3x = 0$$
$$x(x - 3) = 0$$
$$x = 0, \ 3$$

A check shows that 0 is an extraneous solution. The number is 3.

**41.** Let $x =$ the distance between $B$ and $F$.

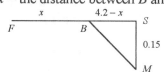

We know that $r \times t = d$, or that time is given by $t = \dfrac{d}{r}$. The time to cover the distance from $M$ to $B$ is given by the distance $\sqrt{(4.2 - x)^2 + 0.15^2}$ (using the Pythagorean Theorem) divided by the rate for this leg of 0.5 mph. The time to cover the distance from $B$ to $F$ is given by the distance $x$ divided by the rate for this leg of 12 mph. These distances are covered in 50 minutes, or 5/6 hour.

$$\frac{\sqrt{(4.2 - x)^2 + 0.15^2}}{0.5} + \frac{x}{12} = \frac{5}{6}$$

$$24\sqrt{(4.2 - x)^2 + 0.15^2} + x = 10$$

$$24\sqrt{(4.2 - x)^2 + 0.15^2} = 10 - x$$

$$\left(24\sqrt{(4.2 - x)^2 + 0.15^2}\right)^2 = (10 - x)^2$$

$$576(4.2 - x)^2 + 12.96 = 100 - 20x + x^2$$

$$576(17.64 - 8.4x + x^2) + 12.96 = 100 - 20x + x^2$$

$$575x^2 - 4818.4x + 10{,}073.6 = 0$$

Using the Quadratic Formula, we find that the solutions are 4 and 4.38. Because 4.38 has no meaning in this context, the athlete rode the bicycle for 4 miles.

**43.** Let $x =$ the number of hours for the amateur to complete the job working alone (the work rate is $1/x$) and $x - 4 =$ the number of hours for the experienced worker working alone (the work rate is $1/(x - 4)$). On the first day, the amateur works a total of 6 hours and the experienced worker works a total of 8 hours. The job is 65% complete at the end of the first day.

$$8\left(\frac{1}{x - 4}\right) + 6\left(\frac{1}{x}\right) = 0.65$$

$$8x + 6(x - 4) = 0.65x(x - 4)$$

$$8x + 6x - 24 = 0.65x^2 - 2.6x$$

$$0.65x^2 - 16.6x + 24 = 0$$

$$x = \frac{-(-16.6) \pm \sqrt{(-16.6)^2 - 4(0.65)(24)}}{2(0.65)}$$
$$= \frac{16.6 \pm \sqrt{213.16}}{1.3} \approx 1.54, \ 24$$

The first value has no meaning in this context. The amateur can do the job alone in 24 hours and the experienced worker can do the job alone in 20 hours. Let $h$ = the number of hours the workers must work together the next day until the job is done. Then their combined work rate multiplied by the number of hours they work should equal 35%, the remaining portion of the job.

$$\left(\frac{1}{24} + \frac{1}{20}\right) \cdot h = 0.35$$
$$\frac{11}{120} h = 0.35$$
$$h \approx 3.82$$

Working together on the next day, they will have to work 3.82 hours, or if they both begin at 9 A.M., until about 12:49 P.M.

**45.** Let $x$ = the number of games played by the junior and $x - 3$ = the number of games played by the senior. The junior's average points per game is $\frac{141}{x}$ and the senior's average points per game is $\frac{174}{x-3}$. The senior scored 5.1 more points per game than the junior.

$$\frac{174}{x-3} - \frac{141}{x} = 5.1$$
$$174x - 141(x - 3) = 5.1x(x - 3)$$
$$174x - 141x + 423 = 5.1x^2 - 15.3x$$
$$5.1x^2 - 48.3x - 423 = 0$$
$$x = \frac{-(-48.3) \pm \sqrt{(-48.3)^2 - 4(5.1)(-423)}}{2(5.1)}$$
$$= \frac{48.3 \pm \sqrt{10{,}962.09}}{10.2} \approx -5.53, \ 15$$

The negative value has no meaning in this context. The junior played in 15 games and the senior played in 12 games.

**47.** $N(x) = -0.025x^2 + 2.71x - 1.76$

Using the Quadratic Formula, we can find the roots of $N$, or equivalently, the $x$-intercepts of the graph of $N(x)$.

$$x = \frac{-2.71 \pm \sqrt{2.71^2 - 4(-0.025)(-1.76)}}{2(-0.025)}$$
$$= \frac{-2.71 \pm \sqrt{7.1681}}{-0.05} \approx 0.65, \ 107.75$$

The $x$-intercepts are about (1, 0) and (108, 0). The interpretation is that no one age 1 or 108 has a credit card.

**49.** Because the percentage who always pay the balance due increases with age, the linear model $y = ax + b$ is the better model. Using (30, 39.6) and (70, 74.9), the slope is $m = \frac{74.9 - 39.6}{70 - 30} = 0.8825$.

The model is
$$y - 39.6 = 0.8825(x - 30)$$
$$y = 0.8825x + 13.125.$$
$$y \approx 0.88x + 13.13$$

**51.** Let $x$ = the number of calculators sold to students and $x + 40$ = the total number of calculators in the shipment. The students' price per calculator was $\frac{10{,}000}{x}$. The price that students would have paid per calculator if the entire shipment would have been sold to students is $\frac{10{,}000}{x+40}$, which is $12.50 less than the price students actually paid.

$$\frac{10{,}000}{x} - \frac{10{,}000}{x+40} = 12.50$$
$$10{,}000(x + 40) - 10{,}000x = 12.5x(x + 40)$$
$$400{,}000 = 12.5x^2 + 500x$$
$$12.5x^2 + 500x - 400{,}000 = 0$$
$$x^2 + 40x - 32{,}000 = 0$$
$$(x + 200)(x - 160) = 0$$
$$x = -200, \ 160$$

The negative value has no meaning in this context. The students purchased 160 calculators total at a cost of $10,000 ÷ 160 = $62.50 each. The total amount collected from nonstudents for the remaining 40 calculators was $62.50 \times 40 = \$2500$. If this was the amount of profit distributed to the 160 students who bought calculators, each student received $\$2500 \div 160 \approx \$15.63$.

53. We know that $AB = 26$. Let $x =$ the distance between $A$ and $C$ and $x + 8 =$ the distance between $A$ and $D$.

$$\frac{AB}{AD} = \frac{AC}{AB}$$
$$\frac{26}{x+8} = \frac{x}{26}$$
$$26^2 = x(x+8)$$
$$676 = x^2 + 8x$$
$$0 = x^2 + 8x - 676$$

$$x = \frac{-8 \pm \sqrt{8^2 - 4(1)(-676)}}{2(1)}$$
$$= \frac{-8 \pm \sqrt{2768}}{2} \approx -30.31,\ 22.31$$

The negative value has no meaning in this context. The spring stretched to $22.31 + 8 = 30.31$ inches long, or $30.31 - 26 = 4.31$ inches from its original position.

## Section 10.6 Quadratic Inequalities

1. For $\leq$ or $\geq$, the $x$-intercepts represent solutions. Then look for the $x$-interval(s) in which the graph is above the $x$-axis (for $>$ or $\geq$) or below the $x$-axis (for $<$ or $\leq$).

3. Let $y = (3-x)(1+x)$. The $x$-intercepts are $(3, 0)$ and $(-1, 0)$. From the graph, we can see that the expression is positive when $x$ is in the interval $(-1, 3)$.

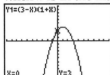

5. Let $y = 5x^2 - 8x = x(5x-8)$. The $x$-intercepts are $(0, 0)$ and $(8/5, 0)$. From the graph, we can see that the expression is positive when $x$ is in the interval $(-\infty, 0) \cup (1.6, \infty)$.

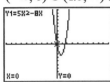

7. Let $y = x^2 + 10 - 7x = (x-5)(x-2)$. The $x$-intercepts are $(2, 0)$ and $(5, 0)$. From the graph, we can see that the expression is negative when $x$ is in the interval $(2, 5)$.

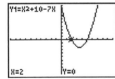

9. Let $y = 3 + x - 2x^2 = (3-2x)(1+x)$. The $x$-intercepts are $(1.5, 0)$ and $(-1, 0)$. From the graph, we can see that the expression is negative when $x$ is in the interval $(-\infty, -1) \cup (1.5, \infty)$.

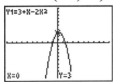

11. Write the inequality in standard form. The function is $f(x) = 2x^2 + 3x + 7$.

13. Graph $y = 9 - x^2$. The $x$-intercepts appear to be $(-3, 0)$ and $(3, 0)$. The solution set of $9 - x^2 < 0$ is $(-\infty, -3) \cup (3, \infty)$.

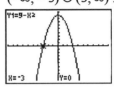 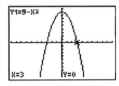

15. Graph $y = 5 - 4x^2 - 8x$. The $x$-intercepts appear to be $(-2.5, 0)$ and $(0.5, 0)$. The solution set of $5 - 4x^2 \geq 8x$ is $[-2.5, 0.5]$.

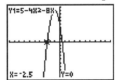 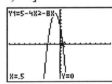

17. Graph $y = x(7-x) - 8$. The $x$-intercepts appear to be approximately $(1.44, 0)$ and $(5.56, 0)$. The solution set of $x(7-x) \geq 8$ is $[1.44, 5.56]$.

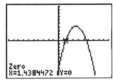 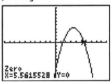

19. $(x-2)(x+5) > 8$
$(x-2)(x+5) - 8 > 0$

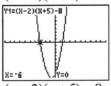

$(x-2)(x+5) - 8 = 0$
$x^2 + 3x - 18 = 0$
$(x+6)(x-3) = 0$
$x = -6, 3$

The solution of the inequality $(x-2)(x+5) > 8$ is $(-\infty, -6) \cup (3, \infty)$.

**21.**
$$x^2 \le 5x$$
$$x^2 - 5x \le 0$$

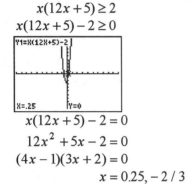

$$x^2 - 5x = 0$$
$$x(x - 5) = 0$$
$$x = 0, 5$$

The solution of the inequality $x^2 \le 5x$ is $[0, 5]$.

**23.**
$$x(12x + 5) \ge 2$$
$$x(12x + 5) - 2 \ge 0$$

$$x(12x + 5) - 2 = 0$$
$$12x^2 + 5x - 2 = 0$$
$$(4x - 1)(3x + 2) = 0$$
$$x = 0.25, -2/3$$

The solution of $x(12x + 5) \ge 2$ is $(-\infty, -0.67] \cup [0.25, \infty)$.

**25.**
$$2x(x - 2) < 7$$
$$2x(x - 2) - 7 < 0$$

$$2x(x - 2) - 7 = 0$$
$$2x^2 - 4x - 7 = 0$$
$$x = \frac{-(-4) \pm \sqrt{(-4)^2 - 4(2)(-7)}}{2(2)}$$
$$= \frac{4 \pm \sqrt{72}}{4} \approx -1.12, 3.12$$

The solution of $2x(x - 2) < 7$ is $(-1.12, 3.12)$.

**27.** "A number does not exceed 5" means that the number is less than or equal to 5. This can be translated as $x \le 5$, which is choice (b).

**29.** "A number is more than 5" can be translated as $x > 5$, which is choice (a).

**31.** "At least 6" means greater than or equal to 6.
$$x(x + 1) \ge 6$$
$$x^2 + x - 6 \ge 0$$
$$(x + 3)(x - 2) \ge 0$$

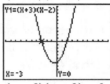

$$(x + 3)(x - 2) = 0$$
$$x = -3, 2$$

The solution to the inequality is $(-\infty, -3] \cup [2, \infty)$.

**33.** "Not more than 6" means less than or equal to 6.
$$x(x - 1) \le 6$$
$$x^2 - x - 6 \le 0$$
$$(x - 3)(x + 2) \le 0$$

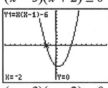

$$(x - 3)(x + 2) = 0$$
$$x = -2, 3$$

The solution to the inequality is $[-2, 3]$.

**35.** If an expression is negative, it is less than 0.
$$x(6x - 5) - 6 < 0$$
$$6x^2 - 5x - 6 < 0$$
$$(2x - 3)(3x + 2) < 0$$

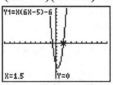

$$(2x - 3)(3x + 2) = 0$$
$$x = -2/3, 3/2$$

The solution to the inequality is $(-0.67, 1.5)$.

**37.** If an expression is not positive, it is less than or equal to 0.

$2x(2-x) + 7 \leq 0$

$4x - 2x^2 + 7 \leq 0$

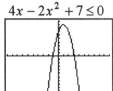

$4x - 2x^2 + 7 = 0$

$x = \dfrac{-(-4) \pm \sqrt{(-4)^2 - 4(2)(-7)}}{2(2)}$

$= \dfrac{4 \pm \sqrt{72}}{4} \approx -1.12, \; 3.12$

The solution to the inequality is $(-\infty, -1.12] \cup [3.12, \infty)$.

**39.** $g(x) = (4-x)(2+x)$

Graph $y = (4-x)(2+x)$. The x-intercepts are (4, 0) and (−2, 0). As the graph shows, $g(x) \geq 0$ for values of x in the interval $[-2, 4]$.

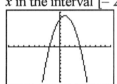

**41.** $g(x) = 4x^2 - 7x = x(4x - 7)$

Graph $y = 4x^2 - 7x$. The x-intercepts are (0, 0) and (1.75, 0). As the graph shows, $g(x) \geq 0$ for values of x in the interval $(-\infty, 0] \cup [1.75, \infty)$.

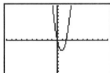

**43.** $f(x) = 16 - x^2 = (4-x)(4+x)$

Graph $y = 16 - x^2$. The x-intercepts are (−4, 0) and (4, 0). As the graph shows, $f(x) \leq 0$ for values of x in the interval $(-\infty, -4] \cup [4, \infty)$.

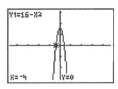

**45.** $g(x) = x(x-8) - 5 = x^2 - 8x - 5$

Graph $y = x(x-8) - 5$. From the graph, the estimated x-intercepts are (−0.58, 0) and (8.58, 0). As the graph shows, $g(x) \leq 0$ for values of x in the interval $[-0.58, 8.58]$.

 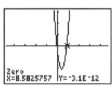

**47.** If the quadratic inequality has no x-intercepts, the possible solution sets are **R** or $\varnothing$.

**49.** Because $x - 2$ is a factor of $ax^2 + bx + c$, (2, 0) is an x-intercept of the graph of $f(x) = ax^2 + bx + c$. But the solutions of $ax^2 + bx + c > 0$ are represented by points of the graph that are above the x-axis. Thus, 2 is not a solution.

**51.** $4x^2 + 23x + 15 = (4x + 3)(x + 5) \geq 0$

Graph $y = 4x^2 + 23x + 15$. The x-intercepts are (−0.75, 0) and (−5, 0). The graph shows that the solution set is $(-\infty, -5] \cup [-0.75, \infty)$.

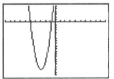

**53.** $5x - x^2 = x(5-x) < 0$

Graph $y = 5x - x^2$. The x-intercepts are (0, 0) and (5, 0). The graph shows that the solution set is $(-\infty, 0) \cup (5, \infty)$.

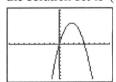

**55.** $x^2 + 3x + 6 < 0$

Graph $y = x^2 + 3x + 6$. The graph shows that the solution set is $\varnothing$.

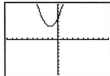

**57.** $4x^2 + 20x + 25 = (2x+5)^2 \leq 0$

Graph $y = 4x^2 + 20x + 25$. The $x$-intercept is $(-2.5, 0)$. The graph shows that the solution set is $\{-2.5\}$.

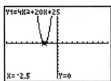

**59.** $x^2 - 4x \geq 4$
$x^2 - 4x - 4 \geq 0$

Graph $y = x^2 - 4x - 4$. From the Quadratic Formula, the $x$-intercepts are $(-0.83, 0)$ and $(4.83, 0)$. The graph shows that the solution set is $(-\infty, -0.83] \cup [4.83, \infty)$.

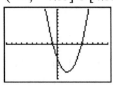

**61.** $x^2 + 9 \leq 6x$
$x^2 - 6x + 9 \leq 0$
$(x-3)^2 \leq 0$

Graph $y = x^2 - 6x + 9$. The $x$-intercept is $(3, 0)$. The graph shows that the solution set is $\{3\}$.

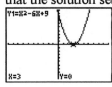

**63.** $2x^2 < 4x + 5$
$2x^2 - 4x - 5 < 0$

Graph $y = 2x^2 - 4x - 5$. From the Quadratic Formula, the $x$-intercepts are $(-0.87, 0)$ and $(2.87, 0)$. The graph shows that the solution set is $(-0.87, 2.87)$.

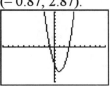

**65.** $x^2 - 6x + 5 = (x-1)(x-5) \geq 0$

Graph $y = x^2 - 6x + 5$. The $x$-intercepts are $(1, 0)$ and $(5, 0)$. The graph shows that the solution set is $(-\infty, 1] \cup [5, \infty)$.

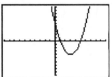

**67.** $0.5x - 0.25x^2 \leq 0.9$
$-0.25x^2 + 0.5x - 0.9 \leq 0$

Graph $y = -0.25x^2 + 0.5x - 0.9$. The graph shows that the solution set is **R**.

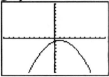

**69.** $x(6x+17) \leq -7$
$x(6x+17) + 7 \leq 0$
$6x^2 + 17x + 7 \leq 0$
$(3x+7)(2x+1) \leq 0$

Graph $y = 6x^2 + 17x + 7$. The $x$-intercepts are $(-7/3, 0)$ and $(-0.5, 0)$. The graph shows that the solution set is $[-2.33, -0.5]$.

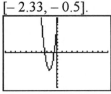

**71.** $x^2 + \dfrac{3}{4}x > \dfrac{5}{2}$
$4x^2 + 3x - 10 > 0$
$(x+2)(4x-5) > 0$

Graph $y = 4x^2 + 3x - 10$. The $x$-intercepts are $(-2, 0)$ and $(1.25, 0)$.

The graph shows that the solution set is $(-\infty, -2) \cup (1.25, \infty)$.

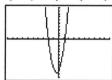

**73.** $3x(x+2) < 4$
$3x^2 + 6x - 4 < 0$
Graph $y = 3x^2 + 6x - 4$. From the Quadratic Formula, the $x$-intercepts are $(-2.53, 0)$ and $(0.53, 0)$. The graph shows that the solution set is $(-2.53, 0.53)$.

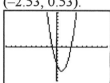

**75.** $(2x+1)(x+2) \le (x-3)(x+2)$
$2x^2 + 5x + 2 \le x^2 - x - 6$
$x^2 + 6x + 8 \le 0$
$(x+4)(x+2) \le 0$

Graph $y = x^2 + 6x + 8$. The $x$-intercepts are $(-4, 0)$ and $(-2, 0)$. The graph shows that the solution set is $[-4, -2]$.

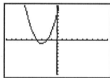

**77.** $2x^2 - 3x + k > 0$
Discriminant:
$b^2 - 4ac = (-3)^2 - 4(2)(k) = 9 - 8k$
For $k > \dfrac{9}{8} = 1.125$, the value of the discriminant is negative and there are no $x$-intercepts. Because $k$ is positive and $(0, k)$ is on the graph of $y = 2x^2 - 3x + k$, the graph is entirely above the $x$-axis and the solution set is the set of real numbers.

**79.** $9 + kx - x^2 > 0$
$-x^2 + kx + 9 > 0$
Discriminant:
$b^2 - 4ac = k^2 - 4(-1)(9) = k^2 + 36$
The value of the discriminant is never negative. There are no values of $k$ such that the solution set is the set of real numbers.

**81.** $x^2 + 5x + k < 0$
Discriminant:
$b^2 - 4ac = 5^2 - 4(1)(k) = 25 - 4k$
For $k \ge \dfrac{25}{4} = 6.25$, the value of the discriminant is nonpositive and there are either no $x$-intercepts or only one $x$-intercept. Because $k$ is positive and $(0, k)$ is on the graph of $y = x^2 + 5x + k$, the graph is entirely above or on the $x$-axis and the solution set is $\varnothing$.

**83.** $kx^2 + 8x - 5 > 0$
Discriminant:
$b^2 - 4ac = 8^2 - 4(k)(-5) = 64 + 20k$
For $k \le -\dfrac{64}{20} = -3.2$, the value of the discriminant is nonpositive and there are either no $x$-intercepts or only one $x$-intercept. Because $(0, -5)$ is on the graph of $y = kx^2 + 8x - 5$, the graph is entirely below or on the $x$-axis and the solution set is $\varnothing$.

**85.** $x^2 + 12x + k \le 0$
Discriminant:
$b^2 - 4ac = 12^2 - 4(1)(k) = 144 - 4k$
For $k = 36$, the value of the discriminant is zero and there is only one $x$-intercept. When $k = 36$, we have $x^2 + 12x + 36 \le 0$ and this results in $(x+6)^2 \le 0$. The only solution is $x = -6$ and the solution set is $\{-6\}$. So, $k = 36$ produces a quadratic inequality such that the real number solution set is a set with only one element.

**87.** $x^2 + kx + 9 \leq 0$
Discriminant:
$b^2 - 4ac = k^2 - 4(1)(9) = k^2 - 36$
For $k = \pm 6$, the value of the discriminant is zero and in each case there is only one x-intercept. So, $k = \pm 6$ produces a quadratic inequality such that the real number solution set is a set with only one element.

**89.** Let $x$ = the smaller number and $x + 5$ = the larger number.
$$x(x+5) \leq 6$$
$$x^2 + 5x - 6 \leq 0$$
$$(x+6)(x-1) \leq 0$$
Graph $y = x^2 + 5x - 6$. The solution set is $[-6, 1]$. The largest possible value of the smaller number is 1.

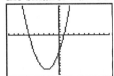

**91.** Let $x$ = the depth of the home. Assume that the front of the home is the minimum length of $x + 35$.

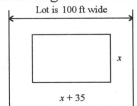

The ground-floor living space is given by $x(x + 35)$. Provision (c) stipulates that the area must be more than 2556 square feet.
$$x(x+35) > 2556$$
$$x^2 + 35x - 2556 > 0$$
$$(x+71)(x-36) > 0$$
Graph $y = x^2 + 35x - 2556$. The x-intercepts are $(-71, 0)$ and $(36, 0)$. The solution set is $(-\infty, -71) \cup (36, \infty)$.

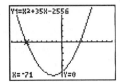

Because the negative values have no meaning in this context, the smallest allowable depth for the house is 36 feet. The minimum length for the front of the house is 71 feet. However, if the lot is 100 feet wide, this leaves a total of only $100 - 71 = 29$ feet, or an average of 14.5 feet, between the house and a property line on either side of the house. This violates provision (b) that no home can be less than 15 feet from any property line.

**93.**

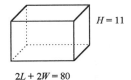

The volume, $LWH$, must be at least 2 cubic feet or 3456 cubic inches.
$$LWH \geq 3456$$
$$L(40-L)(11) \geq 3456$$
$$-11L^2 + 440L - 3456 \geq 0$$
Graph $y = -11x^2 + 440x - 3456$. From the Quadratic Formula, the x-intercepts are $(10.74, 0)$ and $(29.26, 0)$. The graph shows that the solution set is $(10.74, 29.26)$. The length of the package should be between 10.74 inches and 29.26 inches.

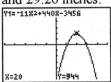

### Section 10.6  Quadratic Inequalities

**95.**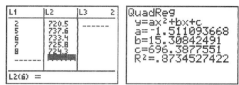

The model function is
$H(x) = -1.5x^2 + 15.3x + 696$.

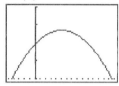

**97.** $-1.5x^2 + 15.3x + 696 = 680$
$-1.5x^2 + 15.3x + 16 = 0$
$$x = \frac{-15.3 \pm \sqrt{(15.3)^2 - 4(-1.5)(16)}}{2(-1.5)}$$
$$= \frac{-15.3 \pm \sqrt{330.09}}{-3} \approx -1, 11$$

**99.**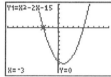

(a) The domain of
$f(x) = \sqrt{x^2 - 2x - 15}$ is the set of all $x$ for which the radicand is nonnegative. Therefore, we estimate the intervals for which the graph of $g(x) = x^2 - 2x - 15$ is on or above the $x$-axis: $(-\infty, -3] \cup [5, \infty)$.

(b) Because $h(x)$ is nonnegative for all $x$, the portion of the graph of $g$ that is below the $x$-axis is inverted to the other side of the $x$-axis.

**101.** $ax^2 + bx > 0$
Because $c = 0$, the discriminant is $b^2 > 0$. That means that there are two $x$-intercepts. Because we have strict inequality, the $x$-values of these $x$-intercepts must be excluded from the solution set and, therefore, the solution set is not the set of all real numbers.

**103.** $9x^2 + kx + 1 \geq 0$
Discriminant:
$b^2 - 4ac = k^2 - 4(9)(1) = k^2 - 36$
For $k^2 - 36 \leq 0$, the value of the discriminant is nonpositive and there is at most one $x$-intercept. Because $(0, 1)$ is on the graph of $y = 9x^2 + kx + 1$, the graph is entirely above or on the $x$-axis and the solution set is the set of real numbers. So if $-6 \leq k \leq 6$, the solution set of $9x^2 + kx + 1 \geq 0$ is **R**.

**105.** $-2x^2 + kx - 2 > 0$
Discriminant:
$b^2 - 4ac = k^2 - 4(-2)(-2) = k^2 - 16$
For $-4 \leq k \leq 4$, then the discriminant is nonpositive and there is at most one $x$-intercept. The graph of $y = -2x^2 + kx - 2$ contains the point $(0, -2)$ and, therefore, if $-4 \leq k \leq 4$, the graph is entirely below the $x$-axis or on the $x$-axis and the solution set for the inequality is $\varnothing$.

**107.** $2kx^2 + x + (k - 1) \leq 0$
Discriminant:
$b^2 - 4ac = 1^2 - 4(2k)(k-1) = 1 - 8k^2 + 8k$
If $-8k^2 + 8k + 1 = 0$, then there will be only one $x$-intercept. Use the Quadratic Formula to find the solutions of this equation.
$$k = \frac{-8 \pm \sqrt{8^2 - 4(-8)(1)}}{2(-8)}$$
$$= \frac{-8 \pm \sqrt{96}}{-16} = \frac{2 \pm \sqrt{6}}{4}$$
Choose either $k = \dfrac{2 + \sqrt{6}}{4}$ or
$k = \dfrac{2 - \sqrt{6}}{4}$ so that the graph is entirely above the $x$-axis and, therefore, the inequality is satisfied by only one value. The graph contains the point $(0, k - 1)$. So, we want $k - 1$ to be greater than

zero. Choosing $k = \dfrac{2+\sqrt{6}}{4}$ makes $k-1$ greater than zero. The value $k = \dfrac{2+\sqrt{6}}{4} \approx 1.11$ is such that the solution set of $2kx^2 + x + (k-1) \leq 0$ is a set with only one element.

## Section 10.7 Quadratic Functions

1. The graph is a parabola. For positive $a$ the graph opens upward, and for negative $a$ the graph opens downward. The graph becomes narrower as $|a|$ increases.

3. $f(x) = ax^2 + bx + c$
   $a = -2$, Vertex: (1, 3)
   Because $a < 0$, the parabola opens downward. The range of the function is $\{y | y \leq 3\}$.

5. $f(x) = ax^2 + bx + c$
   $a = \dfrac{3}{5}$, Vertex: (0, 2)
   Because $a > 0$, the parabola opens upward. The range of the function is $\{y | y \geq 2\}$.

7. (a) Because the axis of symmetry contains the vertex, the equation is $x = h$.
   (b) The first coordinate of the vertex is $a$.

9. Vertex: (0, –1)
   Because the first coordinate of the vertex is 0, the equation of the vertical axis of symmetry is $x = 0$.

11. Vertex: (3, 0)
    Because the first coordinate of the vertex is 3, the equation of the vertical axis of symmetry is $x = 3$.

13. The graph of $f(x) = x^2 - 4$ is the same as the graph of $y = x^2$ with the vertex shifted down 4 units. The function matches graph (e).

15. The graph of $h(x) = -2x^2$ is a parabola that opens down and is a narrower version of $y = -x^2$. The function matches graph (c).

17. The graph of $g(x) = -\dfrac{1}{2}x - 2$ is a line with slope – 0.5 and y-intercept (0, – 2). The function matches graph (b).

19. (a) The graph of $y = -x^2$ has the same vertex and width as $y = x^2$ but opens downward.
    (b) The graph of $y = -2x^2$ has the same vertex as $y = x^2$ but opens downward and is narrower.

21. (a) The graph of $y = (x+3)^2$ is the same width as $y = x^2$ but is shifted left 3 units.
    (b) The graph of $y = (x-3)^2$ is the same width as $y = x^2$ but is shifted right 3 units.

Section 10.7  Quadratic Functions

23. (a) The graph of $y = (x-3)^2 - 2$ is the same width as $y = x^2$ but is shifted right 3 units and down 2 units.

    (b) The graph of $y = (x+1)^2 + 4$ is the same width as $y = x^2$ but is shifted left 1 unit and up 4 units.

25. The x-coordinate is $-\dfrac{b}{2a}$. Determine the y-coordinate by evaluating the function at $-\dfrac{b}{2a}$.

27.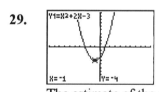

    The estimate of the vertex from the graph is (3, 4). Because $f(x) = (x-3)^2 + 4$ is in the form $(x-h)^2 + k$, the vertex may be found by inspection. The vertex is $(h, k) = (3, 4)$. The minimum of the function is 4.

29.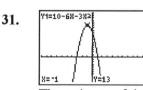

    The estimate of the vertex from the graph is $(-1, -4)$. The vertex of $h(x) = x^2 + 2x - 3$ may be found by evaluating $-\dfrac{b}{2a} = -\dfrac{2}{2(1)} = -1$ and $h(-1) = (-1)^2 + 2(-1) - 3 = -4$. The vertex is $(h, k) = (-1, -4)$. The minimum of the function is $-4$.

31. The estimate of the vertex from the graph is $(-1, 13)$. The vertex of $g(x) = 10 - 6x - 3x^2$ may be found by evaluating $-\dfrac{b}{2a} = -\dfrac{-6}{2(-3)} = -1$ and $g(-1) = 10 - 6(-1) - 3(-1)^2 = 13$. The vertex is $(h, k) = (-1, 13)$. The maximum of the function is 13.

33. Vertex: $(h, k)$. Because the range is $\{y | y \geq 3\}$, $k = 3$. Because the axis of symmetry is $x = 0$ and the axis of symmetry passes through the vertex, $h = 0$. So the vertex is $(0, 3)$.

35. Vertex: $(h, k)$. Because the range is $\{y | y \leq -1\}$, $k = -1$. Because the axis of symmetry is $x = 5$ and the axis of symmetry passes through the vertex, $h = 5$. So the vertex is $(5, -1)$.

37. $f(x) = x^2 - 6x + 9 = (x-3)^2$
    Vertex: $(h, k) = (3, 0)$
    Axis of symmetry: $x = 3$

39. $h(x) = 2x^2 + 4x - 7$
    Evaluate $-\dfrac{b}{2a} = -\dfrac{4}{2(2)} = -1$ and $h(-1) = 2(-1)^2 + 4(-1) - 7 = -9$.
    Vertex: $(h, k) = (-1, -9)$
    Axis of symmetry: $x = -1$

41. $h(x) = 7 - 4x - x^2$
    Evaluate $-\dfrac{b}{2a} = -\dfrac{-4}{2(-1)} = -2$ and $h(-2) = 7 - 4(-2) - (-2)^2 = 11$.
    Vertex: $(h, k) = (-2, 11)$
    Axis of symmetry: $x = -2$

43. Because the domain is **R**, the function is defined for $x = 0$. The y-intercept is $(0, c)$, where $c$ is the constant term of the function.

**45.** $f(x) = x^2 - 3x - 28$

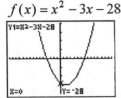

From the graph, the $y$-intercept appears to be $(0, -28)$ and the $x$-intercepts appear to be $(-4, 0)$ and $(7, 0)$.
Algebraically, $y$-intercept:
Because $f(0) = -28$, it is $(0, -28)$.
$x$-intercepts:
$$x^2 - 3x - 28 = 0$$
$$(x - 7)(x + 4) = 0$$
$$x = -4, 7$$
The $x$-intercepts are $(-4, 0)$ and $(7, 0)$.

**47.** $g(x) = 4x^2 - 5x + 9$

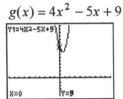

From the graph, the $y$-intercept appears to be $(0, 9)$ and there are no $x$-intercepts.
Algebraically, $y$-intercept:
Because $g(0) = 9$, it is $(0, 9)$.
$x$-intercepts:
$$4x^2 - 5x + 9 = 0$$
Because the discriminant is $b^2 - 4ac = -119$, there are no real solutions and, thus, no $x$-intercepts.

**49.** $f(x) = 5 - 4x - 3x^2$

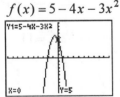

From the graph, the $y$-intercept appears to be $(0, 5)$ and the $x$-intercepts appear to be $(-2, 0)$ and $(1, 0)$.
Algebraically, $y$-intercept:
Because $f(0) = 5$, it is $(0, 5)$.
$x$-intercepts:
$$5 - 4x - 3x^2 = 0$$

$$x = \frac{-(-4) \pm \sqrt{(-4)^2 - 4(-3)(5)}}{2(-3)}$$
$$= \frac{4 \pm \sqrt{76}}{-6} \approx -2.12, \ 0.79$$
The $x$-intercepts are $(-2.12, 0)$ and $(0.79, 0)$.

**51.** $h(x) = 3x^2 + 5x$

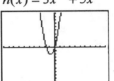

From the graph, the $y$-intercept appears to be $(0, 0)$ and the $x$-intercepts appear to be $(-2, 0)$ and $(0, 0)$.
Algebraically, $y$-intercept:
Because $h(0) = 0$, it is $(0, 0)$.
$x$-intercepts:
$$3x^2 + 5x = 0$$
$$x(3x + 5) = 0$$
$$x = -\frac{5}{3}, \ 0$$
The $x$-intercepts are $(-1.67, 0)$ and $(0, 0)$.

**53.** Because the discriminant is negative, there are no real solutions. Therefore, there are no $x$-intercepts.

**55.** Because the discriminant is positive, there are two real solutions. Therefore, there are two $x$-intercepts. Furthermore, because the discriminant is a perfect square, the $x$-coordinates of these intercepts are rational numbers.

**57.** Because $a = -3 < 0$, the parabola opens downward. Because the vertex $(-1, -2)$ is below the $x$-axis, there are no $x$-intercepts.

**59.** Because $a = \frac{1}{2} > 0$, the parabola opens upward. Because the vertex $(2, -4)$ is below the $x$-axis, there are two $x$-intercepts.

**61.** Because the vertex (2, 0) is on the x-axis, there is one x-intercept.

**63.** Because the vertex (−2, −5) is below the x-axis and the parabola opens upward, the graph has two x-intercepts.

**65.** Because the vertex (−3, 0) is on the x-axis, the graph has one x-intercept.

**67.** Because the maximum value occurs at (−4, 3), the parabola opens downward and the vertex is above the x-axis. Thus, the graph has two x-intercepts.

**69.** $f(x) = x^2 - 6x + 8 = (x-4)(x-2)$

(a) The domain of $f$ is **R**. The x-coordinate of the vertex is
$-\dfrac{b}{2a} = -\dfrac{-6}{2(1)} = 3$. The y-coordinate of the vertex is
$f(3) = (3)^2 - 6(3) + 8 = -1$.
Because the vertex is (3, −1) and $a > 0$, the range is $\{y \mid y \geq -1\}$.

(b) Because $f(0) = 8$, the y-intercept is (0, 8). The factored form of $f(x)$ shows that (2, 0) and (4, 0) are the x-intercepts.

(c) The vertex is (3, −1). The equation of the axis of symmetry is $x = 3$.

**71.** $f(x) = 5 - 4x - x^2 = (5+x)(1-x)$

(a) The domain of $f$ is **R**. The x-coordinate of the vertex is
$-\dfrac{b}{2a} = -\dfrac{-4}{2(-1)} = -2$. The y-coordinate of the vertex is
$f(-2) = 5 - 4(-2) - (-2)^2 = 9$.
Because the vertex is (−2, 9) and $a < 0$, the range is $\{y \mid y \leq 9\}$.

(b) Because $f(0) = 5$, the y-intercept is (0, 5). The factored form of $f(x)$ shows that (−5, 0) and (1, 0) are the x-intercepts.

(c) The vertex is (−2, 9). The equation of the axis of symmetry is $x = -2$.

**73.** $h(x) = 2x^2 + 4x + 9$

(a) The domain of $h$ is **R**. The x-coordinate of the vertex is
$-\dfrac{b}{2a} = -\dfrac{4}{2(2)} = -1$. The y-coordinate of the vertex is
$h(-1) = 2(-1)^2 + 4(-1) + 9 = 7$.
Because the vertex is (−1, 7) and $a > 0$, the range is $\{y \mid y \geq 7\}$.

(b) Because $h(0) = 9$, the y-intercept is (0, 9). Because the vertex is above the x-axis and the parabola opens upward, there are no x-intercepts.

(c) The vertex is (−1, 7). The equation of the axis of symmetry is $x = -1$.

**75.** If $f(x) = (x-h)^2 + k$ is up 3 units compared to $y = x^2$, then $k = 3$, $h = 0$, and $f(x) = (x-0)^2 + 3 = x^2 + 3$.

**77.** If $f(x) = (x-h)^2 + k$ is left 3/5 unit compared to $y = x^2$, then $h = -3/5$, $k = 0$, and $f(x) = \left(x + \tfrac{3}{5}\right)^2 + 0$.

**79.** If $f(x) = (x-h)^2 + k$ is up 2 units and left 4 units compared to $y = x^2$, then $k = 2$, $h = -4$, and $f(x) = (x+4)^2 + 2$.

**81.** If $f(x) = (x-h)^2 + k$ is right 3 units and up 2 units compared to $y = x^2$, then $k = 2$, $h = 3$, and $f(x) = (x-3)^2 + 2$.

**83.** If $f(x) = x^2 + bx + c$ has x-intercepts (5, 0) and (−3, 0) then
$f(x) = (x-5)(x+3) = x^2 - 2x - 15$
and $b = -2$ and $c = -15$.

**85.** If $f(x) = x^2 + bx + c$ has x-intercepts (0, 0) and (−2, 0) then
$f(x) = x(x+2) = x^2 + 2x$ and $b = 2$ and $c = 0$.

87. If $f(x) = x^2 + bx + c$ has only (4, 0) as an x-intercept then
$f(x) = (x-4)^2 = x^2 - 8x + 16$ and $b = -8$ and $c = 16$.

89. If $f(x) = x^2 + bx + c$ has (0, 3) as a y-intercept then $c = 3$. The x-coordinate of the vertex is $-\dfrac{b}{2a} = -\dfrac{b}{2}$. The axis of symmetry, with equation $x = 2$, passes through the vertex. So, $-\dfrac{b}{2} = 2$. Thus, $b = -4$. So $f(x) = x^2 - 4x + 3$.

91. If $f(x) = ax^2 + x + c$ has $(-1, 0)$ on its graph, $f(-1) = 0$ and, therefore, $0 = a - 1 + c$. Because the x-coordinate of the vertex is $-\dfrac{1}{6} = -\dfrac{b}{2a} = -\dfrac{1}{2a}$, $2a = 6$, and, so, $a = 3$. Substituting 3 for $a$ in $0 = a - 1 + c$ gives us that $c = -2$. So, $f(x) = 3x^2 + x - 2$.

93. Let $x =$ one of the numbers and $30 - x =$ the other number. The quadratic function describing their product is
$f(x) = x(30 - x) = -x^2 + 30x$
The maximum value of the function occurs at the vertex. The x-coordinate of the vertex is $x = -\dfrac{b}{2a} = -\dfrac{30}{2(-1)} = 15$.
So, the numbers are 15 and 15.

95. $h = 10 + 30t - 16t^2$
The maximum height of the dive is found at the vertex. The number of seconds after which the peak of the dive occurs is the x-coordinate of the vertex.
$x = -\dfrac{b}{2a} = -\dfrac{30}{2(-16)} = \dfrac{15}{16}$ second

97. Let $x =$ the number of $20 increases in rental price. Then the total income is given by
$f(x) = (140 + 20x)(80 - 5x)$
$= 11{,}200 + 900x - 100x^2$

The greatest income occurs at the vertex of the function's graph.
$x = -\dfrac{b}{2a} = -\dfrac{900}{2(-100)} = 4.5$
The greatest income occurs with 4.5 $20-increases in price. Therefore, the greatest income occurs at a price of $140 + 20(4.5) = \$230$.

99.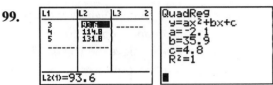
The quadratic model is
$f(x) = -2.1x^2 + 35.9x + 4.8$
Because $a = -2.1 < 0$, the parabola would open downward.

101. $f(x) = -2.1x^2 + 35.9x + 4.8$
Vertex:
$x = -\dfrac{b}{2a} = -\dfrac{35.9}{2(-2.1)} \approx 8.5$
$f(8.5) = -2.1(8.5)^2 + 35.9(8.5) + 4.8$
$\approx 158.2$
For the graph of $f$, the vertex is about (8.5, 158.2), which means that the highest number of accounts, 158.2 million, occurred in 1998. Because the graph is falling from left to right for years after 1998, the model predicts that the number of accounts will decrease.

103. We can use the quadratic regression feature of a calculator to find a model for these points. (Alternatively, we could substitute each point into the equation $f(x) = ax^2 + bx + c$ to obtain three equations in three unknowns, as we did in Chapter 5).
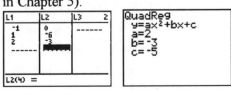
The function is $f(x) = 2x^2 - 3x - 5$.

Vertex:

$$x = -\frac{b}{2a} = -\frac{-3}{2(2)} = 0.75$$

$f(0.75) = 2(0.75)^2 - 3(0.75) - 5 = -6.125$

The vertex is $(0.75, -6.125)$.

**105.** We can use the quadratic regression feature of a calculator to find a model for these points.

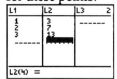 

The function is $f(x) = x^2 + x + 1$.

Vertex:

$$x = -\frac{b}{2a} = -\frac{1}{2(1)} = -0.5$$

$f(-0.5) = (-0.5)^2 + (-0.5) + 1 = 0.75$

The vertex is $(-0.5, 0.75)$.

**107.** $I(x) = (700 + 25x)(30 - x)$
$= 21{,}000 + 50x - 25x^2$

(a) $I(0) = 21{,}000 + 50(0) - 25(0)^2$
$= \$21{,}000$

(b) Vertex:

$$x = -\frac{b}{2a} = -\frac{50}{2(-25)} = 1$$

$I(1) = 21{,}000 + 50(1) - 25(1)^2$
$= 21{,}025$

The vertex is $(1, 21{,}025)$. The income is maximized if one store is vacant.

(c) It makes sense to only consider positive income values from the function. Because the function factors as $(700 + 25x)(30 - x)$, to have only positive incomes, the maximum number of vacancies is 30. So, there must be spaces for 30 stores in the mall.

**109.**
$$f(x) = ax^2 + bx + c$$
$$f(x) - c = ax^2 + bx + c - c$$
$$f(x) - c = ax^2 + bx$$
$$\frac{f(x)}{a} - \frac{c}{a} = \frac{ax^2}{a} + \frac{bx}{a}$$
$$\frac{f(x)}{a} - \frac{c}{a} = x^2 + \frac{bx}{a}$$
$$\frac{f(x)}{a} - \frac{c}{a} + \left(\frac{b}{2a}\right)^2 = x^2 + \frac{bx}{a} + \left(\frac{b}{2a}\right)^2$$
$$\frac{f(x)}{a} - \frac{c}{a} + \frac{b^2}{4a^2} = x^2 + \frac{bx}{a} + \frac{b^2}{4a^2}$$
$$\frac{f(x)}{a} - \frac{4ac}{4a^2} + \frac{b^2}{4a^2} = \left(x + \frac{b}{2a}\right)^2$$
$$\frac{f(x)}{a} - \frac{4ac - b^2}{4a^2} = \left(x + \frac{b}{2a}\right)^2$$
$$\frac{f(x)}{a} = \left(x + \frac{b}{2a}\right)^2 + \frac{4ac - b^2}{4a^2}$$
$$f(x) = a\left(x + \frac{b}{2a}\right)^2 + \frac{4ac - b^2}{4a}$$

With the function written in this form, we see that $h = -\dfrac{b}{2a}$ and $k = \dfrac{4ac - b^2}{4a}$.

## Chapter 10 Project

1. In the table, the column for $Y_1$ gives the target heart rates for women and column $Y_2$ gives the target heart rates for men.

3. (a) A man who is 5 feet 11 inches tall is 71 inches tall.
$$W(h) = 0.077h^2 - 7.61h + 310.11$$
$$W(71) = 0.077(71)^2 - 7.61(71) + 310.11$$
$$\approx 158 \text{ pounds}$$

   (b) Because the ideal weight function for women is $W(h) = 2.95h - 57.32$, the slope of this linear function gives the increase in weight for each additional inch in height. Thus, the ideal weight increases by about 3 pounds per inch of additional height.

5. For women age 60: $C(60) = 1.17(60) + 147.15 = 217.35$
   For men age 60: $C(60) = 0.76(60) + 171.4 = 217$
   At age 60, the cholesterol levels of men and women are about the same.

## Chapter 10 Review Exercises

1. (a) For one graph, write the equation in standard form, graph, and estimate the $x$-coordinates of the $x$-intercepts.
   (b) For two graphs, graph the left and right sides of the equation and estimate the $x$-coordinates of the points of intersection.

3. $13 - 3z^2 = 0$
   Graph $y = 13 - 3x^2$. From the graph the $x$-intercepts appear to be $(-2, 0)$ and $(2, 0)$. The actual $x$-intercepts are approximately $(\pm 2.08, 0)$. The actual solutions are $z = \pm\sqrt{\dfrac{13}{3}} \approx \pm 2.08$.

5. $(d-3)(d+2) = 36$
   $d^2 - d - 6 - 36 = 0$
   $d^2 - d - 42 = 0$
   $(d-7)(d+6) = 0$
   $d - 7 = 0$ or $d + 6 = 0$
   $d = 7$ or $d = -6$

7. $x(x+2) = 15$
   $x^2 + 2x - 15 = 0$
   $(x+5)(x-3) = 0$
   $x + 5 = 0$ or $x - 3 = 0$
   $x = -5$ or $x = 3$

9. $25(b+4)^2 - 16 = 0$
   $25(b+4)^2 = 16$
   $(b+4)^2 = \dfrac{16}{25}$
   $b + 4 = \pm\sqrt{\dfrac{16}{25}}$
   $b = -4 \pm \dfrac{4}{5} = -4.8, -3.2$

11. $(b-2)^2 + 37 = 0$
    $(b-2)^2 = -37$
    $b - 2 = \pm\sqrt{-37}$
    $b = 2 \pm i\sqrt{37} \approx 2 \pm 6.08i$

13. After dividing both sides by 2, we obtain an $x$-coefficient of 5/2. To complete the square, add the square of half the $x$-coefficient to both sides. This number is 25/16.

**15.** 
$$x^2 + 10x - 1 = 0$$
$$x^2 + 10x = 1$$
$$x^2 + 10x + 5^2 = 1 + 5^2$$
$$(x+5)^2 = 26$$
$$x + 5 = \pm\sqrt{26}$$
$$x = -5 \pm \sqrt{26}$$
$$x \approx -10.10, \; 0.10$$

**17.** 
$$x^2 - 6x + 10 = 0$$
$$x^2 - 6x = -10$$
$$x^2 - 6x + 9 = -10 + 9$$
$$(x-3)^2 = -1$$
$$x - 3 = \pm\sqrt{-1}$$
$$x = 3 \pm i\sqrt{1} = 3 \pm i$$

**19.** 
$$x^2 - 5x + 3 = 0$$
$$x^2 - 5x = -3$$
$$x^2 - 5x + \frac{25}{4} = -3 + \frac{25}{4}$$
$$\left(x - \frac{5}{2}\right)^2 = 3.25$$
$$x - \frac{5}{2} = \pm\sqrt{3.25}$$
$$x = \frac{5}{2} \pm \sqrt{3.25} \approx 0.70, \; 4.30$$

**21.** 
$$3 - x - 2x^2 = 0$$
$$(3 + 2x)(1 - x) = 0$$
$$3 + 2x = 0 \quad \text{or} \quad 1 - x = 0$$
$$x = -1.5 \quad \text{or} \quad x = 1$$

**23.** 
$$1 = 2x^2 + 6x$$
$$x^2 + 3x = 0.5$$
$$x^2 + 3x + \frac{9}{4} = 0.5 + \frac{9}{4}$$
$$\left(x + \frac{3}{2}\right)^2 = 2.75$$
$$x + \frac{3}{2} = \pm\sqrt{2.75}$$
$$x = -\frac{3}{2} \pm \sqrt{2.75} \approx -3.16, \; 0.16$$

**25.** All of the equations can be solved with the Quadratic Formula or by completing the square. Parts (i) and (iv) can be solved with the Square Root Property. Parts (i) and (ii) can be solved by factoring.

**27.** 
$$5x^2 - 16 = 0$$
$$a = 5, \; b = 0, \; c = -16$$
$$x = \frac{-0 \pm \sqrt{0^2 - 4(5)(-16)}}{2(5)}$$
$$= \frac{\pm\sqrt{320}}{10} \approx \pm 1.79$$

**29.** 
$$4x^2 = 5x + 3$$
$$4x^2 - 5x - 3 = 0$$
$$a = 4, \; b = -5, \; c = -3$$
$$x = \frac{-(-5) \pm \sqrt{(-5)^2 - 4(4)(-3)}}{2(4)}$$
$$= \frac{5 \pm \sqrt{73}}{8} \approx -0.44, \; 1.69$$

**31.** 
$$2x^2 + 9 = x$$
$$2x^2 - x + 9 = 0$$
$$a = 2, \; b = -1, \; c = 9$$
$$x = \frac{-(-1) \pm \sqrt{(-1)^2 - 4(2)(9)}}{2(2)}$$
$$= \frac{1 \pm \sqrt{-71}}{4} \approx 0.25 \pm 2.11i$$

**33.** 
$$4x = x^2 - 12$$
$$x^2 - 4x - 12 = 0$$
$$a = 1, \; b = -4, \; c = -12$$
Discriminant:
$$b^2 - 4ac = (-4)^2 - 4(1)(-12) = 64$$
Because 64 is positive and a perfect square, there are two rational solutions.

**35.** 
$$6x^2 - 2x + 5 = 0$$
$$a = 6, \; b = -2, \; c = 5$$
Discriminant:
$$b^2 - 4ac = (-2)^2 - 4(6)(5) = -116$$
Because the discriminant is negative, there are two imaginary solutions.

**37.** All are in quadratic form.

(a) $u = \dfrac{1}{x}$; $2u^2 + 3u - 5 = 0$

(b) $u = \sqrt{2x}$; $u^2 + u - 3 = 0$

(c) $u = x + 1$; $u^2 + 2u + 1 = 0$

(d) $u = x^{1/4}$; $u - u^2 + 3 = 0$

**39.** The restricted values are $-1$ and $2$.

$$\dfrac{x}{x+1} = 2 + \dfrac{2}{x-2}$$
$$x(x-2) = 2(x+1)(x-2) + 2(x+1)$$
$$x^2 - 2x = 2x^2 - 2x - 4 + 2x + 2$$
$$0 = x^2 + 2x - 2$$
$$x = \dfrac{-2 \pm \sqrt{2^2 - 4(1)(-2)}}{2(1)}$$
$$= \dfrac{-2 \pm \sqrt{12}}{2} \approx -2.73,\ 0.73$$

**41.**
$$\sqrt{1-2x} = x + 1$$
$$\left(\sqrt{1-2x}\right)^2 = (x+1)^2$$
$$1 - 2x = x^2 + 2x + 1$$
$$0 = x^2 + 4x$$
$$0 = x(x+4)$$
$$x = 0,\ -4$$

A check shows that $-4$ is an extraneous solution. The only solution is 0.

**43.**
$$x - 3x^{1/2} = 10$$
$$u^2 - 3u - 10 = 0 \quad \text{Let } u = x^{1/2}.$$
$$(u - 5)(u + 2) = 0$$
$$u = -2,\ 5$$

$x^{1/2} = -2 \quad$ or $\quad x^{1/2} = 5$
$x = 4 \qquad\qquad\qquad x = 25$

A check shows that 4 is an extraneous solution. The only solution is 25.

**45.**
$$(x^2 + x)^2 + 72 = 18(x^2 + x)$$
$$u^2 - 18u + 72 = 0$$
$$(u - 12)(u - 6) = 0$$
$$u = 6,\ 12$$
$$x^2 + x = 6$$
$$x^2 + x - 6 = 0$$
$$(x + 3)(x - 2) = 0$$
$$x = -3,\ 2$$

or
$$x^2 + x = 12$$
$$x^2 + x - 12 = 0$$
$$(x + 4)(x - 3) = 0$$
$$x = -4,\ 3$$

A check shows that all of these values are valid solutions. The solutions are $-4, -3, 2,$ and $3$.

**47.** Restricted values: $-5, 2$

$$\dfrac{x}{x-2} = \dfrac{2}{x+5}$$
$$x(x+5) = 2(x-2)$$
$$x^2 + 5x = 2x - 4$$
$$x^2 + 3x + 4 = 0$$
$$x = \dfrac{-3 \pm \sqrt{3^2 - 4(1)(4)}}{2(1)}$$
$$= \dfrac{-3 \pm \sqrt{-7}}{2} \approx -1.5 \pm 1.32i$$

**49.**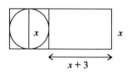

The total area of the two tablecloths is
$$\pi\left(\dfrac{x}{2}\right)^2 + x(x + 3) = 90$$
$$\dfrac{\pi}{4}x^2 + x^2 + 3x - 90 = 0$$
$$\left(\dfrac{\pi}{4} + 1\right)x^2 + 3x - 90 = 0$$
$$x = \dfrac{-3 \pm \sqrt{3^2 - 4\left(\frac{\pi}{4} + 1\right)(-90)}}{2\left(\frac{\pi}{4} + 1\right)}$$
$$\approx -7.99,\ 6.31$$

The dimensions of the necessary piece of fabric are 6.31 by 15.62. Approximately 10.95 square yards of material are needed.

**51.** Let $x$ = the number of children in the family.
$$6 + \frac{60}{x} + \frac{56}{x+3} + 5 = 30$$
$$\frac{60}{x} + \frac{56}{x+3} = 19$$
$$60(x+3) + 56x = 19x(x+3)$$
$$60x + 180 + 56x = 19x^2 + 57x$$
$$19x^2 - 59x - 180 = 0$$
$$x = \frac{-(-59) \pm \sqrt{(-59)^2 - 4(19)(-180)}}{2(19)}$$
$$\approx -1.89, \ 5$$

The negative value has no meaning in this context. There are 5 children in the family.

**53.** Let $x$ = the number of hours for the slower worker to wash the windows alone. Then $x - 5$ = the number of hours it takes the faster worker to wash the windows alone.
$$\frac{1}{x} + \frac{1}{x-5} = \frac{1}{6}$$
$$6(x-5) + 6x = x(x-5)$$
$$6x - 30 + 6x = x^2 - 5x$$
$$x^2 - 17x + 30 = 0$$
$$(x-15)(x-2) = 0$$
$$x = 2, \ 15$$

The value 2 has no meaning in this context. The slower worker can wash the windows in 15 hours when working alone.

**55.** The graph of $f(x) = x^2 + 5x + 6$ is below the $x$-axis in the $x$-interval $(-3, -2)$. Therefore, the graph of $f$ is on or above the $x$-axis for $x \leq -3$ and for $x \geq -2$. Thus the solution set is $(-\infty, -3] \cup [-2, \infty)$.

**57.** $x(x-6) + 9 \leq 0$
$$x^2 - 6x + 9 \leq 0$$
$$(x-3)^2 \leq 0$$

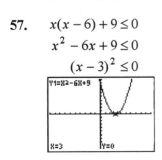

The $x$-intercept is $(3, 0)$
The solution of the inequality $x(x-6) + 9 \leq 0$ is $\{3\}$.

**59.** $x^2 + 5 \geq 3x$
$x^2 - 3x + 5 \geq 0$

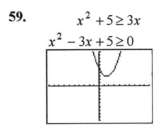

From the graph, we can see that the inequality $x^2 - 3x + 5 \geq 0$ is true for all $x$. Thus, the solution set is **R**.

**61.** $x - x^2 - 1 > 0$

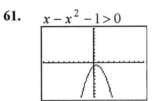

From the graph, we can see that the inequality $x - x^2 - 1 > 0$ is never true for any value of $x$. Thus, the solution set is $\varnothing$.

**63.** $f(x) = 8 + 3x - 2x^2 > 0$

Graph $y = 8 + 3x - 2x^2$. The $x$-intercepts are approximately $(-1.39, 0)$ and $(2.89, 0)$.

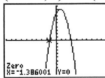 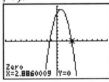

The function is positive for values of $x$ in the interval $(-1.39, 2.89)$.

**65.** $3x^2 + 2x + k > 0$
Discriminant:
$b^2 - 4ac = 2^2 - 4(3)(k) = 4 - 12k$

For $k > \frac{1}{3}$, the value of the discriminant is negative and there are no $x$-intercepts. Because $k$ is positive and $(0, k)$ is on the graph of $y = 2x^2 - 3x + k$, the graph is entirely above the $x$-axis and the solution set is the set of real numbers.

**67.** The graph shows that the domain is **R**. The range can be estimated by first locating the vertex $(h, k)$. If the parabola opens upward, the range is $\{y | y \geq k\}$; if the parabola opens downward, the range is $\{y | y \leq k\}$.

**69.** $f(x) = x^2 - 5x + 9$
Vertex:
$$x = -\frac{b}{2a} = -\frac{-5}{2(1)} = 2.5$$
$$f(2.5) = (2.5)^2 - 5(2.5) + 9 = 2.75$$
The vertex is (2.5, 2.75).
*y*-intercept:
Substitute 0 for *x* and solve for *y*.
$$f(0) = 0^2 - 5(0) + 9 = 9$$
The *y*-intercept is (0, 9).
*x*-intercept:
Set $x^2 - 5x + 9$ equal to 0; solve for *x*.
$$0 = x^2 - 5x + 9$$
$$b^2 - 4ac = (-5)^2 - 4(1)(9) = -11$$
Because the discriminant is negative, there are no real solutions and, thus, no *x*-intercepts.

**71.** $f(x) = 17 - x^2 = -x^2 + 17$
Because $a = -1 < 0$, the parabola opens downward.
$$x = -\frac{b}{2a} = -\frac{0}{2(-1)} = 0$$
$$f(0) = 17 - 0^2 = 17$$
The vertex is (0, 17).
The range is $\{y | y \leq 17\}$.

**73.** $h(x) = 2x^2 - 5x$
Because $a = 2 > 0$, the parabola opens upward.
$$x = -\frac{b}{2a} = -\frac{-5}{2(2)} = 1.25$$
$$h(1.25) = 2(1.25)^2 - 5(1.25) = -3.125$$
The vertex is $(1.25, -3.125)$.
The range is $\{y | y \geq -3.125\}$.

**75.** Let $W$ = the width of the rectangle and $L$ = the length. The perimeter is $2W + 2L = 250$ and $W = 125 - L$.
The area function is:
$$A(L) = L(125 - L) = 125L - L^2$$
The maximum area occurs at the vertex of the graph of this function.
$$L = -\frac{b}{2a} = -\frac{125}{2(-1)} = 62.5$$
The maximum area occurs when the length is 62.5 feet.

## Chapter 10 Looking Back

**1.** $f(x) = \dfrac{x+2}{x-3}$
To find the domain, find the values for which the function is defined. Therefore, find and exclude the values that make the denominator zero.
$$x - 3 = 0$$
$$x = 3$$
The domain of the function is $\{x | x \neq 3\}$.

**3.** If every vertical line intersects the graph at most once, the graph represents a function.

**5.** (a) $x + 3 + x^2 + 4x - 3 = x^2 + 5x$
(b) $3t + 1 - (4 - 6t) = 3t + 1 - 4 + 6t$
$= 9t - 3$

**7.** $f(x) = \sqrt{5 - x}$
$f(-4) = \sqrt{5 - (-4)} = \sqrt{9} = 3$

**9.** $f(x) = 3x - 7$
$f(t + 4) = 3(t + 4) - 7$
$= 3t + 12 - 7 = 3t + 5$

**11.** $g(x) = 0.01x^3 + x$

(a) $g(3) = 3.27$

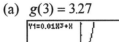

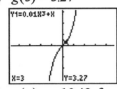

(b) $g(x) = -10.43$ for $x = -7$

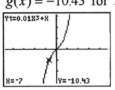

(c) $g(x) = 8.16$ for $x = 6$

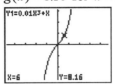

# Chapter 10 Test

**1.**
$$3t^2 + 4t = 1$$
$$t^2 + \frac{4}{3}t = \frac{1}{3}$$
$$t^2 + \frac{4}{3}t + \frac{4}{9} = \frac{1}{3} + \frac{4}{9}$$
$$\left(t + \frac{2}{3}\right)^2 = \frac{7}{9}$$
$$t + \frac{2}{3} = \pm\sqrt{\frac{7}{9}}$$
$$t = -\frac{2}{3} \pm \frac{\sqrt{7}}{3} \approx -1.55, \ 0.22$$

**3.**
$$(2x-1)^2 - 5 = 0$$
$$(2x-1)^2 = 5$$
$$2x - 1 = \pm\sqrt{5}$$
$$2x = 1 \pm \sqrt{5}$$
$$x = \frac{1 \pm \sqrt{5}}{2} \approx -0.62, \ 1.62$$

**5.**
$$(2y+1)(y+3) = 2$$
$$2y^2 + 7y + 3 = 2$$
$$2y^2 + 7y + 1 = 0$$
$$y = \frac{-7 \pm \sqrt{7^2 - 4(2)(1)}}{2(2)}$$
$$= \frac{-7 \pm \sqrt{41}}{4} \approx -3.35, \ -0.15$$

**7.**
$$9x^4 + 7x^2 = 2$$
$$9u^2 + 7u - 2 = 0 \quad \text{Let } u = x^2.$$
$$(9u - 2)(u + 1) = 0$$
$$u = -1, \ 2/9$$

$$x^2 = -1$$
$$x = \pm\sqrt{-1} \quad \text{or}$$
$$= \pm i$$
$$x^2 = \frac{2}{9}$$
$$x = \pm\sqrt{\frac{2}{9}}$$
$$= \pm\frac{\sqrt{2}}{3} \approx \pm 0.47$$

**9.**
$$A = \frac{1}{2}bh$$
$$20 = \frac{1}{2}(h+3)h$$
$$40 = h^2 + 3h$$
$$0 = h^2 + 3h - 40$$
$$0 = (h+8)(h-5)$$
$$h = -8, \ 5$$

The height of the triangle is 5 feet and the base of the triangle is 8 feet.

**11.** We want to maximize the volume. The length is 100 feet, the height is $x$, and the width is $12 - 2x$.
$$V(x) = 100x(12 - 2x)$$
$$= 1200x - 200x^2$$
The volume will be maximized at the vertex of the graph of this function.
$$x = -\frac{b}{2a} = -\frac{1200}{2(-200)} = 3$$
The height should be 3 inches to maximize the capacity.

**13.**
$$6 + x - x^2 \leq 0$$
$$(3 - x)(2 + x) \leq 0$$
Graph $y = 6 + x - x^2$. The $x$-intercepts are $(-2, 0)$ and $(3, 0)$. The solution set is $(-\infty, -2] \cup [3, \infty)$.

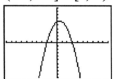

**15.** Using the given graph to answer the questions, we have:
(a) The domain of the function is **R**. The range is $\{y | y \geq -4\}$.
(b) Because the parabola opens upward, we know that the coefficient $a$ positive: $a > 0$.
(c) From the graph, we can estimate that the $x$-intercepts are $(2, 0)$ and $(6, 0)$. Thus, the estimated solutions of the equation $ax^2 + bx + c = 0$ are 2 and 6.

# Chapter 11

# Exponential and Logarithmic Functions

## Section 11.1 Algebra of Functions

1. Evaluate $f$ and $g$ at 1 and add the results, or algebraically add $f$ and $g$ and then evaluate the resulting function at 1.

3. $g(x) = |2x^2 - x - 1|$
   The domain is **R**.

5. $f(x) = \sqrt{1 - 2x}$
   The domain is the set of numbers that make the radicand nonnegative.
   $$1 - 2x \geq 0$$
   $$-2x \geq -1$$
   $$x \leq 0.5$$
   The domain is $(-\infty, 0.5]$.

7. $h(x) = \dfrac{x+4}{x^3 + 3x^2 - 4x} = \dfrac{x+4}{x(x+4)(x-1)}$
   The domain excludes any numbers that make the denominator 0. Thus the domain is **R**, $x \neq -4, 0, 1$.

9. $g(x) = \sqrt{x^2 - 16}$
   The domain is the set of numbers that make the radicand nonnegative.
   $$x^2 - 16 \geq 0$$
   $$(x+4)(x-4) \geq 0$$
   Graph $y = x^2 - 16$. The $x$-intercepts are $(-4, 0)$ and $(4, 0)$.

   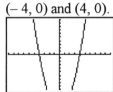

   The domain is $(-\infty, -4] \cup [4, \infty)$.

11. $f(x) = 3x, \ g(x) = x - 1$
    (a) $(f + g)(-1) = f(-1) + g(-1)$
    $= 3(-1) + (-1) - 1 = -5$
    (b) $(f - g)(1) = f(1) - g(1)$
    $= 3(1) - (1 - 1) = 3$
    (c) $(fg)(0) = f(0) \cdot g(0)$
    $= 3(0) \cdot (0 - 1) = 0$
    (d) $\left(\dfrac{f}{g}\right)(0) = \dfrac{f(0)}{g(0)} = \dfrac{3(0)}{0 - 1} = 0$

13. $f(x) = |2x + 1|, \ g(x) = x^2$
    (a) $(f + g)(-1) = f(-1) + g(-1)$
    $= |2(-1) + 1| + (-1)^2$
    $= 2$
    (b) $(f - g)(1) = f(1) - g(1)$
    $= |2(1) + 1| - (1)^2 = 2$
    (c) $(fg)(0) = f(0) \cdot g(0)$
    $= |2(0) + 1| \cdot 0^2 = 0$
    (d) $\left(\dfrac{f}{g}\right)(0) = \dfrac{f(0)}{g(0)}$ is not defined because $g(0)$ is 0.

15. $f(x) = 2^x, \ g(x) = 2^{-x}$
    (a) $(f + g)(-1) = f(-1) + g(-1)$
    $= 2^{-1} + 2^{-(-1)} = 2.5$
    (b) $(f - g)(1) = f(1) - g(1)$
    $= 2^1 - 2^{-(1)} = 1.5$
    (c) $(fg)(0) = f(0) \cdot g(0)$
    $= 2^0 \cdot 2^{-0} = 1$

(d) $\left(\dfrac{f}{g}\right)(0) = \dfrac{f(0)}{g(0)} = \dfrac{2^0}{2^{-0}} = 1$

17. $f(x) = -4x$, $g(x) = 3 - 2x$
    (a) $(f+g)(-3) = f(-3) + g(-3)$
    $= -4(-3) + 3 - 2(-3)$
    $= 21$
    (b) $(f-g)(4) = f(4) - g(4)$
    $= -4(4) - [3 - 2(4)]$
    $= -11$
    (c) $(fg)(0) = f(0) \cdot g(0)$
    $= -4(0) \cdot [3 - 2(0)] = 0$
    (d) $\left(\dfrac{f}{g}\right)(1) = \dfrac{f(1)}{g(1)} = \dfrac{-4(1)}{3 - 2(1)} = -4$

19. $f(x) = |x - 3|$, $g(x) = 2x - 3$
    (a) $(f+g)(2) = f(2) + g(2)$
    $= |2 - 3| + 2(2) - 3 = 2$
    (b) $(f-g)(-2) = f(-2) - g(-2)$
    $= |-2 - 3| - [2(-2) - 3]$
    $= 12$
    (c) $(fg)(0) = f(0) \cdot g(0)$
    $= |0 - 3| \cdot [2(0) - 3] = -9$
    (d) $\left(\dfrac{f}{g}\right)(-1) = \dfrac{f(-1)}{g(-1)} = \dfrac{|-1 - 3|}{2(-1) - 3} = -0.8$

21. $f(x) = x^2 - 2$, $g(x) = \sqrt[3]{x}$
    (a) $(f+g)(0) = f(0) + g(0)$
    $= 0^2 - 2 + \sqrt[3]{0} = -2$
    (b) $(f-g)(1) = f(1) - g(1)$
    $= 1^2 - 2 - \sqrt[3]{1} = -2$
    (c) $\left(\dfrac{f}{g}\right)(27) = \dfrac{f(27)}{g(27)}$
    $= \dfrac{27^2 - 2}{\sqrt[3]{27}} = \dfrac{727}{3}$
    (d) $(fg)(8) = f(8) \cdot g(8)$
    $= (8^2 - 2) \cdot \sqrt[3]{8} = 124$

23. No, 0 is in the domain of $h$, but 0 is not in the domain of $f/g$ because $g(0) = 0$.

25. $f(x) = x^2 - 9$, $g(x) = x + 3$
    (a) $(f-g)(x) = f(x) - g(x)$
    $= x^2 - 9 - (x + 3)$
    $= x^2 - x - 12$, **R**
    (b) $(f+g)(x) = f(x) + g(x)$
    $= x^2 - 9 + x + 3$
    $= x^2 + x - 6$, **R**
    (c) $\left(\dfrac{f}{g}\right)(x) = \dfrac{f(x)}{g(x)}$
    $= \dfrac{x^2 - 9}{x + 3} = x - 3$, **R**, $x \neq -3$
    (d) $(fg)(x) = f(x) \cdot g(x)$
    $= (x^2 - 9)(x + 3)$
    $= x^3 + 3x^2 - 9x - 27$, **R**

27. $f(x) = x^2 + 4$, $g(x) = x^2 - 4$
    (a) $\left(\dfrac{f}{g}\right)(x) = \dfrac{f(x)}{g(x)}$
    $= \dfrac{x^2 + 4}{x^2 - 4}$, **R**, $x \neq -2, 2$
    (b) $\left(\dfrac{g}{f}\right)(x) = \dfrac{g(x)}{f(x)} = \dfrac{x^2 - 4}{x^2 + 4}$, **R**
    (c) $\left(\dfrac{f}{f}\right)(x) = \dfrac{f(x)}{f(x)} = 1$, **R**
    (d) $\left(\dfrac{g}{g}\right)(x) = \dfrac{g(x)}{g(x)}$
    $= \dfrac{x^2 - 4}{x^2 - 4} = 1$, **R**, $x \neq -2, 2$

29. $f(x) = \dfrac{1}{x + 2}$, $g(x) = \dfrac{1}{x - 3}$
    (a) $(f - g)(x) = f(x) - g(x)$
    $= \dfrac{1}{x + 2} - \dfrac{1}{x - 3}$
    $= \dfrac{x - 3 - (x + 2)}{(x + 2)(x - 3)}$
    $= \dfrac{-5}{(x + 2)(x - 3)}$;
    **R**, $x \neq -2, 3$

Section 11.1   Algebra of Functions

(b) $(f+g)(x) = f(x) + g(x)$
$= \dfrac{1}{x+2} + \dfrac{1}{x-3}$
$= \dfrac{x-3+x+2}{(x+2)(x-3)}$
$= \dfrac{2x-1}{(x+2)(x-3)};$
$\mathbf{R}, x \neq -2, 3$

(c) $\left(\dfrac{f}{g}\right)(x) = \dfrac{f(x)}{g(x)}$
$= \dfrac{\dfrac{1}{x+2}}{\dfrac{1}{x-3}}$
$= \dfrac{x-3}{x+2}, \mathbf{R}, x \neq -2, 3$

(d) $(fg)(x) = f(x) \cdot g(x)$
$= \dfrac{1}{x+2} \cdot \dfrac{1}{x-3}$
$= \dfrac{1}{(x+2)(x-3)};$
$\mathbf{R}, x \neq -2, 3$

**31.** $f(x) = x^2 - 4, \ g(x) = x - 1$
$(f - g)(2t) = f(2t) - g(2t)$
$= (2t)^2 - 4 - (2t - 1)$
$= 4t^2 - 4 - 2t + 1$
$= 4t^2 - 2t - 3$

**33.** $f(x) = x^2 - 4, \ g(x) = x - 1$

$\left(\dfrac{f}{g}\right)\left(\dfrac{t}{2}\right) = \dfrac{f\left(\dfrac{t}{2}\right)}{g\left(\dfrac{t}{2}\right)} = \dfrac{\left(\dfrac{t}{2}\right)^2 - 4}{\left(\dfrac{t}{2}\right) - 1} = \dfrac{\dfrac{t^2}{4} - 4}{\dfrac{t}{2} - 1}$

$= \dfrac{4 \cdot \left(\dfrac{t^2}{4} - 4\right)}{4 \cdot \left(\dfrac{t}{2} - 1\right)} = \dfrac{t^2 - 16}{2t - 4}$

**35.** $f(x) = x + 5, \ g(x) = x^2 - 3$
$(f + g)(3t) = f(3t) + g(3t)$
$= 3t + 5 + (3t)^2 - 3$
$= 9t^2 + 3t + 2$

**37.** $f(x) = x + 5, \ g(x) = x^2 - 3$
$(fg)(-2t) = f(-2t) \cdot g(-2t)$
$= (-2t + 5)[(-2t)^2 - 3]$
$= (-2t + 5)(4t^2 - 3)$
$= -8t^3 + 20t^2 + 6t - 15$

**39.** Produce the graphs of $f$ and $g$ and note the intersection of the $x$-values for which $f$ and $g$ are defined.

**41.** $f(x) = \sqrt{2x-3}, \ g(x) = \sqrt{5-x}$
The domain of both (a) $(f+g)(x)$ and (b) $(f-g)(x)$ is the intersection of the domains of $f$ and $g$. The domain is $[1.5, 5]$.

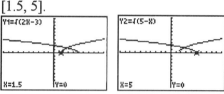

**43.** $f(x) = \sqrt{8-x}, \ g(x) = \sqrt{x-1}$

(a) The domain of $\left(\dfrac{f}{g}\right)(x)$ is the intersection of the domains of $f$ and $g$ and also excludes 1 because that makes the denominator of the combined function 0. The domain is $(1, 8]$.

(b) The domain of $(fg)(x)$ is the intersection of the domains of $f$ and $g$. The domain is $[1, 8]$.

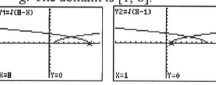

**45.** First evaluate $g(5)$ and then evaluate $f(g(5))$.

**47.** $f(x) = x - 3, \ g(x) = 2 - x^2$

(a) $g(1) = 2 - (1)^2 = 1$
$(f \circ g)(1) = f(g(1)) = f(1)$
$= 1 - 3 = -2$

(b) $g(0) = 2 - (0)^2 = 2$
$(f \circ g)(0) = f(g(0)) = f(2)$
$= 2 - 3 = -1$

(c) $f(3) = 3 - 3 = 0$
$(g \circ f)(3) = g(f(3)) = g(0)$
$\qquad = 2 - 0^2 = 2$
(d) $f(-1) = -1 - 3 = -4$
$(g \circ f)(-1) = g(f(-1)) = g(-4)$
$\qquad = 2 - (-4)^2 = -14$

**49.** $f(x) = x^2,\ g(x) = 2x - 3$
(a) $f(-3) = (-3)^2 = 9$
$(g \circ f)(-3) = g(f(-3)) = g(9)$
$\qquad = 2(9) - 3 = 15$
(b) $g(-0.5) = 2(-0.5) - 3 = -4$
$(f \circ g)(-0.5) = f(g(-0.5)) = f(-4)$
$\qquad = (-4)^2 = 16$
(c) $g(1.5) = 2(1.5) - 3 = 0$
$(g \circ g)(1.5) = g(g(1.5)) = g(0)$
$\qquad = 2(0) - 3 = -3$
(d) $f(-1) = (-1)^2 = 1$
$(f \circ f)(-1) = f(f(-1)) = f(1)$
$\qquad = 1^2 = 1$

**51.** (a) $g(2) = 1$,
$(f \circ g)(2) = f(g(2)) = f(1) = 5$
(b) $f(-1) = 3$,
$(g \circ f)(-1) = g(f(-1)) = g(3) = -2$
(c) $g(-3) = 2$,
$(f \circ g)(-3) = f(g(-3)) = f(2) = -4$

**53.** $f(x) = 3 - x^2,\ g(x) = 2x + 1$
$(f \circ g)(x) = f(g(x)) = f(2x + 1)$
$\qquad = 3 - (2x + 1)^2$
$\qquad = 3 - (4x^2 + 4x + 1)$
$\qquad = -4x^2 - 4x + 2$
$(g \circ f)(x) = g(f(x)) = g(3 - x^2)$
$\qquad = 2(3 - x^2) + 1$
$\qquad = 6 - 2x^2 + 1$
$\qquad = -2x^2 + 7$

**55.** $f(x) = x^2 - 3x + 5,\ g(x) = 3x - 2$
$(f \circ g)(x) = f(g(x)) = f(3x - 2)$
$\qquad = (3x - 2)^2 - 3(3x - 2) + 5$
$\qquad = 9x^2 - 12x + 4 - 9x + 6 + 5$
$\qquad = 9x^2 - 21x + 15$

$(g \circ f)(x) = g(f(x)) = g(x^2 - 3x + 5)$
$\qquad = 3(x^2 - 3x + 5) - 2$
$\qquad = 3x^2 - 9x + 15 - 2$
$\qquad = 3x^2 - 9x + 13$

**57.** $f(x) = x^3,\ g(x) = \sqrt[3]{x - 2}$
$(f \circ g)(x) = f(g(x)) = f(\sqrt[3]{x - 2})$
$\qquad = (\sqrt[3]{x - 2})^3$
$\qquad = x - 2$
$(g \circ f)(x) = g(f(x)) = g(x^3)$
$\qquad = \sqrt[3]{x^3 - 2}$

**59.** $f(x) = 5,\ g(x) = 2x - 3$
$(f \circ g)(x) = f(g(x)) = f(2x - 3)$
$\qquad = 5$
$(g \circ f)(x) = g(f(x)) = g(5)$
$\qquad = 2(5) - 3 = 7$

**61.** $f(x) = 3x - 4,\ g(x) = x + 5$
$(f \circ f)(x) = f(f(x)) = f(3x - 4)$
$\qquad = 3(3x - 4) - 4$
$\qquad = 9x - 16$
$(g \circ g)(x) = g(g(x)) = g(x + 5)$
$\qquad = (x + 5) + 5 = x + 10$

**63.** $f(x) = |x - 3|,\ g(x) = x^2 + 3$
$(f \circ f)(x) = f(f(x)) = f(|x - 3|)$
$\qquad = ||x - 3| - 3|$
$(g \circ g)(x) = g(g(x)) = g(x^2 + 3)$
$\qquad = (x^2 + 3)^2 + 3$
$\qquad = x^4 + 6x^2 + 9 + 3$
$\qquad = x^4 + 6x^2 + 12$

**65.** $f(x) = 2x^2 - 3x,\ g(x) = 2 - x$
$(f \circ f)(x) = f(f(x)) = f(2x^2 - 3x)$
$\qquad = 2(2x^2 - 3x)^2 - 3(2x^2 - 3x)$
$\qquad = 2(4x^4 - 12x^3 + 9x^2)$
$\qquad\quad - 6x^2 + 9x$
$\qquad = 8x^4 - 24x^3 + 12x^2 + 9x$
$(g \circ g)(x) = g(g(x)) = g(2 - x)$
$\qquad = 2 - (2 - x) = x$

Section 11.1   Algebra of Functions

**67.** $f(x) = 3$, $g(x) = 2$
$(f \circ f)(x) = f(f(x)) = f(3) = 3$
$(g \circ g)(x) = g(g(x)) = g(2) = 2$

**69.** $f(x) = 2x + 3$, $g(x) = \dfrac{x-3}{2}$
$(f \circ g)(x) = f(g(x)) = f\left(\dfrac{x-3}{2}\right)$
$= 2\left(\dfrac{x-3}{2}\right) + 3$
$= x - 3 + 3 = x$

**71.** $f(x) = \dfrac{x}{3} + 2$, $g(x) = 3x - 6$
$(f \circ g)(x) = f(g(x)) = f(3x - 6)$
$= \dfrac{3x-6}{3} + 2 = x - 2 + 2 = x$

**73.** $f(x) = x + 3$; $g(x) = 2x$; $h(x) = x^2$
$F(x) = 2x + 6 = 2(x + 3) = 2f(x)$
$= g(f(x)) = (g \circ f)(x)$

**75.** $f(x) = x + 3$; $g(x) = 2x$; $h(x) = x^2$
$H(x) = 4x = 2(2x) = 2g(x)$
$= g(g(x)) = (g \circ g)(x)$

**77.** $f(x) = x + 3$; $g(x) = 2x$; $h(x) = x^2$
$G(x) = x^2 + 3 = h(x) + 3$
$= f(h(x)) = (f \circ h)(x)$

**79.** $f(x) = x + 3$; $g(x) = 2x$; $h(x) = x^2$
$F(x) = 2x^2 = 2h(x) = g(h(x))$
$= (g \circ h)(x)$

**81.** Distance: $d(t) = 54t$; Gals: $g(d) = \dfrac{d}{25}$
Cost per gallon: $c(g) = 1.03g$

(a) Number of gallons in terms of $t$:
$(g \circ d)(t) = g(d(t)) = \dfrac{54t}{25}$

(b) Total fuel cost in terms of $t$:
$c(g(d(t))) = 1.03\left(\dfrac{54t}{25}\right)$

(c) When $t = 12$:
$c(g(d(12))) = 1.03\left(\dfrac{54 \cdot 12}{25}\right) \approx \$26.70$

**83.** (a) The number of people who attend the gathering: $f(x) = x$

(b) Per-person cost: $c(x) = 10 - 0.05x$

(c) Total cost of the gathering:
$T(x) = (fc)(x) = x(10 - 0.05x)$
$= 10x - 0.05x^2$

(d) When $x = 84$:
$T(84) = 10(84) - 0.05(84)^2 = \$487.20$

**85.** The composition of $F$ and $G$ is the number of transactions as a function of the number of years since 1990.

**87.** $F(187) = -0.55(187)^2 + 194(187) - 5879$
$\approx 11{,}166$
The actual entry is 11,160, which is the number (in millions) of ATM transactions when the number of terminals was 187 thousand.

**89.** Let $h(x) = x^2 - 5x + 6$ and $g(x) = \sqrt{x}$.
$f(x) = \sqrt{x^2 - 5x + 6} = \sqrt{h(x)} = g(h(x))$
Thus, $f = (g \circ h)$.

**91.** Let $h(x) = 3x + 2$ and $g(x) = x^4$.
$f(x) = (3x + 2)^4 = [h(x)]^4 = g(h(x))$
Thus, $f = (g \circ h)$.

**93.** Let $h(x) = |2x + 5|$ and $g(x) = x + 8$.
$f(x) = |2x + 5| + 8 = h(x) + 8 = g(h(x))$
Thus, $f = (g \circ h)$.

## Section 11.2   Inverse Functions

1. The set is not necessarily a one-to-one function. If the set has any ordered pairs with the same first coordinate, then the set is not even a function.

3. No, the function is not one-to-one because the ordered pairs (5, 2) and (3, 2) are different ordered pairs with the same second coordinate.

5. Yes, the function is one-to-one because no two ordered pairs have the same second coordinate.

7. No, by the Horizontal Line Test, this is not a one-to-one function because there exist horizontal lines that intersect the graph more than once.

9. No, by the Horizontal Line Test, this is not a one-to-one function because there exist horizontal lines that intersect the graph more than once.

11. Yes, by the Horizontal Line Test, this is a one-to-one function because no horizontal line intersects the graph more than once.

13. $f(x) = x^3 - 3x^2 + 6x + 4$
    Yes, by the Horizontal Line Test, this function is a one-to-one function.
    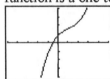

15. $f(x) = |x - 7| - |x + 5|$
    No, by the Horizontal Line Test, this function is not one-to-one.
    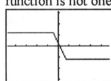

17. $f(x) = x^3 2^x$
    No, by the Horizontal Line Test, this function is not one-to-one.

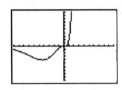

19. $f = \{(1, 2), (3, 4), (5, 6), (7, -1)\}$
    To write the inverse function, interchange the $x$- and $y$-coordinates of each ordered pair.
    $f^{-1} = \{(2, 1), (4, 3), (6, 5), (-1, 7)\}$
    The domain of $f^{-1}$ is $\{2, 4, 6, -1\}$.
    The range of $f^{-1}$ is $\{1, 3, 5, 7\}$.

21. $f = \{(3, 2), (4, 3), (-5, 1), (-7, -2)\}$
    To write the inverse function, interchange the $x$- and $y$-coordinates of each ordered pair.
    $f^{-1} = \{(2, 3), (3, 4), (1, -5), (-2, -7)\}$
    The domain of $f^{-1}$ is $\{2, 3, 1, -2\}$.
    The range of $f^{-1}$ is $\{3, 4, -5, -7\}$.

23. $f(3) = 4$, $f(5) = 7$, $f(-2) = -3$, $f(-4) = -5$;
    Because $f(-2) = -3$, we know that $(-2, -3)$ belongs to $f$. So, $(-3, -2)$ belongs to $f^{-1}$ and $f^{-1}(-3) = -2$.

25. $f(3) = 4$, $f(5) = 7$, $f(-2) = -3$, $f(-4) = -5$;
    Because $f(3) = 4$, we know that $(3, 4)$ belongs to $f$. So, $(4, 3)$ belongs to $f^{-1}$ and $f^{-1}(4) = 3$.

27. $f(x) = 2^x$
    From the graph, $f(6) = 64$. So, $f^{-1}(64) = 6$.

## Section 11.2  Inverse Functions

**29.** $f(x) = 2^x$

From the graph, $f(-4) = 0.0625$. So,

$f^{-1}(0.0625) = -4$.

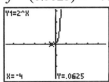

**31.** $f(x) = 2^{-x}$

From the graph, $f(-5) = 32$. So,

$f^{-1}(32) = -5$.

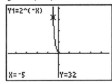

**33.** $f(x) = 2^{-x}$

From the graph, $f(3) = 0.125$. So,

$f^{-1}(0.125) = 3$.

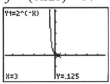

**35.** No; the $x^{-1}$ key returns the reciprocal of $x$, not the inverse function. In this case, $f(x) = f^{-1}(x) = x$.

**37.** Graph $y_1 = f(x) = 2x + 1$ and

$y_2 = g(x) = \dfrac{x-1}{2}$ simultaneously. Yes,

it is reasonable to believe that the functions may be inverses.

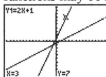

**39.** Graph $y_1 = f(x) = x^2 + 1$ and

$y_2 = g(x) = 1 - x^2$ simultaneously. No, it is not reasonable to believe that the functions are inverses.

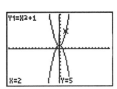

**41.** Graph $y_1 = f(x) = x^3 - 2$ and

$y_2 = g(x) = \sqrt[3]{x+2}$ simultaneously. Yes, it is reasonable to believe that the functions may be inverses.

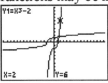

**43.**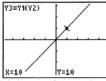

Yes, because $(f \circ g)(x) = x$ for each $x$, $f$ and $g$ are inverses.

**45.**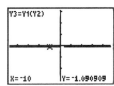

No, because $(f \circ g)(x) \neq x$ for each $x$, $f$ and $g$ are not inverses.

**47.**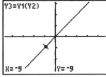

Yes, because $(f \circ g)(x) = x$ for each $x$, $f$ and $g$ are inverses.

**49.** Graph $y = 2^x$ with a calculator. Then sketch the graph of the inverse function by using the fact that the line $y = x$ is the line of symmetry.

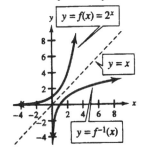

**51.** Graph $y = x^3 - 1$ with a calculator. Then sketch the graph of the inverse function by using the fact that the line $y = x$ is the line of symmetry.

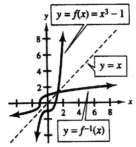

**53.** The graph of a nonconstant linear function is a nonvertical line that passes the Horizontal Line Test. Therefore, the function has an inverse.

**55.** $f(x) = 2x$, $g(x) = -2x$
$(f \circ g)(x) = f(g(x)) = f(-2x)$
$= 2(-2x) = -4x$
Because $(f \circ g)(x) \neq x$, $f$ and $g$ are not inverses.

**57.** $f(x) = \sqrt[3]{x+1}$, $g(x) = x^3 - 1$
$(f \circ g)(x) = f(g(x))$
$= f(x^3 - 1)$
$= \sqrt[3]{(x^3 - 1) + 1}$
$= \sqrt[3]{x^3} = x$
$(g \circ f)(x) = g(f(x))$
$= g(\sqrt[3]{x+1})$
$= (\sqrt[3]{x+1})^3 - 1$
$= x + 1 - 1 = x$
Because $(f \circ g)(x) = (g \circ f)(x) = x$, $f$ and $g$ are inverses.

**59.** $f(x) = \dfrac{3}{x}$, $g(x) = \dfrac{3}{x}$
$(f \circ g)(x) = f(g(x))$
$= f\left(\dfrac{3}{x}\right)$
$= \dfrac{3}{\frac{3}{x}} = x$

$(g \circ f)(x) = g(f(x))$
$= g\left(\dfrac{3}{x}\right)$
$= \dfrac{3}{\frac{3}{x}} = x$
Because $(f \circ g)(x) = (g \circ f)(x) = x$, $f$ and $g$ are inverses.

**61.** $f(x) = 3x + 5$
$y = 3x + 5$
$x = 3y + 5$
$x - 5 = 3y$
$\dfrac{x-5}{3} = y$
$f^{-1}(x) = \dfrac{x-5}{3}$
Verification:
$(f \circ f^{-1})(x) = 3\left(\dfrac{x-5}{3}\right) + 5 = x$
$(f^{-1} \circ f)(x) = \dfrac{(3x+5)-5}{3} = x$

**63.** $f(x) = -2x$
$y = -2x$
$x = -2y$
$-0.5x = y$
$f^{-1}(x) = -0.5x$
Verification:
$(f \circ f^{-1})(x) = -2(-0.5x) = x$
$(f^{-1} \circ f)(x) = -0.5(-2x) = x$

**65.** $f(x) = \dfrac{x}{3} + 5$
$y = \dfrac{x}{3} + 5$
$x = \dfrac{y}{3} + 5$
$x - 5 = \dfrac{y}{3}$
$3(x - 5) = y$
$f^{-1}(x) = 3(x - 5)$
Verification:
$(f \circ f^{-1})(x) = \dfrac{3(x-5)}{3} + 5 = x$

Section 11.2   Inverse Functions

$(f^{-1} \circ f)(x) = 3\left[\left(\dfrac{x}{3}+5\right)-5\right] = x$

**67.**
$$f(x) = \dfrac{1}{x}$$
$$y = \dfrac{1}{x}$$
$$x = \dfrac{1}{y}$$
$$xy = 1$$
$$y = \dfrac{1}{x}$$
$$f^{-1}(x) = \dfrac{1}{x}$$
Verification:
$$(f \circ f^{-1})(x) = \dfrac{1}{1/x} = x$$
$$(f^{-1} \circ f)(x) = \dfrac{1}{1/x} = x$$

**69.**
$$f(x) = \dfrac{1}{x+2}$$
$$y = \dfrac{1}{x+2}$$
$$x = \dfrac{1}{y+2}$$
$$xy + 2x = 1$$
$$xy = 1 - 2x$$
$$y = \dfrac{1-2x}{x}$$
$$f^{-1}(x) = \dfrac{1}{x} - 2$$
Verification:
$$(f \circ f^{-1})(x) = \dfrac{1}{\dfrac{1}{x}-2+2} = x$$
$$(f^{-1} \circ f)(x) = \dfrac{1}{\dfrac{1}{x+2}} - 2 = x$$

**71.**
$$f(x) = \dfrac{x+1}{x-2}$$
$$y = \dfrac{x+1}{x-2}$$
$$x = \dfrac{y+1}{y-2}$$
$$xy - 2x = y + 1$$
$$xy - y = 2x + 1$$
$$y(x-1) = 2x + 1$$
$$y = \dfrac{2x+1}{x-1}$$
$$f^{-1}(x) = \dfrac{2x+1}{x-1}$$
Verification:
$$(f \circ f^{-1})(x) = \dfrac{\dfrac{2x+1}{x-1}+1}{\dfrac{2x+1}{x-1}-2} = x$$
$$(f^{-1} \circ f)(x) = \dfrac{2 \cdot \left(\dfrac{x+1}{x-2}\right)+1}{\left(\dfrac{x+1}{x-2}\right)-1} = x$$

**73.**
$$f(x) = x^3 + 1$$
$$y = x^3 + 1$$
$$x = y^3 + 1$$
$$x - 1 = y^3$$
$$\sqrt[3]{x-1} = y$$
$$f^{-1}(x) = \sqrt[3]{x-1}$$
Verification:
$$(f \circ f^{-1})(x) = \left(\sqrt[3]{x-1}\right)^3 + 1 = x$$
$$(f^{-1} \circ f)(x) = \sqrt[3]{(x^3+1)-1} = x$$

**75.**
$$f(x) = x^{-3}$$
$$y = x^{-3}$$
$$x = y^{-3}$$
$$x^{-1/3} = y$$
$$f^{-1}(x) = x^{-1/3}$$
Verification:
$$(f \circ f^{-1})(x) = (x^{-1/3})^{-3} = x$$
$$(f^{-1} \circ f)(x) = (x^{-3})^{-1/3} = x$$

**77.**
$$f(x) = \frac{\sqrt{x}}{2}$$
$$y = \frac{\sqrt{x}}{2}$$
$$x = \frac{\sqrt{y}}{2}$$
$$2x = \sqrt{y}$$
$$4x^2 = y, \quad x \geq 0$$
$$f^{-1}(x) = 4x^2, \quad x \geq 0$$
Verification:
$$(f \circ f^{-1})(x) = \frac{\sqrt{4x^2}}{2} = x$$
$$(f^{-1} \circ f)(x) = 4 \cdot \left(\frac{\sqrt{x}}{2}\right)^2 = x$$

**79.**
$$f(x) = \sqrt[3]{2-x}$$
$$y = \sqrt[3]{2-x}$$
$$x = \sqrt[3]{2-y}$$
$$x^3 = 2 - y$$
$$y = 2 - x^3$$
$$f^{-1}(x) = 2 - x^3$$
Verification:
$$(f \circ f^{-1})(x) = \sqrt[3]{2-(2-x^3)} = x$$
$$(f^{-1} \circ f)(x) = 2 - \left(\sqrt[3]{2-x}\right)^3 = x$$

**81.**
$$f(x) = \sqrt[5]{x} + 3$$
$$y = \sqrt[5]{x} + 3$$
$$x = \sqrt[5]{y} + 3$$
$$x - 3 = \sqrt[5]{y}$$
$$(x-3)^5 = y$$
$$f^{-1}(x) = (x-3)^5$$
Verification:
$$(f \circ f^{-1})(x) = \sqrt[5]{(x-3)^5} + 3 = x$$
$$(f^{-1} \circ f)(x) = (\sqrt[5]{x} + 3 - 3)^5 = x$$

**83.** The inverse of $f$ multiplies $x$ by 3 and then subtracts 2.

**85.** The inverse of $f$ takes the cube root of $x$ and then subtracts 1.

**87.** The pairs given in (i) and (ii) are not inverses because $(f \circ g)(x) \neq x$. In (iii), we have
$$(f \circ g)(x) = (g \circ f)(x) = \frac{5}{5/x} = x,$$
so $f(x) = \frac{5}{x}$ and $g(x) = \frac{5}{x}$ are inverses.

**89.** The first is a one-to-one function, but the second function is not.

**91.**
$$C(t) = \tfrac{5}{9}(t - 32)$$
$$y = \tfrac{5}{9}(t - 32)$$
$$t = \tfrac{5}{9}(y - 32)$$
$$\tfrac{9}{5}t = y - 32$$
$$\tfrac{9}{5}t + 32 = y$$
$$C^{-1}(t) = \tfrac{9}{5}t + 32$$
This inverse function converts Celsius temperatures to Fahrenheit.

**93.** A function that gives the value in dollars of $p$ British pounds is
$$d(p) = 1.50p$$
The inverse function is
$$p = \frac{d}{1.50},$$
which gives the value $p$ in pounds of $d$ dollars.

**95.** $C(s) = 1.3s - 23$

(a) From the graph, we can see that $C$ is a one-to-one function.

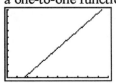

(b) $\dfrac{C + 23}{1.3} = s$. This inverse function represents the number of subscribers (in millions) as a function of the number of cell sites (in thousands).

Section 11.3  Exponential Functions

**97.**
$y(p) = 17.5p - 108$

**99.**
The functions appear to be inverse functions.

**101.**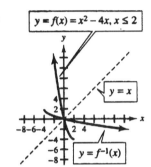

**103.**
$f(x) = x^2 + 1, \quad x \geq 0$
$y = x^2 + 1$
$x = y^2 + 1$
$x - 1 = y^2$
$\sqrt{x-1} = y, \quad x \geq 1$
$f^{-1}(x) = \sqrt{x-1}, \quad x \geq 1$
Verification:
$(f \circ f^{-1})(x) = \left(\sqrt{x-1}\right)^2 + 1 = x$
$(f^{-1} \circ f)(x) = \sqrt{(x^2+1) - 1} = x$

**105.** $f(x) = -|x+5|, \; x \leq -5$
Because $x \leq -5$, we have
$|x+5| = -(x+5)$ and
$-[-(x+5)] = x+5, \; x \leq -5$. The graph is in Quadrant III.
$y = x + 5$
$x = y + 5$
$x - 5 = y, \quad x \leq 0$
$f^{-1}(x) = x - 5, \quad x \leq 0$
Verification:
$(f \circ f^{-1})(x) = -|x - 5 + 5| = -(-x) = x$
$(f^{-1} \circ f)(x) = (x+5) - 5 = x$

## Section 11.3  Exponential Functions

**1.** (a) The graph rises from left to right if $b > 1$.
(b) The graph falls from left to right if $0 < b < 1$.

**3.** $f(x) = 2^{3x-1}$
$f(2) = 2^{3(2)-1} = 2^{6-1} = 2^5 = 32$

**5.** $g(x) = \left(\frac{1}{3}\right)^{2x}$
$g(1) = \left(\frac{1}{3}\right)^{2(1)} = \left(\frac{1}{3}\right)^2 = \frac{1}{9}$

**7.** $g(x) = \left(\frac{1}{3}\right)^{2x}$
$g(0) = \left(\frac{1}{3}\right)^{2(0)} = \left(\frac{1}{3}\right)^0 = 1$

**9.** $f(x) = 2^{3x-1}$
$f\left(\frac{2}{3}\right) = 2^{3(2/3)-1} = 2^{2-1} = 2^1 = 2$

**11.** $g(x) = e^{-2x}$
$g(2) = e^{-2(2)} = e^{-4} \approx 0.02$

**13.** $g(x) = e^{-2x}$
$g(-3) = e^{-2(-3)} = e^6 \approx 403.43$

**15.**
$h(x) = \left(\dfrac{1}{5}\right)^x$

$h\left(-\dfrac{3}{2}\right) = \left(\dfrac{1}{5}\right)^{-3/2} \approx 11.18$

**17.** $g(x) = e^{-2x}$

$g(-2) = e^{-2(-2)} = e^4 \approx 54.60$

**19.** $f(x) = 4e^{3x}$

$f(2) = 4e^{3(2)} = 4e^6 \approx 1613.72$

**21.** $g(x) = \left(\sqrt{3}\right)^{2x}$

$g(5) = \left(\sqrt{3}\right)^{2(5)} = (3^{1/2})^{10} = 3^5 = 243$

**23.** $f(x) = (0.23)^{3x-1}$

$f(-2) = (0.23)^{3(-2)-1}$

$= (0.23)^{-7} = 29{,}370.08$

**25.** $g(x) = -3e^{2x} + 4$

$g(-2) = -3e^{2(-2)} + 4 = -3e^{-4} + 4 \approx 3.95$

**27.** Because $3^{-x} = \left(\dfrac{1}{3}\right)^x = \dfrac{1}{3^x}$, functions (ii), (iii), and (iv) all have the same graph.

**29.** $f(x) = -e^x - 2$
Domain: **R**; Range: $\{y \mid y < -2\}$

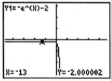

**31.** $f(x) = 2 - 3^{-x}$
Domain: **R**; Range: $\{y \mid y < 2\}$

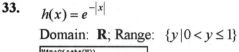

**33.** $h(x) = e^{-|x|}$
Domain: **R**; Range: $\{y \mid 0 < y \leq 1\}$

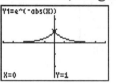

**35.**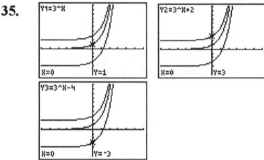

For $c > 0$ the graph is shifted up $c$ units, and for $c < 0$ the graph is shifted down $|c|$ units.

**37.** (a) The graph of $f(x) = 2^x + 1$ is the graph of $h(x) = 2^x$ shifted up 1 unit.

(b) The graph of $g(x) = 2^x - 3$ is the graph of $h(x) = 2^x$ shifted down 3 units.

**39.** (a) The graph of $f(x) = \left(\dfrac{1}{3}\right)^x + 3$ is the graph of $h(x) = \left(\dfrac{1}{3}\right)^x$ shifted up 3 units.

(b) The graph of $g(x) = \left(\dfrac{1}{3}\right)^{x+3}$ is the graph of $h(x)$ shifted left 3 units.

**41.** (a) The graph of $g(x) = b^x + 3$ is the graph of $f(x) = b^x$ shifted up 3 units. This matches graph (B).

(b) The graph of $h(x) = 4 - b^x$ is the graph of $f(x) = b^x$ reflected in the $x$-axis and shifted up 4 units. This matches graph (A).

**43.** From the graph, $h(x) = (2.3)^x = 0.76$ is true when $x \approx -0.33$.

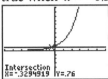

**45.** From the graph, $h(x) = (2.3)^x = 7.51$ is true when $x \approx 2.42$.

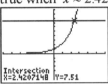

**47.** From the graph, we see that $f(x) = e^x = -3.45$ is never true.

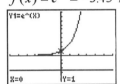

**49.** $e^x = 4$

Graph $y_1 = e^x$ and $y_2 = 4$. From the graph, we can estimate the solution of the equation as $x \approx 1.39$.

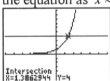

**51.** $(0.27)^x = 6$

Graph $y_1 = (0.27)^x$ and $y_2 = 6$. From the graph, we can estimate the solution of the equation as $x \approx -1.37$.

**53.** $e^x = -2x + 1$

Graph $y_1 = e^x$ and $y_2 = -2x + 1$. From the graph, we can estimate the solution of the equation as $x = 0$.

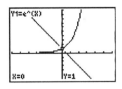

**55.** $e^x = x$

Graph $y_1 = e^x$ and $y_2 = x$. As we can see from the graph, the graphs do not intersect. Therefore, there is no solution to the equation $e^x = x$.

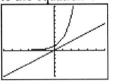

**57.** Both sides of the equation $4^x = 8$ can be written with the same base, and then the exponents can be equated. The two sides of $4^x = 12$ cannot be written with the same integer base.

**59.** $5^{-x} = 125$
$5^{-x} = 5^3$
$-x = 3$
$x = -3$

**61.** $\left(\dfrac{2}{5}\right)^x = \dfrac{25}{4}$

$\left(\dfrac{5}{2}\right)^{-x} = \dfrac{5^2}{2^2}$

$\left(\dfrac{5}{2}\right)^{-x} = \left(\dfrac{5}{2}\right)^2$

$-x = 2$
$x = -2$

**63.** $e^{3x+1} = e^{x-2}$
$3x + 1 = x - 2$
$2x = -3$
$x = -1.5$

**65.**
$$\frac{1}{9^x} = 27$$
$$\frac{1}{(3^2)^x} = 3^3$$
$$\frac{1}{3^{2x}} = 3^3$$
$$3^{-2x} = 3^3$$
$$-2x = 3$$
$$x = -1.5$$

**67.**
$$49^x = 343$$
$$(7^2)^x = 7^3$$
$$7^{2x} = 7^3$$
$$2x = 3$$
$$x = 1.5$$

**69.**
$$16 = 8^{x-2}$$
$$2^4 = (2^3)^{x-2}$$
$$2^4 = 2^{3(x-2)}$$
$$4 = 3(x-2)$$
$$4 = 3x - 6$$
$$10 = 3x$$
$$x = \frac{10}{3}$$

**71.**
$$3 \cdot 8^x = 12$$
$$3 \cdot 8^x = 3 \cdot 4$$
$$8^x = 4$$
$$(2^3)^x = 2^2$$
$$2^{3x} = 2^2$$
$$3x = 2$$
$$x = \frac{2}{3}$$

**73.**
$$9 \cdot 3^x = \frac{1}{3}$$
$$3^2 \cdot 3^x = 3^{-1}$$
$$3^{2+x} = 3^{-1}$$
$$2 + x = -1$$
$$x = -3$$

**75.**
$$2^{-x-3} = -4$$
Because an exponential function is never negative, there is no solution.

**77.**
$$\frac{5^{x+1}}{5^{1-x}} = \frac{1}{25^x}$$
$$5^{x+1-(1-x)} = 25^{-x}$$
$$5^{2x} = (5^2)^{-x}$$
$$5^{2x} = 5^{-2x}$$
$$2x = -2x$$
$$4x = 0$$
$$x = 0$$

**79.**
$$e^x + 4 = 3$$
$$e^x = -1$$
Because an exponential function is never negative, there is no solution.

**81.**
$$\left(\frac{1}{3}\right)^x 27^{x+1} = 9^{3-2x}$$
$$3^{-x} \cdot (3^3)^{x+1} = (3^2)^{3-2x}$$
$$3^{-x} \cdot 3^{3(x+1)} = 3^{2(3-2x)}$$
$$3^{-x+3(x+1)} = 3^{2(3-2x)}$$
$$-x + 3(x+1) = 2(3-2x)$$
$$-x + 3x + 3 = 6 - 4x$$
$$6x = 3$$
$$x = 0.5$$

**83.**
$$32^{1-x} = 8^{x^2-x}$$
$$(2^5)^{1-x} = (2^3)^{x^2-x}$$
$$2^{5(1-x)} = 2^{3(x^2-x)}$$
$$5(1-x) = 3(x^2-x)$$
$$5 - 5x = 3x^2 - 3x$$
$$0 = 3x^2 + 2x - 5$$
$$0 = (x-1)(3x+5)$$
$$x = -\frac{5}{3}, 1$$

**85.**
$$(0.25)^{2x+1} = 8^{2-x}$$
$$\left(\frac{1}{4}\right)^{2x+1} = 8^{2-x}$$
$$(2^{-2})^{2x+1} = (2^3)^{2-x}$$
$$2^{-2(2x+1)} = 2^{3(2-x)}$$
$$-2(2x+1) = 3(2-x)$$
$$-4x - 2 = 6 - 3x$$
$$-x = 8$$
$$x = -8$$

Section 11.3 Exponential Functions

**87.**
$$e^{x^2-1} = 7^0$$
$$e^{x^2-1} = 1$$
$$e^{x^2-1} = e^0$$
$$x^2 - 1 = 0$$
$$x^2 = 1$$
$$x = \pm\sqrt{1} = \pm 1$$

**89.**
$$(\sqrt{3})^{2-2x} = 9^{x^2}$$
$$(3^{1/2})^{2-2x} = (3^2)^{x^2}$$
$$3^{1-x} = 3^{2x^2}$$
$$1 - x = 2x^2$$
$$0 = 2x^2 + x - 1$$
$$0 = (2x-1)(x+1)$$
$$x = -1,\ 0.5$$

**91.** $t = 2$ years, $r = 0.09$, $A = \$5974.16$, $n = 4$

$$A = P\left(1 + \frac{r}{n}\right)^{nt}$$
$$5974.16 = P\left(1 + \frac{0.09}{4}\right)^{4(2)}$$
$$5974.16 = P(1 + 0.0225)^8$$
$$5974.16 = P(1.0225)^8$$
$$\frac{5974.16}{(1.0225)^8} = P$$
$$5000 \approx P$$

The person received $5000 in severance pay.

**93.** For the first scholarship fund:
$P = \$15{,}000$, $r = 0.10$, $n = 365$

$$A = P\left(1 + \frac{r}{n}\right)^{nt}$$
$$A = 15{,}000\left(1 + \frac{0.10}{365}\right)^{365t}$$

For the second scholarship fund:
$P = \$20{,}000$, $r = 0.06$, $n = 4$

$$A = P\left(1 + \frac{r}{n}\right)^{nt}$$
$$A = 20{,}000\left(1 + \frac{0.06}{4}\right)^{4t}$$
$$A = 20{,}000(1.015)^{4t}$$

Graph each equation to find the year in which the value of the funds is the same. We can estimate that the funds will have the same value (approximately $30,500) after about 7.12 years.

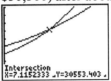

**95.** (a)

| Year | n | Expenditure |
|---|---|---|
| 1997 | 0 | 98.69 |
| 1998 | 1 | $98.69(1.06)^1$ |
| 1999 | 2 | $98.69(1.06)^2$ |
| 2000 | 3 | $98.69(1.06)^3$ |
| 2001 | 4 | $98.69(1.06)^4$ |
| 2002 | 5 | $98.69(1.06)^5$ |

(b) $E(n) = 98.69(1.06)^n$

(c) For 2002, $n = 5$:
$E(5) = 98.69(1.06)^5 \approx \$132.07$
For 2003, $n = 6$:
$E(8) = 98.69(1.06)^6 \approx \$139.99$

(d) From the graph, it appears that the average annual expenditure will reach $175 when $n \approx 10$, or approximately 2007.

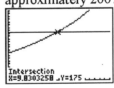

**97.**

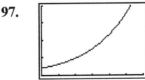

The graph rises rapidly from left to right.

**99.** Half the federal budget would be $500 billion. From the graph, we can estimate that the cost of Medicare would represent half of the federal budget in the year 2009.

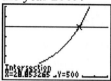

**101.** $6x \cdot 3^x = x + 1$

Graph $y_1 = 6x \cdot 3^x$ and $y_2 = x + 1$. From the graph, we can estimate the solutions of the equation as $x \approx -2.19$ and $x \approx 0.16$.

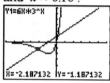 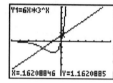

**103.** $5xe^x = -1$

Graph $y_1 = 5xe^x$ and $y_2 = -1$. From the graph, we can estimate the solutions of the equation as $x \approx -2.54$ and $x \approx -0.26$.

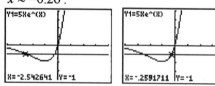

**105.** $e^x < e^{-x}$

Graph $y_1 = e^x$ and $y_2 = e^{-x}$. From the graph, we can see that the graphs intersect at (0, 1). The graph of $y_1 = e^x$ is below the graph of $y_2 = e^{-x}$ to the left of (0, 1). The solution set is $(-\infty, 0)$.

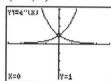 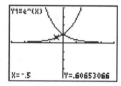

**107.** Graph $y = x^2 e^x$. The local minimum appears to be at (0, 0). The local maximum appears to be at $(-2, 0.54)$.

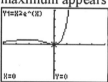 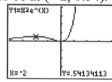

**109.** Graph $y = e^{|x|} + 3$. The local minimum appears to be at (0, 4).

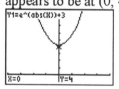

## Section 11.4  Logarithmic Functions

**1.** The logarithmic function is the inverse of the exponential function, which is not defined for $b \leq 0$ or for $b = 1$.

**3.** The equation $\log_7 49 = 2$ is equivalent to $7^2 = 49$ in exponential form.

**5.** The equation $\log 0.001 = -3$ is equivalent to $10^{-3} = 0.001$ in exponential form.

**7.** The equation $\ln \sqrt[3]{e} = \dfrac{1}{3}$ is equivalent to $e^{1/3} = \sqrt[3]{e}$ in exponential form.

**9.** The equation $\log_m n = 2$ is equivalent to $m^2 = n$ in exponential form.

**11.** The equation $4^3 = 64$ is equivalent to $\log_4 64 = 3$ in logarithmic form.

Section 11.4  Logarithmic Functions

**13.** The equation $10^{-5} = 0.00001$ is equivalent to $\log 0.00001 = -5$ in logarithmic form.

**15.** The equation $\left(\dfrac{1}{2}\right)^{-3} = 8$ is equivalent to $\log_{1/2} 8 = -3$ in logarithmic form.

**17.** The equation $P = e^{rt}$ is equivalent to $\ln P = rt$ in logarithmic form.

**19.**
$$f(x) = e^x$$
$$y = e^x$$
$$x = e^y$$
$$\ln x = y$$
$$f^{-1}(x) = \ln x$$

**21.**
$$f(x) = 4^{-x}$$
$$y = 4^{-x}$$
$$x = 4^{-y}$$
$$x = \left(\dfrac{1}{4}\right)^y$$
$$\log_{1/4} x = y$$
$$f^{-1}(x) = \log_{1/4} x$$

**23.**
$$f(x) = \log x$$
$$y = \log x$$
$$x = \log y$$
$$10^x = y$$
$$f^{-1}(x) = 10^x$$

**25.**
$$f(x) = \log_{1.5} x$$
$$y = \log_{1.5} x$$
$$x = \log_{1.5} y$$
$$1.5^x = y$$
$$f^{-1}(x) = 1.5^x$$

**27.** Let $y = \log_4 16$.
$$4^y = 16$$
$$4^y = 4^2$$
$$y = 2$$
Therefore, $\log_4 16 = 2$.

**29.** Let $y = \ln 1$.
$$e^y = 1$$
$$e^y = e^0$$
$$y = 0$$
Therefore, $\ln 1 = 0$.

**31.** Let $y = \log_9 243$.
$$9^y = 243$$
$$(3^2)^y = 3^5$$
$$3^{2y} = 3^5$$
$$2y = 5$$
$$y = 2.5$$
Therefore, $\log_9 243 = 2.5$.

**33.** Let $y = \log 1000$.
$$10^y = 1000$$
$$10^y = 10^3$$
$$y = 3$$
Therefore, $\log 1000 = 3$.

**35.** $\ln(-3)$ is not defined. The domain of a logarithmic function is $(0, \infty)$.

**37.** Let $y = \log_{1/9} 3$.
$$\left(\dfrac{1}{9}\right)^y = 3$$
$$(3^{-2})^y = 3$$
$$3^{-2y} = 3^1$$
$$-2y = 1$$
$$y = -0.5$$
Therefore, $\log_{1/9} 3 = -0.5$.

**39.** Let $y = \log_2 8^{1/3}$.
$$2^y = 8^{1/3}$$
$$2^y = (2^3)^{1/3}$$
$$2^y = 2^1$$
$$y = 1$$
Therefore, $\log_2 8^{1/3} = 1$.

**41.** Let $y = \log_3 \sqrt{3^5}$.
$$3^y = \sqrt{3^5}$$
$$3^y = 3^{5/2}$$
$$y = \frac{5}{2}$$
Therefore, $\log_3 \sqrt{3^5} = \frac{5}{2}$.

**43.** log 0 is not defined. The domain of a logarithmic function is $(0, \infty)$.

**45.** Let $y = \log_{2/3}\left(\frac{3}{2}\right)$.
$$\left(\frac{2}{3}\right)^y = \frac{3}{2}$$
$$\left(\frac{2}{3}\right)^y = \left(\frac{2}{3}\right)^{-1}$$
$$y = -1$$
Therefore, $\log_{2/3}\left(\frac{3}{2}\right) = -1$.

**47.** Let $y = \log_b b^3$.
$$b^y = b^3$$
$$y = 3$$
Therefore, $\log_b b^3 = 3$.

**49.** Let $y = \log_a \sqrt{a}$.
$$a^y = \sqrt{a}$$
$$a^y = a^{1/2}$$
$$y = \frac{1}{2}$$
Therefore, $y = \log_a \sqrt{a} = 0.5$.

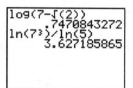

**Figure for Exercises 51, 53, and 55**

**51.** $\ln 4 \approx 1.39$ (see first entry)

**53.** $\log_4 7 = \frac{\ln 7}{\ln 4} \approx 1.40$ (see second entry)

**55.** $\ln\left(\frac{\sqrt{3}}{2}\right) \approx -0.14$ (see third entry)

**Figure for Exercises 57 and 59**

**57.** $\log\left(7 - \sqrt{2}\right) \approx 0.75$ (see first entry)

**59.** $\log_5 7^3 = \frac{\ln 7^3}{\ln 5} \approx 3.63$
(see second entry)

**61.** Write the equation in exponential form.

**63.** $\log_3 x = 5$
$$3^5 = x$$
$$243 = x$$

**65.** $\ln x = 1$
$$e^1 = x$$
$$e = x$$

**67.** $\log_8 t = -\frac{2}{3}$
$$8^{-2/3} = t$$
$$\left(\sqrt[3]{8}\right)^{-2} = t$$
$$2^{-2} = t$$
$$\frac{1}{4} = t$$

**69.** $\log_x 81 = 2$
$$x^2 = 81$$
$$x = \sqrt{81} = 9$$
Because $x$ is defined as being positive, we consider only the positive square root.

Section 11.4   Logarithmic Functions

**71.**
$$\log_x 25 = \frac{2}{3}$$
$$x^{2/3} = 25$$
$$(x^{2/3})^{3/2} = 25^{3/2}$$
$$x = \left(\sqrt{25}\right)^3$$
$$x = 5^3 = 125$$

**73.** $\log_x 1 = 0$
$$x^0 = 1$$
The solution is any $x$ such that $x > 0$ and $x \neq 1$.

**75.**
$$\log_2 \sqrt{x} = \frac{1}{2}$$
$$2^{1/2} = \sqrt{x}$$
$$2^{1/2} = x^{1/2}$$
$$2 = x$$

**77.** $\log_6 6^x = 3$
$$6^3 = 6^x$$
$$3 = x$$

**79.** $\log_4 (x + 1) = 2$
$$4^2 = x + 1$$
$$16 = x + 1$$
$$15 = x$$

**81.** $\log_5 x^2 = 4$
$$5^4 = x^2$$
$$625 = x^2$$
$$x = \pm\sqrt{625} = \pm 25$$

**83.** (a) The graph rises from left to right if $b > 1$.
(b) The graph falls from left to right if $0 < b < 1$.

**85.** $\ln(5 - 2x)$
Graph $y_1 = \ln(5 - 2x)$. The estimate from the graph agrees with the actual domain of $(-\infty, 2.5)$.

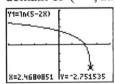

**87.** $\ln|x - 3|$
Graph $y_1 = \ln|x - 3|$. The estimate from the graph agrees with the actual domain of **R**, $x \neq 3$.

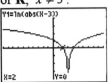

**89.** $g(x) = \log x = 1.751$
Graph $y = \log$. From the graph, after several zoom-and-trace cycles, we have that if $g(x) = 1.751$, then $x \approx 56.36$.

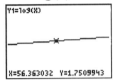

**91.** $f(x) = \ln x = 3.807$
Graph $y = \ln x$. From the graph, after several zoom-and-trace cycles, we have that if $f(x) = 3.807$, then $x \approx 45.02$.

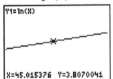

**93.** (a) The graph of $g(x) = 3 + \ln x$ is the graph of $f(x) = \ln x$ shifted up 3 units.
(b) The graph of $h(x) = \ln(3 + x)$ is the graph of $f(x) = \ln x$ shifted left 3 units.

**95.** (a) The graph of $g(x) = \ln(-x)$ is the graph of $f(x) = \ln x$ reflected in the $y$-axis.
(b) The graph of $h(x) = -\ln x$ is the graph of $f(x) = \ln x$ reflected in the $x$-axis.

**97.** (a) The graph of $g(x) = \log_b (x+3)$ is the graph of $f(x) = \log_b x$ shifted left 3 units. This matches graph (B).
(b) The graph of $h(x) = 3 + \log_b x$ is the graph of $f(x) = \log_b x$ shifted up 3 units. This matches graph (D).

**99.** $G = 10 \log (P_o / P_i)$
(a) When $P_o = 16$ and $P_i = 0.004$,
$G = 10 \log (16 / 0.004)$
$\quad = 10 \log (4000)$
$\quad \approx 36.02$
The power gain is approximately 36.02 decibels.
(b) When $G = 50$ and $P_i = 0.006$,
$50 = 10 \log (P_o / 0.006)$
$5 = \log (P_o / 0.006)$
$10^5 = \dfrac{P_o}{0.006}$
$100{,}000 = \dfrac{P_o}{0.006}$
$600 = P_o$
The power output is 600.

**101.** $pH = -\log H^+$
(a) When $H^+ = 0.005$,
$pH = -\log 0.005 \approx 2.3$
The pH value is approximately 2.3.
(b) When $pH = 7$,
$7 = -\log H^+$
$-7 = \log H^+$
$10^{-7} = H^+$
The hydrogen ion concentration is $10^{-7}$.

**103.** $N(x) = 0.93 + 0.63 \ln x$
$A(x) = 1.32 + 0.54 \ln x$
(a) When $x = 12$,
$N(12) = 0.93 + 0.63 \ln(12) \approx 2.50$
$A(12) = 1.32 + 0.54 \ln(12) \approx 2.66$
(b) Graph $y_1 = 0.93 + 0.63 \ln x$ (Nat'l) and $y_2 = 1.32 + 0.54 \ln x$ (Am).
From the graph, we can conclude that the American League average will be higher than the National League average, but the difference in the averages will decrease.

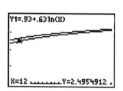

**105.** 

Both appear to be good models.

**107.** $N_2$ predicts that the number of telecommuters will level off at about 17.5 million.

**109.** Because $\ln \tfrac{1}{2} < 0$, reverse the inequality symbol to obtain $a\left(\ln \tfrac{1}{2}\right) < b\left(\ln \tfrac{1}{2}\right)$.

**111.** $e^{\ln 5 - \ln 2} = \dfrac{e^{\ln 5}}{e^{\ln 2}} = \dfrac{5}{2}$

**113.** The function $\ln |x|$ is positive for $|x| > 1$. So $\ln |x|$ is positive for $x$ belonging to the set $(-\infty, -1) \cup (1, \infty)$.

**115.** $\log_3 x < \log_4 x$
We can graph $\log_3 x$ as $y_1 = \dfrac{\ln x}{\ln 3}$ and $\log_4 x$ as $y_2 = \dfrac{\ln x}{\ln 4}$. We can see from the graphs that the graph of $\log_3 x$ is above the graph of $\log_4 x$ for $x > 1$. At $x = 1$, the values are equal. For $0 < x < 1$, the graph of $\log_3 x$ is below the graph of $\log_4 x$. Thus, the solution set of the inequality is $(0, 1)$.

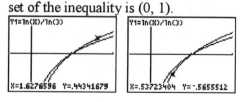

## Section 11.5 Properties of Logarithms

1. What exponent on 9 results in 3? The answer is 1/2.

3. Because, in general, $\log_b b^x = x$, we have $\log_5 5^7 = 7$.

5. Because, in general, $\log_b 1 = 0$, we have $\log_{1.2} 1 = 0$.

7. Because, in general, $b^{\log_b x} = x$, we have $7^{\log_7 2} = 2$.

9. $\log_9 \sqrt{3} = \log_9 9^{1/4} = \dfrac{1}{4} = 0.25$

11. $9^{-\log_3 2} = (3^2)^{-\log_3 2} = 3^{-2\log_3 2}$
    $= (3^{\log_3 2})^{-2} = 2^{-2} = \dfrac{1}{4} = 0.25$

13. $e^{2\ln 3} = (e^{\ln 3})^2 = 3^2 = 9$

15. $\log_3 (\log_4 4) = \log_3 1 = 0$

17. $\log_3 (\log_2 8) = \log_3 (\log_2 2^3)$
    $= \log_3 3 = 1$

19. $\log_7 3x = \log_7 3 + \log_7 x$

21. $\log_7 \dfrac{5}{y} = \log_7 5 - \log_7 y$

23. $\log x^2 = 2 \log x$

25. $\log_9 \sqrt{n} = \log_9 n^{1/2} = \dfrac{1}{2} \log_9 n$

27. $\log_3 \left(\dfrac{5}{6}\right)^3 = 3 \log_3 \dfrac{5}{6}$
    $= 3(\log_3 5 - \log_3 6)$
    $= 3 \log_3 5 - 3 \log_3 6$
    $= 3 \log_3 5 - 3 \log_3 (2 \cdot 3)$
    $= 3 \log_3 5 - 3(\log_3 2 + \log_3 3)$
    $= 3 \log_3 5 - 3(\log_3 2 + 1)$
    $= 3 \log_3 5 - 3 \log_3 2 - 3$
    $= 3B - 3A - 3$
    This matches choice (c).

29. $\log_3 \dfrac{6\sqrt{5}}{5}$
    $= \log_3 6 + \log_3 \sqrt{5} - \log_3 5$
    $= \log_3 (2 \cdot 3) + \log_3 5^{1/2} - \log_3 5$
    $= \log_3 2 + \log_3 3 + \dfrac{1}{2} \log_3 5 - \log_3 5$
    $= \log_3 2 + 1 - \dfrac{1}{2} \log_3 5$
    $= A + 1 - \dfrac{1}{2} B$
    This matches choice (a).

31. $\log_3 \dfrac{5\sqrt{10}}{9}$
    $= \log_3 5 + \log_3 \sqrt{10} - \log_3 9$
    $= \log_3 5 + \dfrac{1}{2} \log_3 10 - \log_3 3^2$
    $= \log_3 5 + \dfrac{1}{2} \log_3 (5 \cdot 2) - 2$
    $= \log_3 5 + \dfrac{1}{2} \log_3 5 + \dfrac{1}{2} \log_3 2 - 2$
    $= \dfrac{1}{2} A + \dfrac{3}{2} B - 2$
    This matches choice (g).

33. $\log_3 \dfrac{\sqrt[4]{96}}{12}$
    $= \log_3 \sqrt[4]{96} - \log_3 12$
    $= \dfrac{1}{4} \log_3 96 - \log_3 (2^2 \cdot 3)$
    $= \dfrac{1}{4} \log_3 (2^5 \cdot 3) - (\log_3 2^2 + \log_3 3)$
    $= \dfrac{1}{4} (\log_3 2^5 + \log_3 3) - \log_3 2^2 - 1$
    $= \dfrac{1}{4} (\log_3 2^5 + 1) - 2 \log_3 2 - 1$
    $= \dfrac{1}{4} \log_3 2^5 + \dfrac{1}{4} - 2 \log_3 2 - 1$
    $= \dfrac{5}{4} \log_3 2 + \dfrac{1}{4} - 2 \log_3 2 - 1$
    $= -\dfrac{3}{4} A - \dfrac{3}{4}$
    This matches choice (d).

**35.** Begin with the Power Rule for Logarithms to obtain $n \log (AB)$. In the Order of Operations, exponents have priority over products.

**37.** $\log_9 x^2 y^{-3} = \log_9 x^2 + \log_9 y^{-3}$
$= 2\log_9 x - 3\log_9 y$

**39.** $\log \dfrac{(x+1)^2}{x+2} = \log (x+1)^2 - \log (x+2)$
$= 2\log (x+1) - \log (x+2)$

**41.** $\ln \left(\dfrac{2x+5}{x+7}\right)^5 = 5[\ln(2x+5) - \ln(x+7)]$
$= 5\ln(2x+5) - 5\ln(x+7)$

**43.** $\log \dfrac{\sqrt{10}}{x^2} = \log \sqrt{10} - \log x^2$
$= \dfrac{1}{2}\log 10 - 2\log x$
$= \dfrac{1}{2} - 2\log x$

**45.** $\log_5 \sqrt[3]{xy^2} = \dfrac{1}{3}\log_5 xy^2$
$= \dfrac{1}{3}(\log_5 x + \log_5 y^2)$
$= \dfrac{1}{3}\log_5 x + \dfrac{2}{3}\log_5 y$

**47.** $\log_2 \dfrac{\sqrt{x+1}}{32} = \log_2 \sqrt{x+1} - \log_2 32$
$= \dfrac{1}{2}\log_2 (x+1) - \log_2 2^5$
$= \dfrac{1}{2}\log_2 (x+1) - 5$

**49.** $\log_2 \dfrac{x^2 \sqrt{y}}{\sqrt[3]{x}}$
$= \log_2 x^2 + \log_2 \sqrt{y} - \log_2 \sqrt[3]{x}$
$= 2\log_2 x + \dfrac{1}{2}\log_2 y - \dfrac{1}{3}\log_2 x$
$= \dfrac{5}{3}\log_2 x + \dfrac{1}{2}\log_2 y$

**51.** $\log (4x^2 + 4x + 1) = \log (2x+1)^2$
$= 2\log (2x+1)$

**53.** $\ln 1000 e^{0.08t} = \ln 1000 + \ln e^{0.08t}$
$= \ln 10^3 + 0.08t$
$= 3\ln 10 + 0.08t$

**55.** $\log_4 (x+4)$ cannot be expanded

**57.** $\ln Pe^{rt} = \ln P + \ln e^{rt} = \ln P + rt$

**59.** Use the Distributive Property to obtain $(2+1)\log 3 = 3\log 3$. For $2\log x + \log y$, $x \neq y$, the Distributive Property does not apply.

**61.** $\log_4 3 + \log_4 5 = \log_4 (3 \cdot 5) = \log_4 15$

**63.** $\log_3 10 - \log_3 5 = \log_3 \dfrac{10}{5} = \log_3 2$

**65.** $3\log_2 y = \log_2 y^3$

**67.** $\dfrac{1}{2}\log x = \log x^{1/2} = \log \sqrt{x}$

**69.** $\log_4 60 - \log_4 15 = \log_4 \dfrac{60}{15} = \log_4 4 = 1$

**71.** $\log_5 50 + 2\log_5 10 - \log_5 40$
$= \log_5 50 + \log_5 10^2 - \log_5 40$
$= \log_5 \dfrac{50 \cdot 100}{40} = \log_5 125 = \log_5 5^3 = 3$

**73.** $\log_8 3 + \log_8 x - \log_8 y$
$= \log_8 (3x) - \log_8 y = \log_8 \dfrac{3x}{y}$

**75.** $2\log_7 x - \log_7 y = \log_7 x^2 - \log_7 y$
$= \log_7 \dfrac{x^2}{y}$

**77.** $\dfrac{1}{2}\log x - \dfrac{2}{3}\log y = \log x^{1/2} - \log y^{2/3}$
$= \log \dfrac{x^{1/2}}{y^{2/3}} = \log \dfrac{\sqrt{x}}{\sqrt[3]{y^2}}$

**79.** $3\log y - 2\log x - 4\log z$
$= \log y^3 - \log x^2 - \log z^4$
$= \log y^3 - (\log x^2 + \log z^4)$
$= \log y^3 - \log x^2 z^4 = \log \dfrac{y^3}{x^2 z^4}$

**81.** $2\log(x+3) - 3\log(x+1)$
$= \log(x+3)^2 - \log(x+1)^3$
$= \log \dfrac{(x+3)^2}{(x+1)^3}$

**83.** $\dfrac{1}{2}[\log x - \log(x+1)] = \dfrac{1}{2}\log\dfrac{x}{x+1}$
$= \log\left(\dfrac{x}{x+1}\right)^{1/2} = \log\sqrt{\dfrac{x}{x+1}}$

**85.** $5 + 2\log_3 x = \log_3 3^5 + \log_3 x^2$
$= \log_3 243 + \log_3 x^2 = \log_3 243x^2$

**87.** $rt + \ln P_0 = \ln e^{rt} + \ln P_0 = \ln P_0 e^{rt}$

**89.** Because $\log_8 \dfrac{3}{5} = \log_8 3 - \log_8 5$ by the Quotient Rule for Logarithms, $\dfrac{\log_8 3}{\log_8 5} \neq \log_8 \dfrac{3}{5}$. The statement is false.

**91.** Because $\dfrac{\log_3 81}{\log_3 9} = \dfrac{\log_3 3^4}{\log_3 3^2} = \dfrac{4}{2} = 2$, the statement is true.

**93.** Because $\log 6 = \log(3 \cdot 2) = \log 3 + \log 2$ by the Product Rule for Logarithms, $(\log 3)(\log 2) \neq \log 6$. The statement is false.

**95.** Because $\log_2 x^x = x\log_2 x$ by the Power Rule for Logarithms, the equation is always true.

**97.** $\log_5 x^2 y = \log_5 x^2 + \log_5 y$
$= 2\log_5 x + \log_5 y$
The equation is always true.

**99.** Because $\log_3(x+10)$ cannot be simplified, $\log_3(x+10) = \log_3 x + \log_3 10$ is false.

**101.** $P(t) = \dfrac{10}{\log 5 + \log(t+2)}$

(a) After $t = 0$ hours of travel:
$P(0) = \dfrac{10}{\log 5 + \log(0+2)} = 10$.
After $t = 4$ hours of travel:
$P(4) = \dfrac{10}{\log 5 + \log(4+2)} \approx 6.77$.
The percent decline is
$\dfrac{6.77 - 10}{10} \times 100\% = -32.3\%$

(b) The denominator becomes
$\log 5 + \log(t+2) = \log[5(t+2)]$
$= \log(5t+10)$
The function is $P(t) = \dfrac{10}{\log(5t+10)}$

**103.**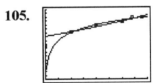
The graph of $N$ does not contain a point for 1970 because $\ln 0$ is not defined.

**105.**

Both appear to be good models.

**107.** Let $\log_b M = r$ and $\log_b N = s$. Then $b^r = M$ and $b^s = N$. Therefore, the quotient $\dfrac{M}{N} = \dfrac{b^r}{b^s} = b^{r-s}$. But $b^{r-s} = \dfrac{M}{N}$ implies that $\log_b \dfrac{M}{N} = r - s = \log_b M - \log_b N$.

**109.** $2 + \log_3 x = \log_3 3^2 + \log_3 x$
$= \log_3 9 + \log_3 x$
$= \log_3 9x$

**111.** Using the Change-of-Base Formula:
$$\log_2(xy) = \frac{\ln xy}{\ln 2} = \frac{\ln x + \ln y}{\ln 2}$$

## Section 11.6   Logarithmic and Exponential Equations

**1.** If $b$ can be written as a power of 3, then we can equate exponents instead of writing the equation in logarithmic form.

**3.**
$$3^{5x} = \frac{1}{27}$$
$$3^{5x} = 3^{-3}$$
$$5x = -3$$
$$x = -0.6$$

**5.**
$$7^{1-x} = 49$$
$$7^{1-x} = 7^2$$
$$1 - x = 2$$
$$x = -1$$

**7.**
$$3^x = 5$$
$$\ln 3^x = \ln 5$$
$$x \ln 3 = \ln 5$$
$$x = \frac{\ln 5}{\ln 3} \approx 1.46$$

**9.**
$$3^{x-3} = 2^{x+2}$$
$$\ln 3^{x-3} = \ln 2^{x+2}$$
$$(x-3)\ln 3 = (x+2)\ln 2$$
$$x \ln 3 - 3 \ln 3 = x \ln 2 + 2 \ln 2$$
$$x \ln 3 - x \ln 2 = 2 \ln 2 + 3 \ln 3$$
$$x(\ln 3 - \ln 2) = 2 \ln 2 + 3 \ln 3$$
$$x = \frac{2 \ln 2 + 3 \ln 3}{\ln 3 - \ln 2}$$
$$x \approx 11.55$$

**11.**
$$e^{20k} = 0.6$$
$$\ln e^{20k} = \ln 0.6$$
$$20k = \ln 0.6$$
$$k = \frac{\ln 0.6}{20} \approx -0.03$$

**13.**
$$4e^{2x-1} = 5$$
$$e^{2x-1} = 1.25$$
$$\ln e^{2x-1} = \ln 1.25$$
$$2x - 1 = \ln 1.25$$
$$x = \frac{\ln 1.25 + 1}{2} \approx 0.61$$

**15.**
$$\left(\frac{5}{3}\right)^{-x} = 20$$
$$\ln\left(\frac{5}{3}\right)^{-x} = \ln 20$$
$$-x \ln\left(\frac{5}{3}\right) = \ln 20$$
$$x = -\frac{\ln 20}{\ln\left(\frac{5}{3}\right)} \approx -5.86$$

**17.**
$$3 + e^x = 7$$
$$e^x = 4$$
$$\ln e^x = \ln 4$$
$$x = \ln 4 \approx 1.39$$

**19.**
$$5^x + 7 = 2$$
$$5^x = -5$$
Because an exponential function is never negative, this equation has no solution.

**21.**
$$600(1 + 0.02)^{4t} = 1000$$
$$(1 + 0.02)^{4t} = \frac{5}{3}$$
$$\ln(1 + 0.02)^{4t} = \ln(5/3)$$
$$4t \ln(1 + 0.02) = \ln(5/3)$$
$$t = \frac{\ln(5/3)}{4 \ln(1 + 0.02)} \approx 6.45$$

Section 11.6  Logarithmic and Exponential Equations

**23.**
$$5^{-x^2} = 0.2$$
$$5^{-x^2} = \frac{1}{5}$$
$$5^{-x^2} = 5^{-1}$$
$$-x^2 = -1$$
$$x^2 = 1$$
$$x = \pm\sqrt{1} = \pm 1$$

**25.**
$$5^{|x+1|} = 2$$
$$\ln 5^{|x+1|} = \ln 2$$
$$|x+1|\ln 5 = \ln 2$$
$$|x+1| = \frac{\ln 2}{\ln 5}$$

$x + 1 = \dfrac{\ln 2}{\ln 5}$   $x + 1 = -\dfrac{\ln 2}{\ln 5}$

$x = \dfrac{\ln 2}{\ln 5} - 1$  or  $x = -\dfrac{\ln 2}{\ln 5} - 1$

$x \approx -0.57$    $x \approx -1.43$

**27.**  $|2 - 3^x| = 1$

$2 - 3^x = 1$      $2 - 3^x = -1$
$-3^x = -1$        $-3^x = -3$
$3^x = 1$    or   $3^x = 3$
$3^x = 3^0$        $3^x = 3^1$
$x = 0$            $x = 1$

**29.** No, the base must be the same to equate arguments.

**31.**
$$\log_5 x = \log_5(2x - 3)$$
$$x = 2x - 3$$
$$x = 3$$

**33.**
$$\log_5(x - 1) = 2$$
$$\log_5(x - 1) = \log_5 5^2$$
$$x - 1 = 5^2$$
$$x - 1 = 25$$
$$x = 26$$

**35.**
$$\log_3 \sqrt[3]{3x - 5} = 1$$
$$\log_3 \sqrt[3]{3x - 5} = \log_3 3$$
$$\sqrt[3]{3x - 5} = 3$$
$$3x - 5 = 3^3$$
$$3x = 32$$
$$x = \frac{32}{3}$$

**37.**
$$\frac{1}{3}\log_8(x - 7) = \log_8 3$$
$$\log_8(x - 7)^{1/3} = \log_8 3$$
$$(x - 7)^{1/3} = 3$$
$$x - 7 = 3^3$$
$$x = 34$$

**39.**
$$\log_2(x^2 - 7x) = 3$$
$$\log_2(x^2 - 7x) = \log_2 2^3$$
$$x^2 - 7x = 8$$
$$x^2 - 7x - 8 = 0$$
$$(x - 8)(x + 1) = 0$$
$$x = -1, 8$$

**41.**
$$\log_4(5 - 3x)^3 = 6$$
$$\log_4(5 - 3x)^3 = \log_4 4^6$$
$$(5 - 3x)^3 = 4^6$$
$$5 - 3x = 4^{6/3}$$
$$-3x = 4^2 - 5$$
$$x = -\frac{11}{3}$$

**43.**
$$\frac{1}{2}\log_9(6 - x) = \log_9 x$$
$$\log_9(6 - x)^{1/2} = \log_9 x$$
$$(6 - x)^{1/2} = x$$
$$6 - x = x^2$$
$$0 = x^2 + x - 6$$
$$0 = (x + 3)(x - 2)$$
$$x = -3, 2$$

A check shows that $-3$ is an extraneous solution. The only solution is 2.

**45.** 
$$2 - \log_9 x^2 = 1$$
$$1 = \log_9 x^2$$
$$\log_9 9 = \log_9 x^2$$
$$9 = x^2$$
$$x = \pm\sqrt{9} = \pm 3$$

**47.** 
$$\log_4 6x - \log_4(x-5) = 2$$
$$\log_4 \frac{6x}{x-5} = \log_4 4^2$$
$$\frac{6x}{x-5} = 16$$
$$6x = 16(x-5)$$
$$-10x = -80$$
$$x = 8$$

**49.** 
$$\log x + \log(7-x) = 1$$
$$\log x(7-x) = \log 10^1$$
$$x(7-x) = 10$$
$$7x - x^2 = 10$$
$$x^2 - 7x + 10 = 0$$
$$(x-5)(x-2) = 0$$
$$x = 2, 5$$

**51.** 
$$1 - \log(x-5) = \log\left(\frac{x}{5}\right)$$
$$1 = \log\left(\frac{x}{5}\right) + \log(x-5)$$
$$\log 10^1 = \log \frac{x(x-5)}{5}$$
$$10 = \frac{x(x-5)}{5}$$
$$50 = x^2 - 5x$$
$$0 = x^2 - 5x - 50$$
$$0 = (x-10)(x+5)$$
$$x = -5, 10$$

A check shows that $-5$ is an extraneous solution. The only solution is 10.

**53.** 
$$\log_5(2x-3) - \log_5 x = 1$$
$$\log_5 \frac{2x-3}{x} = \log_5 5^1$$
$$\frac{2x-3}{x} = 5$$
$$2x - 3 = 5x$$
$$-3x = 3$$
$$x = -1$$

A check shows that $-1$ is an extraneous solution. There is no solution.

**55.** 
$$\log_5 x(x+2) = 0$$
$$\log_5 x(x+2) = \log_5 5^0$$
$$x(x+2) = 1$$
$$x^2 + 2x - 1 = 0$$
$$x = \frac{-2 \pm \sqrt{2^2 - 4(1)(-1)}}{2(1)} \approx -2.41, \ 0.41$$

**57.** 
$$\log x + \log(4+x) = \log 5$$
$$\log x(4+x) = \log 5$$
$$x(4+x) = 5$$
$$x^2 + 4x - 5 = 0$$
$$(x+5)(x-1) = 0$$
$$x = -5, 1$$

A check shows that $-5$ is an extraneous solution. The only solution is 1.

**59.** 
$$\log(x+6) - 2\log x = \log 12$$
$$\log(x+6) - \log x^2 = \log 12$$
$$\log \frac{(x+6)}{x^2} = \log 12$$
$$\frac{(x+6)}{x^2} = 12$$
$$x + 6 = 12x^2$$
$$12x^2 - x - 6 = 0$$
$$(4x-3)(3x+2) = 0$$
$$x = -0.67, \ 0.75$$

A check shows that $-0.67$ is an extraneous solution. The only solution is 0.75.

**61.** 
$$|2 - \log_2(x-1)| = 3$$
$$2 - \log_2(x-1) = 3 \ \text{or}$$
$$2 - \log_2(x-1) = -3$$
$$2 - \log_2(x-1) = 3$$
$$\log_2(x-1) = -1$$
$$\log_2(x-1) = \log_2 2^{-1}$$
$$x - 1 = 0.5$$
$$x = 1.5$$
or
$$2 - \log_2(x-1) = -3$$
$$\log_2(x-1) = 5$$
$$\log_2(x-1) = \log_2 2^5$$
$$x - 1 = 32$$
$$x = 33$$

### Section 11.6 Logarithmic and Exponential Equations

**63.** If $b$ were permitted to be 1, then $1^3 = 1^8$, but $3 \neq 8$.

**65.**
$(\log x)^2 + \log x^3 - 4 = 0$
$(\log x)^2 + 3\log x - 4 = 0$
$u^2 + 3u - 4 = 0 \quad$ Let $u = \log x$.
$(u+4)(u-1) = 0$
$u = -4,\ 1$

Now, replace $u$ with $\log x$.

$\log x = -4 \qquad\qquad \log x = 1$
$\log x = \log 10^{-4} \quad$ or $\quad \log x = \log 10^1$
$x = 10^{-4} \qquad\qquad x = 10$

**67.** $\log_4 x - 3\sqrt{\log_4 x} + 2 = 0$

Let $u = \sqrt{\log_4 x}$.

$u^2 - 3u + 2 = 0$
$(u-2)(u-1) = 0$
$u = 1,\ 2$

Now, replace $u$ with $\sqrt{\log_4 x}$.

$\sqrt{\log_4 x} = 1 \qquad \sqrt{\log_4 x} = 2$
$\log_4 x = 1 \quad$ or $\quad \log_4 x = 4$
$4^1 = x \qquad\qquad 4^4 = x$
$x = 4 \qquad\qquad x = 256$

**69.**
$e^{2x} + 6 = 5e^x$
$e^{2x} - 5e^x + 6 = 0$
$u^2 - 5u + 6 = 0 \quad$ Let $u = e^x$.
$(u-3)(u-2) = 0$
$u = 2,\ 3$

Now, replace $u$ with $e^x$.

$e^x = 2 \qquad\qquad e^x = 3$
$\ln e^x = \ln 2 \quad$ or $\quad \ln e^x = \ln 3$
$x = \ln 2 \qquad\qquad x = \ln 3$

**71.**
$5^{2x} = 5^x + 42$
$5^{2x} - 5^x - 42 = 0$
$u^2 - u - 42 = 0 \quad$ Let $u = 5^x$.
$(u-7)(u+6) = 0$
$u = -6,\ 7$

Now, replace $u$ with $5^x$.

$5^x = 7 \qquad\qquad\qquad 5^x = -6$
$\ln 5^x = \ln 7 \qquad\qquad$ No solution
$x \ln 5 = \ln 7 \quad$ or
$x = \dfrac{\ln 7}{\ln 5}$

**73.** $R = \log I$
$I = 10^R$

**75.**
$L = 10\log\left(\dfrac{I}{I_0}\right)$
$L = \log\left(\dfrac{I}{I_0}\right)^{10}$
$10^L = \left(\dfrac{I}{I_0}\right)^{10}$
$\sqrt[10]{10^L} = \dfrac{I}{I_0}$
$I_0\sqrt[10]{10^L} = I$

**77.**
$P = P_0 e^{rt}$
$\dfrac{P}{P_0} = e^{rt}$
$\ln\dfrac{P}{P_0} = \ln e^{rt}$
$\ln\dfrac{P}{P_0} = rt$
$\dfrac{1}{t}\ln\dfrac{P}{P_0} = r$

**79.**
$2 = \left(1+\dfrac{r}{n}\right)^{nt}$
$\ln 2 = \ln\left(1+\dfrac{r}{n}\right)^{nt}$
$\ln 2 = nt\ln\left(1+\dfrac{r}{n}\right)$
$\dfrac{\ln 2}{n\ln\left(1+\dfrac{r}{n}\right)} = t$

**81.** Let $P$ = the initial investment, $A = 2P$, and $t = 14$.

$$A = Pe^{rt}$$
$$2P = Pe^{14r}$$
$$2 = e^{14r}$$
$$\ln 2 = \ln e^{14r}$$
$$\ln 2 = 14r$$
$$r = \frac{\ln 2}{14} \approx 0.0495$$

The bank is offering 4.95% interest.

**83.** $P(t) = 14e^{0.03t}$

(a) For 1990, $t = 0$:
$$P(0) = 14e^{0.03(0)} = 14 \text{ million}$$

(b) For 2010, $t = 20$:
$$P(20) = 14e^{0.03(20)} \approx 25.51 \text{ million}$$

(c) For 2020, $t = 30$:
$$P(30) = 14e^{0.03(30)} \approx 34.43 \text{ million}$$

**85.** Use $P(t) = P_0 e^{rt}$. The half-life is $t_h$, where $P(t_h) = \frac{1}{2} P_0$. We are given that $t_h = 5730$. Use this fact to determine $r$.

$$\frac{1}{2} P_0 = P_0 e^{5730r}$$
$$0.5 = e^{5730r}$$
$$\ln 0.5 = \ln e^{5730r}$$
$$\ln 0.5 = 5730r$$
$$r = \frac{\ln 0.5}{5730}$$

To determine how many years it will take for the remaining amount to be 30% of the original, we solve $0.3 P_0 = P_0 e^{rt}$ for $t$ and substitute the value just found for $r$.

$$0.3 P_0 = P_0 e^{rt}$$
$$0.3 = e^{rt}$$
$$\ln 0.3 = \ln e^{rt}$$
$$\ln 0.3 = rt$$
$$\frac{\ln 0.3}{r} = t$$
$$t = \frac{\ln 0.3}{\left(\frac{\ln 0.5}{5730}\right)} \approx 9953$$

It takes approximately 9953 years for carbon-14 to decay to the point where only 30% of the original amount remains.

**87.** Use $V(t) = V_0 e^{-0.19t}$ with $t = 7$ and $V(t) = \$22{,}480.57$. Solve for $V_0$.

$$V(t) = V_0 e^{-0.19t}$$
$$22{,}480.57 = V_0 e^{-0.19(7)}$$
$$V_0 = \frac{22{,}480.57}{e^{-0.19(7)}} \approx 85{,}000$$

The original cost of the bulldozer was $85,000.

**89.** Use $T = T_m + D_0 e^{kt}$, where $T_m = 60\,°F$. At 8:00 A.M., the temperature of the body was $88\,°F$. So, $D_0 = 88 - 60 = 28\,°F$. After 45 minutes, $T = 85\,°F$. Use this information to determine $k$.

$$T = T_m + D_0 e^{kt}$$
$$85 = 60 + 28 e^{45t}$$
$$25 = 28 e^{45t}$$
$$\ln \frac{25}{28} = 45k$$
$$\frac{1}{45} \ln \frac{25}{28} = k$$

Now, determine $t$ for $T = 98.6\,°F$.

$$98.6 = 60 + 28 e^{kt}$$
$$38.6 = 28 e^{kt}$$
$$\frac{38.6}{28} = e^{kt}$$
$$\ln \frac{38.6}{28} = kt$$
$$t = \frac{\ln \frac{38.6}{28}}{k}$$
$$t = \frac{\ln \frac{38.6}{28}}{\frac{1}{45} \ln \frac{25}{28}} \approx -127$$

The murder occurred at approximately 127 minutes before 8:00 A.M., or at approximately 5:53 A.M.

**91.**
$R = \log I$
$6.9 = \log I$
$I = 10^{6.9}$
The intensity of the Armenian earthquake was $10^{6.9}$. An earthquake that is 10 times as intense would be
$10(10^{6.9}) = 10^{1+6.9} = 10^{7.9}$
$I = 10^{7.9}$
$7.9 = \log I$
$7.9 = R$
The Richter scale reading would be 7.9.

**93.** (a) $51.59(1.045)^x = 340$

(b) By solving the equation in part (a), we can estimate the year in which the horse population is projected to reach 300.
$$51.59(1.045)^x = 340$$
$$(1.045)^x = \frac{340}{51.59}$$
$$\ln(1.045)^x = \ln\frac{340}{51.59}$$
$$x\ln(1.045) = \ln\frac{340}{51.59}$$
$$x = \frac{1}{\ln(1.045)}\ln\frac{340}{51.59}$$
$$x \approx 43$$
The population will reach 340 in about 2003.

**95.**
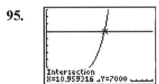
Using the graphing method, we can estimate the year in which there will be 7 million point-of-sale terminals as 2001 ($x \approx 11$).

**97.**
$A(x) = \dfrac{1000P(x)}{T(x)} = \dfrac{190{,}600e^{0.26x}}{50.5e^{0.45x}}$
$\approx 3774e^{(0.26-0.45)x}$
$= 3774e^{-0.19x}$
A graph of $A(x)$ shows that the number of transactions per terminal is declining.

**99.** $f(x) = \log x$, $g(x) = \ln x$
$(g \circ f)(x) = (g(f(x))$
$= g(\log x)$
$= \ln(\log x)$
Solving the given equation, we have
$\ln(\log x) = 1$
$e^1 = \log x$
$x = 10^e$

**101.** (a) $2\log_3 x + \log_3(10 - x^2) = 2$
$\log_3 x^2 + \log_3(10 - x^2) = \log_3 3^2$
$\log_3 x^2(10 - x^2) = \log_3 9$
$x^2(10 - x^2) = 9$
$x^4 - 10x^2 + 9 = 0$
$(x^2 - 9)(x^2 - 1) = 0$
$x = -3, -1, 1, 3$

A check shows that only 1 and 3 are solutions.

(b) $\log_3 x^2 + \log_3(10 - x^2) = \log_3 3^2$
$\log_3 x^2(10 - x^2) = \log_3 9$
$x^2(10 - x^2) = 9$
$x^4 - 10x^2 + 9 = 0$
$(x^2 - 9)(x^2 - 1) = 0$
$x = -3, -1, 1, 3$

In (a), negative numbers are not in the domain of $\log_3 x$, but in (b), negative numbers are in the domain of both logarithmic functions.

**103.** The graph of the left side of the equation cannot be produced, which implies that the equation has no solution. The domain of $\ln(3 - 4x)$ is $(-\infty, 0.75)$, while the domain of $\ln(x - 2)$ is $(2, \infty)$. The intersection of these two domains is $\varnothing$.

**105.** Graph $y_1 = 2^x$ and $y_2 = x^2$. From the graph, it appears that the solutions are $-0.77$, 2, and 4.

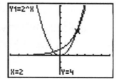
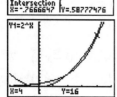

## Chapter 11 Project

**1.** The rate of increase is increasing.

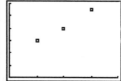

**3.** The graph models the data well.

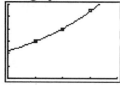

**5.** In the long run, the rate of increase is decreasing.

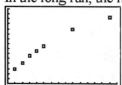

## Chapter 11 Review Exercises

**1.** $f(x) = x^2 - 3x$, $g(x) = 3x + 4$
$(f + g)(x) = f(x) + g(x)$
$\phantom{(f + g)(x)} = x^2 - 3x + 3x + 4$
$\phantom{(f + g)(x)} = x^2 + 4$

**3.** $f(x) = x^2 - 3x$, $g(x) = 3x + 4$
$\left(\dfrac{f}{g}\right)(3.7) = \dfrac{f(3.7)}{g(3.7)}$
$\phantom{\left(\dfrac{f}{g}\right)(3.7)} = \dfrac{(3.7)^2 - 3(3.7)}{3(3.7) + 4}$
$\phantom{\left(\dfrac{f}{g}\right)(3.7)} = \dfrac{2.59}{15.1} \approx 0.17$

5.  $f(x) = x^2 - 3x$, $g(x) = 3x + 4$
$$(f \circ g)(x) = f(g(x))$$
$$= f(3x + 4)$$
$$= (3x + 4)^2 - 3(3x + 4)$$
$$= 9x^2 + 24x + 16 - 9x - 12$$
$$= 9x^2 + 15x + 4$$

7.  $f(x) = x^2 - 3x$, $h(x) = \sqrt{x + 1}$
$$(f \circ h)(3.56) = f(h(3.56))$$
$$= f(\sqrt{3.56 + 1})$$
$$= (\sqrt{3.56 + 1})^2 - 3(\sqrt{3.56 + 1})$$
$$= 3.56 + 1 - 3\sqrt{3.56 + 1}$$
$$\approx -1.85$$

9.  No, $g(-1) = 0$, but $f(0)$ is not defined.

11. $f(x) = \dfrac{1}{x + 2}$, $g(x) = |2x + 5|$
$$(f \circ g)(x) = f(g(x))$$
$$= f(|2x + 5|)$$
$$= \dfrac{1}{|2x + 5| + 2}$$
$$(g \circ f)(x) = g(f(x))$$
$$= g\left(\dfrac{1}{x + 2}\right)$$
$$= \left|2\left(\dfrac{1}{x + 2}\right) + 5\right|$$
$$= \left|\dfrac{2}{x + 2} + 5\right|$$

13. Use the Vertical Line Test to determine if a relation is a function. Use the Horizontal Line Test to determine if a function is one-to-one.

15. $f(x) = x^3 + x^2 - 4x - 3$
Because the graph fails the Horizontal Line Test, the function is not one-to-one.

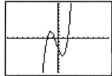

17. For the one-to-one function in (b), the inverse function is
$g^{-1} = \{(2, 7), (6, -3), (-1, -2), (4, 0)\}$
Domain: $\{-1, 2, 4, 6\}$
Range: $\{-3, -2, 0, 7\}$

19. $f(x) = \dfrac{1}{x} + 1$, $g(x) = \dfrac{1}{x - 1}$
$$(f \circ g)(x) = f(g(x)) = f\left(\dfrac{1}{x - 1}\right)$$
$$= \dfrac{1}{\left(\dfrac{1}{x - 1}\right)} + 1 = x - 1 + 1 = x$$
$$(g \circ f)(x) = g(f(x)) = g\left(\dfrac{1}{x} + 1\right)$$
$$= \dfrac{1}{\left(\dfrac{1}{x} + 1\right) - 1} = \dfrac{1}{\left(\dfrac{1}{x}\right)} = x$$
The functions are inverses.

21. $f(x) = x^3 - 5$
$y = x^3 - 5$
$x = y^3 - 5$
$x + 5 = y^3$
$\sqrt[3]{x + 5} = y$
$f^{-1}(x) = \sqrt[3]{x + 5}$
Verification:
$(f \circ f^{-1})(x) = f(\sqrt[3]{x + 5})$
$= (\sqrt[3]{x + 5})^3 - 5 = x$

23. The exponent $x$ can be any real number, but $b > 0$ and $b \neq 1$.

25. (a) $f(x) = -8e^{-2x}$
$f(1) = -8e^{-2(1)} = -\dfrac{8}{e^2}$

(b) $g(x) = \left(\dfrac{9}{16}\right)^{3x}$
$g(1/2) = \left(\dfrac{9}{16}\right)^{3/2} = \left(\dfrac{3}{4}\right)^3 = \dfrac{27}{64}$

**27.** Graph $y_1 = \pi^x$ and $y_2 = 5$. From the graph, we can estimate the solution as about 1.41.

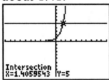

**29.**
$$4^x = 2048$$
$$(2^2)^x = 2^{11}$$
$$2^{2x} = 2^{11}$$
$$2x = 11$$
$$x = 5.5$$

**31.**
$$\left(\frac{1}{27}\right)^{3x} = 9^{4x-5}$$
$$(3^{-3})^{3x} = (3^2)^{4x-5}$$
$$3^{-9x} = 3^{2(4x-5)}$$
$$-9x = 2(4x-5)$$
$$-9x = 8x - 10$$
$$-17x = -10$$
$$x = \frac{10}{17}$$

**33.** For growth, the function should be increasing: $b > 1$.

**35.** The equation $8^{2/3} = 4$ can be written in logarithmic form as $\log_8 4 = \frac{2}{3}$.

**37.** The equation $\log_{25} \frac{1}{5} = -\frac{1}{2}$ can be written in exponential form as $25^{-1/2} = \frac{1}{5}$.

**39.** The function $y = 1^x$ is not one-to-one. Thus it does not have an inverse.

**41.** $\ln e^{3.23} = 3.23$

**43.** $\log_{1/4} 32 = \log_{1/4} \left(\frac{1}{4}\right)^{-5/2} = -\frac{5}{2}$

**45.** $\ln x = 5.2$
$$x = e^{5.2} \approx 181.27$$

**47.** $\log_{343} x = \frac{2}{3}$
$$x = 343^{2/3} = \left(\sqrt[3]{343}\right)^2 = 7^2 = 49$$

**49.** $\log_4 32 = x$
$$4^x = 32$$
$$(2^2)^x = 2^5$$
$$2^{2x} = 2^5$$
$$2x = 5$$
$$x = 2.5$$

**51.** $f(x) = \log_b x$
(a) The $x$-intercept is $(1, 0)$.
(b) There is no $y$-intercept.
(c) The domain is $(0, \infty)$.
(d) The range is **R**.

**53.** $\log_{21} 21^{41} = 41$

**55.** $\log_{21} 21 = 1$

**57.** Because $\log 49 = 2 \log 7$, we estimate $\log 49$ as $2(0.845) = 1.69$.

**59.**
$$\log \frac{\sqrt[4]{x^2 y^3}}{x^3 z^4}$$
$$= \log \sqrt[4]{x^2 y^3} - \log x^3 z^4$$
$$= \log (x^2 y^3)^{1/4} - \log x^3 z^4$$
$$= \frac{1}{4} \log x^2 y^3 - (\log x^3 + \log z^4)$$
$$= \frac{1}{4}(\log x^2 + \log y^3) - 3 \log x - 4 \log z$$
$$= \frac{1}{2} \log x + \frac{3}{4} \log y - 3 \log x - 4 \log z$$
$$= \frac{3}{4} \log y - \frac{5}{2} \log x - 4 \log z$$

**61.** $\dfrac{1}{4}[3\log(x+4) - \log y - \log 2z]$

$= \dfrac{1}{4}[3\log(x+4) - (\log y + \log 2z)]$

$= \dfrac{1}{4}[\log(x+4)^3 - \log 2yz]$

$= \dfrac{1}{4}\log\dfrac{(x+4)^3}{2yz}$

$= \log\left[\dfrac{(x+4)^3}{2yz}\right]^{1/4}$

$= \log\sqrt[4]{\dfrac{(x+4)^3}{2yz}}$

**63.** $\log_2 24 + 2\log_2 5 - \log_2 75$

$= \log_2 24 + \log_2 5^2 - \log_2 75$

$= \log_2(24 \cdot 25) - \log_2 75$

$= \log_2 \dfrac{24 \cdot 25}{75}$

$= \log_2 8$

$= \log_2 2^3 = 3$

**65.** In the first equation, both sides can be written as powers of 3. In the second equation, 4 cannot easily be written as a power of 3. The equation can be solved with logarithms.

**67.** $3^x = 7$

$\ln 3^x = \ln 7$

$x \ln 3 = \ln 7$

$x = \dfrac{\ln 7}{\ln 3} \approx 1.77$

**69.** $\log_4(x+3) + \log_4(x-3) = 2$

$\log_4(x+3)(x-3) = \log_4 4^2$

$(x+3)(x-3) = 16$

$x^2 - 9 = 16$

$x^2 = 25$

$x = \pm\sqrt{25} = \pm 5$

A check shows that $-5$ is an extraneous solution. The only solution is 5.

**71.** $\ln(x-4) - \ln(3x-10) = \ln\dfrac{1}{x}$

$\ln\dfrac{x-4}{3x-10} = \ln\dfrac{1}{x}$

$\dfrac{x-4}{3x-10} = \dfrac{1}{x}$

$x(x-4) = 3x - 10$

$x^2 - 4x = 3x - 10$

$x^2 - 7x + 10 = 0$

$(x-5)(x-2) = 0$

$x = 2, 5$

A check shows that 2 is an extraneous solution. The only solution is 5.

**73.** $a = 6.9e^{-2.3b}$

$0.45 = 6.9e^{-2.3b}$

$\dfrac{0.45}{6.9} = e^{-2.3b}$

$\ln\dfrac{0.45}{6.9} = \ln e^{-2.3b}$

$\ln\dfrac{0.45}{6.9} = -2.3b$

$-\dfrac{1}{2.3}\ln\dfrac{0.45}{6.9} = b$

$b \approx 1.19$

**75.** Use $N = ke^{0.04t}$ where $N = 5500$ and $t = 11$.

$N = ke^{0.04t}$

$5500 = ke^{0.04(11)}$

$5500 = ke^{0.44}$

$k = \dfrac{5500}{e^{0.44}} \approx 3542$

After 11 minutes, there were 3542 bacteria.

## Chapter 11 Test

**1.** $f(x) = x^2 - 4, \quad g(x) = x + 2$
$(f+g)(x) = f(x) + g(x)$
$\phantom{(f+g)(x)} = x^2 - 4 + x + 2$
$\phantom{(f+g)(x)} = x^2 + x - 2$

**3.** $f(x) = x^2 - 4, \quad g(x) = x + 2$
$(f \circ g)(x) = f(g(x))$
$\phantom{(f \circ g)(x)} = f(x+2)$
$\phantom{(f \circ g)(x)} = (x+2)^2 - 4$
$\phantom{(f \circ g)(x)} = x^2 + 4x + 4 - 4$
$\phantom{(f \circ g)(x)} = x^2 + 4x$

**5.** $f(x) = \dfrac{x-5}{3}, \quad g(x) = 3x + 5$
$(f \circ g)(x) = f(g(x))$
$\phantom{(f \circ g)(x)} = f(3x+5)$
$\phantom{(f \circ g)(x)} = \dfrac{(3x+5)-5}{3} = x$
$(g \circ f)(x) = g(f(x))$
$\phantom{(g \circ f)(x)} = g\left(\dfrac{x-5}{3}\right)$
$\phantom{(g \circ f)(x)} = 3\left(\dfrac{x-5}{3}\right) + 5 = x$
Thus, $f$ and $g$ are inverses.

**7.**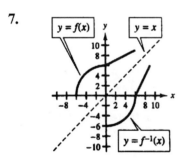

**9.** $\log_2 4^{3/4} = \log_2 (2^2)^{3/4} = \log_2 2^{3/2} = \dfrac{3}{2}$

**11.** $\left(\dfrac{1}{16}\right)^{x-1} = 8$
$(2^{-4})^{x-1} = 2^3$
$2^{-4(x-1)} = 2^3$
$-4(x-1) = 3$
$-4x + 4 = 3$
$-4x = -1$
$x = 0.25$

**13.** $e^{2x} = 10$
$\ln e^{2x} = \ln 10$
$2x = \ln 10$
$x = \dfrac{\ln 10}{2} \approx 1.15$

**15.** $\log \dfrac{A^2}{\sqrt{B}} = \log A^2 - \log \sqrt{B}$
$\phantom{\log \dfrac{A^2}{\sqrt{B}}} = 2 \log A - \dfrac{1}{2} \log B$

**17.** $2\log_6 x + \log_6 (x-1) - 1$
$= \log_6 x^2 + \log_6 (x-1) - \log_6 6^1$
$= \log_6 x^2(x-1) - \log_6 6^1$
$= \log_6 \left[\dfrac{x^2(x-1)}{6}\right]$

**19.** $\log x + \log(x+3) = 1$
$\log x(x+3) = \log 10^1$
$x(x+3) = 10$
$x^2 + 3x - 10 = 0$
$(x+5)(x-2) = 0$
$x = -5, 2$
A check shows that $-5$ is an extraneous solution. The only solution is 2.

**21.** $2 + \ln x = 0$
$\ln x = -2$
$x = e^{-2} \approx 0.14$

**23.** $D(t) = 2 + 5(0.89)^t$

(a) When the ad campaign began, $t = 0$.
$D(0) = 2 + 5(0.89)^0 = 7$ thousand cases.

(b) The graph shows that as time $t$ increases, demand $D$ decreases. The ad campaign is ineffective.

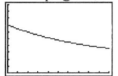

(c) Half of the initial demand is 3.5 thousand cases.
$$3.5 = 2 + 5(0.89)^t$$
$$1.5 = 5(0.89)^t$$
$$0.3 = 0.89^t$$
$$\ln 0.3 = \ln 0.89^t$$
$$\ln 0.3 = t \ln 0.89$$
$$t = \frac{\ln 0.3}{\ln 0.89} \approx 10.3$$

After about 10.3 weeks, the demand for this brand of soap had fallen to half the initial demand.

**25.** Use $T = T_m + D_0 e^{kt}$ and $T_m = 32$, $T = 40$, $D_0 = 90 - 32 = 58$, and $k = -1.32$.
$$T = T_m + D_0 e^{kt}$$
$$40 = 32 + 58e^{-1.32t}$$
$$8 = 58e^{-1.32t}$$
$$\frac{8}{58} = e^{-1.32t}$$
$$\ln \frac{8}{58} = \ln e^{-1.32t}$$
$$\ln \frac{8}{58} = -1.32t$$
$$-\frac{1}{1.32} \ln \frac{8}{58} = t$$
$$t \approx 1.5$$

It will take about 1.5 hours for the soda to cool to 40°F.

# Cumulative Test for Chapters 9-11

**1.**
(a) $\sqrt{y^8} = \sqrt{(y^4)^2} = |y^4| = y^4$

(b) $\sqrt{y^{10}} = \sqrt{(y^5)^2} = |y^5|$

**3.**
(a) $\dfrac{a^{-2}b^{1/3}}{a^{1/2}b} = \dfrac{1}{a^{5/2}b^{2/3}}$

(b) $(x^{-1}y^2)^{-1/2} = x^{1/2}y^{-1} = \dfrac{x^{1/2}}{y}$

(c) $\left(\dfrac{x^{-1}}{y^2}\right)^{-3} = \dfrac{x^3}{y^{-6}} = x^3 y^6$

**5.**
(a) $\sqrt{20x^9} = \sqrt{4x^8}\sqrt{5x} = 2x^4\sqrt{5x}$

(b) $\sqrt[14]{y^{21}} = y^{21/14} = y^{3/2} = y\sqrt{y}$

**7.** The conjugate of $3 - \sqrt{5}$ is $3 + \sqrt{5}$.
Product: $(3-\sqrt{5})(3+\sqrt{5}) = 9 - 5 = 4$

**9.**
(a) $\sqrt{2x+1} = x\sqrt{3}$
$$2x + 1 = 3x^2$$
$$3x^2 - 2x - 1 = 0$$
$$(3x+1)(x-1) = 0$$
$$x = -\frac{1}{3}, 1$$

A check shows that $-1/3$ is an extraneous solution. The only solution is 1.

(b) $(2x-3)^{2/3} = 9$
$(2x-3)^2 = 9^3$
$(2x-3)^2 = 729$
$2x - 3 = \pm\sqrt{729}$
$2x = 3 \pm 27$
$x = \dfrac{3 \pm 27}{2} = -12, 15$

11. Graphing can be used to estimate solutions. Algebraic methods include factoring, using the Square Root Property, and using the Quadratic Formula. The Quadratic Formula always works.

13. Let $x$ = the number.
$x - 6 = \dfrac{7}{x}$
$x(x - 6) = 7$
$x^2 - 6x - 7 = 0$
$(x - 7)(x + 1) = 0$
$x = -1, 7$
The number could be $-1$ or $7$.

15. $a + a^{1/2} = 12$
$a + a^{1/2} - 12 = 0$
$u^2 + u - 12 = 0$   Let $u = a^{1/2}$.
$(u + 4)(u - 3) = 0$
$u = -4, 3$
Now, replace $u$ with $a^{1/2}$.
$a^{1/2} = -4$     or    $a^{1/2} = 3$
$a = (-4)^2 = 16$         $a = 3^2 = 9$
A check shows that 16 is an extraneous solution. The only solution is 9.

17. (a) $p(x) = x^2 + 2x - 24 = (x + 6)(x - 4)$
Graph $y = x^2 + 2x - 24$. The $x$-intercepts are $(-6, 0)$ and $(4, 0)$. We can see from the graph that the solution set of $p(x) > 0$ is $(-\infty, -6) \cup (4, \infty)$.

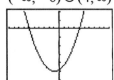

(b) $p(x) = x^2 + 2x - 24 < -25$. Graph $y = x^2 + 2x + 1$. The $x$-intercept is $(-1, 0)$. We can see from the graph that the solution set of $p(x) < -25$ is $\varnothing$.

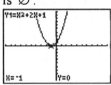

(c) $p(x) = x^2 + 2x - 24 > -26$. Graph $y = x^2 + 2x + 2$. There are no $x$-intercepts. We can see from the graph that the solution set of $p(x) > -26$ is $\mathbf{R}$.

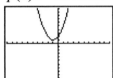

19. $P(x) = 8x - x^2$
(a) The profit is maximized at the vertex of the graph of the parabola. That is, when
$x = -\dfrac{b}{2a} = -\dfrac{8}{2(-1)} = 4$ units are produced.
(b) The maximum profit is
$P(4) = 8(4) - 4^2 = 32 - 16 = 16$.

21. $f(x) = |x - 1|$, $g(x) = 2x + 1$

(a) $(f \circ g)\left(-\dfrac{1}{2}\right) = f\left(g\left(-\dfrac{1}{2}\right)\right)$
$= f(0) = |0 - 1| = 1$

(b) $\left(\dfrac{f}{g}\right)\left(-\dfrac{1}{2}\right) = \dfrac{f\left(-\dfrac{1}{2}\right)}{g\left(-\dfrac{1}{2}\right)} = \dfrac{\left|-\dfrac{1}{2} - 1\right|}{\left|-\dfrac{1}{2}(2) + 1\right|}$
$= \dfrac{\left|-\dfrac{1}{2} - 1\right|}{0}$ is undefined

23. Both have a $y$-intercept of $(0, 1)$ and no $x$-intercept. The graph of $f$ rises and the graph of $g$ falls from left to right.

**25.**

(a) $\ln \dfrac{x^2 y}{\sqrt{z}} = \ln x^2 + \ln y - \ln \sqrt{z}$

$\qquad = 2\ln x + \ln y - \dfrac{1}{2}\ln z$

(b) $3\log(x-1) - \log(x+1)$

$\qquad = \log(x-1)^3 - \log(x+1)$

$\qquad = \log \dfrac{(x-1)^3}{x+1}$

**27.** $P = \$1200,\ A = \$2187,\ t = 10$:

$$A = Pe^{rt}$$
$$2187 = 1200e^{10r}$$
$$\dfrac{2187}{1200} = e^{10r}$$
$$\ln \dfrac{2187}{1200} = \ln e^{10r}$$
$$\ln \dfrac{2187}{1200} = 10r$$
$$\dfrac{1}{10}\ln \dfrac{2187}{1200} = r \approx 0.06$$

The interest rate should be approximately 6%.